ADHESIVES, SEALANTS, AND COATINGS FOR SPACE AND HARSH ENVIRONMENTS

POLYMER SCIENCE AND TECHNOLOGY

Recent volumes in the series:

Volume 25 NEW MONOMERS AND POLYMERS
Edited by Bill M. Culbertson and Charles U. Pittman, Jr.

Volume 26 POLYMER ADDITIVES
Edited by Jiri E. Kresta

Volume 27 MOLECULAR CHARACTERIZATION OF COMPOSITE INTERFACES
Edited by Hatsuo Ishida and Ganesh Kumar

Volume 28 POLYMER LIQUID CRYSSTALS
Edited by Alexandre Blumstein

Volume 29 ADHESIVE CHEMISTRY
Edited by Lieng-Huang Lee

Volume 30 MICRODOMAINS IN POLYMER SOLUTIONS
Edited by Paul Dubin

Volume 31 ADVANCES IN POLYMER SYNTHESIS
Edited by Bill M. Culbertson and James E. McGrath

Volume 32 POLYMERIC MATERIALS IN MEDICATION
Edited by Charles G. Gebelein and Charles E. Carraher, Jr.

Volume 33 RENEWABLE-RESOURCE MATERIALS: New Polymer Sources
Edited by Charles E. Carraher, Jr., and L. H. Sperling

Volume 34 POLYMERS IN MEDICINE II: Biomedical and Pharmaceutical Applications
Edited by E. Chiellini, P. Giusti, C. Migliaresi, and L. Nicolais

Volume 35 ADVANCES IN BIOMEDICAL POLYMERS
Edited by Charles G. Gebelein

Volume 36 FOURIER TRANSFORM INFRARED CHARACTERIZATION OF POLYMERS
Edited by Hatsuo Ishida

Volume 37 ADHESIVES, SEALANTS, AND COATINGS FOR SPACE AND HARSH ENVIRONMENTS
Edited by Lieng-Huang Lee

A Continuation Order Plan is available for this series. A continuation order will bring delivery of each new volume immediately upon publication. Volumes are billed only upon actual shipment. For further information please contact the publisher.

ADHESIVES, SEALANTS, AND COATINGS FOR SPACE AND HARSH ENVIRONMENTS

Edited by

Lieng-Huang Lee
Xerox Corporation
Webster, New York

PLENUM PRESS • NEW YORK AND LONDON

Library of Congress Cataloging in Publication Data

International Symposium on Adhesives, Sealants, and Coatings for Space and Harsh Environments (1987: Denver, Colo.)
Adhesives, sealants, and coatings for space and harsh environments / edited by Lieng-Huang Lee.
p. cm.—(Polymer science and technology; v. 37)
"Proceedings of an International Symposium on Adhesives, Sealants, and Coatings for Space and Harsh Environments, sponsored by the Division of Polymeric Materials: Science and Engineering of the American Chemical Society, held April 7–9, 1987, in Denver, Colorado"—T.p. verso.
Includes bilbiographical references and index.
ISBN 0-306-42989-6
1. Adhesives—Congresses. 2. Sealing compounds—Congresses. 3. Coatings—Congresses. I. Lee, Lieng-Huang, 1924– II. American Chemical Society. Division of Polymeric Materials: Science and Engineering. III. Title. IV. Series.
TP967.I57 1987 88-21002
668′.3—dc19 CIP

Proceedings of an International Symposium on Adhesives, Sealants, and Coatings for Space and Harsh Environments, sponsored by the Division of Polymeric Materials Science and Engineering of the American Chemical Society, held April 7–9, 1987, in Denver, Colorado

A Division of Plenum Publishing Corporation
233 Spring Street, New York, N.Y. 10013

Printed in the United States of America

PREFACE

New technologies constantly generate new demands for exotic materials to be used in severe environments. The rapid developments of aerospace industries during the last two decades have required new materials to survive extreme high and low temperatures and various radiations. The exploration of new energy sources, e.g., solar and geothermal, has led us to develop new solar collectors and geothermal devices. Even the search for new oils has demanded that we study the corrosive environment of oil fields.

In the telecommunication industries, optical fibers have been adopted broadly to replace metallic conductors. However, none of the optical fibers can survive abrasion or corrosion without the application of a coating material. For microelectronics, protection in terms of coatings and encapsulants is deemed necessary to prevent corrosion. One of the major causes of corrosion has been shown to be water which appears to be abundant in our earthly environments.

Water can attack the bulk adhesive (or sealant), the interface, or the adherend. Water can also cause delamination of coating film, and it is definitely the major ingredient in causing cathodic or anodic corrosion. Thus, water becomes the major obstacle in solving durability problems of various materials in harsh environments.

As an intent to solve some of the above problems, an International Symposium on Adhesives, Sealants, and Coatings for Space and Harsh Environments was organized for the Division of Polymeric Materials and Engineering of the American Chemical Society. The Symposium was held between April 7 and 9, 1987 in the Radisson Hotel, Denver, Colorado. The meeting was well attended by those from academia and industries.

A total of 34 papers was presented to six sessions. The papers, which were properly refereed and revised, are now included in this volume of Proceedings, which also contains several contributions submitted after the Symposium and the discussions. This volume is divided into six parts:

I. Environmental Exposure

II. Stress and Interface

III. Adhesives for Space and Harsh Environments

IV. Sealants for Space and Harsh Environments

V. Coatings for Corrosive Environments

VI. Coatings for Electronic and Optical Environments

As Chairman of the Symposium, I would like to thank our contributors; their biographies are attached at the end of this volume. I am also indebted to our Session Chairmen: Professor C.E. Rogers, Dr. W.D. Bascom, Dr. R.M. Evans and Dr. J.W. Holubka, for their assistance. I would like to thank Dr. J. Scott Thornton, for his talk during the Symposium Banquet. I sincerely appreciate the skill of Ms. D.M. Costenoble in retyping all manuscripts for this volume. I appreciate the support of Xerox Webster Research Center for me to perform this task. I would like to thank Mr. F. Belli and his group of the Technical Information Center in assisting me to prepare author and subject indexes.

Acknowledgment is also made to the donors of the Petroleum Research Fund, administered by the American Chemical Society. Their funding partially assisted several overseas speakers to attend the symposium.

Lieng-Huang Lee
April, 1988

CONTENTS

PART SIX: COATINGS FOR ELECTRONIC AND OPTICAL ENVIRONMENTS

PART ONE:

ENVIRONMENTAL EXPOSURE

Introductory Remarks

Lieng-Huang Lee

Webster Research Center
Xerox Corporation
Webster, New York 14580

Water is one of the factors that cause the failure of adhesives, sealants, and coatings. During this Session, we shall address the water problem from different approaches. I shall present an overview about various problems related to adhesives, sealants, and coatings with the emphasis on water being the culprit.

Professor C.E. Rogers will provide us with a lecture about the fundamentals of permeation, diffusion, and solubility. Specifically, he will discuss the effects of environment exposure on sorption and transport of penetrants in polymeric materials. In reducing the water absorption, fluoropolymers have been used to replace non-fluorinated counterparts. Specifically, Dr. S.J. Shaw will discuss mechanical and water absorption behavior of fluoroepoxy resins. The properties of fluoroepoxy will be compared with those of non-fluorinated polymers.

In the marine environments, the durability of adhesive bonds is generally adversely affected by salt and water. Dr. D.A. Dillard will present a paper* on the durability of rubber-to-metal bonds in the marine environments. Sacrificial anodes are often attached to marine structures. While this practice can protect the metal surface cathodically from being corroded, it may cause severe deterioriation of adhesives or coatings that are bonded to the structure. The mechanisms of degradation will be discussed with the support of ESCA data.

It is a miracle how the inside of a human body functions in an adverse environment full of water. Generally, it is not easy to place a polymer in the body. An interesting approach of some medical adhesives is to use water as a catalyst to polymerize a monomer anionically into an adhesive polymer. A good example of this type of adhesive is cyanoacrylates. This is our first opportunity to hear a paper presented by Mr. Ruan from the People's Republic of China. In China, they have been using cyanoacrylates to seal human tubular tissues for sterilization and nonsurgical treatments. We can now hear their success story about using this medical adhesive to perform birth control and other treatments.

Editor's note: Dr. Dillard's paper was published elsewhere. In its place, a paper by Xie et al. on the underwater adhesives used in the PRC is included.

We also include an interesting paper by Xia Wen-gan of the PRC on the inorganic adhesive used in China around 207 B.C. This may be one of the earliest adhesives invented by man. We are encouraged to see that the environmental exposure of this inorganic adhesive for such a long time has not caused serious degradation.

ADHESIVES, SEALANTS, AND COATINGS FOR SPACE AND HARSH ENVIRONMENTS

Lieng-Huang Lee

Webster Research Center
Xerox Corporation
Webster, NY 14580

ABSTRACT

This review discusses new developments in adhesives, sealants, and coatings for space and harsh environments. Several papers presented to the Symposium on the subject matter held in Denver, Colorado in April, 1987 will be included in this review. Other developments reported through 1987 will be briefly mentioned.

The purpose of this review is not only to enumerate problems associated with harsh environments but also to highlight the breakthroughs in solving some of the problems. Neither this review nor this entire volume addressing to the subject matter can be inclusive about the state-of-the-art. However, we sincerely hope that we can somewhat point to the directions for future research in materials needed for aerospace, underwater, corrosive environments, optical and electronic environments.

INTRODUCTION

The durability of synthetic materials is affected by environmental factors. In harsh environments, synthetic materials often degrade and fail to perform. This paper reviews new developments designed to overcome environmental problems, such as moisture (or water), temperature, radiation, and corrosion.(1)

The impact of these problems on various technologies, e.g., space exploration, transportation, energy development, telecommunications, and microelectronics, will also be discussed briefly. The synthetic materials involved are adhesives, sealants,(2) and coatings. Our discussion will start with the most detrimental environmental factor, water, the problems it causes, and some solutions. Then we shall discuss other factors and solutions for environmentally caused durability problems. We will also examine some protective coatings for space, corrosive, optical, and electronic environments. In contrast, we shall also mention new developments in elastomers and ceramic coatings for extreme situations.

ENVIRONMENTAL EXPOSURE TO WATER

Unlike metals, most polymers are permeable to moisture or to liquid water. With adhesives and sealants, water attacks the materials, the interface, or the adherend, e.g., metal. Water can also permeate coatings and attack the substrate underneath. The permeability coefficient P of a polymer[3] is a product of the diffusion coefficient D and the solubility coefficient S:

$$P = D \times S.$$

This general subject of penetrants in polymeric materials has been reviewed by Rogers.[4] Cassidy et al.[5,6] studied water permeation through elastomer laminates and concluded that the laminates showed directional dependence of permeation rate. Thomas and Muniandy[7] demonstrated that the presence of a small amount of hydrophilic impurities in vulcanized rubbers influenced strongly their water absorption and desorption behavior.

The diffusion of water into epoxy adhesives has been reported by several authors.[8,9] If the surface concentration of the epoxy film is constant at all times and the mass transport process is completely controlled by diffusion, then the sorption process is said to be Fickian and exhibits the following characteristics:

1. Both sorption and desorption curves are linear in the initial stage.

2. Above the linear portion both sorption and desorption curves are concave to the abscissa.

3. Reduced sorption (desorption) curves of specimens of different thicknesses are superposable and this is called film-thickness scaling.

However, recently Wong and Broutman[10] found that non-Fickian sorption of water by an epoxy resin could be caused by molecular relaxation, insufficient curing or oxidation of the resin. The authors[11] later proposed a model for the diffusion of water in a glassy epoxy resin to consist of two regions in which water molecules possess different mobilities. Despite the mechanism of diffusion, the net effect of water on the epoxy adhesive[12] is plasticization of the polymer and the weakening of adhesive bond.

In order to lower the permeability P of a polymer by water, both the diffusivity D and the solubility S should be low.[13] In general, fluorine-containing and silicon-containing polymers, such as fluoroacrylic, fluorosilicone, are low in water permeability (Table 1). However, no polymer has been found to be entirely hermetic or hydrophobic.[14]

Fluoroepoxy polymers developed by Griffith[15] have been studied with continuing interest.[16] Fluorepoxides[17,18] have been demonstrated to absorb less water than their nonfluorinated counterparts. Blocking the reactive groups of a nonfluorinated epoxide with fluorinated aromatic

Table 1. Moisture Solubility, Diffusivity and Permeation Constants of "Hydrophobic" Polymers

Material	Diffusivity D, cm^2/sec	Solubility S, $\frac{cm^3 H_2O}{cm^3 \text{ polymer-cm Hg}}$	Permeability (P = D x S)
Siloxyimides	$1.5–9.4 \times 10^{-6}$	0.069–0.100	$1.035–9.21 \times 10^{-7}$
Fluorosilicone	Not calculable	due to	anomalies
Epoxy Novolac	3.8×10^{-8}	0.50	1.9×10^{-8}
Fluoroacrylic	3×10^{-7}	0.06	5.4×10^{-8}
Phenylated Silicone	3.8×10^{-5}	0.025	9.5×10^{-7}
Silastyrene	No	absorption	Detected

Rf. B.L. Rathbun and P.wh.Schuessler, in "Adhesive Chemistry-Developments and Trends," Ed.L.H. Lee, 785, Plenum, N.Y. 1984.

reagents, e.g., pentafluorobenzoyl chloride,[19] can also reduce the moisture absorption by as much as 75%.

A recent paper by Nakamura and Maruno[20] reveals new applications of fluoroepoxides as optical adhesives by lowering the refractive index to match that of quartz (1.46), beside having low absorption of water. The refractive index of epoxy resin (e.g., 1.5777) can be lowered to 1.415 by increasing the fluorine content of the resin. The fluoroepoxides can be used to adhere optical fibers and form a transparent product.

Though water has been the worst enemy for manmade adhesives, interestingly water has presented little problem to some natural adhesives secreted by barnacles or mussels. Recently, Waite[21] described the structure of the repeated decapeptide sequence of the mussel polyphenolic protein (Fig. 1). The use of this type of protein in various industrial and medical applications is being investigated. Biopolymers, Inc. of Farmington, Conn. has developed the protein as a biocompatible, nonspecific adhesive for cell and tissue culture. At present, one of the problems is the adequate supply of this type of adhesive or protein. Some companies, e.g., Genex Corp. of Gaithersburg, Md., even consider using the recombinant DNA technology to make this type of adhesive.

Salt water is more detrimental than water to adhesives, sealants, and coatings. In salt water, additional corrosion problems accelerate debonding. In general, salt water has a lower diffusivity D through elastomers than does distilled water.[5] For one elastomer, natural rubber, a recent report[22] indicates that after 42 years of immersion in sea water at 80 ft (34.3 m), only 5% of the water was absorbed, based on the dry rubber mass. Surprisingly, the rubber showed essentially no loss of physical properties.

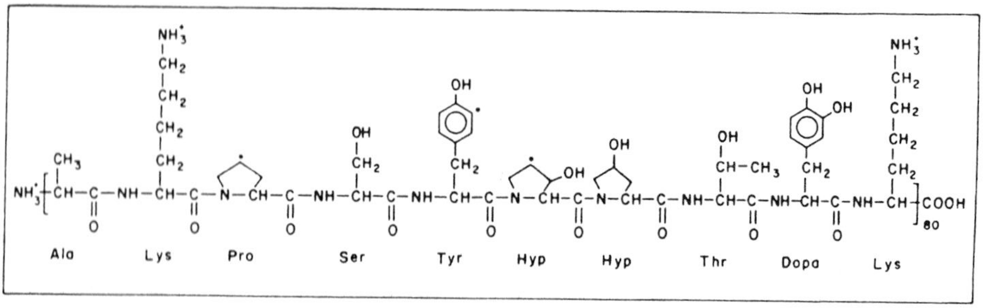

Ala Lys Pro Ser Tyr Hyp Hyp Thr Dopa Lys

Fig. 1. The dots represent positions of additional hydroxylation Ref. J.H. Waite, Int. J. Adhesion and Adhesives 7, No. 1,9. January 1987.

In the absence of electrochemically active conditions, the rubber-metal bonds between carbon steel and titanium, respectively, were found[23] to be highly durable, despite a wide range of water absorption, e.g., natural rubber and acrylonitrile rubber (1-4%) and polychloroprene (35%). However, in the presence of electrochemically active conditions, such as sea water, bond failure occurred at a relatively rapid rate.

Coincidentally, polychloroprene has been used widely in underwater acoustic systems.[24] The sonar-dome windows for many surface ships are constructed by using a heavy steel wire-reinforced polychloroprene (Neoprene®). In this application, the rubber-steel bond is critical to the integrity of the dome. The sonar device is constantly exposed to attack by sea water, temperature extremes, pressure cycling, and mechanical stress. The search for a durable adhesive system for this application has been a continuing effort.

The effect of electrochemical potential on the durability of rubber/metal in sea water was reported by Stevenson.[25] In electrochemically inert system, the bond was stable even after three years immersion in sea water. However, bond failure was detected within days in the presence of an electrical pair due to the presence of contact potential.

The effect of salt on the adhesion of brass-plated steel core to rubber[26] is similar to that of sea water. Here, an actual corrosion of the brass takes place due to the dezincification and dissolution of brass in the plating. In this case, the effect of the contact potential of the metallic pair could have accelerated the galvanic corrosion.

In the application of medical adhesives, both silicones and cyanoacrylates perform various functions in the water-dominant environments, e.g., the human body. Water is also a problem for electronic prostheses implanted in the human body. Silicone encapsulants[27] have been used successfully to control corrosion and current leakage. In China, cyanoacrylate has been used to seal tubular tissues for sterilization and non-surgical purposes[28] because this adhesive polymerizes instantaneously in the presence of water or other anionic species. In the latter case, water the adversary has been properly utilized as beneficial catalyst for the polymerization.

STRESS AND INTERFACE

Stress can affect the diffusion of water in an epoxy resin.[29] In a bent epoxy bar, more water is picked up at the tension side than at the compression side because the diffusivity of water is increased at the tension side. In theory, the diffusion constant should increase exponentially with stress; however, the water uptake through sorption does not follow the same pattern.[30]

In the case of adhesives, whether stress affects the joint strength after exposure to humid environments depends greatly on the type of adhesive. Cotter[31] found that epoxide-polyamide in all climates and vinyl phenolics in hot/wet surroundings were sensitive to stress, while nitrile-phenolic, epoxy-Novolak and epoxide-phenolic were not.

Water can also attack the interface. As a result, adhesive debonds, coating delaminates, and metal corrodes. To fortify interfacial barriers, various coupling agents[32] have been employed for nonmetal and metal surfaces. In the last few years, new coupling agents based on chelates have been reported, but are not widely used. Polyfunctional mercapto-esters,[33,34] were used for the epoxy-metal system; a β-diketone chelate[35] was developed, and organofunctional zircoaluminates[36] were claimed as new coupling agents. A fumaro coordination compound,[37,38] Volan 82, has been shown to improve adhesion between polyethylene and aluminum under wet conditions.

Two potential coupling agents have been synthesized by Sounik and Kenney.[39] These coupling agents are $HOSiPcOSi(CH_3)_2CH_2CH_2C_6H_4CH_2I$-m and -p and $HOSiPcOSi(CH_3)_2$-$CH_2CH_2C_6H_4CH_2NHCH_2CH_2NH_2$-m and-p (Pc=phthalocyanato ligand). These agents have the tendency to form an oxygen bridge with the glass surface.

A new family of coupling agents based on azidosilane have been reported by Kolpak.[40] The structure of azidosilanes is:

$$N_3SO_2\text{-}R\text{-}Si(OR')_3$$

The reaction of azidosilanes with polymers can be expressed as follows:

$$\text{—R—}SO_2N_3 \xrightarrow{\text{heat}} \text{—R—}SO_2N: + N_2$$

$$\text{—R—}SO_2N: + \text{H—C—} \longrightarrow \text{—R—}SO_2\text{—}\underset{\substack{|\\ H}}{N}\text{—}\overset{|}{\underset{|}{C}}\text{—}$$

Silane coupling agents have not only been used to improve adhesion in composites but also as adhesion promoters for adhesives.[41] Among them, the epoxy silane, γ-glycidoxypropyltrimethoxy silane, was found to be a very efficient adhesion promoter for epoxy-based adhesives.

Besides coupling agents, various surface treatment methods have been used to strengthen the interfacial barriers for metal adherends. Recently Clearfield, McNamara and Davis[42] reviewed various surface treatment methods for aluminum, titanium and steel adherends. Venables[43] summarized that the morphology and the stability of metal oxides at the interface were responsible for the initial bond strength. In the case of aluminum, in the presence of moisture, aluminum oxide can transform into aluminum hydroxide to cause the deterioration of bond strength. In the case of titanium, the oxides are more stable than the oxides of aluminum. However, under severe environmental conditions, titanium oxides can undergo a polymorphic transformation leading to the degradation of adhesive bond.

According to Brockmann et al.,[44] the failure of bonded aluminum joints can be of two mechanisms: (1) an alkaline (anionic) mechanism of destruction caused by the adverse effect of moisture on the aluminum oxide, and (2) an acidic (cationic) degradation, which can only be accompanied by electrochemical (or bondline) corrosion of the metal. The alkaline mechanism is presumably caused by the conversion of aluminum oxide into aluminum hydroxide. Thus, the new hydration inhibitor based on nitrilotrismethylene phosphonic acid (NMTP) reported by Hardwick, Ahearn, and Venables[45] has been found efficient in preventing the adhesive bond from degrading.[46-48] However, this type of inhibitor may not be effective in inhibiting the acidic degradation due to the bondline corrosion.

For cathodic corrosion, zinc-rich primers,[49] such as zinc phosphate, can provide some protection for steel. Zinc-rich primers can form a relatively impermeable barrier and retard or inhibit the corrosion process. For the zinc-primer system, a new additive, polyacrylic acid has been introduced by Sugama et al.[50,51,52] to enhance the bond durability between the primer and the metal.

Besides zinc primers, other inorganic primers,[53,54,55] such as amorphous aluminum oxide generated from an aluminum alkoxide or chelate, have been used for the protection of aluminum. These primers could eliminate the use of relatively more toxic chromium-containing primers. Preliminary studies indicate that the inorganic primers are effective with short-time, room temperature metal surface treatments which are adoptable to field repair situations.

ADHESIVES FOR SPACE AND HARSH ENVIRONMENTS

Water is also a problem for re-entry spacecraft, such as shuttle orbiters. For non-reentry spacecraft and missiles, the major problems are heat, particle bombardment, and radiation. The temperatures of the space shuttle's outer skin can reach 2300° F, while those for the proposed Orient Express reach 3000° F.[56] (The Orient Express is the transatmospheric craft designed to take off from a standard runway, rise through the atmosphere, and cruise at a great speed in space.) This craft's dual propulsion system will involve ramjets for atmospheric flight and scramjets for space flight. The engine temperature of the scramjets can be as high as 3500° F. The hydrogen fuel will be stored in tanks with a skin temperature of -423° F. This is a temperature extreme that no organic material can, by any chance, survive.

For high temperature applications, the current upper limit appears to be 1000-1400° F for organic materials. High temperature polymers have been arbitrarily divided into three ranges (Table 2).[57]

Examples for Range I polymers are polybenzimidazoles (PBIs) and polyquinoxalines (PQs). The PBI/carbon composites retain about 18% of

Table 2: Three Ranges of High Temperature Polymers

Range	Use Temperature, °C(°F)	Use Time
I	538-760 (1000-1400)	Seconds to minutes
II	288-371 (550-700)	Hundreds of hours
III	177-232 (350-450)	Thousands of hours

their room temperature flexural strength after a 3-minute exposure to 1200° F.

For Range II applications, polymers, such as polyimides (PIs)[58] and polyphenylquinoxalines (PPQs), have been used as adhesives and composite matrices. Both PMR-15, developed by NASA-Lewis Research Center, and LARC-160, developed by NASA-Langley Research Center, have been used to form graphite composites that retain high strengths up to 600° F. These PIs are generally cured at high temperature under pressure.

The acetylene-terminated (AT) counterparts of PIs and PPQs[59] belong to Range III because of the lowering of their thermal stability by the presence of the acetylene-terminal groups. Both ATI and ATPQ can be crosslinked without giving off gases. Currently, ATI is available from National Starch and Chemical as Thermid 600.

One of the fast growing PI groups is the bismaleimides (BMIs), primarily due to their epoxy-like cure at much lower temperatures (e.g., 350° F) than other PIs. Polyamino-bismaleimides (PABMs), e.g., Rhone-Poulenc Kerimide 601, are stable at 500° F for thousands of hours. In general, most BMIs are Range III polymers. Currently, there are at least ten suppliers of different BMIs worldwide.

Recent progress in high-temperature polymers and adhesives have been reviewed.[60-63] New semicrystalline polyimides[61] containing carbonyl and ether connecting groups exhibit Tgs of 215 to 247°C and Tm's of 350 to 442°C, and good film properties. Polyarylene ethers derived from 9,9-bis(4-hydroxyphenyl)fluorene (Tgs of 223 to 280°C) also have good adhesive properties. PPQ's containing pendant phenyl-ethynl groups have been used to form adhesive bonds with good retention of strength to a temperature as high as 316°C.

The modification of polyimide LARC-TPI with 5 mol% of 4,4'-oxydianiline (ODA) has been shown by Mitsui Toatsu to be a good film adhesive[64] for bonding titanium alloys. Moreover, the LARC-TPI/ODA has a better flow than the unmodified LARC-TPI.

To avoid toxic and tenacious solvents, other forms of high temperature adhesives have been introduced by NASA-Langley Research Center. Burks and St.Clair[65] prepared the first hot-melt adhesive BDSDA/APB. Later, another hot-melt polyimide with a flexible m-phenylene diamine unit (MPD) by the code number of 422[66] was found to function near 200°C. It has a very low viscosity to ensure good wetting of the adherends. On the other hand, the introduction of a water-base LARC-TPI adhesive[67,68] certainly opened up a new possibility using low organic volatile dispersions containing high-temperature adhesives.

Silicone polymers are rather stable for space applications. Therefore, the adhesives for the ceramic tiles protecting space orbiters are based on silicone. Recently, PIs have been used to form silicone copolymers. One hot-melt adhesive introduced by the General Electric Company is a silicone-imide copolymer (SSAX-SPI), which is suitable for hot/wet environments and is stable to engine oil and jet fuel. Another silicone-modified PI (M&T 4605-40) has been a useful adhesive for aerospace and electronic applications. A siloxane-modified polyethersulfideimide has been prepared by Burks and St.Clair[69] to improve processability.

A recent interesting product based on the semi-interpenetrating network (semi-IPN) has been introduced jointly by NASA-Langley Research Center and Kentron International. According to Hanky and St. Clair,[70] the linear polyimidesulfone is dispersed in the matrix of the crosslinked acetylene-terminated polyimidesulfone to yield a flexible, tough network. This processable material is a potential high temperature adhesive with a high strength retention up to 1000° F.

SEALANTS FOR SPACE AND HARSH ENVIRONMENTS

Unlike adhesives, there have not been many new sealants developed for special environmental requirements. Sealants may be divided into four categories: 1) low performance, 2) medium performance, 3) high performance and 4) advanced performance. Low performance sealants usually do not cure and are based on drying oils.[71] They tend to crack. Medium performance sealants are solvent-based acrylics and butyl rubber. Their drawbacks are joint shrinkage due to solvent loss. Recently some water-based sealants based on poly(vinyl acetate) and acrylics have been made. Shrinkage, damages due to freezing, and mildew are new problems confronting the water-based sealants.

High performance sealants are polysulfides,[72] polyurethanes, and silicones.[73] All three, which are the curing type, have little or no shrinkage and possess excellent joint movement. Styrene block copolymers[74] are gradually joining this group of sealants, especially for hot-melt applications.[75] In practice, there have been different hydrids of them. Today acrylic latex has also been formulated to be one of the high performance sealants providing transparency and weatherability of the products.

The effect of equatorial environment and water immersion conditions on high performance sealants has been studied by Gan, Ong and Tan.[76] The shear strengths of lap joints were measured up to one year. According to the findings, the shear strengths of the acrylic/Al and the polysulfide/Al lap joints were much less than those of the silicone/Al. However, their short-term durabilities under the equatorial environment were better than that of the latter. Both the acrylic and the silicone systems, but not the polysulfide, suffered varying degrees of strength deterioration through water immersion for one year. Previously, Hanhele et al.[77,78] studied the water immersion of polysulfide sealants, and found that the dichromate-cured materials were more effective sealants in hot water immersion than those cured with manganese oxide.

For about four decades, polysulfide sealants have been successfully used for aircraft[79,80,81] and construction[82] partially due to their fuel and solvent resistance, good low-temperature performance and weatherability.

During the last several years, a new class of high performance polysulfide polymers was developed by Singh[83] by reacting Thiokol LP polysulfides with dithiols. The products, known as Permapol P-5 polymers manufactured by Products Research & Chemicals Corp., are formed through the following reactions:

$$HS\text{-}[CH_2\text{-}CH_2\text{-}O\text{-}CH_2\text{-}O\text{-}CH_2\text{-}CH_2\text{-}(S)_x]_n\ CH_2\text{-}CH_2\text{-}O\text{-}CH_2\text{-}O\text{-}CH_2\text{-}CH_2\text{-}SH + HS\text{-}R\text{-}SH$$

LP polymer   Dithiol

$$\xrightarrow{\text{Catalyst}} HS\text{-}[CH_2\text{-}CH_2\text{-}O\text{-}CH_2\text{-}CH_2\text{-}CH_2\text{-}S_x]_m CH_2\text{-}CH_2\text{-}O\text{-}CH_2\text{-}O\text{-}CH_2\text{-}CH\text{-}SH$$

LP polymer

$$HS\text{-}[CH_2\text{-}CH_2\text{-}O\text{-}CH_2\text{-}CH_2\text{-}CH_2\text{-}(S)_x]_{n-m} R\text{-}SH$$

Dithiol modified LP polymer

n is the sulfur rank of the polymers with the value of 2 to 3

x is 5 to 25

m is 0 to 24

This class of modified polysulfide polymers exhibits low viscosity and better compatibility with plasticizers, pigments, and fillers. The sealants derived from them are more thermally stable, more chemical resistant than the unmodified polysulfide sealants based on Thiokol LP polymers.

For a moist environment, a new one-component sealant[84] can cure with moisture as catalyst. The chemistry involves the reaction of a 1,3-dipole (a difunctional nitrile oxide) and a dipolarophile (an unsaturated polymer); a difunctional hydroximoyl chloride, which yields a nitrile oxide by the reaction with a latent base (e.g., barium oxide), reacts with the polymer upon exposure to moisture by forming isoxazoline crosslinks.

$$Ba(OH)_2 + Ar(COC(Cl){=}NOH)_2 \longrightarrow Ar(COC{\equiv}\overset{+}{N}\text{-}\overset{-}{O})_2 + BaCl_2 + 2H_2O$$

$$\downarrow\ {>}C{=}C{<}$$

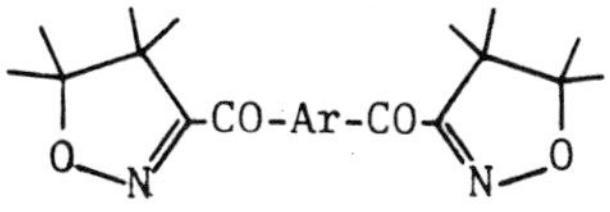

-CO-Ar-CO-

ISOXAZOLINE CROSSLINKS

For aerospace applications, advanced sealants are required to sustain extreme high and low temperatures. The problem has been that there are not many commercially available products. For advanced sealants, FASIL and phosphazenes are two materials that have a low servicing

Table 3. Test Run for Advanced Sealants
(Temp.:-20°C→200°C; Pres: 20 psi)

Sealant	Cycle to Failure
DC94011 (or LS77135), Poly[trifluoropropyl(methyl)siloxane]	5
GE651: Poly[ethylcyano(methyl)siloxane]	15 (but not re-injectable)
LS/FCS 210 (plus particles with primed surface) MW = 268,000	No failure after 20
PNF (plus particles)	No failure after 20
Fasil	No failure after 20

From B. Boutevin and Y. Pietrasanta, Prog. Org. Coatings 13, 319 (1985). (Ref. 88)

temperature between -50° F and -80° F. FASIL, fluoroalkylarylenesiloxanylene, which was developed by the Air Force Materials Laboratory,[85] has an operating range of -54°→250°C (-66°→500° F). The phosphazene,[86] which was originally made by Firestone as PNF®, is now available from Ethyl as EYPEL-F™. It is a phosphonitrilic fluoroelastomer, which has been used for low temperature seals, such as O-rings. The temperature range is -65° and 175°C (-85° and 347° F). Most other commercially available fluoroelastomers, including duPont's Kalrez® perfluoroelastomers,[87] have glass transition temperatures (Tgs) higher than those of EYPEL-F. However, in the laboratory, polyfluorovinyl ether terpolymers have shown lower Tgs than those of EYPEL-F or fluorosilicones.[88] Figure 2 shows the relationship between fluid resistance and effective temperature range for related

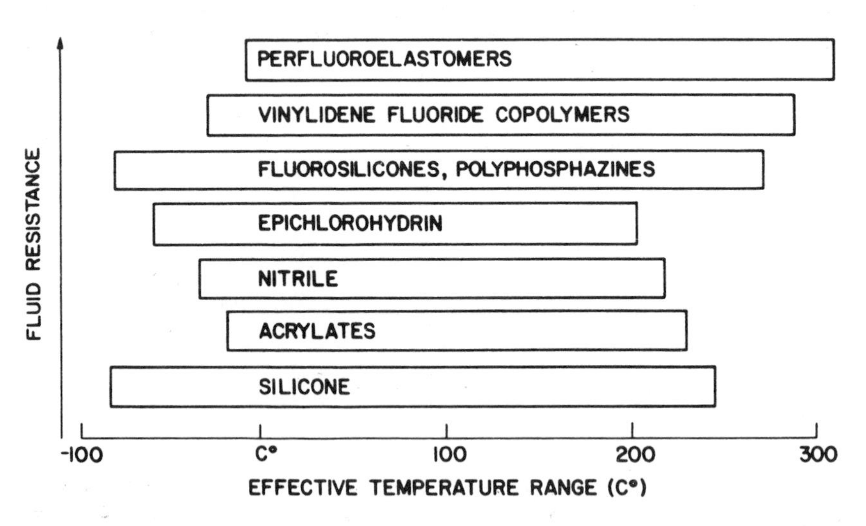

Fig. 2. Dependence of fluid resistance on effective temperature range. From R.E. Uschold, Polymer J. (Japan) 17 (1), 253 (1985). (Ref. 87)

polymers.[87] Among them, fluorosilicones and polyphosphazenes appear to cover the widest temperature ranges. In Table 3, the test runs for various advanced sealants are compared in terms of cycles to failure.[88] Among them, LS/FCS210, PNF or EYPEL-F, and Fasil showed no failure after 20 cycles.

COATINGS FOR SPACE ENVIRONMENTS

The extreme heat generated during the ascent or descent of a spacecraft can cause the ablation of organic polymers (such as phenolic resins, epoxy resins, PTFE, and polyimides) into chars. These chars may serve as a protective shell for a part of the spacecraft. An interesting example of ablative coatings[89] is based on the isocyanurate spray-on foam insulation (SOFI) that insulates the cryogenic liquid oxygen or liquid hydrogen tank from boiling-off on the space shuttle launcher. The temperature of these materials changes from -430° F before launch to greater than 1,000° F during ascent. At the latter temperature, the polymeric foam would char into an ablative coating.

Much like ablative coatings, intumescent coatings[89] have been employed in aircraft or building structures to protect against heat and flame without alteration of the original properties of the underlying materials. These coatings generally consist of a polyhydroxy compound, a dehydrating chemical, a blowing agent, and a resin binder. Upon ablation, these coatings can form a charry layer 200 times the original thickness.

Ablative materials, or ablators, have been the main protective shells for past spacecraft including the Apollos. However, since 1981, ceramic tiles[90] made of rigid silica fibers have been the most important thermal protective system (TPS) for shuttle orbiters. On the upper surface of the orbiter, the temperature is about 700° F, but lower than the rest of surface; thus a flexible, coated nylon felt has been used for protection. The ceramic tiles consist of those with high temperature, reusable surface insulation (HRSI) and those with low temperature, reusable surface insulation (LRSI). The HRSI tiles are coated with a black borosilicate glass, the LRSI with a white material. Both are then overcoated with a water repellent. The nylon felt of the flexible reusable surface insulation (FRSI) is then coated with a thin silicone elastomeric film.

Besides heat, the spacecraft will be bombarded by atoms, ions or molecules in the low earth orbit (LEO) environments. Both atomic oxygen and O^+ have been speculated[91] as the major bombarding particles. Upon bombardment, many organic polymers have been shown to suffer a loss of surface integrity.[92] However, reports have[93] claimed that silicone polymers are more atomic oxygen resistant than others (see Table 4). A copolymer of poly(siloxane-imide) has been reported by Yilgor(94) to be resistant to atomic oxygen.

For protection, sputtered coatings have been developed by NASA-Lewis Research Center[95] for Kapton® polyimide on the spacecraft. The ion-beam sputtered Al_2O_3 and SiO_2 and a co-deposited mixture of SiO_2 containing a small amount of polytetrafluoroethylene (PTFE), 4%, were found to be effective. The mass loss rate of the coated Kapton® was reduced to 0.2% compared to the rate of the uncoated.

Table 4. Reaction efficiencies of selected materials with atomic oxygen in low earth orbit (LEO)

Material	Reaction Efficiency, cm^3/atom
Kapton	$3x10^{-24}$
Mylar	3.4
Tedlar	3.2
Polyethylene	3.7
Polysulfone	2.4
Graphite/epoxy 1034C 5208/T300	 2.1 2.6
Epoxy	1.7
Polystyrene	1.7
Polybenzimidazole	1.5
25% Polysiloxane/45% Polyimide	0.3
Polyester 7% Polysilane/93% Polyimide	0.6
Polyester	Heavily attacked
Polyester with Antioxidant	Heavily attacked
Silicones	0.2*
Perfluorinated polymers Teflon, TFE Teflon, FEP	 <0.05 <0.05
Carbon (various forms)	0.9-1.7
Silver (various forms)	Heavily attacked
Osmium	0.026

Data Source-NASA Johnson Space Center (Cited in Ref. 93)

A simulated test(96) on ground was carried out to determine the extent of atomic oxygen damage on protective coatings for Galileo spacecraft. As a result, an ITO (indium-tin-oxide)-coated polyester was chosen for the thermal blankets for the Galileo spacecraft because of low weight loss.

In the LEO environment, the best protective coating has been shown to be inorganic materials. The graphite/epoxy tubular structures for a manned Space Station truss assembly(97) have been protected by anodized aluminum foil with or without the sputtered SiO_2 on the surface. The sputtered foil has poorer adhesion to the composite than the unsputtered foil. The anodized aluminum foil offers optical tailorability, ease of manufacturing and excellent handling properties. It is also relatively low cost.

COATINGS FOR CORROSIVE ENVIRONMENTS

Though organic coatings have long been employed to protect metals from corrosion, there are still many unsolved problems[98] with corrosion mechanisms, adhesion, and practical testing. Water is generally regarded as the major ingredient causing corrosion. The consensus is that under an organic coating a layer of water builds up. (Fig. 3) When a cathodic corrosion takes place, water, with the aid of oxygen and electrons, can produce hydroxide ions according to the following equation.[99]

$$H_2O + \frac{1}{2}O_2 + 2e^- = 2OH^-.$$

This reaction is facilitated by cation counterions and a catalytic surface of metal oxide. According to the above equation, we can speculate on some of the measures that should help prevent or deter the corrosion (cathodic) process and the subsequent loss of adhesion of the coating. We could:

1) Modify the polymer structure and achieve lower water permeability. For example, the fluorinated epoxy resin absorbs much less water (about 85% less), than the unmodified epoxy.[17]

2) Reduce oxygen permeability through coating the metal by applying a certain primer as discussed in an earlier section.

3) Reduce the conductivity of the oxide layer. A recent work by Jain et al.[100,101] describes a layered semiconductor/insulator structure to prevent corrosion that is presumably similar to this approach, but depends on the formation of an active electronic barrier.

4) Incorporate a cation-exchange material at the interface. Leidheiser and Wang[98] demonstrated that the rate of the cation counterion diffusion through the organic coating can control the hydroxide formation. Thus, Leidheiser[102] proposed the use of a cation-exchanger at the interface. However, in reality, this proposal is not easy to carry out.

5) Use a corrosion inhibitor to suppress the catalytic activity of the metal oxide. This is perhaps the most common way to prevent corrosion. In recent years, with aluminum, a corrosion inhibitor, NMTP, nitrilostris (methylene) phosphoric acid ($N[CH_2P(O)(OH)_2]_3$, has been used to inhibit the hydration of Al_2O_3 into bayerite $Al(OH)_3$[45] and to prevent the adsorption of water on the Al_2O_3 surface. With copper, polyvinylimidazole[103] has been found to be an effective inhibitor at elevated temperatures. With metal oxide surfaces, polyacrylic acid[50] has been adsorbed to enhance adhesion and also, presumably, to suppress the oxide activity. A list of corrosion inhibitors for other metals has been compiled by Trabanelli and Carassiti.[104]

Even in the absence of corrosion, water can cause disbondment of the coating[105] from the adherend (Fig. 4). Water can diffuse into the coated film due to a concentration gradient. The direct interaction of water with the adhesive bond at the interface can lead to disbondment or delamination. Thus, in some cases, the addition of a coupling agent [106] to a coating system can improve adhesion and reduce the chances of disbondment.

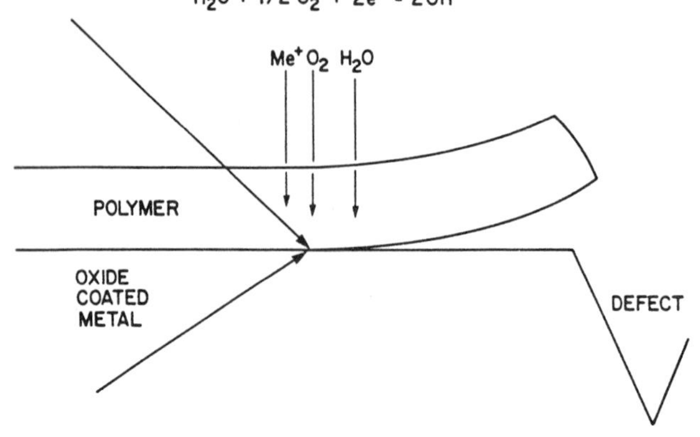

Fig. 3. Simplified model of the mechanism of cathodic delamination in an electrolyte

From H. Leidheiser, Jr. and W. Wang, J. Coat. Technol. 53, No. 672, 77, 1987 (Ref. 99).

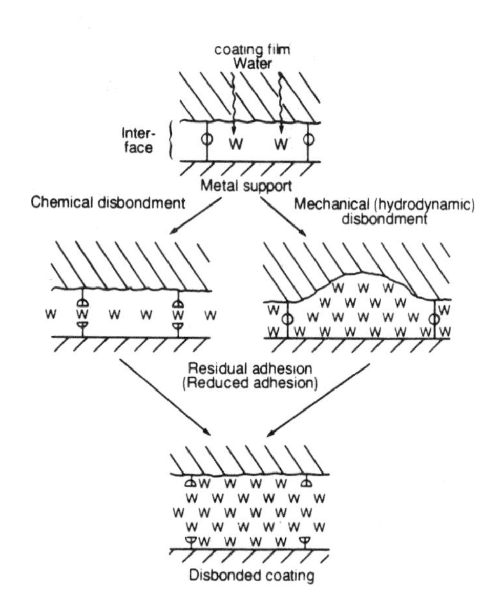

Fig. 4. Proposed mechanism for the disbonding of coating film by water.

From H. Leidheiser, Jr. and W. Funke, J. Oil and Col. Chem. Assoc. 70(5), 121) (1987) (Ref. 105).

Two papers on the corrosion of steel in a salt solution[107] and in an acidic solution[108] have been published. In a salt solution, the loss of adhesion has been attributed to oxygen depolarization, while in an acid solution, the delamination or blister-formation is caused by the evolution of hydrogen.

In marine environments, corrosion by salt water has always been a severe problem. Aerospace missiles[109] on ships are equally susceptible to attack by salt and water. A finish system of a cathodic base coat with an epoxy or polyurethane topcoat was demonstrated to provide exceptional corrosion resistance for the motor cases of missiles in severe marine environments.

For a wet rusty surface, a new coating material has been developed to displace water. The polymer[110] is a product of polyetheramine D400 and p-benzoquinone(2:3).

In marine environments, salt water is not the only problem. The marine life, such as barnacles and mussels, tend to form hard and soft fouling colonies on ship hulls. In a previous section, we mentioned the powerful water-based adhesives[21] secreted by them, and those adhesives cause the strong attachment of marine life to the ship hulls. To prevent the building of a thick layer of uncoated, marine colonies, antifouling paints[111,112] have been used to coat the ship hulls. Those paints generally contain toxic biocides, such as cuprous oxide[112] or tin-based chemicals. The problem about those antifouling paints is the leaching of biocides into water to poison or contaminate other marine life.

For environmental protection purposes, new alternatives[113] to antifouling paints have been experimented with. Low surface energy polymers,[114] such as fluorourethane and silicone rubbers, have been actually applied as coating to prevent fouling colonies from attaching to the ship hulls. An effective fluorourethane coating contains 24% PTFE (polytetrafluoroethylene) by volume and the NRL fluoropolyol, Desodur® N-75 (an aliphatic polyisocyanate), titanium dioxide and solvents. This is a more durable coating than those containing a higher percentage of PTFE. One silicone elastomer from General Electric Company, which is essentially a high-strength silicone release coating, has been tested and shows promising results.

ELASTOMERS AND SEALS FOR CORROSIVE AND HARSH ENVIRONMENTS

In a very severe, corrosive environment common organic coatings might fail; however fluoropolymers[115] could provide some protection. For example, the PFA elastomer, which is a copolymer of tetrafluoroethylene with a perfluoroalkoxy pendant side chain, has excellent flex life, chemical inertness, and melt-processability. PFA can be used to protect machine components and electrical parts. Some fluoropolymers, e.g., CM-X, which has unusual mechanical properties, can be applied by electrostatic coating. In addition, fluorosilicone elastomers[116] have also performed well under severe environments. Most of these polymers are excellent seal materials[117] or sealants. Some of the fluoroelastomers have also been used for the automotive underhood environments.[118,119]

The failure of Challenger was caused by a leak at a joint between segments of one of the two booster rockets. Superheated gases and flame shot past two synthetic rubber (presumably fluoroelastomer) O-rings and touched off the explosion of the huge external fuel tank. The Rogers Commission report said that the cold weather, 36°F at 11:38 a.m. liftoff on January 28, 1986, contributed to the accident by robbing the O-rings of their resiliency, preventing them from sealing the joint properly. A salient question is "could a lower temperature elastomeric seal prevent such a catastrophe?" After we reviewed this volume of literature, we wonder whether research on this subject matter deserves attention. For our future space program, we urgently need to develop more low-temperature sustaining polymers and rubbers.

One of the most corrosive environments is that of an oilfield in which elastomers(120) have been used as components for deep-well recovery. In general, deep-well drilling equipment must withstand continuous temperatures of 400° F with approximately 16-hour excursions around 600° F and pressures in excess of 20,000 psi in the presence of H_2S, CO_2, CH_4, saline, and acidic and corrosive chemicals, plus amine inhibitors. One of the problems is that most base corrosion inhibitors attack conventional fluoroelastomers.(121) Recently, two families of organic elastomers have been reported to survive under these corrosive environments. The first is a family of fluoroelastomers(122) containing the terpolymers of TFE/olefin/FVE (TFE=tetrafluoroethylene and FVE=perfluoroalkyl vinyl ether); the second a family of saturated nitrile rubbers.(123) The latter also found applications for the automotive underhood environments.(124)

The saturated (or hydrogenated) nitrile rubber (HSN or HNBR) can be represented by the following formula:

$$-\!\!\left(CH_2\text{-}CH_2\text{-}CH_2\text{-}CH_2\text{-}CH_2\text{-}\underset{\underset{CN}{|}}{CH}\right)\!\!-$$

The trade names of HSN are Therban (Bayer AG and Mobay Corp.) and Zetpol (Nippon Zeon). Major processing conditions and properties for Therban(118) are as follows:

Processing

- easy mixing
- very good storage stability of the compounds
- very good processing safety
- easy processing on standard industry equipment

Properties

- excellent physical properties, even at elevated temperatures
- outstanding abrasion resistance even at elevated temperatures
- good low temperature flexibility
- low compression set values
- very good hot air resistance
- outstanding resistance to weathering and ozone
- good resistance to additive-containing oils and cooling fluids
- good resistance to crude oil even in the presence of hydrogen sulfide and amines (corrosion inhibitors)

One of the problems for the HSN is the wet H_2S exposure. The HSN can still be degraded by H_2S in the presence of water.(125) However, the application temperature of 350° F for HSN[39] is significantly higher than that (250° F) for NBR.

CERAMIC COATINGS FOR EXTREME ENVIRONMENTS

When organic coatings and fluoropolymers all fail to perform at extreme high temperatures or harsh application conditions, ceramics take over. The three most important features of ceramic coatings[126] are abrasion resistance, resistance to high temperatures, and electrical insulation. The coatings can be resistant to oxidizing agents, acids, and alkalis. For abrasion resistance, chromium oxide, aluminum oxide, and aluminatitania can be used. For thermal resistance, zirconium oxide (m.p. 4,500° F), magnesium zirconate (m.p. 3,900° F) and yittria-stabilized zirconium oxide (m.p. 4,800° F) are the top choices. For electrical insulation, aluminum oxide and magnesium aluminate should be adequate. These ceramic coatings can be applied by one of the four methods: oxygen acetylene powder, oxygen acetylene rod, plasma, and detonation gun.

For gas turbines[127] a common ceramic coating system consists of a plasma-sprayed ZrO_2-(6%-8%) Y_2O_3 layer over an MCr ALY (M≡Ni, Co or NiCo) bond-coated layer plasma-sprayed at low pressure. Nearly all thermal barrier coatings for engines have been concentrated on ZrO_2-based materials applied by plasma-spraying.[128]

COATINGS FOR ELECTRONIC AND OPTICAL ENVIRONMENTS

Water is not only a major ingredient for metal corrosion, it is also a key culprit for the corrosion and degradation of electronic and optical devices. In the case of optical fibers, nascent silica fibers must be protected by a polymer coating to prevent abrasion and microbending loss. A good coating should have the following attributes:[129]

- the ability to be stripped for splicing and connecting,
- thermal, oxidative, and hydrolytic stability,
- the ability to bond in cable structures,
- resistance to compounds such as water and gasoline, and
- handling characteristics (e.g., low surface tack, toughness, abrasion resistance).

The fatigue properties of the fibers depend greatly upon the water permeability and adhesion of the polymer coating.[130] Most optical fibers are coated with UV-curable polymers, e.g., epoxy acrylate[131,132] or silicone.[133,134] The silicone coating exhibits relatively lower water absorption during 21 days immersion at room temperature (0.3% vs about 4% for UV-cured acrylates tested under the same conditions). Other polymeric materials, e.g., fluoropolymers,[20] polyamides, polyester, ethylene-vinyl acetate etc.[135] have also been used for coating optical fibers.

For solar reflector applications, protective coatings are also needed for the back surfaces of mirrors and sealants on the edges of solar heliostats.[136] The primary reason for mirror failure is silver corrosion from moisture. Thus, an intimate bonding is essential to reduce moisture on the silvered surface. With good adhesion, a Kraton® styrene-butadiene block copolymer was found to provide a superior back surface protection against corrosion for at least 12 months. An ultraviolet-stabilized butyl rubber appeared to be the best edge seal. Silicone was not suitable for the latter application because of its higher water permeability. For antireflection coatings of transparent polymeric materials, surface fluorination was shown to lower permeability and to improve bondability and thermal stability.[137] Optical measurements also revealed a substantial improvement in specular transmittance after fluorination.

All polymer glazings for silver mirrors have a common environmental problem--UV degradation. A suitable stabilizer against UV radiation is incorporated with each polymer to improve its weatherability.[138,139]

For electronic or microelectronic devices, corrosion can cause catastrophic failures.[140] To achieve reliability, corrosion-resistant materials, such as protective coatings or encapsulants,[141] and corrosion-preventing agents, such as inhibitors, are commonly used. The corrosion can be atmospheric, galvanic, or electrolytic. Atmospheric corrosion results from the interactions of moisture, oxygen, ionic impurities, and low molecular weight organic compounds, while galvanic corrosion is caused by the contact of dissimilar materials. Electrolytic corrosion of conductors is induced by the potentials normally applied to an electronic circuit.

Oxidation of metals or semiconductors can take place spontaneously. However, not all oxidation processes are beneficial, as in the case of the passivation of Si by SiO_2 or of Al by Al_2O_3. For many electronic devices, low temperature oxidation can be detrimental. A thin layer of oxide film in the range of 100 Å can result in an open circuit, due to the increase in resistance.[142]

The magnetic or magneto-optical properties of amorphous thin films[143] derived from Fe, Co, Tb, or Gd could be severely affected by oxidation or corrosion. These materials are potential storage media in high-density mass memories, if not oxidized.

Moisture can also affect the rate of oxidation and the structure and electrical properties of the product.[144] Interactions of moisture with ionic contaminants tend to affect electrolytic corrosion. High levels of chloride and humidity can totally corrode the positively biased aluminum bonding pads in devices.[145] Therefore, recent reductions of the chloride content in encapsulants from several hundred ppm to 5 ppm have improved overall device reliability by factors of 10 to 100.[146]

CONCLUSIONS

In this review, we have pointed out the importance of environmental factors, such as moisture or water, oxygen, temperature, and contaminants, in the corrosion of metals and organic materials. Among them, water is the major detrimental factor.

In space environments, temperature extremes require exotic polymers as adhesives and sealants for spacecraft, and bombardment by particles, e.g., molecules, ions, and atoms, requires special coatings on the surface of spacecraft.

Several commonly known measures that prevent cathodic corrosion of metals were briefly discussed. For optical and electronic materials, water was also shown to be the major problem for both cathodic (atmospheric) and electrolytic corrosions, but not for the galvanic corrosion.

REFERENCES

1. L.H. Lee, Ed. Adhesives, Sealants, and Coatings for Space and Harsh Environments, Plenum, New York (1988).
2. L.H. Lee, "Adhesives and Sealants for Severe Environments," Int. J. Adhes. and Adhes., 7(2), 81 (1987).
3. J. Comyn, Ed., Polymer Permeability, Elsevier Applied Science Publishers, New York and London, 1985.
4. C.E. Rogers, "Effect of Environmental Exposure on Sorption and Transport of Penetrants in Polymeric Materials," in Ref. 1.
5. P.E. Cassidy, T.M. Aminabhavi and J.C. Brunson, "Water Permeation Through Elastomer. Laminates I. Neoprene/EPDM," Rubber Chem. and Technol., 56, 357 (1983).
6. P.E. Cassidy and T.M. Aminabhavi, "Water Permeation Through Elastomer Laminates II. SBR/EPDM," Rubber Chem. and Technol., 59, 779 (1980).
7. A.G. Thomas and K. Muniandy, "Absorption and Desorption of Water in Rubbers," Polymer, 28, 408 (1987).
8. J. Crank, The Mathematics of Diffusion, Ch. 12, Clarendon Press, Oxford (1956).
9. H.L. Frisch, "Anomalous Polymer-Penetrant Permeation," J. Chem. Phys., 37, 2408 (1962).
10. T.C. Wong and L.J. Broutman, "Water in Epoxy Resins Part I. Non-Fickian Sorption Processes," Polym. Eng. Sci., 25 (9), 521 (1985).
11. T.C. Wong and L.J. Broutman, "Water in Epoxy Resins Part II. Diffusion Mechanism," Polym. Eng. Sci., 25 (9), 529 (1985).
12. D.M. Brewis, J. Comyn and R.J.A. Shalash, "The Effect of Water and Heat on the Properties of an Epoxide Adhesive in Relation to the Performance of Single Lap Joints," Polym. Comm., (24), 67 (1983).
13. B.L. Rathbun and P. Wh. Schuessler, "Moisture Permeation of Polymer Sealants and Interface Modifying Films," in Adhesive Chemistry-Developments and Trends, Ed. L.H. Lee p.785, Plenum, New York (1984).
14. P. Wh. Schuessler, "Hydrophobic Resins," NASC Phase I Report, May, 1981.
15. J.R. Griffith, "Epoxy Resins Containing Fluorine," CHEMTECH 12, 290 (1982).
16. S.J. Shaw, D.A. Tod and J.R. Griffith, "The Mechanical and Water Absorption Behavior of Fluoroepoxy Resins," in Ref. 1.
17. P. Johncock and G.F. Tudgey, "Epoxy Systems with Improved Water Resistance, and the Non-Fickian Behavior of Epoxy Systems During Water Ageing," Brit. Polym. J., 15, 14 March 1983.
18. S. Sasaki and K. Nakamura, "Syntheses and Properties of Cured Epoxy Resins Containing the Perfluorobutenyloxy Group I. Epoxy Resins Cured with Perfluorobutenyloxyphthalic Anhydride," J. Polym. Sci. Polym. Chem., 22, 831 (1984).
19. H.P. Hu, R.D. Gilbert and R.E. Fornes, "Chemical Modification of Cured MY720/DDS Epoxy Resins Using Fluorinated Aromatic Compounds to Reduce Moisture Sensitivity," J. Polym. Sci., Polym. Chem., A25, 1235 (1987).
20. K. Nakamura and T. Maruno, "Recent Progress in Organic Materials for Optical Communications--Application of Fluorinated Epoxy Resin as Optical Adhesive," Technol. Japan, 20 (1), 8 (1987).
21. J.H. Waite, "Nature's Underwater Adhesive Specialist," Int. J. Adhes. & Adhes. 7 (1), 9 (1987), also CHEMTECH, 17 692 Nov. 1987.
22. K. Ab-Malek and A. Stevenson, "The Effect of 42 Years Immersion in Seawater on Natural Rubber," J. Mater. Sci., 21, 147 (1986).

23. A. Stevenson, "On the Durability of Rubber/Metal Bonds in Seawater," Int. J. Adhes. & Adhes. , 5 (2), 81 (1985).
24. R.Y. Ting, "A New Approach to the Quality Analysis of Transducer Elastomers," Elastomerics, 117, 29, Aug. 1985.
25. A. Stevenson, "The Effect of Electrochemical Potentials on the Durability of Rubber/Metal Bonds in Sea Water," J. Adhes., 21, 313 (1987).
26. Y. Ishikawa and S. Kawakami, "Effects of Salt Corrosion on the Adhesion of Brass-Plated Steel Core to Rubber," Rubber Chem. and Technol., 59, 1 (1986).
27. P.R. Troyk, M.J. Watson and J.J. Poyezdala, "Humidity Testing of Silicone Polymers for the Corrosion Control of Implanted Medical Electronic Prosthesis," Polym. Mater. Sci. Eng., 53, 457, Sept. 1985.
28. C.-L. Ruan, J.-H. Liu and Z.-Y. Qin, "Sealing of Human Tissues with High Alkyl Cyanoacrylate Adhesive XKM-2 and Its Effects--A New Sterilization and Nonsurgical Technique," in Ref. 1.
29. A.A. Fahmy and J.C. Hurt, "Stress Dependence of Water Diffusion in Epoxy Resin," Polym. Compos., 1, 77 (1980).
30. O. Gillat and L.J. Broutman, ASTM STP 685, p.61, American Society for Testing and Materials (1978).
31. J.L. Cotter, "The Durability of Structural Adhesives" in Developments in Adhesives 1., Ed. W.C. Wake, Chapt. 1, Applied Science Publishers, London, U.K. (1977).
32. E.P. Plueddemann, "Silane Coupling Agents", Plenum, New York, (1982).
33. A.J. DeNicola, Jr. and J.P. Bell, "Polyfunctional Chelating Agents for Improved Durability of Epoxy Resin to Steel," Org. Coat. and Appl. Polym. Sci. Prepr., 46, 489 (1982).
34. R.G. Schmidt and J.P. Bell, "Investigation of Steel/Epoxy Adhesion Durability Using Polymeric Coupling Agents I. Synthesis and Characterization of EME Coupling Agents," in Ref. 1.
35. A.J. DeNicola, Jr. and J.P. Bell, "Synthesis and Testing of β-Diketone Coupling Agents for Improved Durability of Epoxy Adhesion to Steel," in Adhesion Aspects of Polymer Coatings, Ed. K.L. Mittal, Plenum, New York (1982) p.443.
36. L.B. Cohen, "Zircoaluminate Coupling Agents as High Performance Adhesion Promoters," presented to 59th ACS Colloid and Surface Chemistry Symposium, June, 1985.
37. Q.-X. Yang and Q.-L. Zhou, "ESCA and AES Studies of the Interfacial Chemical Bonding Between Aluminum and Chromium (III) Fumarato-Coordination Compound," in Adhesive Chemistry-Developments and Trends, Ed. L.H. Lee, Plenum, New York (1984), p.799.
38. C.Q. Yang, T.W. Rausch, S.P. Clough and W.G. Fately, "The Bonding and Hydrolysis of Cr(III) Fumarato Coupling Agent on Aluminum Surfaces Studied by Auger Electron Spectroscopy and X-ray Photoelectron Spectroscopy," in Ref. 1.
39. D.F. Sounik and M.E. Kenney, "New Potential Silane Coupling Agents," Polym. Compos., 6 (3) 151 (1985).
40. F.J. Kolpak, "Applications of Azidosilane Coupling Agents in Reinforced Thermoplastic Composites," SAMPE Q., 18 (1) 21, Oct., 1986.
41. W. Thiedman, F.C. Tolan, P.J. Pearce and C.E.M. Morris, "Silane Coupling Agents as Adhesion Promoters for Aerospace Structural Film Adhesives," J. Adhes., 22, 197 (1987).
42. H.M. Clearfield, D.K. McNamara and G.D. Davis, "Adherend Surface Preparation for Structural Adhesive Bonding," in Fundamentals of Adhesion, Ed. L.H. Lee, to be published by Plenum, New York (1988).

43. J.D. Venables, "Review-Adhesion and Durability of Metal-Polymer Bonds, "J. Mater. Sci., 19, 2431 (1984).
44. W. Brockmann, O.-D. Hennemann, H. Kollek and C. Matz, "Adhesion in Bonded Aluminum Joints for Aircraft Construction," Int. J. Adhes. & Adhes., 6 (3) 115 (1986).
45. D.A. Hardwick, J.S. Ahearn and J.D. Venables, "Environmental Durability of Aluminum Adhesive Joints Protected with Hydration Inhibitors," J. Mater. Sci., 19, 223 (1984).
46. G.D. Davis, J.S. Ahearn, L.J. Matienzo and J.D. Venables, "Use of Hydration Inhibitors to Improve Bond Durability of Aluminum Adhesive Joints, "J. Mater. Sci., 20, 975 (1985).
47. D.A. Hardwick, J.S. Ahearn, A. Desai and J.D. Venables, "Environmental Durability of Phosphoric Acid Anodized Aluminum Adhesive Joints Protected with Hydration Inhibitors," J. Mater. Sci., 21, 179 (1986).
48. L.J. Matienzo, D.K. Shaffer, W.C. Moshier and G.D. Davis, "Environmental and Adhesive Durability of Aluminum-Polymer Systems Protected with Organic Corrosion Inhibitors," J. Mater. Sci., 21, 1601 (1986).
49. D. Finch, "Zinc-rich Primers Protect Metal Substrates," Mater. Eng., 103, 41 (1986).
50. T. Sugama, L.E. Kukacka and N. Carciello, "Nature of Interfacial Interaction Mechanisms Between Polyacrylic Acid Macromolecules and Oxide Metal Surfaces," J. Mater. Sci., 19, 4045 (1984).
51. T. Sugama, L.E. Kukacka, N. Carciello and J.B. Warren, "Chemisorption Mechanism and Effect of Polyacrylic Acid on the Improvement in Bond Durability of Zinc Phosphate-to-Polymer Adhesive Joints," Informal Report BNL 37808, Brookhaven National Laboratory, January, 1986.
52. T. Sugama, "Adhesion Properties of Polyelectrolytes--Chemisorbed Zinc Phosphate Conversion Coatings," in Ref. 1.
53. R.A. Pike, "Inorganic Primers in Bonded Joints," Int. Adhes. and Adhes., 5 (1), 3 (1985).
54. R.A. Pike, "Inorganic Adhesive Primers: Effect of Metal Surface Treatments and Formation Temperature," Int. Adhes. and Adhes. 6 (1), 21 (1986).
55. R.A. Pike, "Inorganic Primers for Adhesive Bonding: A Status Report," in Ref. 1.
56. M.A. Steinberg, "Materials for Aerospace," Sci. Am., 255, 66, Oct. 1986.
57. Reliability of Adhesive Bonds Under Severe Environments, National Materials Advisory Board, National Research, Council, Dec. 1984.
58. L.K. English, "Premium Performance From Polyimides," Mater. Eng., 103, 14, Jan. 1986.
59. P. Hergenrother, "Status of High Temperature Adhesives," in Adhesive Chemistry-Developments and Trends, Ed., L.H. Lee, Plenum, New York (1984) p.447.
60. P. Hergenrother, "High Performance Thermoplastics," Angew. Makromol. Chem., 145/146, 323 (1986).
61. P. Hergenrother, "Recent Advances in High Temperature Polymers," Polymer J., (Japan) 19, (1), 73 (1987).
62. P.E. Cassidy, "An Overview of Polymers for Harsh Environments: Aerospace, Geothermal and Undersea," in Ref. 1.
63. Chang Chih-Ching, "High-Temperature Organic Adhesives--A Review" in Ref. 1.
64. D.J. Progar, "Evaluation of Polyimide Films as Adhesives,"--J. Adhes. Sci. Tech., 1 (1) 53 (1987).
65. H.D. Burks and T.L. St. Clair, "Synthesis and Characterization of a Melt Processable Polyimide," NASA Technical Memo 84494 (1982).

66. D.J. Progar and T.L. St. Clair, "Evaluation of a Novel Thermoplastic Polyimide for Bonding Titanium," Int. J. Adhes. and Adhes., 6 25, Jan. 1986.
67. T.L. St. Clair, "Adhesive Development of NASA Langley" presented to Program Review/Workshop, Virginia Polytechnic Institute and State University, Blacksburg, VA, May 1, 1984.
68. D.J. Progar, "Adhesive Evaluation of LARC-TPI and a Water-soluble Version of LARC-TPI," Int. J. Adhes and Adhes., 6, 12 (1986).
69. H.D. Burks and T.L. St. Clair, "Siloxane-modified Polyether Sulfideimide," J. Appl. Polym. Sci., 34, 351 (1987).
70. A.O. Hanky and T.L. St. Clair, "Semi-2-interpenetrating Polymer Networks of High Temperature Systems," presented to 30th National SAMPE Symposium, March, 1985.
71. V.R. Foster, "Polymers in Caulking and Sealant Materials," J. Chem. Edu., 64 (10), 861 (1987).
72. D.B. Paul, P.J. Hanhela and R.H.E. Huang, "Effects of Environment on Performance of Polysulfide Sealants," in Ref. 1.
73. M.J. Owen and J.M. Klosowski, "Durability of Silicone Sealants," in Ref. 1.
74. G. Holden, "Styrenic Block Copolymers in Sealants," Elastomerics, 119 (3), 26 (1987).
75. S.G. Chu, "Hot Melt Sealants Based on Thermoplastics Elastomers," in Ref. 1.
76. L.M. Gan, H.W.K. Ong and T.L. Tan, "Durability of Acrylic, Polysulfide and Silicone Sealants Bonded to Aluminum," Durability Build. Mater., 3, 225 (1986).
77. P.J. Hanhela, R.H.E. Huang and D.B. Paul, "Water Immersion of Polysulfide Sealants. 1. Effect of Temperature on Swell and Adhesion," Ind. Eng. Chem. Prod. Res. Dev., 25, 31 (1986).
78. P.J. Hanhela, R.H.E. Huang, D.B. Paul and T.E.F. Symes," Water Immersion of Polysulfide Sealants II. An Interpretation of the Influence of Curing Systems on Water Resistance," J. Appl. Polym. Sci., 32, 5415 (1986).
79. R.E. Meyer, "Polysulfide Sealants for Aerospace Part I. Theory and Background," SAMPE J. 6, Nov./Dec., 1982.
80. R.E. Meyer, "Polysulfide Sealants for Aerospace Part II. Application and Handling," SAMPE J., 7, March/April, 1983.
81. J. Fasold, "Polysulfide Helps Prevent Icing on Vanes of B-1B Aircraft," Adhes. Age, 29, (11) 20 (1986).
82. J.R. Panek and J.P. Cook, Construction Sealants and Adhesives, 2nd Edition, Wiley, New York (1984).
83. H. Singh, "A New Class of High Performance Polysulfide Polymers," Rubber World, 196 (5), 30 Aug., 1987.
84. D.S. Breslow, K. Brack and H. Boardman, "A One-component Sealant Based on 1,3-Dipoles," J. Appl. Polym. Sci., 32, 4657 (1986).
85. H. Rosenberg and E.-W. Choe, "Methyl- and 3,3,3-Trifluoropropyl-substituted m-xylylenesiloxanylene Polymers," Org. Coat. and Plast. Prepr, 37, No. 1, 166 (1977).
86. "EYPEL™ Performance Polymers-Expanding The Universe of Polymer Technology"--Ethyl Corporation (1985).
87. R.E. Uschold, "Fluoroelastomers: Today's Technology and Tomorrow's Polymers," Polym. J. (Japan) 17 (1), 253 (1985).
88. B. Boutevin and Y. Pietrasanta, "The Synthesis and Applications of Fluorinated Silicones, Notably in High-Performance Coatings," Prog. Org. Coat, 13, 297 (1985).
89. L.K. English, "Intumescent Coatings: First Line of Defense Against Fires," Mater. Eng., 103 (2) 39, Feb. 1986.
90. R.L. Dotts, D.M. Curry and D.J. Tillian, CHEMTECH, 14 616, Oct. 1984.

91. A. Garton, P.D. McLean, W. Wiebe and R.J. Densley, "Exposure of Cross-linked Epoxy Resins to the Space Environment," J. Appl. Polym. Sci., 32, 3941 (1986).
92. A. Garton, W.T.K. Stevenson and P.D. McLean, "The Stability of Polymers in Low Earth Orbit," Mater. Sci. Design, I (6), Nov./Dec., 1986.
93. L.K. English, "Atomic Oxygen: Achilles' Heel of Man in Space," Mater. Eng., 104 (8) 39, Aug., 1987.
94. I. Yilgör, "Synthesis and Characterization of Atomic Oxygen Resistant Poly(siloxane-imide) Coatings," in Ref. 1.
95. B.A. Banks, M.J. Mirtich, S.K. Rutledge and D.M. Swec, "Sputtered Coatings for Protection of Spacecraft Polymers," Thin Solid Films, 127, 107 (1985).
96. F.L. Bouquet and C.R. Maag, "Ground Radiation Tests and Flight Atomic Oxygen Tests of ITO Protective Coatings for Galileo Spacecraft," IEEE Trans. on Nucl. Sci., NS-33, (6), 1408 (1986).
97. H.W. Dursch and C.L. Hendricks, "Protective Coatings for Composite Tubes in Space Applications," SAMPE Q. 14, Oct. 1987.
98. W. Funke, et al., "Unsolved Problems of Corrosion Protection by Organic Coatings: A Discussion," J. Coat. Technol., 58 (741), 79 Oct. 1986.
99. H. Leidheiser and W. Wang, "Some Substrate and Environmental Influences on the Cathodic Delamination of Organic Coatings," J. Coat.Technol., 53 (672), 77, Jan. 1981.
100. F.C. Jain, J.J. Rosato, K.S. Kalona and V.S. Agarwala, "Formation of an Active Electronic Barrier at Aluminum/Semiconductor Interfaces: A Novel Approach in Corrosion Prevention," Corrosion (Houston), 42 (12), 700 (1986).
101. F.C. Jain, J.J. Rosato, K.S. Kalonia and V.S. Agarwala, "Corrosion Prevention in Metals Using Layered Semiconductor/Insulator Structures Forming an Interfacial Electronic Barrier," in Ref. 1.
102. H. Leidheiser, Jr., "Solid-State Chemistry as Applied to Cathodically Driven or Corrosion--Induced Delamination of Organic Coatings," Ind. Eng. Chem. Prod. Res. Dev., 20 (3), 547 (1981).
103. F.P. Eng and H. Ishida, "New Corrosion Inhibitors for Copper-Polyvinylimidazoles," Polym. Mater. Sci., Eng. 53, 725, Sept. 1985.
104. G. Trabanelli and V. Carassitti, in Advances in Corrosion Science and Technology, Vol. 1, Plenum, New York (1970), p.147.
105. H. Leidheiser, Jr. and W. Funke, "Water Disbondment and Wet Adhesion of Organic Coatings on Metals: A Review and Interpretation," J. Oil and Col. Chem. Asso. 70 (5), 121 (1987).
106. E.P. Plueddemann, "Silane Adhesion Promoters in Coatings," Prog. Org. Coat., 11, 297 (1983).
107 M. Knaster and J. Parks, "Mechanism of Corrosion and Delamination of Painted Phosphated Steel During Accelerated Corrosion Testing," J. Coat. Technol., 58 (738), 31, July, 1986.
108. M.L. White, H. Vedage, R.D. Granata and H. Leidheiser, Jr., "Failure Mechanisms for Organic Coatings Subjected to 0.1 M. Sulfuric Acid," Ind. Eng. Chem. Prod. Res. Dev., 25, 129 (1986).
109. W.B. Tolley and T.A. Eckler, "Corrosion Prevention for Aerospace Missile Environments," 31st Int. SAMPE Symposium, April 7-10 (1986), p.124.
110. K. Kaleem, F. Chertok and S. Erhan, "A Novel Coating Based on Poly(etheraminequinone) Polymers," Prog. Org. Coat., 15, 63 (1987).
111. C.A. Giúdice, B.d. Amo and V.J.D. Rascio, "Influence of Composition and Film Thickness on Anti-fouling Paints Bioactivity Containing Caster Oil as the Thixotropic Agent," in Ref. 1.

112. C.A. Giúdice, B.d. Amo and J.C. Benítez, "Determination of Metallic Copper, Cuprous Oxide and Cupric Oxide During the Manufacture and Storage of Anti-fouling Paints," J. Oil Col. Chem. Assoc., 64, 1 (1981).
113. N.A. Ganem, N.I. El-Awady, W.S. El-Hamouly and M.M. El-Awady, "New Approaches to Non-Toxic Anti-fouling Coatings for Ship-Hull Protection," J. Coat. Technol., 54 (684), 83 (1982).
114. R.F. Brady, J.R. Griffith, K.J. Love and D.E. Field," Nontoxic Alternatives to Anti-fouling Paints," J. Coat. Technol., 59 (755) 113 (1987).
115. L.K. English, "Fluoropolymers: First Family in Engineering Resins," Mater. Eng., 105 (1), 27 (1988).
116. R.B. Bush, "Flourosilicone Elastomers: Exceptional Performers for Severe Environments," Mater. Eng., 104, (7), 29 (1987).
117. R.A. Brullo and A.M. Sohlo, "New Considerations in the Selection of Fluorocarbon Elastomers for Seal Applications," Rubber World, 195 35, Oct. 1986.
118. D.D. Early, "Innovative Products Find Niches in Underhood Applications," Elastomerics, 118 (2), 13 (1986).
119. Editor, "Automotive Underhood Elastomers Adapt to Harsh Environments," Elastomerics, 119 (9), 27, (1987).
120. L.K. English, "Tougher Elastomers Tackle New Applications," Mater. Eng. 103, 37, July, 1986.
121. T.W. Ray and C.E. Ivey, "Effects of Organic Amine Inhibitors on Elastomers," Elastomerics, 118 (9) 24 (1986).
122. A.L. Moore, "Base-Resistant Fluoroelastomers Developed for Severe Environments," Elastomerics, 118 (9) 14 (1986).
123. J. Thoermer, J. Mirza and N. Shoen, "Therben Hydrogenated NBR in Oilfield Applications," Elastomerics, 118 (9) 28 (1986).
124. D.D. Early, "New Hydrogenated Materials Bridge 'Performance Gap'," Elastomerics, 117 (11), 26 (1985).
125. L.A. Peters and D.E. Cain, "Saturated Nitrile (HSN) vs Nitrile (NBR) in Temperature and Compatibility Tests," Elastomerics 118 (6), 28 (1986).
126. E.S. Hamel, "Ceramic Coatings: More Than Just Wear Resistant," Mater. Eng., 103, 30, Aug. 1986.
127. R.A. Miller, "Current Status of Thermal Barrier Coatings--An Overview," Surf. and Coat. Technol., 30, 1 (1987).
128. W.J. Lackey, D.P. Stinton, G.A. Gerny, A.C. Schaffhauser and L.L. Fehrenbacher, "Ceramic Coatings for Advanced Heat Engines--A Review and Projection," Adv. Ceram. Mater., 2 (1) 24 (1987).
129. L.L. Blyler, Jr. and C.J. Aloisio, "Polymer Coatings for Optical Fibers," CHEMTECH, 680, Nov. 1987.
130. T.-S. Wei, "Effects of Polymer Coatings on Strength and Fatigue Properties of Fused Silica Optical Fibers," Adv. Ceram. Mater., 1 (3) 237 (1986).
131. T.T. Wang, H.N. Vazirani, H. Schonhorn and H.M. Zupko, "Effects of Water and Moisture on Strengths of Optical Glass (Silica) Fibers Coated with a UV-cured Epoxy Acrylate," J. Appl. Polym. Sci., 23, 887-892 (1979).
132. A.H. Rodas, R.E.S. Bretas and A. Reggianni, "Ultraviolet-curing Blends for Coating of Optical Fibres," J. Mater. Sci., 21, 3025 (1986).
133. W.E. Dennis and D.W. Burke, "Elastomer Protects Optical Fibers," Res. and Dev. 70, Jan., 1986.
134. W.E. Dennis, "Optical Fiber Coatings" in Ref. 1.
135. T.-S. Wei, "Polymer Materials for Optical Fibers," in Ref. 1.
136. K.B. Wischmann, "Protective Coatings and Sealants for Solar Applications," in Polymers in Solar Energy Utilization, ACS Symposium Series, Washington, D.C. (1983) p.7.

137. G. Jorgensen and P. Schissel, "Effective Antireflection Coatings of Transparent Polymeric Materials by Gas-Phase Surface Fluorination," Solar Energy Mater., 12, 491 (1985).
138. H.H. Neidlinger and P. Schissel, "Stabilized Acrylic Glazings for Solar Reflectors," in Ref. 1.
139. H.H. Neidlinger and P. Schissel, "Polymer Glazings for Silver Mirrors," Solar Energy Mater., 14, 327 (1986).
140. R.B. Comizzoli, R.P. Frankenthal, P.C. Milner and J.D. Sinclair, "Corrosion of Electronic Materials and Devices," Science, 234, 340, Oct. 1986.
141. R.N. Gounder, J. Maiden, S. Keck and S. Carrle, "Advanced Materials Technologies for Spacecraft Systems," RCA Engineer, 29-5, 18, Sept./Oct. 1984.
142. R. Holm, Electrical Contacts: Theory and Applications, Ed. 4, Springer-Verlag, New York, 1967, Chap. 30.
143. R.B. vanDover, E.M. Gyorgy, R.P. Frankenthal, M. Hong and D.J. Siconolfi, "Effect of Oxidation on the Magnetic Properties of Unprotected TbFe Thin Films," J. Appl. Phys., 59 (4) 1291 (1986).
144. F.P. Fehler, Low Temperature Oxidation, The Role of Vitreous Oxides, Wiley, New York, 1986, Chapts. 8 and 9.
145. S.C. Kolesar, Annu Proc. Reliab. Phys., 12, 155 (1974).
146. M.T. Goosey, unpublished results (cited in Ref. 135).

EFFECTS OF ENVIRONMENTAL EXPOSURE ON SORPTION AND TRANSPORT OF PENETRANTS IN POLYMERIC MATERIALS

Charles E. Rogers

Center for Adhesives, Sealants and Coatings
Case Western Reserve University
Cleveland, Ohio 44106

ABSTRACT

The exposure of polymeric materials to hostile environments and/or conditions leading to mechanical deformation usually have marked effects on sorption and transport behavior in the materials. These effects can be interpreted, and in some cases predicted, by consideration of the chemical and physical changes in the materials resulting from exposure and deformation.

INTRODUCTION

The sorption, diffusion and permeation of low molecular weight substances (penetrants) in polymeric materials is governed by the relative physiochemical natures of the polymer and the penetrant.[1,2] Changes in the chemical composition, structure, or morphology of the polymer due to environmental exposure factors usually have pronounced effects on sorption and transport behavior. Chemical changes strongly affect sorption, which in turn causes changes in diffusion due to plasticization. Changes in void or free volume contents due to mechanical deformations or other causes can lead to drastic effects on transport behavior.

The effects of environmental factors can be interpreted, and in some cases predicted, in terms of the induced composition and structural changes with consideration of the resultant interactions with penetrant and modification of transport magnitudes and mechanisms. The effects of imposed hydrostatic pressure, mechanical deformations, chemical modifications (including thermal- and photo-degradation), and swelling induced structural changes on sorption and transport will be considered. Special attention will be given to changes in void and free volume content and in chemical composition in terms of their domain size and temporal and spatial distributions as functions of exposure conditions and time.

SORPTION AND TRANSPORT IN POLYMERS

The simplest relationship describing the unidirectional sorption-diffusion process through a membrane of thickness ℓ is:

$$J = Q/At = -D\,(\partial c/\partial x) = P\,\Delta c/\ell\ , \qquad (1)$$

where J is the flux, Q is the total amount of penetrant which has passed through area A during t and $(\partial c/\partial x)$ is the penetrant concentration gradient within the membrane at any given point or plane. The permeability constant $\underline{P}$ is defined as $\underline{P}$ = DS, where D is the diffusion coefficient and S is the solubility coefficient relating the internal sorbed concentration, c, to the ambient concentration (or pressure), c, S = c/c. More rigorous treatments, based on irreversible thermodynamics, consider that the driving force is the gradient of chemical potential, $\partial\mu/\partial x$.

For inert gases, D and S are usually constants at temperatures well above the critical temperature of the penetrant. For penetrants below their critical temperature, D and S are usually functions of the sorbed concentration with the dependence magnitude increasing as the sorbed concentration increases. D is generally more dependent on c than is S.

In polymers with high internal viscosities (e.g., glassy polymers) the sorption and transport processes are often accompanied by penetrant-induced swelling processes which are governed by the relaxation processes of the swelling polymer. Hence, the rates of sorption and transport, leading to the equilibrium permeation steady state of flow, may deviate considerably from simple Fickian scaling parameters (e.g., $t^{\frac{1}{2}}/\ell$). Such "anomalous" sorption-diffusion processes also are common in polymers with significant polymer-polymer interactions which are labile to disruption by penetrant on a time-scale comparable to or greater than the sorption-diffusion unit process timescale. This situation often arises due to chemical or other modifications caused by environmental exposure of the polymer. It is often accompanied by a thickness dependence of D and S caused by a gradient of chemical composition of the polymer due to the environmental exposure. A gradient of stress through the polymer also will affect D and S.

A major factor affecting the sorption-diffusion process is the "mode-of-sorption" relating to the nature of the interactions between the components as they affect the mobility of the sorbed penetrant molecules. Clustering of penetrant sorbed within the polymer or strong penetrant-polymer interactions effectively immobilizes a portion of the total sorbed penetrant content, removing it from participation in the overall transport process. This phenomenon, usually now analyzed as "dual-mode sorption" with a fraction immobilized and the remainder free to diffuse, again often is observed to develop during environmental exposure of certain polymers due to void formation favoring clustering and/or the formation of strongly interacting groups on the polymer which leads to both immobilization of penetrant and reduction of chain segmental mobilities.

EFFECTS OF MECHANICAL DEFORMATION

Polymeric materials used in many applications are subjected to various kinds and magnitudes of mechanical deformation either as a normal aspect of the application or as the result of some excessive applied stress. These deformations may increase the free volume in the system and/or cause other changes in polymer structure and morphology such as orientation, shear banding, yielding (plastic deformation), voiding, crazing and fracture. The sorption and, especially, transport of gases, vapors and liquids are very sensitive to even minor changes in polymer free volume and structure.[1,2] The marked dependence of these processes on deformation has been reported for a few polymers such as polyethylene[3-5], polystyrene and polycarbonate.[6,7]

Mechanical deformation of polymeric materials is defined here as involving either elastic, plastic, or voiding structural changes or some combination of those changes. Pure elastic response is reversible and does not involve any change in volume of the structure. Plastic deformation involves a yielding of structure with a more-or-less uniform volume increase throughout the deformed region. The deformation may be totally reversible but most often it will not be, exhibiting only a partial recovery toward the original structure by a relatively long-term relaxation process. Voiding includes the formation of interval voids, which may coalesce to form pore-like structures, crazes, cracks and fracture. Again, with the exception of cracks and fracture, these structures may partially recover but regions of permanent damage usually will remain.

In many polymers, orientation or strain-induced crystallization may occur during any of the types of deformation. The formation of shear bands apparently involves both localized orientation with concurrent void formation.[8] In filled polymers, deformation may result in loss of polymer-filler adhesion, a form of voiding. Cyclic deformations often give fatigue effects on structure which are different from those observed in singly deformed samples.[9]

Some major variables affecting deformation include the composition and initial morphology of the material, the deformation mode, rate, magnitude and temperature. A combination of these factors govern the nature of a material's response to deformation in terms of the balance between various structural changes (e.g., brittle-ductile transition, crazing-shear banding transition, etc.). Deformation modes of major interest include uniaxial tensile, biaxial tensile (concurrent or sequential), and shear. Cyclic deformations in these modes, leading to fatigue effects, are of particular interest since they relate to many application conditions.

Changes in S and D due to deformation can be attributed to stress, per se, as it affects the chemical potential (μ) of the system and hence the driving force for diffusion.[10] If a constitutive relation between stress (σ) and strain (ϵ) is known and if $\mu(\sigma)$ is known or can be assumed, then D, S and $\underline{P}$ can be obtained as a function of σ and ϵ. This approach must be tempered by consideration that not all deformations give stresses which affect S and D.

A more convenient, although somewhat less rigorous, approach is to consider changes in the volume of the penetrant-polymer system caused by deformation. It is expected that the formation of excess free volume by strain-induced structural changes will increase the magnitude of S, D and $\underline{P}$ for gases, vapors and liquids in polymeric materials. The formation of voids (not considered as free volume) also will increase D and P, with generally a lesser effect on S. The magnitude of voiding effects will increase as the connectivity between defect regions increases. Actual pores, cracks, or fracture through the material will allow bulk convective flow of penetrant to occur.

Such voiding or other dilative changes are not generally expected in elastomeric materials. Hence, the progressive orientation of chain structure with increasing deformation should cause a decrease in D with larger decreases expected with increasing penetrant molecular size and shape. This behavior becomes very apparent when elongation is sufficient to cause crystallization in self-reinforcing elastomers such as natural rubber. However, comparable effects on D, although of lower magnitude, also should be observed in non-crystallizing elastomers such as silicone rubber due to the restraint imposed upon segmental motions as the chains are extended.

The formation of oriented or crystalline regions by deformation will reduce or entirely eliminate sorption and diffusion within those regions.[1-5] However, the formation of those regions will almost always involve concurrent voiding or related dilative structural changes in neighboring regions in a bulk polymer material.

The time between deformation and exposure to penetrant, and the time of exposure to penetrant, will be of significance due to recovery and relaxation of the deformed structure. The sorption of vapors or liquids which swell the polymeric material will enhance the recovery/relaxation process. Sufficient swelling may effectively "heal" the deformed regions but at the expense of a more rapid swelling (higher initial sorption rate due to the presence of defect structures, free volume, etc.) and the probability of higher D and P due to plasticization of the polymer (concentration dependent S, D and P behavior).[1,2]

Effects of Deformation of Elastomers on PDS It has been reported (e.g., Refs. 1-5,11), that in the absence of crystallization, there is little, if any, effect of mechanical strain on the permeability of inert gases (e.g., N_2, CO_2) or plasticizing vapors (e.g., n-butane) in natural rubber. This is nominally in agreement with predictions based on free volume concepts. Since the initial value of Poisson's ratio is nearly 0.5 for the elastomer, there is little, if any, change in its free volume upon simple deformation. However, there are changes induced in the chain conformations of the elastomer which should affect S and D. The absence of any observed change in P implies that any induced change in S is compensated by opposite changes in D.

However, it has been shown both by theory and by experiment[12,13] that the effect of simple tensile stress on the equilibrium degree of swelling of a rubbery material is to increase the amount of swelling. For swelling confined to only the thickness of the membrane, the stress in the x-direction, σ_x, is given[13] approximately as:

$$\sigma_x = (RT/V_1)[ln\,\phi_1 + \phi_2 + \chi_1\phi_2^2 + (P_2V_1/M_c\phi_2)], \quad (2)$$

where ϕ_1 and ϕ_2 are the volume fractions of penetrant and polymer, V_1 is the molar volume of penetrant in the mixture, χ_1 is the Flory-Huggins solution theory parameter, ρ_2 is the density of dry polymer, and M_c is the average molecular weight between crosslinks. This equation predicts, and experiments confirm, that the effect of applied stress becomes more pronounced as the molar volume of the sorbed and diffusing penetrant becomes larger. In the case of a good swelling agent the effect is quite large. The corresponding change (probable decrease) in D is not expected to be necessarily equal to the change in S so that there may be a net change in P (P = DS), especially for the transport of larger penetrants.

In recent study[14] to investigate the above expectation has been made of methanol transport through several different silicone rubber samples with various loading levels of treated fumed silica. The samples were cured with a platinum catalyst, then subjected to uniaxial elongations and clamped securely into a permeation cell. P, D and S were determined using a carrier sweep apparatus, with GC detection of the permeated vapor.

The results under elongations from zero to 400% showed that P and D decreased with increasing strain while S increased. The increase in S was somewhat greater than that predicted by the above theory. The excess increase in S was attributed to an additional contribution due to sorption sites created by reversible voiding around filler. The decrease in D more than compensated for the increase in S giving a net decrease in their product, P. The cause of this decrease in D, in the absence of any significant free volume change with elongation, was attributed to restrictions imposed upon chain segmental motions related to the extended chain conformations under elongated conditions.

Effects of Deformation on PDS in Semicrystalline and Glassy Polymers Several studies have been made using gaseous penetrants (e.g., N_2, CO_2, Ar, etc.) in deformed semicrystalline polymers[3-5] and amorphous glassy polymers.[6,7] There is generally an increase in S, D and P which is ascribed to dilational strain; an increase in free volume. In polyaxially deformed polyethylene the increase levels off as the strain increases. In uniaxially deformed samples, S continues to level off but D and P go through a maximum and then decrease to values even below that for the initial undeformed sample due to structural yielding to form oriented regions.

It was observed[6,7] as shown in Table 1, that P and D increase with strain essentially independent of the size of the penetrant molecule indicating that, at low strains, the size distribution of free volume regions is not distorted by the strain. The subsequent decrease of P

Table 1. Strain and Time Dependence of P an D in Biaxially-Oriented Polystyrene at 50°C (6,7)

Gas	$d^2 (Å)^2$	$\Delta P/P_o\epsilon$	$\Delta D/D_o\epsilon$	$-(1/P)dP/d\log t$	$-(1/D)dD/d\log t$
Ar	11.7	32	19	0.058	0.058
Kr	12.2	31	21	----	----
N_2	13.6	31	19	----	----
CO_2	16.0	25	15	0.067	0.037
Xe	16.4	30	22	0.114	0.085

and D with time with the samples held at a constant strain was more rapid for larger gas penetrant molecules (Xe) than for smaller (CO_2). This indicates that large free volume regions decrease in size faster than small ones.

Smith and Adam[7] have presented a semiquantitative theory for the dependence of D on the free volume change induced by small deformations within the elastic limit of the polymers. The excess volume was taken as:

$$v - v_o = v_f + \Delta v_c. \qquad (3)$$

where the free volume fraction, $f = v_f/v$, is that part of $(v - v_o)$ that can be redistributed without an increase in energy.

The dependence of the diffusion coefficient on f is then given as:

$$D = D_o \exp[-(B/f + E^*/RT)], \qquad (4)$$

where B is a constant (as in the Doolittle and WLF equations), E^* is an activation energy, and D_o is a weak function of temperature. It is assumed that f is essentially independent of temperature and time below T_g and that B, D_o, and E^* are independent of strain. Then:

$$(1/D)\, dD/d\varepsilon = (B/f^2)\, df/d\varepsilon \approx (B/f_o^2)\, df/d\varepsilon. \qquad (5)$$

The isothermal compressibilities of a sample and of its free volume are:

$$\beta = -(1/v)dv/dp; \qquad (6a)$$

$$\beta_f = -(1/v)dv_f/dp. \qquad (6b)$$

In simple uniaxial tension, the fractional increase in volume is

$$(1/\nu)d\nu/d\epsilon = 1 - 2\nu, \qquad (7)$$

where ν is Poisson's ratio, defined in elasticity theory for infinitesimal strains. It then follows that:

$$df/d\epsilon = (\beta_f/\beta - f)\,(1 - 2\nu). \qquad (8)$$

Since f is usually small compared with β_f/β, which probably is near to unity, then:

$$(1/D)\, dD/d\epsilon \approx (B/f^2_o)\,(\beta_f/\beta)\,(1 - 2\nu). \qquad (9)$$

The utility of this expression is illustrated by the theoretical prediction of $(1/D)dD/d\epsilon$ for CO_2 in polycarbonate of 13.1 which is in good agreement with the experimental value of 15. The use of more rigorous expressions for the free volume dependence, and especially the dependence of ν on ϵ,[14] should enhance the utility of expressions based upon this concept.

The number of polymers for which ν values have been determined (and published) is very limited. However, values of ν can be estimated by the expression:[15]

$$\ln(B/\rho)=8.3-4\nu \qquad (10)$$

where B is the bulk modulus, ρ is density, and $B/\rho = (U/V)^6$. U (the Rao function relating to molar sound velocity) and V (the molar volume) both can be calculated by group contribution methods.[15] For example, for amorphous poly (bisphenol carbonate), the calculated value of ν of 0.406 is in nearly exact agreement with the experimental value.

Effects of Larger Deformations on PDS A prediction of the effects of larger deformations on sorption and transport of penetrants in polymers must consider several factors which affect deformation and others which affect sorption and transport.[14,15] Criteria exist which define the onset of yielding (e.g., von Mises criterion), crazing, and fracture (e.g., modified Griffith equation). If deformation is limited to strains which do not produce voiding deformations, it can be assumed as a first approximation that the strain produces only an increase in free volume which will vary from zero in elastomers to large values in various plastics. This strain limitation is reasonable in terms of material selection and development design practices which would avoid application conditions leading to voiding deformations which could lead to material failure. The limitation is not as reasonable for high-impact materials where controlled crazing, shear banding, and/or yielding are essential features of the impact energy dissipation process.

The presence of microcrazes would be expected to increase the apparent diffusion and permeability coefficients due to a contribution of essentially porous flow to the total flux. The occurrence of concurrent convective and activated flow processes has been considered.[16] The presence of shear bands on diffusion is more complex. It might be expected that the oriented structure within the shear banded region would impede the diffusion process. However, it has been observed,[17] for the diffusion of methylene chloride in polystyrene, that the diffusion coefficient during sorption into a shear band region is greater than that in the undeformed region by a factor of 2.2 in the shear plane direction and 1.4 in the tensile plane direction. These differences can be attributed to concurrent voiding in the shear band region, but the evidence also suggests that energy stored in the region by deformation enhances the sorption process. Comparable phenomena have been noted in polystyrene samples molded under very high pressures.[18]

Effects of Mechanical Deformation-Summary In summary, mechanical deformation of a polymeric material can be expected to cause changes in its transport properties due to stress per se, changes in free volume, voiding and orientation. Even in the absence of free volume changes, diffusive motions can be hindered by restrained chain segmental motions due to deformation-induced chain conformational states.

Fairly simple free volume relationships can serve as reasonable representations of changes in transport behavior provided that the deformation is not too great (no extensive yielding, voiding, or orientation). A value of Poisson's ratio can be calculated by the group contribution method, but it must be remembered that it is only applicable, as such, for the case of small deformations. The general predictive procedure is being extended to include voiding deformation modes, the use of swelling penetrants, and more rigorous expressions relating S, D and $\underline{P}$ to free volume or other characteristic material parameters.

Relaxation recovery of the deformed material affects transport behavior, especially for larger size penetrant molecules. This size dependence also should be apparent for transport in materials with deformation-induced voiding or orientation. Diffusive motion is very dependent on the size and shape of the migrating molecule relative to the effective dimensions of the voids or the pathways along or across oriented regions.

EFFECTS OF CHEMICAL MODIFICATIONS

The effects of modification of the chemical structure of a polymeric material on its $\underline{P}$DS behavior are always very pronounced. Solubility in any material is governed primarily by the relative chemical composition of the penetrant and the polymer. The introduction of other chemical groups into a polymeric material, especially if the groups are polar or ionic, causes major changes in penetrant solubilities. If the solubilities are increased, there is the possibility (probability) of plasticization of the structure increasing D. This effect may be countered by a concurrent increase in polymer chain segmental interactions decreasing segmental motion which decreases D. The net effect on $\underline{P}$ depends on the system and conditions.

In such cases, it is very common to observe a change in sorption-diffusion kinetics from Fickian to non-Fickian behavior. The increasing sorption magnitude imposes swelling stresses on the polymeric matrix in which the chain segmental mobility is decreasing. The consequent increase in segmental relaxation times relative to the diffusion process correlation time leads to progressively anomalous sorption behavior with increasing magnitude of compositional modification.

If the chemical modification process is diffusion-controlled or otherwise has a spatial variation through the thickness of the polymer sample, it will lead to the formation of a gradient in chemical composition in the sample. As an example, photo-oxidation of polyethylene films of moderate (~20 mil) thicknesses leads to the formation of a gradient of oxidized material due to radiation attenuation and oxygen diffusion effects. The presence of such chemical composition gradients has been shown (e.g., 1,2) to have marked effects on S and D behavior.

The effects of chemical modification on sorption and transport behavior are particularly important for materials exposed to moisture.[1,14,19] The presence of oxidation or photodegradation products can lead to enhanced water sorption and diffusion with resultant swelling. This poses a severe challenge to exposed adhesives, sealants and coatings.

Mode-of-Sorption Effects In chemically degrading systems, "mode-of-sorption" effects, as discussed above, often become dominant factors affecting sorption and transport behavior. Illustrations of these effects are given in Tables 2 and 3.

Table 2 gives the moles of water sorbed per mole of functional group in the polymer at 90% relative humidity calculated from the data of McLaren and Rowen[20] given by Barrie.[21] These data fall into two fairly well defined categories (a plot of moles of polar groups versus moles of water sorbed gives two nearly straight lines). One category, with essentially a one-to-one relation between moles of water sorbed versus moles of groups is for (acidic) hydrogen-donors (hydroxyl, carboxyl and peptide). The other category, with an average ratio of about 0.15 moles of water/mole of groups, is for (basic) hydrogen-acceptors (nitrile, carbonyl, ester, ether).

When a polymer such as polyethylene undergoes progressive photo-oxiation, the sorption and transport processes reflect the changing modes of sorption for different types of penetrants as the composition of the polymer changes due to photodegradation product formation.[14] In Table 3, the solubility, diffusion and permeation coefficients are given for nitrogen and for water vapor as a function of carbonyl content. The degradation products of polyethylene are primarily carbonyl formation with some trans-unsaturation followed by the appearance of hydroxyl which increases significantly above about 15 moles/cm^3x10^{-5} carbonyl concentration.

Table 2. Mode-of-Sorption (Hydrogen Bonding) Effects for Water Sorption at 90% R.H.

Polymer	Functional Group	Moles of sorbed water / Mole of group
Poly(vinyl alcohol)	hydroxyl	0.93
Poly(glycine-alanine)	peptide	1.10
Poly(methacrylic acid)	carboxyl	0.78
Polyacrylonitrile	nitrile	0.22
Poly(methyl vinyl ketone)	carbonyl	0.20
Poly(vinyl butyral)	----	0.13
Poly(vinyl acetate)	ester	0.15
Poly(ethylene-vinyl acetate)	ester	0.17
Poly(vinyl benzoate)	ester	0.15
Poly(isobutyl methacrylate)	ester	0.11
Poly(vinyl isobutyl ether)	ester	0.06

Table 3. Nitrogen and Water Vapor Sorption and Transport in Low Density Polyethylene as Carbonyl Content Increases due to Photodegradation (14)

Carbonyl conc ($moles/cm^3 x 10^5$)	4.9	13.1	17.8
Nitrogen			
$\underline{P} \times 10^{10}$	1.7	1.2	0.86
$D \times 10^7$	3.7	2.3	1.5
$S \times 10^4$	4.6	5.3	5.6
Water Vapor			
$\underline{P} \times 10^9$	1.7	2.3	3.8
$D \times 10^7$	2.2	1.7	1.3
$S \times 10^2$	0.8	1.3	2.9

$\underline{P}$ units: $cm^3 cm/cm^2 sec cmHg$
D units: cm^2/sec
S units: $cm^3/cm^3 cmHg$

As seen in Table 3, the increase in carbonyl (and hydroxyl) groups leads to a decrease in D for both nitrogen and water. The increase in interactions between chains lowers their segmental mobilities to decrease the diffusion rate. The increase in the degradation products increases the solubility of both penetrants with a marked increase in water solubility upon the appearance of hydroxyl groups. The increase in the degradation groups effectively increases the polymer's interaction with the penetrants (essentially an increase in polymer solubility parameter), with a more specific mode-of-sorption effect (hydrogen bonding) operative for water. These changes in S and D are reflected in the $\underline{P}$ data. The effect of photodegradation is to lower nitrogen permeability and to drastically increase water vapor permeability.

Polymer Composition Gradient Effects As an example of the effects of a gradient in chemical modification of a polymer on its transport behavior we considered[22] the effects of hydrolysis of a miscible polyblend of poly(methyl methacrylate) (70%) and poly(vinylidene fluoride) (PVF_2) by concentrated sulfuric acid at 40°C. The hydrolysis leads to the formation of poly(methacrylic acid) and the consequent phase separation of crystalline PVF_2.

As shown in Table 4, the separation factor, $\alpha = \underline{P}(CO_2)/\underline{P}(CH_4)$, for the permeation of a 50:50 mixture of carbon dioxide and methane through the polyblend membrane jumps from 0.8 for the unhydrolyzed film to 14 for only 6% hydrolysis. It continues to increase with increasing hydrolysis. Since the sulfuric acid treatment is highly diffusion-controlled, the hydrolyzed region probably corresponds to a laminate or gradient structure.

Table 4. Permeation and Separation of CO_2-CH_4 Mixtures (50:50) in Hydrolyzed Blends of Poly(vinylidene fluoride) (30%) and Poly(methyl methacrylate) (22)

Reaction time (minutes)	% Hydrolysis	$\underline{P}(CO_2)$	$\underline{P}(CH_4)$	α
0	0	0.8	1.0	0.8
2	6	59	4.1	14
5	10	51	2.9	18
15	18	56	2.9	19
30	26	56	3.0	19
60	38	99	4.8	21

$\alpha = \underline{P}(CO_2)/\underline{P}(CH_4)$
$\underline{P}$ in units of cc(STP)cm/cm^2 seccmHg x 10^{10}

The membrane structure and the mechanism for this transport separation enhancement are under further study. It is significant to note, however, that the increase in α is due mainly to a selective increase in $\underline{P}(CO_2)$, which increases by a factor of 124 as the percent hydrolysis goes from zero to 38, whereas $\underline{P}(CH_4)$ only increases by a factor of about five. This selectivity is tentatively attributed to a specific mode-of-sorption for carbon dioxide in the hydrolyzed membrane. This results in an effective increase in both rejection of methane and flux of carbon dioxide. This is a very desirable occurrence for membrane separation technology on the one hand, but it is of significance also for considerations of the effects of environmental exposure of adhesives, sealants, coatings and other polymeric materials in which the occurrence may not be so favorable in the context of the application.

Effects of Chemical Modifications-Summary In summary, chemical degradation of a polymeric material can be expected to cause major changes in its transport properties due mainly to changes in sorption behavior relating to the development of specific modes-of-sorption. As degradation proceeds, the sorption rate may change from one governed by diffusion-control to one governed by relaxation-control as the segmental response of the polymer to swelling decreases due to increased segmental interactions and crosslinking. The sorption rate curves change from "Fickian" through "anomalous" to "Case II" types.

The formation of degradation products can be expected to be as a gradient into the polymer due to diffusion-controlled reaction of the environmental agent. In other cases, the gradient will be as a sharp front advancing into the material, thereby resembling a laminate structure. This front may advance linearly with exposure time, as does the swollen-region front in Case II sorption; the phenomena are closely related and mathematically nearly identical. In any case, these gradient/laminate structures affect flux, selectivity, swelling, mechanical deformation, etc., since they introduce a gradient in the chemical potential of the polymer which affects the driving force for transport (in sense of irreversible thermodynamics), or may be considered to affect the functional dependence of the sorption and diffusion processes on experimental parameters such as concentration and pressure as a function of the thickness parameter.

CONCLUSION

Both mechanical deformations and chemical modifications can be expected to cause significant changes in the sorption and transport behavior of adhesives, sealants, coatings and other polymeric materials. These changes are to be anticipated as natural results of application exposure to various environmental conditions. The prediction of their magnitude and consequences for performance lifetime should be a prime consideration in the selection and development of prospective polymeric materials for specific applications. An appreciation of the factors and phenomena involved is an essential prerequisite for establishing such predictive procedures.

ACKNOWLEDGEMENT

The support of research programs concerned with this general topic by grants from the U.S. Army Research Office, the Jet Propulsion Laboratory, the National Science Foundation and the CWRU Center for Adhesives, Sealants and Coatings are gratefully acknowledged.

NOMENCLATURE

A	Membrane area (cm^2)
B	Constant in the Doolittle and WLF equations
B	Bulk modulus
D	Diffusion coefficient (cm^2/sec)
D_o	Preexponential diffusion coefficient (value of D at $T \rightarrow \infty$).
E^*	Activation energy
J	Flux (cm^3STP/cm^2sec)
M_c	Average polymer molecular weight between crosslinks (g/mol)
$\underline{P}$	Permeability coefficient (cm^3STP cm/cm^2seccmHg)
Q	Permeated penetrant amount (cm^3STP)
R	Gas constant
S	Solubility coefficient (cm^3STP/cm^3cmHg)
T	Temperature (°K)
U	Rao function
V_1	Molar volume of penetrant (cc^3/mol)
X_1	Flory-Huggins solution theory parameter
c	Ambient penetrant concentration (cm^3STP/cm^3)
c	Sorbed penetrant concentration (cm^3STP/cm^3)
d	Diameter of gas molecule (Å)
f	Free volume fraction
f_o	Free volume fraction in dry polymer
ℓ	Membrane thickness (cm)
t	Time (sec)
v	Total volume of polymer
v_o	Reference atomically occupied volume
v_f	Defined free volume
v_c	Unavailable excess volume
x	Membrane thickness coordinate (cm)
α	Separation factor ($\underline{P}(CO_2)/\underline{P}(CH_4)$)
β	Isothermal compressibility
ϵ	Strain
μ	Chemical potential (partial molar free energy)
ν	Poisson's ratio
ρ_2	Polymer density (dry) (g/cm^3)
σ	Stress
Φ_1, Φ_2	Volume fraction of penetrant and polymer, respectively, in the membrane

REFERENCES

1. C.E. Rogers, in Polymer Permeability, J. Comyn, ed., Applied Science Publishers Ltd., London, Chapter 2 (1985).
2. C.E. Rogers, in Physics and Chemistry of the Organic Solid State, Volume II, Interscience Publications, New York, Chapter 6 (1965).
3. H. Yasuda, V. Stannett, H.L. Frisch and A. Peterlin, Makromol. Chem., 73, 188 (1964).
4. A. Peterlin, Makromol. Chem. Sup., 3, 215 (1979).
5. J.C. Phillips and A. Peterlin, Polym. Eng. Sci., 23, 734 (1983).
6. G. Levita and T.L. Smith, Polym. Eng. Sci., 21, 936 (1981).
7. T.L. Smith and R.E. Adam, Polymer, 22, 299 (1981).
8. S.P. McCarthy and C.E. Rogers, Polym. Eng. Sci., in press.
9. D. Benachour and C.E. Rogers, ACS Symposium Series, 220, 307 (1983).
10. P. Neogi, M. Kim and Y. Yang, AIChE J., 32 1146 (1986).
11. J.A. Barrie and B. Platt, J. Polym. Sci., 49, 479 (1961); ibid, 54, 261 (1961).
12. L.R.G. Treloar, The Physics of Rubber Elasticity, 3rd Edition, Oxford University Press, (1975).
13. H.R. Brown, Polymer, 19, 1186 (1978).
14. C.E. Rogers, N. Mitchell and A.N. Dudek Shine, to be published.
15. D.W. van Krevelen, Properties of Polymers, 2nd Edition, Elsevier Science Publication Company, Chapter 13, (1976).
16. H.L. Frisch, J. Phys. and Chem., 60, 1177 (1956).
17. S.P. McCarthy and C.E. Rogers, to be published.
18. W. Dale and C.E. Rogers, AIChE J., 19, 445 (1973).
19. C.E. Rogers, ACS Symposium Series, 220, 231 (1983).
20. A.D. McLaren and J.W. Rowen, J. Polym. Sci., 7, 289 (1951).
21. J.A. Barrie, "Water in Polymers," in Diffusion in Polymers, J. Crank and G.S. Park, eds., Academic Press, London, Chapter 8, p. 264 (1968).
22. R. Duran, L. Kim and C.E. Rogers, to be published.

THE MECHANICAL AND WATER ABSORPTION BEHAVIOUR OF FLUOROEPOXY RESINS

S.J. Shaw*, D.A. Tod* and J.R. Griffith+

*RARDE, Waltham Abbey, Essex, UK
+Naval Research Laboratory, Washington DC, 20375 USA

ABSTRACT

The absorption of water by the organic polymeric materials employed in adhesive and surface coating systems is generally considered harmful. Disadvantageous effects resulting from exposure to warm/moist environments can include plasticization, lowering of glass temperature and disruption of the interfacial region between substrate and organic phase. One approach to alleviate such effects concerns the use of polymers having a high degree of hydrophobicity. In recent years fluorinated epoxy resins have been developed at the Naval Research Laboratory (NRL) offering this capability. This paper briefly discusses the underlying principles of fluoroepoxy resin chemistry followed by a discussion concerning the water absorption characteristics of various NRL developed formulations. An attempt is made to correlate this behavior with changes in formulation variables. The mechanical and fracture properties of the various fluoroepoxy systems are also discussed and compared with properties generally found with non-fluorinated epoxy resins.

INTRODUCTION

Epoxy resins currently form the basis of the majority of structural adhesive and matrix resin formulations where advantage is taken of favourable attributes such as high modulus, low creep and reasonable elevated temperature performance.

Unfortunately epoxy resins can absorb considerable quantities of water, the precise amount and rate of absorption depending upon resin and curing agent structure together with external environmental factors such as temperature and relative humidity. Deleterious effects which can result from exposure to a warm/moist environment include plasticiza-tion[1-4] lowering of glass temperature[2,4,5,6,7] and disruption of the interfacial region between substrate/reinforcement and the organic phase.[8,9,10]

In recent years halogenated epoxy resins having substantial hydrophobic characteristics[11-18] have been developed by several researchers. In particular, fluorinated epoxy resins together with compatible curing agents were developed by Griffith and co-workers in the 1970s.[11-15] Although the chemistry of these systems has been discussed, no attempt has previously been made to conduct detailed mechanical property and water absorption studies with a view to correlating such properties with formulation variables. Such a study is currently underway and the findings obtained to date form the basis of this paper.

FLUOROEPOXY RESIN CHEMISTRY

Synthesis of the fluoroepoxy resins employed in this study, the molecular structure of which are shown below, have been outlined by Griffith.[15]

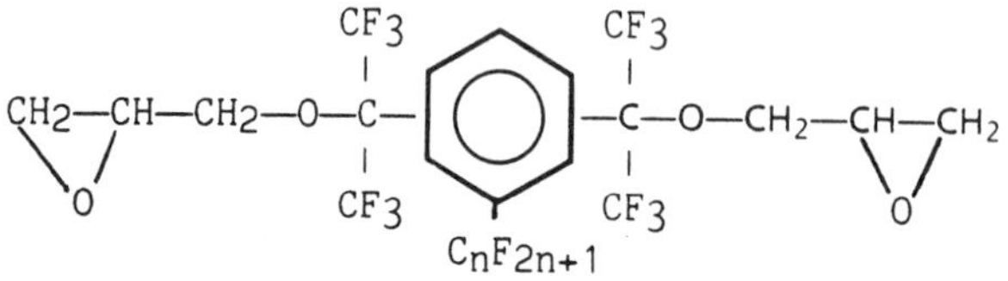

CF_3 CF_3

$CH_2—CH—CH_2—O—C—C_6H_3—C—O—CH_2—CH—CH_2$

O CF_3 CF_3 O

C_nF_{2n+1}

The synthetic process developed has been capable of producing resins with n in the C_nF_{2n+1} group ranging from zero (in effect a hydrogen atom on the 5 position of the benzene ring) to numbers in excess of 10.

All members of the fluoroepoxy resin series are clear, colourless syrups (except for n=o, which is a crystalline solid). As well as hydrophobic character, further advantages of these systems quoted have included[15]:

(i) They react like typical diglycidyl ether resins since the fluorine-containing groups are fairly remote from the reactive epoxy groups.

(ii) Every carbon atom containing fluorine is totally fluorinated, resulting in maximum stability to the degrading effects of the environment.

(iii) The fluorine is present in aliphatic form. This is particularly important because fluoroaromatic compounds are expensive and fluorine is susceptible to nucleophilic displacement by, for example, amine curing agents.

(iv) The epoxy groups are so spaced as to allow the development of reasonable strength and toughness properties in the crosslinked product

(v) In the uncured liquid state, they are extremely good wetting agents; thorough wetting of a polytetrafluoroethylene surface (0° contact angle) by some fluoroepoxy resins has been demonstrated.

One major disadvantage which was realized early in their development concerned compatibility with common curing agents. The incorporation of large quantities of fluorine, although introducing hydrophobic behaviour, was sufficient to make the resins incompatible with many of the commonly employed curing agents, such as the aliphatic amines. Two main types of curing agent have been developed in an attempt to overcome this problem, these being silicone-amines and fluoroanhydrides.[15]

The silicone-amines have the following general structure where n can be varied from 1 to high numbers.

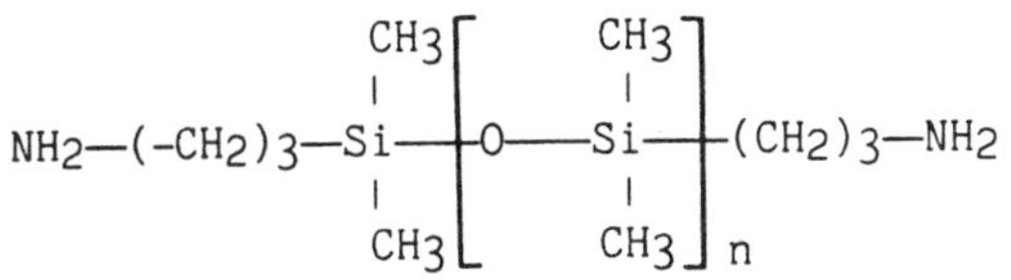

$$NH_2-(-CH_2)_3-\underset{CH_3}{\overset{CH_3}{\underset{|}{\overset{|}{Si}}}}-\left[O-\underset{CH_3}{\overset{CH_3}{\underset{|}{\overset{|}{Si}}}}\right]_n-(CH_2)_3-NH_2$$

An initial two-phase mixture transforms into a single-phase prior to gelation. It is particularly interesting and relevant to note that the silicone-amines resemble fluoroepoxies in many respects, particularly hydrophobocity. Thus the use of these curing agents would not be expected to significantly affect the hydrophobic characteristics of a particular resin.

Variation in the value of n can influence substantially the final properties of the crosslinked product. Consequently, a series of products can be envisaged having properties ranging from rigid plastics ($n<5$) to elastomers ($n>5$).

A number of different fluoroanhydride curing agents have been developed having sufficient compatibility with the fluoroepoxies.[15] The structures of the two fluoroanhydrides used in this work are shown in Fig. 1.

In all cases elevated temperature cures are required together with catalysts such as tertiary amines so as to accelerate the cure reactions.

EXPERIMENTAL PROCEDURES

Materials: The fluoroepoxy resin formulations employed in this study together with cure conditions are shown in Table 1, where formulations are identified by both resin type and curing agent: eg., $C_6/1SA=C_6F_{13}$ (pendant C_nF_{2n+1} group on the resin) resin + n=1 silicone-amine, and $C_8/FA=C_8F_{17}$ resin + a fluoroanhydride curing agent. In this study two

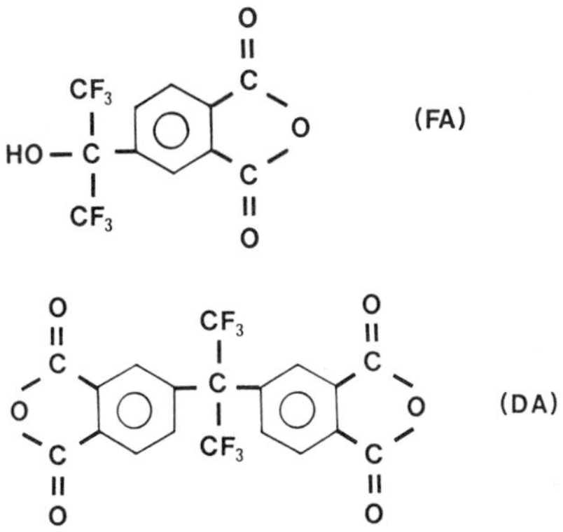

(FA)

(DA)

Fig. 1. Fluoroanhydride curing agents.

types of fluoroanhydride were investigated. In the majority of cases the fluoroanhydride based upon phthalic anhydride referred to as FA in Fig. 1 and text was employed However the last formulation in Table 1 contained a fluoroanhydride mixture based upon 90% FA and 10% of a dianhydride referred to as DA. Two types of cure catalyst were employed with the fluoroanhydrides; namely, cetyltrimethyl ammonium bromide (CAB) and dimethylbenzylamine (DMB).

All silicone-amine systems were cured using a stoichiometric concentration of curing agent. The anhydride curing agents were employed at 80% of theoretical stoichiometry.

Water Absorption Studies: Disc shaped specimens, approximately 24 mm in diameter with thicknesses ranging from approximately 2 to 3 mm, were employed for water absorption studies. Prior to examination all specimens were placed in a desiccator over predried molecular sieve for a period of one month. They were then weighed and immersed in water baths at temperatures of 20, 40, 60 and 90°C. At various time intervals samples were removed, dried of surface water with blotting paper, weighed on an analytical balance and returned to the bath.

Fracture and Mechanical Property Studies: Cured sheets, approximately 6 mm thick, were obtained by a casting technique.

Room temperature fracture properties were determined using compact tension specimens[19] machined from the cast sheets (Fig. 2). A sharp crack was formed at the base of the slot indicated in Fig. 2, by carefully tapping a fresh razor blade into the base at 20°C which caused a natural crack to grow for a short distance ahead of the blade. The specimens were then mounted in a tensile testing machine and fractured at 20°C at a constant crosshead displacement rate of 1 mm min-1.

Table 1. Fluoroepoxy Resin/Curing Agent Formulations and Cure Conditions

Formulation	Cure Conditions hr/°C
C_0/1SA	16/20, 3/110
C_6/1SA	16/20, 3/110
C_6/7.5SA	16/20, 3/110
C_8/1SA	16/20, 3/110
C_8/FA/CAB CAB = Cetyltrimethylammonium Bromide, 0.3%	16/120
C_8/FA/DMB DMB=Dimethylbenzylamine, 1%	2/75, 1/100, 5/150
C_8/FA/DA/DMB DA=Dianhydride	2/150, 2/180

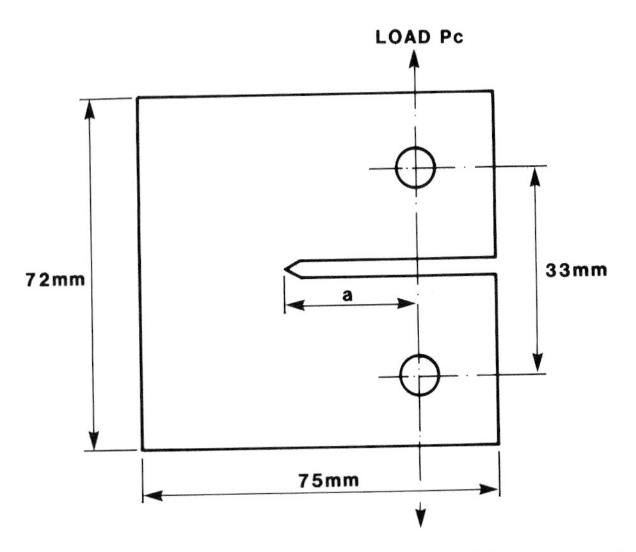

Fig. 2. Compact tension specimen; all dimensions in mm.

Values of stress intensity factor, K_{Ic}, were calculated from the expression,

$$K_{Ic} = \frac{P_c}{BW^{\frac{1}{2}}} Q, \tag{1}$$

where P_c is the load at crack initiation; W the specimen width as indicated in Fig. 2; B the specimen thickness and Q a geometry dependent factor given by,

$$Q = 29.6\left(\frac{a}{w}\right)^{\frac{1}{2}} - 185.5\left(\frac{a}{w}\right)^{3/2} + 655.7\left(\frac{a}{w}\right)^{5/2} - 1017\left(\frac{a}{w}\right)^{7/2} + 638.9\left(\frac{a}{w}\right)^{9/2}, \tag{2}$$

where a is crack length.

The critical stress intensity factor, K_{Ic} values were converted to fracture energy, G_{Ic} values using the equation,

$$G_{Ic} = \frac{K^2_{Ic}}{E}(1-\nu^2), \tag{3}$$

where E is Young's modulus and ν is Poisson's ratio (assumed to be 0.35).

Values of flexural modulus, strength and strain at failure were obtained from flexural bending experiments conducted on rectangular bars according to ASTM D790-71[20] at 20°C and 1 mm min^{-1} crosshead displacement.

Dynamic mechanical studies were conducted on a Rheometrics Mechanical Spectrometer. Rectangular specimens, measuring 85 x 10 x 6 mm were mounted vertically in the spectrometer and clamped securely at both ends. The upper fixture was subjected to torsional sinusoidal oscillations at a frequency of 1 Hz actuated by a voltage signal from a generator. The resultant torque in the specimens was measured at the lower fixture by a transducer system. These two signals were analyzed to provide values of storage shear modulus, G′ and loss shear modulus G″. Values of the loss tangent tan δ, were calculated from the equation.

$$tan\ \delta = \frac{G''}{G'} . \qquad (4)$$

Measurements were taken at approximately 5°C intervals between -160°C and 150°C. Specimens were allowed to reach equilibrium for five minutes at each temperature with a heating rate between test temperature of approximately 5°C per minute.

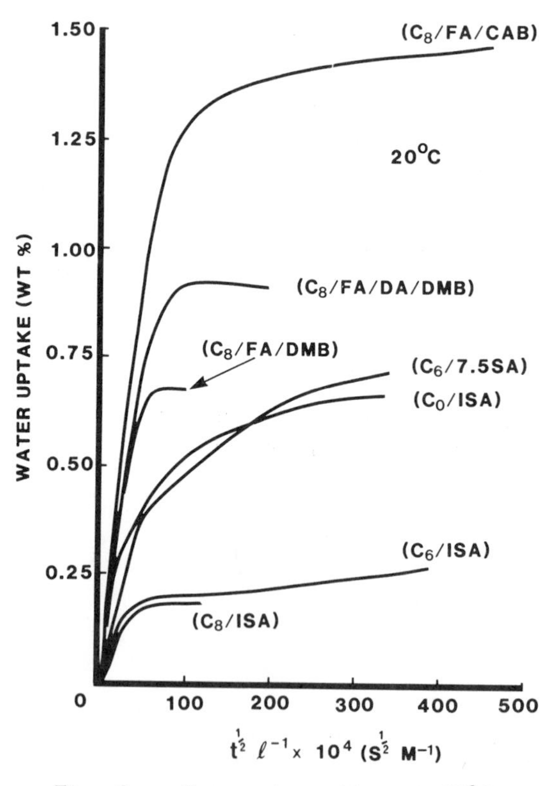

Fig. 3. Water absorption at 20°C.

RESULTS AND DISCUSSION

Water Absorption Behaviour

Water absorption data for immersion conditions of 20, 40, 60 and 90°C are shown in Figs. 3 to 6 respectively as plots of water uptake as a function of square root time, normalized for specimen thickness.

Dealing firstly with room temperature immersion, as indicated in Fig. 3 many of the formulations show deviations from Fickian behaviour,[21] this being particularly so for the C_o/1SA and C_6/7.5SA systems. Also clearly indicated is the fact that variations in formulation (resin and curing agent) have substantial effects on the water absorption characteristics. Immediately apparent are the comparatively high water absorption values obtained with the anhydride cured formulations in comparison to the silicone-amine cured systems.

With the anhydride systems it is particularly interesting to note the substantial differences in absorption behaviour which occur by mere change in accelerator (compare C_8FA/CAB and C_8/FA/DMB); a change from 0.3% cetyltrimethylammonium bromide to 1% dimethylbenzylamine reducing total water absorption by approximately 50%. Although cure temperature employed for the dimethylbenzylamine accelerated system was comparatively high, such increases in cure temperature are often found to increase levels of absorption due to free volume related effects.[22] Consequently, if true for this sytem, a reduction in cure temperature would be expected to reduce the level of water absorbed still further. Clearly, further experimentation is required to determine the extent and reasons for the apparent hydrophobic enhancement provided by the dimethylbenzylamine accelerator.

It seems likely that the reduced hydrophobicity exhibited by the fluoroanhydride systems could be associated with the hydroxyl group on anhydride FA remaining accessible to water molecules after cross-linking. Consequently, substitution of a proportion of this anhydride with the dianhydride DA should reduce water absorption. Fig. 3 shows this not to be the case; such substitution appears to increase total water absorption. Once again, however, differences in cure conditions exist, with the C_8/FA/DA/DMB system being cured some 30°C higher than the C_8/FA/DMB formulation. This could have resulted in reduced hydrophobicity due to the free volume related effects mentioned above, which could have exceeded any improvements provided by a reduced hydroxyl concentration. Although data is clearly lacking at present, it is reasonable to assume that removal of the hydroxyl group on anhydride FA Could lead to enhanced hydrophobic behaviour. Experiments are currently in progress to determine the validity of this proposal.

With the silicone-amine cured systems, two major points are of interest. Firstly, increasing resin fluorine content increases the hydrophobic characteristics of the cured resin. This can be clearly seen by comparing the C_o/1SA, C_6/1SA and C_8/1SA systems in Fig. 3, where a detailed analysis indicates a linear relationship between the amount of water absorbed and fluorine content. Second, increasing the size of the silicone-amine molecule (increasing the value of n) produces a cured product having reduced hydrophobicity. Although an increase in n would increase the number of species present in the system capable of exhibiting hydrophobic behaviour, this would almost certainly be counteracted by a reduction in crosslink density which, due to increased free volume, could allow a greater degree of water absorption. Furthermore it is of interest to note that the C6/7.5SA system was a moderately flexible material at room temperature in comparison to the rigid-like state exhibited by the C6/1SA system. This would indicate substantial differences in T_g (see later) which could clearly influence differences in the amount of water absorbed by the two systems, particularly if the C6/7.5SA system were to exhibit a T_g close to the immersion temperature.

Water absorption data obtained at 40°C is shown in Fig. 4. The data indicates general similarity to the trends previously discussed for the 20°C absorption. The exception is of course the C6/7.5SA data. At this absorption temperature this system is now the least hydrophobic formulation. Bearing in mind the above discussion concerning the reduced crosslink density and flexibility of the C6/7.5SA system, it would seem likely that this substantial change in behaviour be due to the T_g of this system being somewhat lower than the immersion temperature. Unfortunately a 'dry' glass temperature is not available, but it would be expected to be significantly lower than the C6/1SA system, which had a 'dry' T_g of 61°C (discussed later).

Figure 5 shows water absorption data obtained at 60°C, where the major trends are similar to those discussed previously for 40°C immersion. Figure 6 shows results obtained at an immersion temperature of 90°C, where some unusual and dramatic effects are demonstrated. With the n=1 silicone-amine cured systems, as found at the lower immersion temperatures, increasing resin fluorine content increases the general level of hydrophobicity. However it is interesting to note

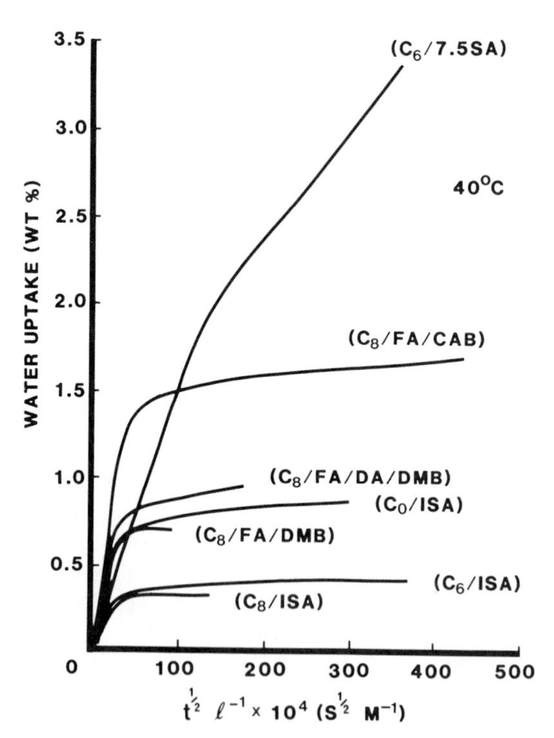

Fig. 4. Water absorption at 40°C.

that the size and indeed the presence of the pendant C_nF_{2n+1} group on the resin molecule influences substantially the shape of the uptake curve. The C_8/1SA system, as shown in Fig. 6, simply shows a gradual plateau effect followed by a gentle decline in weight, so that eventually a net <u>loss</u> in specimen weight is obtained. The C_6/1SA system initially shows similar behaviour but eventually shows a small, but sharp peak followed by a fairly rapid decline. Once again a net loss in weight eventually occurs. The C_0/1SA system (a hydrogen atom on the 5 position of the benzene ring) behaves initially in a similar way, but in this case, the secondary water uptake stage is quite dramatic, showing at present a water uptake of approximately 140% with no indications of a maximum followed by a precipitous decline. The important point to note here is that although a decrease in the size of the pendant C_nF_{2n+1} group increases the degree of water absorption, it also allows a second uptake stage to occur which is dramatic in the case of the C_0/1SA system. Reasons for this behaviour are currently unknown. The C_6/7.5SA system, although having an uptake curve in certain respects similar to the C_0/1SA system, was found to suffer disintegration shortly after attaining a water absorption value of about 27%. This system is currently the only silicone-amine cured system to have suffered this fate.

As before the three anhydride cured systems show similar absorption trends with greatly different magnitudes. The C_8/FA/CAB system shows a steep sigmoidal uptake curve, peaking at approximately 8-9%, followed by a pronounced loss in weight eventually resulting in disintegration of the specimen. Rather surprisingly the C_8/FA/DMB system shows substantially different behaviour to its cetyltrimethyl ammonium bromide catalyzed counterpart. In this case, a comparatively modest degree of absorption is observed (together with no second uptake stage) followed by a gentle weight decline. As mentioned previously, further studies will be necessary to determine the reasons for these substantial changes in absorption behaviour. To date the C_8/FA/CAB

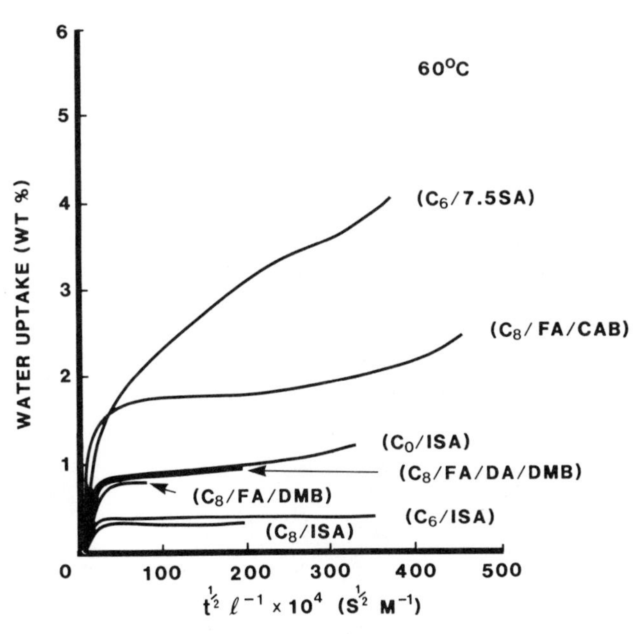

Fig. 5. Water absorption at 60°C.

formulation is the only fluoroanhydride system to have suffered complete disintegration at this absorption temperature. Although not clear at present, it would seem likely that the fluoroanhydride-cured systems would have a substantial number of ester groups in the crosslinked product which could result in susceptibility to hydrolytic degradation.[23]

Figure 7 shows maximum levels of water uptake as a function of immersion temperature for the seven formulations studied so far (it is important to point out that non-Fickian absorption characteristics could well lead to later revision of some of the data shown in this figure). It is of interest to note that increased temperature generally results in an increased level of water absorption.

With the C_6/1SA, C_8/1SA and particularly the C_8/FA/DMB and C_8/FA/DA/DMB systems, the temperature effect can be regarded as virtually insignificant at temperatures up to approximately 60°C. At higher temperatures, however, an upturn does occur, albeit a relatively minor one. With the C_0/1SA and C_8/FA/CAB systems some greater temperature dependence exists at temperatures up to 60°C with a substantial upturn in water absorption behaviour occurring at higher temperatures. The C_6/7.5SA system, however, exhibits a different trend. In this case, increasing temperature from 20 to 40°C causes a substantial upturn in water absorption. It would seem likely that

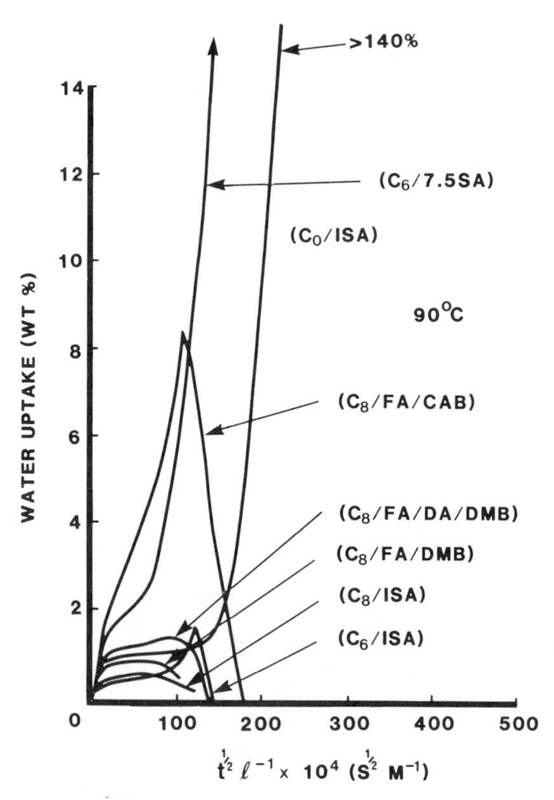

Fig. 6. Water absorption at 90°C.

these substantial upturns in water absorption behaviour could be associated with T_g related effects; the glass temperature in each case eventually falling below the exposure temperature and thus producing marked changes in absorption behaviour.

Values of diffusion coefficient were calculated using Eq. 5, derived from Fick's second law of diffusion.

$$\frac{M_t}{M_\infty} = \frac{4}{\ell}\left(\frac{Dt}{\pi}\right)^{\frac{1}{2}}, \qquad (5)$$

where M_t is water uptake at time t; M_∞ the equilibrium water uptake; ℓ specimen thickness; t absorption time and D the diffusion coefficient. In an attempt to compensate for the somewhat anomolous absorption characteristics exhibited in a number of cases, values of M_∞ were estimated by extrapolating the data in the 'near plateau' regions to the y-axis. M_∞ was simply chosen as the value halfway between the intercept on the y-axis and its meeting point with the curve. Diffusion coefficients obtained from the seven formulations, at each of the temperatures investigated are shown in Table 2. It is interesting to note that a comparison of this data with the water absorption values shown in Figs. 3 to 6 reveals an apparent inverse correlation between the amount of water absorbed by a particular formulation and its diffusion coefficient; the most hydrophobic systems generally exhibiting the highest diffusion coefficients. Conflicting trends of this kind would clearly be of importance when considering such systems for use as protective coatings.

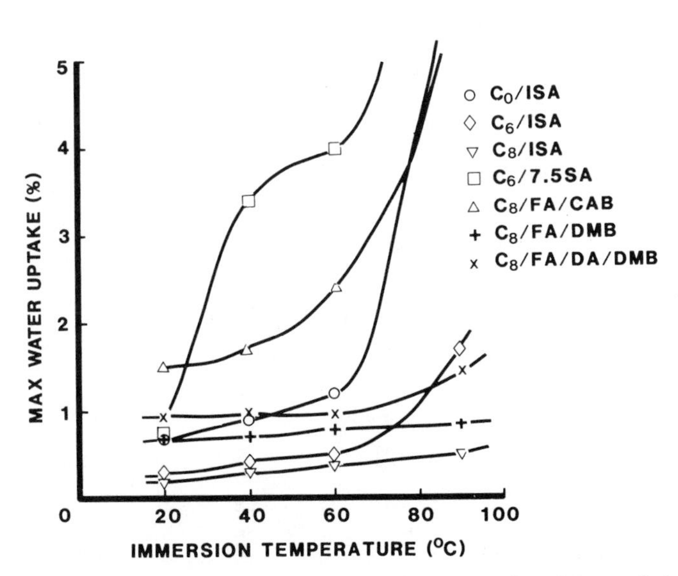

Fig. 7. Maximum levels of water uptake as a function of immersion temperature.

Table 2. Diffusion Coefficients (m^2s^{-1})

Formulation	20°C ($\times10^{-13}$)	40°C ($\times10^{-12}$)	60°C ($\times10^{-12}$)	90°C ($\times10^{-11}$)
C_0/1SA	5.77	2.42	5.78	>3.01
C6/1SA	19.8	4.02	8.01	>3.53
C6/7.5SA	+	87.7*	2.89*	0.47*
C8/1SA	7.66	6.21	13.7	>6.95
C8/FA/CAB	4.72	1.61	3.95	1.72
C8/FA/DMB	16.0	5.00	6.38	>1.77
C8/FA/DA/DMB	7.17	3.78	4.44	1.59

+ Gross non-Fickian absorption prevents calculation of diffusion coefficient.

* Inaccuracy likely due to deviations from Fickian behavior.

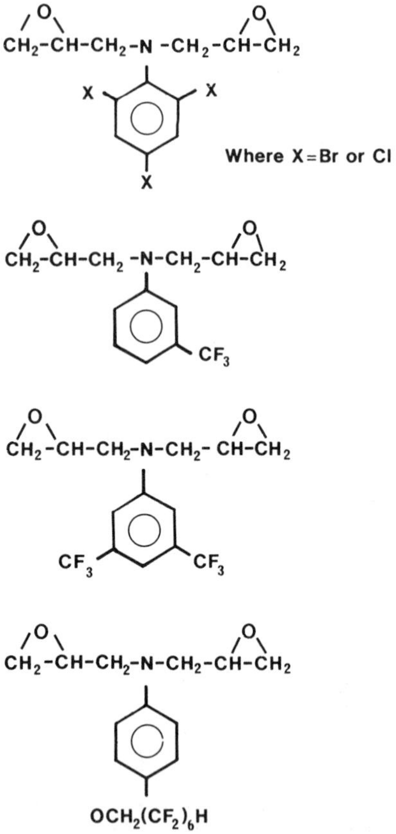

CH_2-CH-CH_2-N-CH_2-CH-CH_2 (X, X, X)

Where X=Br or Cl

CH_2-CH-CH_2-N-CH_2-CH-CH_2 (CF_3)

CH_2-CH-CH_2-N-CH_2-CH-CH_2 (CF_3, CF_3)

CH_2-CH-CH_2-N-CH_2-CH-CH_2 ($OCH_2(CF_2)_6H$)

Fig. 8. Halogenated diglycidylamine structures (after Johncock et al.[17,18]).

As expected, the data in Table 2 shows increasing values of diffusion coefficient with increasing immersion temperature, trends found in many other investigations using a wide variety of epoxy formulations. The values obtained, ranging from approximately 4×10^{-13} to 3×10^{-11} m^2s^{-1} are fairly typical for a wide range of epoxy systems.[24]

It is worthwhile noting at this point that other attempts have been made at promoting hydrophobicity by halogen modification. Halogenated epoxy resins have been developed and studied by Johncock et al.[17,18] Initial studies were concerned with halogen containing diglycidylamines having the structures shown in Fig. 8, all cured with 1,3-diaminobenzene. Various degrees of halogenation were found to result in significant increases in hydrophobic character, with both bromine and chlorine modification having greater affects than fluorination. The most hydrophobic fluorine-modified diglycidylamine was found to exhibit a water absorption at 20°C in excess of 1.3%, which is substantially greater than most of the fluoroepoxy formulations investigated in this programme. However, diffusion coefficients were comparatively low, being roughly equivalent to the current fluoroepoxy formulations having the lowest D values. It is particularly interesting to note that Johncock and Tudgey[17] also observed an inverse correlation between water affinity and diffusion coefficient with their fluorine-modified formulations.

Johncock and co-workers have also conducted investigations on the halogenation of tetraglycidyl-4, 4'-diaminodiphenyl methane (TGDDM) cured with 4,4'-diaminodiphenylsulphone (DDS), the currently preferred resin for many composite aerospace applications.[17,18] The resin structures examined are shown in Fig. 9 together with water absorption and diffusion coefficient values obtained from liquid water immersion experiments conducted at 20°C. As indicated, halogenation increases resin hydrophobicity quite significantly, with the trifluoromethyl modification having the greatest effect. Interestingly, only the trifluoromethyl modification has a marked effect on the diffusion coefficient. Further experiments conducted in moist atmospheres of various activities have shown an order of decreasing diffusion coefficient of CF_3>Cl>Br> unmodified, which was the order of increasing

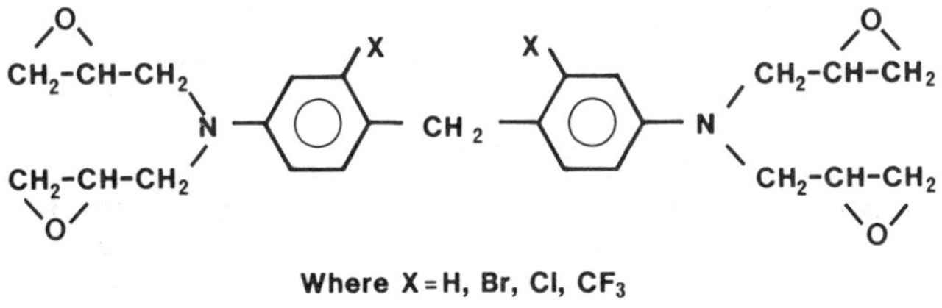

Where X=H, Br, Cl, CF_3

Substituent	Water absorption (%)	$Dm^2s^{-1} \times 10^{-13}$
H	5.5	0.6
Br	3.3	0.5
Cl	3.7	0.6
CF_3	2.9	2.0

Fig. 9. Halogenated TGDDM structure (after Johncock et al.[17,18]).

affinity for water. Once again an apparent inverse correlation between affinity and diffusion coefficient exists. Although the current fluoroepoxy formulations could not be considered equivalent to the halogenated TGDDMs (substantially different glass temperatures) the current systems clearly exhibit far greater hydrophobicity; from 0.19 to 1.5% in comparison to 2.9%

Goobich and Marom[16] have investigated the effects of introducing bromine into epoxy resin formulations by employing a brominated co-reactant with TGDDM/DDS systems. This type of modification indicates that bromine addition significantly reduced the maximum level of moisture absorption whilst having only small effects on the high temperature properties and glass temperatures of the dry resins. For example a bromine content of approximately 25% reduced the level of moisture absorption (liquid immersion at 98°C) from approximately 7% down to 3%; a particularly impressive reduction bearing in mind the limited effect on T_g (228° down to 195°C).

With regard to non-fluorinated epoxy resins, clearly the amount of water absorbed will depend upon many factors, including in particular the types of resin and curing agent employed. Wright, in an excellent review[24], has indicated that water absorption may differ by a factor of ten between different resin types and by a factor of three with the same resin, but different curing formulations. A cursory examination of the literature indicates that water absorptions for various types of non-halogenated resins can vary enormously between, for example, 1% to in excess of 10%, with 2.0 to 6.0% being typical for a wide range of systems under many temperature/humidity conditions. The current fluoroepoxy formulations can therefore be considered as substantial improvements over the majority of 'conventional' epoxy resin systems with regard water absorption capacity. It is however necessary, for many potential applications to take into account other, perhaps equally desirable attributes. The next section considers and discusses some other aspects of the fluoroepoxy resins.

Mechanical Properties

Results obtained from the dynamic mechanical studies conducted on five of the fluoroepoxy formulations are shown in Figs. 10 and 11. Unfortunately due to supply limitations no mechanical characterizations have so far been possible with the C_o/1SA and C_6/7.5SA systems.

The tan δ values obtained from the two silicone-amine cured systems (C_6/1SA and C_8/1SA) plotted as functions of temperature (Figure 10) show two major transitions at approximately -85°C and 60°C. The latter, major transition can be regarded as the glass temperature, which at 60°C can be considered low. As indicated, both systems exhibit this transition at the same approximate temperature. Sub-zero temperature transitions have been observed with a wide range of amine-cured unmodified non-halogenated epoxies. In particular, the most frequently observed transition, generally occurring in the -50° to -80°C temperature range, has been referred to as the β-relaxation.[25-29] Its presence has been attributed to crankshaft rotations of glyceryl units, $-CH_2-CH(OH)-CH_2-O-$, in the epoxy matrix.[25,26,29] Although occurring at a somewhat lower temperature then generally found, it seems likely that the transition centered at approximately -85°C could also be associated with glyceryl unit relaxations. However, more detailed studies would be required to reinforce this view.

Tan δ - temperature plots for the three fluoroanhydride-cured formulations (Figure 11) similarly show two major transitions. Once again a low-temperature transition exists at approximately -90°C for all three systems. In this case only limited dynamic mechanical studies have been conducted on anhydride cured epoxies from which comparisons can be made. Unfortunately from the data which does exist conflicting trends have been found. For example, May and Weir[30] found no evidence of low-temperature transitions in a range of anhydride-cured epoxies. They attributed this to an absence of three or four unrestricted carbon atoms in the crosslinked network which they considered necessary for sufficient rotation to result in a low temperature transition. However, Cuddihy and Moacanin[31] found different trends with hexahydrophthalic anhydride and pyromellitic dianhydride cured diglycidyl ether based epoxies, with, in particular, the latter curing agent producing a β-transition at approximately -95°C. Molecular rotations involving smaller numbers of atoms than that previously proposed by May and Weir were outlined and it seems reasonable from the limited data available that the low-temperature transition found for the current fluoroanhydride cured systems could be associated with similar rotations.

The high temperature relaxations (Figure 11), which once again can be attributed to the glass transition temperature, show some significant formulation dependence. These will be briefly discussed later.

Table 3 shows room temperature mechanical property data obtained from the five fluoroepoxy formulations studied. Although limited at this stage in the program, the data does provide indications as to the effect of formulation variables on mechanical behaviour.

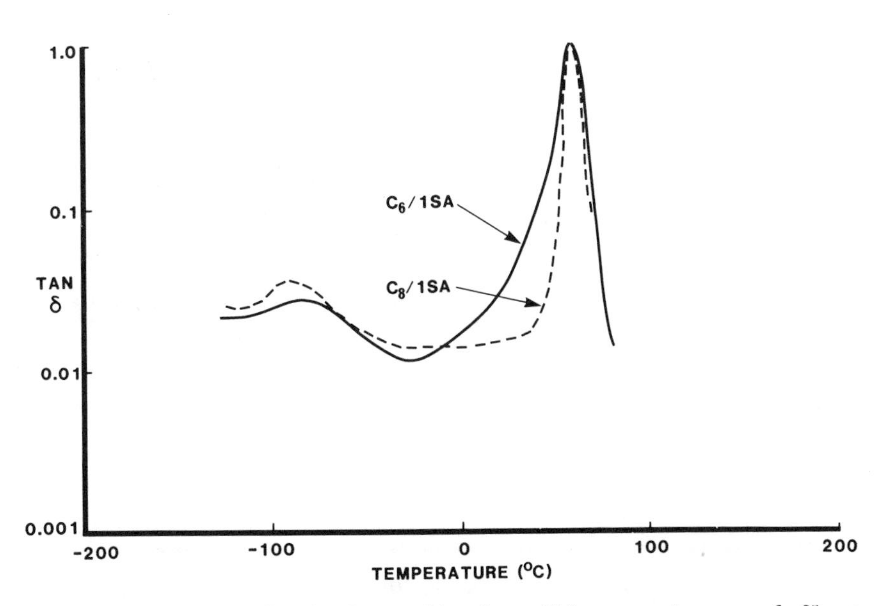

Fig. 10. Dynamic mechanical results for silicone-amine cured fluoro-epoxy resins.

Table 3. Mechanical Property Values for Five Fluoroepoxy Resin Formulations

Formulation	K_{IC} ($MNm^{-3/2}$)	G_{Ic} (Jm^{-2})	E (GPa)	σ_f (MPa)	e_f (%)	T_g (°C)
C6/1SA	0.488	120.8	1.74	42.79	2.6	61
C8/1SA	0.548	160.0	1.66	41.72	2.7	60
C8/FA/CAB	0.426	71.1	2.24	69.07	3.6	142
C8/FA/DMB	0.455	76.3	2.38	40.72	1.7	116
C8/FA/DA/DMB	0.495	96.2	2.23	26.90	1.2	112

Firstly, by comparing the C6/1SA and C8/1SA formulations it can be seen that an increase in resin fluorine content (with the same silicone-amine curing agent) indicates, albeit from limited information, a reduction in modulus together with an increase in toughness in terms of both fracture toughness K_{Ic} and fracture energy G_{Ic}. Since the difference in fluorine content between these two formulations are due entirely to the size of the pendant C_nF_{2n+1} group, it would seem likely that the above trends could be associated with variations in crosslink density resulting from steric effects, rather than an increased degree of flexibility within the network structure. Clearly with the present fluoroepoxy resins, main chain flexibility will derive from the hexafluoroisopropylidene groups. Since they are common to all the resins studied, main chain flexibility, resulting from the resin component, will remain constant.

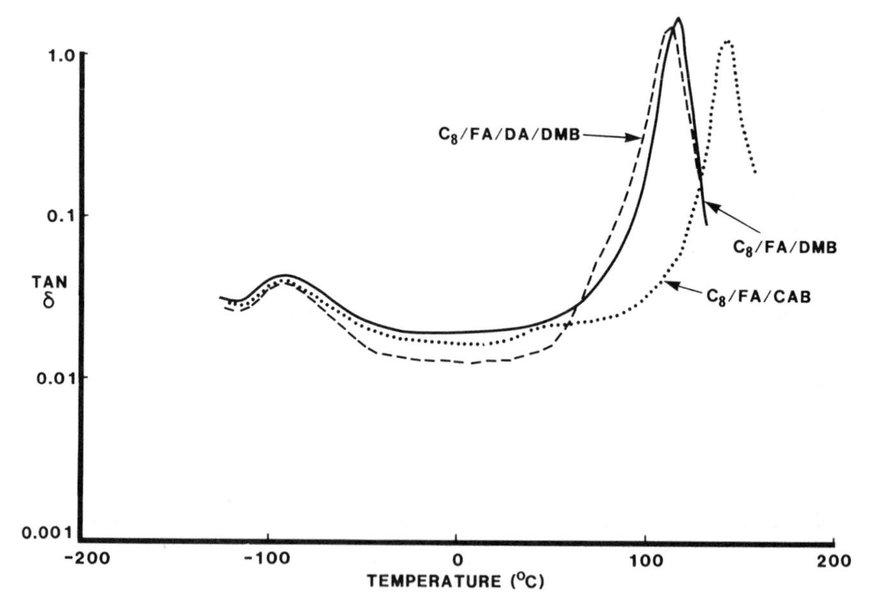

Fig. 11. Dynamic mechanical results for fluoroanhydride cured fluoro-epoxy resins.

Increasing resin fluorine content appears to have a negligible effect on the glass temperature. It is however important to note that the difference in fluorine content in this case is small, wider variations possibly having an influence on T_g. The results for the two silicone-amine cured systems suggest that improvements in modulus could be obtained by reducing the size of the pendant C_nF_{2n+1} group on the fluoroepoxy molecule (and hence a reduction in fluorine content). However as previously shown in Figs. 3 to 6 and Table 3, a reduction in toughness and hydrophobic character would also result.

Unfortunately it has not so far been possible to conduct mechanical and fracture experiments on the C_0/1SA formulation. Such data would clearly be able to confirm the above trends concerning resin fluorine content. It is hoped that this data will eventually be forthcoming.

Since no mechanical data is available for the C_6/7.5SA system, it is not possible to discuss in any reasonable depth the effects of silicone-amine molecular weight on mechanical behaviour. However as mentioned previously, the C_6/7.5SA system was a moderately flexible material at room temperature, in comparison to the rigid-like state exhibited by the C_6/1SA system. This cursory observation would suggest a reducing modulus and glass temperature with increasing silicone-amine molecular weight.

Table 3 clearly shows that the type of curing agent employed has a profound influence on the mechanical behaviour of the cured product, with the anhydride-cured systems exhibiting substantially higher modulus and T_g values than their equivalent silicone-amine cured counterparts. These trends can clearly be attributed to major differences in structure between the two types of systems. The silicone-amine curing agents would clearly impart molecular flexibility to the crosslinked network and thus influence mechanical properties in the manner described. This enhanced flexibility could also contribute to a greater toughness which could thus account for the marginally increased values of K_{Ic} experienced by the silicone-amine cured systems. Conversion of K_{Ic} values for fracture energy, G_{Ic}, using Eq. 3 shows a greater apparent difference between the two cure regimes, this being primarily due to major differences in modulus between the two broad types of formulations.

It is of particular interest to note from Table 3 the major differences in mechanical properties which occur by simply changing the fluoroanhydride accelerator (C_8/FA/CAB in comparison to C_8/FA/DMB). This is perhaps not too surprising bearing in mind the major differences in water absorption behaviour previously discussed. The major property differences which should be noted are that the C_8/FA/CAB system exhibits significantly greater values of flexural strength, σ_f, strain at failure, e_f, and T_g than its dimethylbenzylamine accelerated counterpart (C_8/FA/DMB). The precise reasons for these intriguing differences are presently unknown but are possibly associated with major differences in structure between the two systems. Although, as mentioned previously, the differences in cure conditions could be expected to impart property variations, it is particularly interesting to note that the C_8/FA/CAB system, which was cured at the lower temperature, had the highest T_g.

Rather surprisingly partial substitution of the dianhydride (DA) for the fluoroanhydride (FA), has no significant effect on mechanical properties and Tg. The reasons for this are currently unclear.

Table 4. Glass Transition Temperatures, T_g, and Water Absorption Values of Various Halogenated Epoxy Formulations

Formulation	T_g (°C)	H_2O absorption %	Formulation	T_g (°C)	H_2O absorption (%)
C_6/1SA	61	0.27	Trichlorinated Diglycidylamine (structure A, Fig. 8)	137	1.14
C_8/1SA C_8/FA/CAB	60 142	0.19 1.49	Tribrominated Diglycidylamine (structure A, Fig. 8)	152	0.82
C_8/FA/DMB	116	0.70	Fluorinated TGDDM (see Fig. 9)	217	2.90
C8/FA/DA/DMB	112	0.94	Chlorinated TGDDM (Fig. 9)	265	3.70
Fluorinated Diglycidylamines (structure B, Fig. 8)	157	2.60	Brominated TGDDM (Fig. 9)	238	3.30
Structure C, Fig. 8	143	1.52			
Structure D, Fig. 8	118	1.20			

Of further interest in Table 3 are indications that both fracture toughness, K_{Ic}, and modulus are rather low in comparison to many-non-fluorinated epoxies. The latter property is particularly true for the two silicone-amine cured systems. Although difficult to generalize, K_{Ic} values of approximately 0.65 to 0.90 MN $m^{-3/2}$ could be expected for a wide range of untoughened, non-fluorinated epoxies derived from diglycidyl based resins, together with modulus values of about 3 GPa.[32-37] Furthermore, the T_g values for both the silicone-amine cured systems can be considered low in comparison to many non-fluorinated epoxies, this of course being primarily due to the silicone groups within the crosslinked structure. Although the hexafluoroisopropylidence groups present in all of the resin systems studied here, would be expected to enhance molecular flexibility, it is interesting to observe that the fluoroanydride-cured systems exhibit T_g values similar, if not identical to that found for a wide range of non-fluorinated epoxies. This indicates that cured products obtained from the currently studied fluoroepoxy resins need not necessarily have very low glass temperatures.

Table 5. 'Dry' and 'Wet' T_g Values for Various Halogenated Epoxy Formulations

Formulation	T_g 'Dry' (°C)	T_g 'Wet' (°C)	Formulation	T_g 'Dry' (°C)	T_g 'Wet' (°C)
C_6/1SA	61	56	Trichlorinated Diglycidylamine Structure A, Fig. 8)	137	114
C_8/1SA C_8/FA/CAB	60 142	56 112	Tribrominated Diglycidylamine (Structure A, Fig. 8	152	136
C_8/FA/DMB	116	102	Fluorinated TGDDM (Fig. 9)	217	159
C8/FA/DA/DMB	112	93	Chlorinated TGDDM	265	191
Fluorinated Diglycidylamine s			Brominated TGDDM	238	172
Structure B, Fig. 8)	157	105			
Structure C, Fig. 8	143	113			
Structure D, Fig. 8	118	94			

Comparisons with other halogenated systems is virtually impossible due to a general lack of mechanical property data determined by other workers. However, comparison of glass temperature values are possible. Table 4 compares values obtained for the present systems with formulations developed by other workers.[17,18] For further information, water absorption values obtained at 20°C (full immersion) are also included. The data indicates that the present fluorinated systems exhibit glass temperatures which, broadly speaking, are somewhat lower than the equivalent diepoxide-based systems. Some of the data shown in Table 4 was obtained from tetraglycidyl-based systems, which would clearly result in comparatively high T_g values. As discussed previously, the relatively low T_g values obtained from the current systems can be attributed to molecular flexibilization resulting from the silicone-amine curing agent (when used) and the hexafluoroisopropylidence groups present in all the resins studied. However it is of interest to note that for the anhydride-cured systems, T_g values do begin to resemble those obtained from the other diepoxides shown in Table 4 (Particularly for the C_8/FA/CAB system).

One of the greatest problems associated with water absorption concerns glass temperature reduction. This could have serious implications in service situations where a combination of elevated temperatures and moist atmospheres are experienced. A reduction in T_g below the operating temperature could have serious consequences with avoidance of this situation being particularly necessary. A reduction in T_g of 20°C for each 1% of water absorbed is a generally accepted rule of thumb.[24] Table 5 shows both 'dry' and 'wet' T_g values for each of the systems previously shown in Table 4, with the 'wet' Tg values calculated using this approximate approach. As indicated, although most of the halogenated diglycidylamines developed by Johncock et al.[17,18] have superior 'dry' T_g values, their lower hydrophobic characteristics result in 'wet' T_g values which can be considered broadly similar to the current fluoroanhydride-cured fluoroepoxy systems (particularly so for the fluorinated diglycidylamines).

CONCLUSIONS

The results obtained so far in this program allow us to draw the following conclusions.

1. Fluoroepoxy resins have been developed which, when cured with silicone-amines or fluoroanhydrides, exhibit substantial hydrophobicity.

2. Increasing resin fluorine content increases hydrophobic characteristics and influences mechanical properties.

3. The silicone-amine cured systems exhibit greater hydrophobic characteristics than their fluoroanhydride cured counterparts. However this is counteracted by relatively low modulus and glass temperature values.

4. The catalyst employed with the fluoroanhydride-cured systems has a major influence on both water affinity and mechanical behaviour, particularly flexural strength and T_g.

5. The enhanced water affinity of the fluoroanhydride cured systems is possibly associated, to a certain degree at least, with the presence of hydroxyl groups that remain accessible to water after crosslinking. The removal of these hydrophilic sites would be expected to result in cured formulations exhibiting improved hydrophobic characteristics.

6. Some of the mechanical properties obtained, such as fracture toughness, K_{Ic} and modulus, are low in comparison to many non-halogenated epoxies.

ACKNOWLEDGEMENTS

The authors wish to express their appreciation to Mrs. M. Corthine, RARDE, Waltham Abbey, UK for experimental assistance.

NOMENCLATURE

a	Crack length
e_f	Strain at failure
l	Specimen thickness (water uptake)
t	Time
B	Specimen thickness (fracture)

D Diffusion coefficient
E Young's modulus
G_{Ic} Fracture energy (Mode I)
G' Storage shear modulus
G'' Loss shear modulus
K_{Ic} Stress intensity factor for crack initiation (Mode I)
M_t Water uptake at time t
M_∞ Equilibrium water uptake
P_c Load at crack initiation
Q Geometry factor
T_g Glass temperature
W Specimen width
σ_f Flexural strength
δ Loss angle
ν Poisson's ratio

REFERENCES

1. C.E. Browning, Polym. Eng. Sci., 18, 16, 1978.
2. C. Carfagna, A. Apicella and L. Nicolais, J. Appl. Polym. Sci., 27, 105, 1982.
3. D.M. Brewis, J. Comyn and R.J.A. Shalash, Int. J. Adhes. and Adhes., 215, Oct. 1982.
4. D.M. Brewis, J. Comyn and R.J.A. Shalash, Polym. Comm., 24, 67, 1983.
5. E.L. McKague, J.D. Reynolds and F.E. Halkais, J. Appl. Polym. Sci., 22, 1643, 1978.
6. T.S. Ellis and F.E. Karasz, Polymer, 25, 664, 1984.
7. J. Mijovic and S.A. Weinstein, Polym. Comm., 26, 237, 1985.
8. R.A. Gledhill and A.J. Kinloch, J. Adhes., 6, 315, 1974.
9. M.K. Antoon and J.L. Koenig, J. Macromol. Sci.,-Rev. Macromol. Chem., C19(1), 135, 1980.
10. W.C. Wake, Adhesion and the Formulation of Adhesives, Applied Science, London, p.174, 1982.
11. J.R. Griffith and J.E. Quick, Adv. Chem. Ser., 92, 8, 1970.
12. J.R. Griffith, A.G. Sands and J. Cowling, Adv. Chem. Ser., 99, 471, 1971.
13. J.R. Griffith, J.G. O'Rear and S.A. Reines, CHEMTECH, 311, 1972.
14. D.L. Hunston, J.R. Griffith and R.C. Bowers, Ind. Eng. Chem. Prod. Res. Dev., 17, 10, 1978.
15. J.R. Griffith, CHEMTECH, 290, 1982.
16. J. Goobich and G. Marom, Polym. Eng. Sci., 22, (16), 1052, 1982.
17. P. Johncock and G.F. Tudgey, Br. Polym. J., 15, 14, 1983.
18. J.A. Barrie, P.S. Sagoo and P. Johncock., Polymer, 26, 1167, 1985.
19. J.F. Knott, Fundamentals of Fracture Mechanics, Butterworths, London, 1973.
20. ASTM Standard D790-71, 1973.
21. H. Fujita, Adv. Polym. Sci., 3, 1, 1961.
22. J.B. Enns and J.K. Gillham, J. Appl. Polym. Sci., 28, 2831, 1983.
23. W.G. Potter, Epoxide Resins, Illiffe, London, 1970.
24. W.W. Wright, Composites, 12, 201, 1981.
25. F.R. Dammont and T.K. Kwei, J. Polym. Sci., 5, 761, 1967.
26. O. Delatycki, J.C. Shaw and J.G. Williams, J. Polym. Sci., 7, 753, 1969.
27. R.G.C. Arridge and J.H. Speake, Polymer, 13, 44, 3 1972.
28. T. Hirai and D.E. Kline, J. Appl. Polym. Sci., 17, 31, 1973.
29. J.G. Williams, J. Appl. Polym. Sci., 23, 3433, 1979.
30. C.W. May and F.E. Weir, S.P.E. Trans., 2, (3), 201, 1962.
31. E.F. Cuddihy and J. Moacanin, J. Polym. Sci., A2, 8, 1637, 1970.

32. R.J. Young, Developments in Polymer Fracture-1, E.H. Andrews Ed., Applied Science, London, p. 183, 1979.
33. S. Yamini and R.J. Young, J. Mater Sci., 14, 1609, 1979.
34. R.J. Young, Developments in Reinforced Plastics-1, G. Pritchard Ed., Applied Science, London, p. 257, 1980.
35. A.J. Kinloch and R.J. Young, Fracture Behaviour of Polymers, Applied Science, London. p. 293, 1983.
36. A.J. Kinloch, S.J. Shaw, D.A. Tod and D.L. Hunston, Polymer, 24, 1341, 1981.
37. S.J. Shaw, The Fracture of Epoxy Resins and the Effects of Rubber Inclusions, Ph.D Thesis, City University, London, 1984.

CHARACTERIZATION OF A DURABLE INORGANIC METAL-TO-METAL ADHESIVE FOUND IN A CHINESE QIN's CHARIOT BUILT AROUND 207 B.C.

Xia Wen-gan

Central Laboratory
The Huang-He Machinery Factory
Xi'an, The People's Republic of China

ABSTRACT

A white mass for the metal-to-metal bonding of the gap between the chariot axle pin and the sleeve was identified to be an inorganic adhesive with a possible composition of $Ca_5(PO_4)_3OH$. The chariot was build around 207 B.C. for Emperor Qin Shi Huang. We analyzed the white mass with an infrared spectrometer and an X-ray diffractometer.

We also obtained some known mineral containing $Ca_5(PO_4)_3OH$ for comparison. However, the analytical results of the mineral did not completely agree with those of the adhesive. Despite the disagreement, we believe that we have identified the first metal-to-metal inorganic adhesive used around 207 B.C. in China.

INTRODUCTION

In December, 1980 in Xi'an, China, two bronze coloured chariots and horses, located to the west of the tomb of Emperor Qin (221-207 B.C.), were unearthed. They displayed not only exquisite craftsmanship, but also were of great value for scientific research. Of them, chariot No. 2 was broken into 1555 small pieces (Fig. 1), while another chariot into many more pieces than No. 2, when unearthed. Since then we have restored them, mainly by using adhesive bonding method, and since October 1, 1983 the restored chariot No. 2 (Fig. 2) has been opened to the public as an exhibit. Its total weight is 1244 kg; overall length 3.17 m; width 1.43 m and height 1.06 m. Of the total weight, gold is 3033 g, and silver 4342.1 g. During the course of restoring we found some white solid mass stuck to the gap between the axle pin and the sleeve, which were made of silver and attached to the axle shaft at both ends. The outside diameter of the sleeve at the thin end was 25 mm, and the inside diameter 19.2 mm; at the thick end its outside diameter 40.5 mm, and inside diameter 21 mm. At the thick end there was a rectangular hole with 9.2 mm in length and 4.5 mm in width. The axle pin had one end with the shape like sheep head, and the remainder with the rectangular form, its height was 41 mm, length 9 mm, width 4.3 mm. The rectangular part of the axle pin was inserted into the

Fig. 1. Chariot No. 2 broken into 1555 small pieces.

rectangular hole of the sleeve. After carefully taking, with difficulty, the axle pin out of the sleeve hole, the above-mentioned white solid mass still strongly stuck to the place as shown in Fig. 3.

In this paper we describe the analyses of the white mass to be an inorganic adhesive used for the metal-to-metal bonding.

Fig. 2. Restored chariot.

ANALYTICAL RESULTS

We obtained the infrared absorption spectrum for the sampled mass by means of KBr pressed sheet (Fig. 4). Figure 4 shows the white solid mass to be an inorganic compound: 1450cm^{-1} and 1000–1100cm^{-1} are the carbonate and phosphate absorption peaks respectively. The result suggests that the adhesive may be a carbonate-and-phosphate composite. At the same time, we analyzed the white coating of the chariot. The IR spectrum (Fig. 5) shows that there is a large number of the phosphate in the white coating. At 1450cm^{-1} the carbonate absorption peak of Fig. 5 is lęss distinct than that in Fig. 4. Furthermore, the spectrum shows that calcium orthophosphate, $Ca_3(PO_4)_2$ (plus $ca(OH)_2$ impurity) and a small amount of Si, Mg and Pb are contained in this coating. Having consulted the book entitled "The Standard Infrared Spectrum Graphics of Inorganic Compounds," we found this coating agreed with Graphic No. 238 in this book (Fig. 6). With the aid of X-ray diffraction, we confirmed that it was a hexagonal $Ca_5(PO_4)_3{\cdot}OH$, the natural mineral mass. From a comparison between Fig. 4 and Fig. 5, we can conclude that the white solid mass for bonding the axle and the sleeve resembled $Ca_5(PO_4)_3{\cdot}OH$ containing a large amount of carbonate.

DISCUSSION

We are certain that the white solid mass was not a part of the soil for burying the chariot. The chariot was buried 7-8 m deep under ground. After more than 2,000 years have passed, the downfall of earth took place in the course of long history, finally the whole body of the chariot was buried. When archaeologists unearthed them, they found the covering soil in different deep layers with different colours, such as the black soil, the yellow soil and the sand were mixed as a sort of multi-colour soil, and the remainder was the grey soil and the red soil.[2] Those soils possibly contained SiO_2, but did not contain $Ca_5(PO_4)_3{\cdot}OH$, and the colour of the white mass was white which is greatly different from those of the above mentioned solids. Thus, it is clear that the white solid mass was placed into the gap for bonding purpose only during the time of assembly.

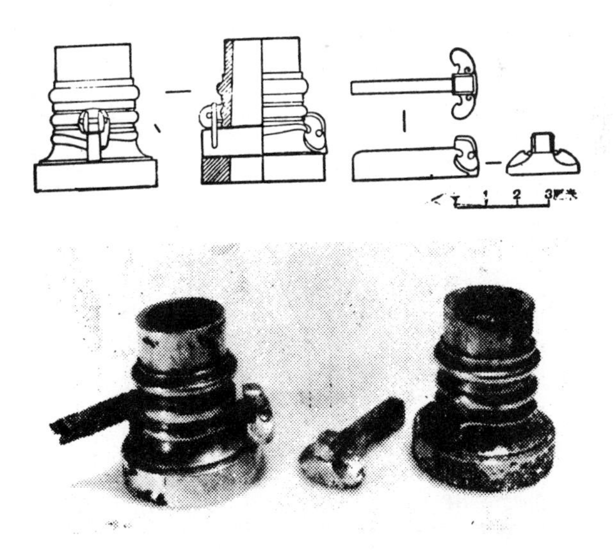

Fig. 3. Solid mass adhering strongly in place.

The white solid mass was identified to be an inorganic adhesive originally used for bonding metals. According to the archaeological study report[2], bronze chariots and horses, which were one half of the life-size, copied strictly after the genuine chariots available for Qin Shi Huang (221-207B.C.). He was the first Emperor who brought all of China under one rule. In ancient time one can imagine that the machine work was generally very poor; it was difficult to tightly fit the axle

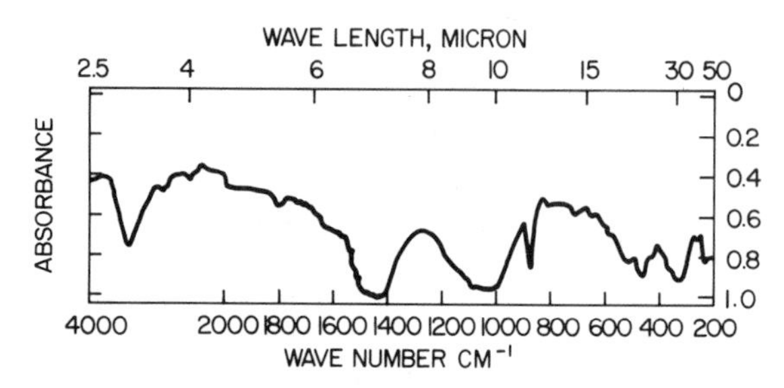

Fig. 4. White solid mass shown to be an inorganic compound.

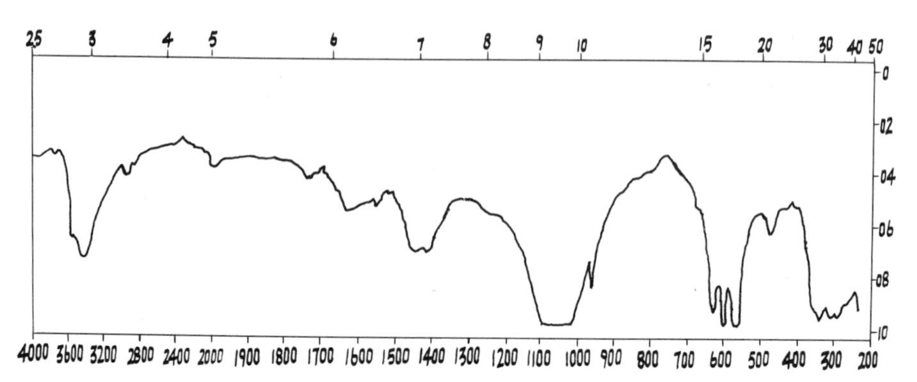

Fig. 5. Showing a large number of phosphate in the white coating.

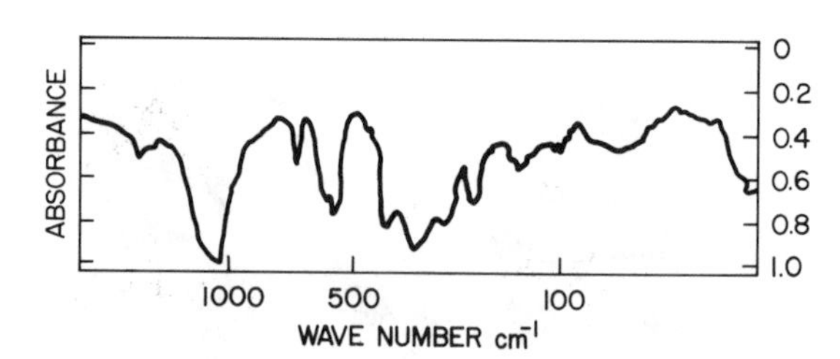

Fig. 6. Abundance of P, Ca and a small amount of Si, Mg and Pb are contained in this coating.

pin with the sleeve. In order to better fit, there was a 0.2 mm tolerance remaining between them. But with 0.2 mm tolerance the axle pin would drop down while the axle shaft was running; or even if, with a close fit between them, the axle pin also would drop down due to the vibrations when the chariot was traveling on the road. For reasons given above, it was necessary for chariots to use such white solid mass serving as an adhesive for bonding the axle pin and the sleeve for getting a reasonable construction design and meeting the requirements.

The general formula for a modern phosphate inorganic adhesive is supposed to be $MeO \cdot P_2O_5 \cdot H_2O$; Me mainly denotes Ca, Cu, Mg. etc. Of them Ca gives the best watertight. The phosphate curing agent belongs mainly to the oxides of Mg, Ca, etc.[3] This inorganic phosphate adhesive agrees with that of the white solid mass analyzed. The chemical composition of $Ca_5(PO_4)_3 \cdot OH$ is mainly CaO and P_2O_5, and a trace of F, Cl, H_2O, etc.[1] It can be concluded that the white solid mass of the chariot must be some sort of inorganic adhesive. At that time, this inorganic adhesive possibly was composed of a natural phosphate and CaO, then water was added onto them to form a paste. For example, after $Ca_5(PO_4)_3 \cdot OH$ was ground into powder, it was mixed with CaO and water with the following reaction:

$$CaO + H_2O \rightarrow Ca(OH)_2$$

In China, we have been using $Ca(OH)_2$ as a wall painting material since the ancient time. When $Ca(OH)_2$ absorbs CO_2 from air, it becomes a white hard solid mass of $CaCO_3$. We tried to verify the analytical results. So we obtained one kind of mineral rock of $Ca_5(PO_4)_3 \cdot OH$ from Xi'an Mineral Institute, and mixed it with CaO, after solidification we obtained an IR spectrum; however its spectrum did not fully agree with the above results. This disagreement could have been caused by one of the following reasons:

1. There are a great variety of $Ca_5(PO_4)_3 \cdot OH$ from different sources. Their compositions are somewhat different depending upon the sources. Consequently, their infrared absorption spectra somewhat disagree with one another.

2. The curing time was too short.

3. Modern method of mixing could be different from that of the ancient time.

Where was the source of $Ca_5(PO_4)_3 \cdot OH$ for the chariot in the Qin dynasty? What was the technology used for formulating such an adhesive? All these problems remain to be investigated and studied.

CONCLUSION

An inorganic adehesive for metal-to-metal bonding has been found in a China's historical relics.

"Using adhesive for bonding metals has only about 25 year's history" said Mr. S. Semerdjiev.[4] However our finding shows that the adhesive for metal bonding can be dated back to 200 B.C. in the Qin Dynasty in China. Chariots unearthed near Qin Shi Huang's tomb have given us a clear proof for the first known adhesive in history.

This adhesive could be one of the earliest inventions known in our history, but we have not found any written record about such an invention in the ancient time.

ACKNOWLEDGEMENT

We appreciate the technical support provided by Messrs. Chao Yu-Din and Ye Gin-Yen for obtaining X-ray diffraction data and infrared spectra for the adhesive obtained from the unearthed chariot.

REFERENCES

1. Geological Department of Nanjing University Crystallography and Mineralogy (in Chinese). Geological Publishing House, (1978).
2. "Qin Tomb's Bronze Chariot No. 2," Archaeology and Cultural Relics (in Chinese) No. 1, (1983).
3. Wang Shi-An, Adhesive and Its Application (in Chinese). Shanghai Technical Publishing House (1981).
4. S. Semerdjiev, Metal-to-Metal Adhesive Bonding, p.4. Business Books Limited, England, (1970).

THE INVESTIGATION OF UNDERWATER ADHESIVES IN THE PEOPLE'S REPUBLIC OF CHINA

Xie Ju-niang, Zhong Song-hui, Zou Xiao-ping, Chen Ru and Lu Sang-heng

Gangzhou Institute of Chemistry
Academia Sinica
P.O. Box 1122, Guangzhou
People's Republic of China

ABSTRACT

This paper introduces the investigations of underwater adhesives since 1970 in the P.R.C. We discuss the adhesives studied by Guangzhou Institute of Chemistry, the Hydraulic Research Institute in Beijing, and Changsha Institute of Chemical Industry, etc. At present there are four different kinds of polymeric materials used as underwater adhesives; namely, epoxy resin, polyurethane, ethylenic-terminated polyurethane and epoxy-acrylate adhesives. The adhesives which give good bond strength and aging resistance in water have been successfully used in repairing sunken ships, treating concrete structures in water and bonding plastizized PVC sheet on wet concrete floors.

INTRODUCTION

Underwater adhesive is an adhesive which can be directly used for repairing underwater structures, such as underwater buildings, pipelines and submerged portion of ships, etc. Common adhesives cannot be used in ordinary ways to repair the damaged part which is in the presence of water or is situated underwater. Thus, an underwater adhesive can be applied directly in water; the submerged substrate can be wetted with it and it can be cured in the presence of water. It is of utmost importance that it should have good bond strength and long service life by being stable in water.

Since 1970 underwater adhesives have been studied in Guangzhou Institute of Chemistry, Academia Sinica. The need of salvaging a sunken ship had put a new problem before us. In the meantime adhesive workers in our country have paid more attention to the study of underwater adhesives. Apart from Guangzhou Institute of Chemistry, Tianjing Institute of Synthetic Materials[1], the Hydraulic Research Institute[2] and Changsha Institute of Chemical Industry[3] have been studying underwater adhesives in succession too. For the present, in China, there are four different kinds of polymeric materials used as underwater adhesives; namely, epoxy resin, polyurethane, ethylenic-terminated polyurethane and epoxy-acrylate adhesives.

EPOXY RESIN UNDERWATER ADHESIVES

Epoxy resin adhesives were our initial work on underwater adhesives. We found that when an epoxy resin was mixed with asphalt or water absorbent additives, or cured with a curing agent which could react with water and release an amine, then, such an epoxy resin could be used in the presence of water or as underwater adhesive. We infer that the underwater adhesive should be composed of either a water-repellent or a water-activated or a water-absorbent material. Thus, we studied epoxy resins as well as other polymeric adhesives to determine whether they could serve as underwater adhesives.

The epoxy resin we used is made of bisphenol A - epichlorohydrin having a molecular weight in the range of 380 to 470, and an epoxy equivalent in the range of 184 to 270. Gypsum, metallic oxide and an alkali-earth metallic oxide can be used as the water-absorbent additives. Diethylene triamine (DETA), polyethylene polyamine, phenolaldehyde-modified amine, ketimine, carbamate etc. can be used as curing agents. An example of a suitable composition of underwater adhesive is given in Table 1.[4] The difference between the underwater adhesive in Table 1 and a common adhesive is the presence or absence of the water-absorbent CaO. The adhesive in Table 1 was successfully used in repairing the propeller blade of a sunken ship and in sealing the leakage of axle at the tail of the ship. The formula can be used only for relatively short-term, such as three months or so, because of the decreasing bond strength during aging.

In order to decrease the toxicity of amines, some adhesive workers tried to modify it with phenolaldehyde, ketone, etc. and improve the aging property of the adhesive. We modified the amines with cyano-substituted alkene. We gave the new curing agent a code name AC. The adhesives cured with AC showed better bond strength and aging life than those cured with DETA, (Tables 2 and 3).[5] After the adhesives were stored for one year in two components the bond strengths did not decrease (Table 4).

Table 1. The Formula of Epoxy Underwater Adhesive

	Parts
Epoxy resin 618 or 634	100
Polyester 702	10-20
Calcium oxide	50
DETA	10
DMP-30	3-5

Table 2. Comparison of Bond Strengths of Epoxy Resin Cured With Various Curing Agents (Underwater Operation, steel)

	Curing Agent	Shear strength(kg/cm^2)
DETA	Chemical reagent	161-184
AC	Guangzhou Institute of Chemistry	250-300
810	Changsha Institute of Chemical Industry	80-90
Dongfeng	Guangzhou Dongfeng Chemical Plant	50-70
301N	Japan Ashahibond	70-75
377	Japan Ashahibond	125-140

Table 3. The Influence of Aging on Bond Strength of AC Hardened Epoxy Underwater Adhesive (Steel; 10-30°C. in water)

Aging time (days)	Shear strength (kg/cm^2)	Aging time (days)	Shear strength kg/cm^2)
1	224	100	295
3	224	150	200
6	243	180	210
11	269	200	200
13	270	220	190
24	290	240	160
30	290	260	180
37	293	280	192
50	290	300	190

Table 4. Bond Strength of Epoxy Underwater Adhesive After Storing In Two-Components At Room Temperature. (Hardened With AC).

Storing time	Shear strength (kg/cm^2.)
Fresh adhesive	300-310
2 months	290-200
4 months	300-310
6 months	280-290
1 year	290-300

Huang Zheng-yan et al.[3] of Changsha Institute of Chemical Industry, reported that they used a curing agent with a code name of 810 to formulate underwater adhesives, and Mai Shu-fang et al.[2] of the Hydraulic Research Institute, Beijing, mentioned that they used a curing agent made of polyol and polyamine to prepare underwater adhesives. Their adhesives have also been shown to be useful for the treatment of concrete structures underwater or under wet conditions.

POLYURETHANE UNDERWATER ADHESIVES

Polyurethanes can be made into a hydroxyl-terminated prepolymer or an isocyanate-terminated prepolymer. The isocyanate-terminated prepolymer can react with water and release carbon dioxide:

$$—NCO + H_2O \rightarrow —NH_2 + CO_2$$

$$—NCO + —NH_2 \rightarrow —NHCONH—$$

Provided that we can treat the released carbon dioxide with a special process, the polyurethane NCO-terminated prepolymer can be used in the presence of water or used as underwater adhesive. The properties of polyurethane underwater adhesives shown in Table 5[6] indicate that our previous inference was correct. Polyurethane adhesive 201[7] was successfully used for bonding the plasticized PVC on wet concrete floor. The properties of Adhesive 201 are shown in Table 6 and Fig. 1.

Yang Chao-xiong[8] used the polyurethane adhesive as an underwater pressure-sensitive adhesive. She reported that the 180° peel strength (dry) of the adhesive was 0.5 to 4.0 kg./cm. After the immersion in water or sea water for 300 days, the peel strength did not decrease.

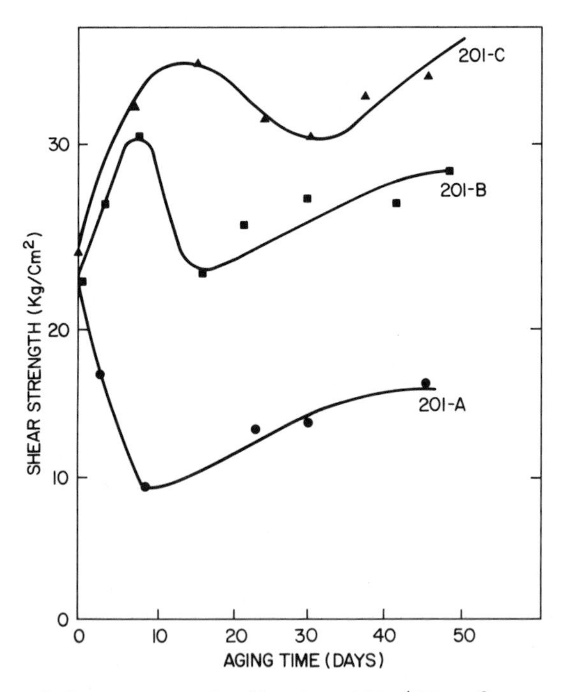

Fig. 1. Aging curve of adhesive 201 (55$\pm$ 1°C; R.H.>95%).

Table 5. Mechanical Properties of Various Polyurethane Underwater Adhesives (Underwater Operating).

Adhesives	Strengths		
	Shear* (kg/cm^2)	Peel** (kg/cm.)	Heterogeneous Tensile (kg/cm^2)
A-2	143-164	1.3-1.4	16-18
GS-1	150-180	1.0-1.5	20-25
GS-2	170-200	1.5-1.8	25-30
GS-3	200-250	2.2-2.5	40-42

*: Steel 45 **: Steel/Canvas

ETHYLENIC-TERMINATED POLYURETHANE UNDERWATER ADHESIVES

Ethylenic-terminated polyurethane is the product of NCO-terminated polyurethane prepolymer reacted with a carboxylic acid. The NCO-groups have already reacted with the carboxyl, so they cannot react with water any more. The active groups are the unsaturated double bonds. If we add a water absorbent additive or a suitable water repellent diluent to the polymer and use an oxidation-reduction system to initiate the polymerization of the unsaturated bonds, it can be used in the presence of water as an underwater adhesive. The properties shown in Tables 7 and 8[9] again support our previous inference.

EPOXY-ACRYLATE UNDERWATER ADHESIVES

Epoxy-acrylate is the reaction product of an epoxy resin and acrylic acid. It functions similarly as epoxy resin. It contains neither a water absorber nor a water reactive material. The difference between them is the curing process: one depends on its epoxy groups, and the other, its unsaturated bonds. If a water absorbing additive is mixed with the epoxy-acrylate, we believe the adhesive we made can be used in the presence of water or underwater conditions. The epoxy-acrylate modified with TDI gives better bond strengths than the unmodified adhesive (Table 9).[10]

SUMMARY

The underwater adhesives should be composed of either a water repellent or a water-activated or a water absorbent material as in the cases of four different kinds of polymeric materials. The underwater adhesives we made gave good bond strengths and aging resistance in water and have successfully been used in repairing sunken ships, treating concrete structures in water and bonding plasticized PVC sheet on wet concrete floors. The advantages and disadvantages of the adhesives are as follows:

Table 6. Bond Strengths Adhesive 201

Adhesive	Substrate	Peel Strength (kg/cm.)		Substrate	Shear Strength (kg/cm^2.)		Substrate	Tensile Strength (kg/cm^2.) wet
		dry	wet		dry	wet		
201-A	a	1.0-3.7	----	b	49-54	23-40	g	11-24
	b	2.4-3.0	1.4-2.1	c	42-47	20-38		
	c	2.8-3.2	1.9-2.7	e	122-130	106-131		
	d	1.9-3.0	2.9-3.6	f	98-122	111-126		
201-B	a	1.0-4.0	----				g	16-31
	b	3.0-3.4	1.2-1.9					
	c	2.8-3.0	2.5-2.9					
201-C	a	1.8-3.0	----	b	56-59	52-64	g	14-29
	b	2.9-3.5	3.3-3.8	c	45-48	46-58		
	c	3.0-3.6	2.9-3.3	e	63-78	66-68		
	d	----	2.5-2.9	f	71-82	61-69		

a: PVC/PVC;
b: PVC/Steel;
c: PVC/Stainless steel;
d: PVC/Wood;
e: Steel/Steel;
f: Stainless steel;
g: PVC/Mortar

Table 7. Shear Strength of Ethylenic-Terminated Polyurethane Underwater Adhesives (Steel; in water; room temp.)

Adhesive	Shear strength (kg/cm^2)	Adhesive	Shear Strength (kg/cm2)
103	174	107	47
104	173	109	55
105	138	106	65

Table 8. Aging Properties of Ethylenic-Terminated Polyurethane Under-water Adhesives (Steel, 55±1°C., RH. 95%) (shear strength kg/cm^2).

Adhesives	Aging Time (days)						
	1	7	14	21	28	35	42
103	91	186	197	154	142	225	219
104	95	205	192	160	143	210	207

Table 9. The Comparison of Epoxy-acrylate (EA) and TDI-Modified Epoxy-acrylate (EAT) Underwater Adhesives In Shear Strengths. (kg/cm^2.)

Day	EA	EAT
1	80-100	180-200
3	100-130	195-225

Underwater Adhesive	Advantage	Disadvantage
Epoxy Resin	It can be cured with many kinds of curing agents for various usage of underwater adhesives.	Its curing time is comparatively slow, so it cannot be used at the vertical plane without supporting.
Polyurethane	It can be used at low temperature without difficulty, and can bond plasticized PVC to wet concrete.	Because of the water-sensitivity of the prepolymer, its shelf-life is comparatively short.
Ethylenic-Terminated Polyurethane	It can be adjusted with the oxidation-reduction initiation system to make fast curing adhesive, so it can be used to seal the pipe-line with flowing water.	Because of the short curing time, it cannot be used in large scale.
Epoxy-acrylate	identical	identical

In order to understand the mechanism of adhesion under wet condition, we have been studying the underwater adhesives from various angles, for example, thermodynamic factor of adhesive adhering in water; the diffusion of water at the interface between substrate and adhesive by using radioisotope, etc. We wish to report our other findings in the future.

REFERENCES

1. Tianjing Institute of Synthetic Materials. He Cheng Cai Liao (in Chinese) 2, 51 (1974).
2. Mai Shu-fang et al., Paper No. 1-48 of the First Annual Meeting of Chinese Adhesion Society (in Chinese).
3. Huang Zheng-yan et al., Society Paper No. 1-21 of the First Annual Meeting of Chinese Adhesion Society (in Chinese).
4. "Epoxy Resin Underwater Adhesion,"Hua Xue Tung Xun, (in Chinese) 32-38 (1976).
5. "Epoxy Resin Underwater Adhesive No. 2" (in Chinese), Research Paper of Guangzhou Institute of Chemistry, Academia Sinica.
6. "Polyurethane Underwater Adhesive" (in Chinese), Research Paper of Guangzhou Institute of Chemistry, Academia Sinica.
7. Xie Ju-niang et al. Hua Xue Tung Xun (in Chinese), 1, 39-51 (1983).
8. Yang Chao-xiong, Hua Xue Tung Xun (in Chinese), 2, 50-61 (1979).
9. "Ethylenic-Terminated Polyurethane Underwater Adhesive" (in Chinese), Research Paper of Guangzhou Institute of Chemistry, Academia Sinica.
10. "Epoxy-acrylate Underwater Adhesive" (in Chinese), Research Paper of Guangzhou Institute of Chemistry, Academia Sinica.

SEALING OF HUMAN TUBULAR TISSUES WITH HIGH ALKYL CYANOACRYLATE ADHESIVE XKM-2 AND ITS EFFECTS--A NEW STERILIZATION AND NONSURGICAL TECHNIQUE

Ruan Chuan-liang*, Liu Jin-hui+ and Qin Zhao-yin**

*Xi'an Adhesive Technology Institute
No. 21, Youyixi Road, Xi'an

+Xi'an No. 4 Hospital
Jiefang Road, Xi'an

**Xi'an Medical University
Zhuquennan Road, Xi'an
The People's Republic of China

ABSTRACT

Sealing of human tubular tissues, such as oviduct, with adhesives has a good developing prospect for human birth control and treatments of some diseases. This paper discusses an ideal adhesive for sealing tubular tissues, explores operation methods of sealing, investigates the interaction of solidifed adhesive with the inside tube wall and evaluates clinical effects of sealing. The operation procedure is to first insert XKM-2 with a special appliance into the narrow part of the oviduct. After its inside wall is sealed, a bacteria-free inflammation appears there, then a foreign-matter granulation tumor is formed, finally the tube cavity is sealed by the proliferous fibre structure together with XKM-2 polymer for a long time. The curative success rate of sterilization is up to 95%. The longest time for sealing has been 15 years. Besides, sealing of pancreatic duct with XKM-2 has obtained satisfactory results as well.

INTRODUCTION

Owing to the continuous growth of modern polymer chemistry and its progressive permeation into medical domain, some original medical technologies have changed a great deal. As a result of the application of adhesive technology in the medical domain, the suturing of human skin, hemostasis of wound, adhering of tooth and skeleton and sealing of tubular tissues have entered a new stage of adhesive bonding. Sealing, with adhesives, of oviduct, spermatic duct, pancreatic duct and blood vessels in certain positions has a good developing prospect for human birth control and treatments of some diseases.

This paper reports an ideal medical adhesive, high alkyl cyanoacrylate adhesive XKM-2, for sealing human tubular tissues, explores operation methods of the sealing, investigates the interaction of solidified adhesive with the inside duct wall and evaluates the clinical effects of sealing.

ADHESIVE FOR SEALING HUMAN TUBULAR TISSUES

As there is usually a great deal of weak-base solution in human tissues to be sealed with adhesives and their surface can not be treated beforehand, epoxy resin, polyurethane, rubber, acrylic resin and organosilicone adhesives to seal tubular tissues are unsatisfactory. We specially formulated, through repeated testing, one kind of high alkyl cyanoacrylate adhesive XKM-2 which can adhere to the inside surface of tubular tissues rapidly. When this adhesive is injected into the inside narrow cavity with the diameter of about 1 mm of the oviduct, under the catalytic reaction of the mucous in the cavity, the adhesive is rapidly solidified and bonded with the mucous membrane of the duct wall to prevent it from overflowing or outflowing to the unnecessary adhering parts.

It has been proved by testing with bacteria culture that this XKM-2 adhesive itself is bacteria-free and has distinct effects of inhibition to many bacteria. Proved by animal tests, the acute toxicity of the medical adhesive XKM-2 has been rated low; after through giving medicine many times, the examinations of the body weight, the weight of liver and kidney and their functions, blood parameters and tissue shapes of the animals tested, show no adverse affects. No growth factor of transplanting malignant tumors have been promoted; no distortion of cell chromosome of big mouse marrow has aroused. Furthermore, the test of either Ames, distortiogen or carcinogen, is negative. Therefore, it is further considered that the application of the adhesive to human body is safe.

The XKM-2 adhesive also has a capacity of preventing X-rays from penetration. When the sealed position is examined with X-rays, the sealed conditions and effective length of the adhesive in the cavity could be revealed clearly on the photographic plate, thus at any time after sealing, the sealing conditions and result could be checked simply and conveniently with X-rays.

OPERATION METHODS FOR SEALING HUMAN TUBULAR TISSUES WITH XKM-2 ADHESIVE

How to inject XKM-2 adhesive accurately into the duct cavity to be sealed is the key as to whether the sealing is successful or not. Since human oviduct is located at both ends of uterine bottom, the sealing operation can not be done under direct vision. Therefore, we first used rabbits' and women's isolated uteri to do exploratory tests, then on the basis of the experiences obtained, we progressively determine better methods for sealing human oviduct with XKM-2 adhesive. The method for sealing one side of the oviduct is as follows: for a woman who is a volunteer to be sterilized, tally with the sterilization conditions and free of any contraindications, make a routine disinfection first, lightly insert a special appliance for sealing the oviduct into the uterine cavity (Fig. 1). After finding the mouth of the oviduct, inject a 0.1 ml of XKM-2 adhesive with an injector, then remove the special appliance. Sealing of the opposite side is the same as mentioned above (Fig. 2). Finally, take an X-ray photograph, if the

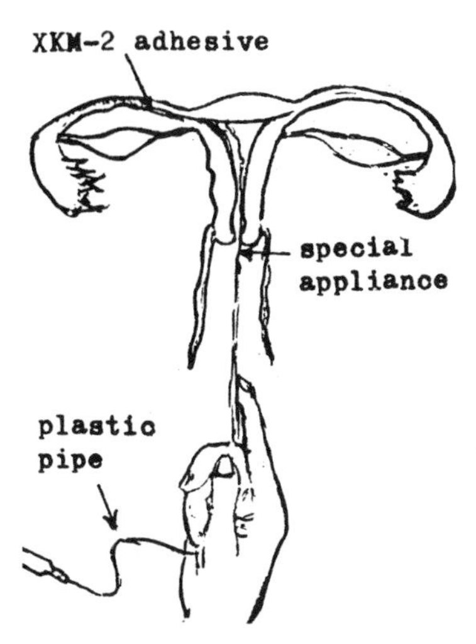

Fig. 1. Insertion of a special appliance with XKM-2 adhesive into the uterine cavity.

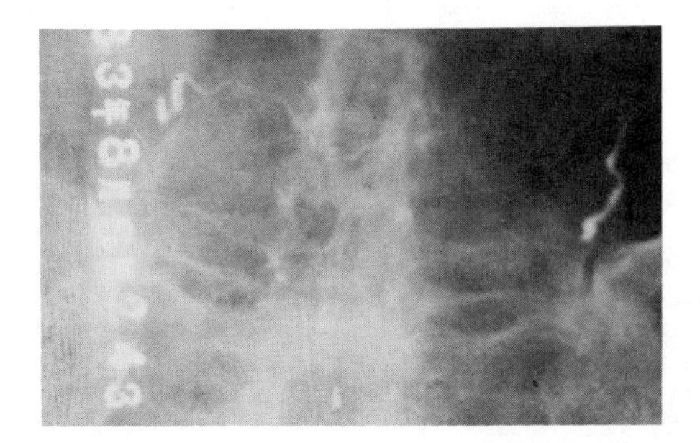

Fig. 2. Sealing the opposite side.

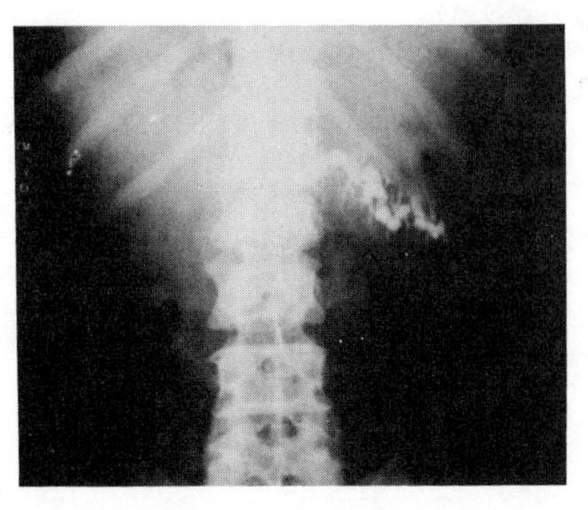

Fig. 3. Injection of XKM-2 adhesive into the pancreatic duct.

developments of both sides of the oviduct are at the very narrow positions, it indicates that the sealing operation is successful.

For sealing a pancreatic duct, we first dissociate the broken end of the duct (the pancreas has been cut off) out a little, insert an equithick disinfected plastic pipe into it, inject with a dry injector 4-5 ml of XKM-2 adhesive into the pancreatic duct, then remove the plastic pipe, finally sew up the tubercle of the pancreas remains (Fig. 3).

For sealing blood vessel and spermatic duct, we generally adopt the method similar to the injection of intravenous puncture.

INTERACTION OF THE SOLIDIFIED XKM-2 ADHESIVE AND THE INSIDE WALL OF THE OVIDUCT

After XKM-2 adhesive has been injected into the narrow part of the oviduct, the adhesive will immediately solidify in a short time and adhere to the mucous membrane on the contact part. At this very moment, the sealed part can withstand a hydraulic pressure higher than 120 mmHg without being peeled off.

It is generally known that the mucous membrane itself containing a great deal of moisture will gradually lower the adhesive strength of the XKM-2 adhesive. Secondly, a living oviduct, unlike a dead one, after being stimulated by the foreign matter (XKM-2 adhesive), will immediately create abnormal wriggle and swing, and try hard to squeeze the foreign matter out for recovering its original function. Moreover, the mucous membrane of the oviduct adhered by XKM-2 may, after a certain period of time, peel off and lose its adhering function. For example, after a certain period of time injecting XKM-2 adhesive into the oviduct cavity, the solidified XKM-2 may often give rise to longitudinal movement due to the external longitudinal thrust. For this reason, only relying on the temporary adhering force between XKM-2 and the mucous membrane, it is difficult to permanently seal the duct cavity in full. But in addition, XKM-2 will inevitably produce mechanical, physical and chemical stimulations to the inside wall of the duct cavity to cause the cavity undergoing an effective pathological change for complete sealing. We have observed the pathological changes in the rabbits' and women's oviducts injected with XKM-2, and have confirmed that both of the changing courses are basically the same and fall roughly into the following three stages:

1. The mechanical, physical and chemical stimulations of XKM-2 cause the mucous membrane in the duct cavity bacteria-free inflammation, hyperaemia, slight oedema and local necrosis, and a great number of neutrophil leucocyte appears. Figure 4 shows the pathological change in the early sealing period of a woman's oviduct.

2. Then macrophage of foreign matter wraps the XKM-2 polymer and forms a granulation tumor of the foreign matter (Fig. 5).

3. Finally, the fibroid hyperplasia is a small beam in shape, extending into the duct cavity, twisting with the XKM-2 polymer and firmly sealing the cavity forever (Fig. 6).

It should be seen from the afore-mentioned figures that owing to the interaction of the XKM-2 polymer with the mucous membrane in the cavity

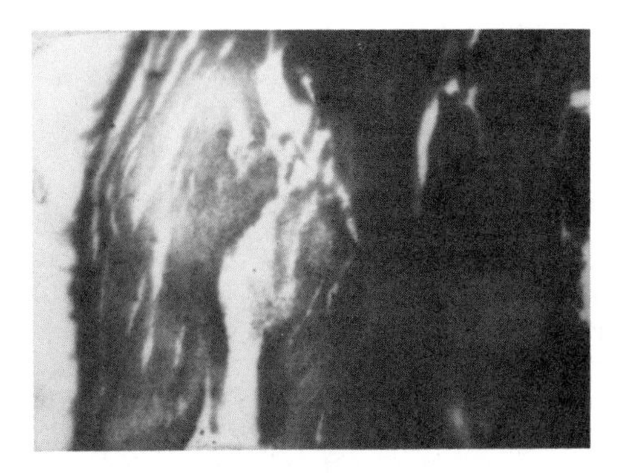

Fig. 4. Pathological change in the earling sealing period of a woman's oviduct.

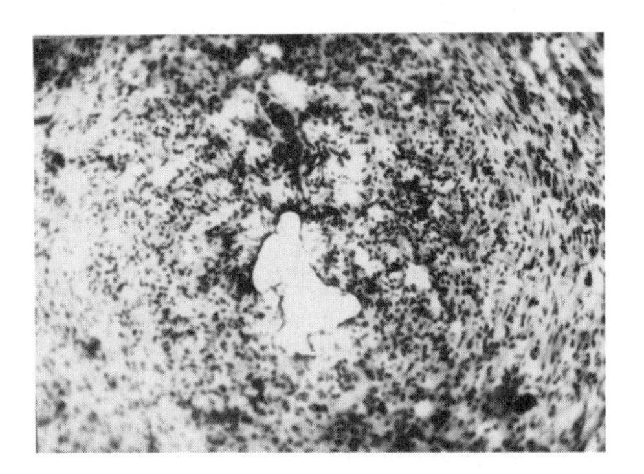

Fig. 5. A granulation tumor wrapping XKM-2 adhesive.

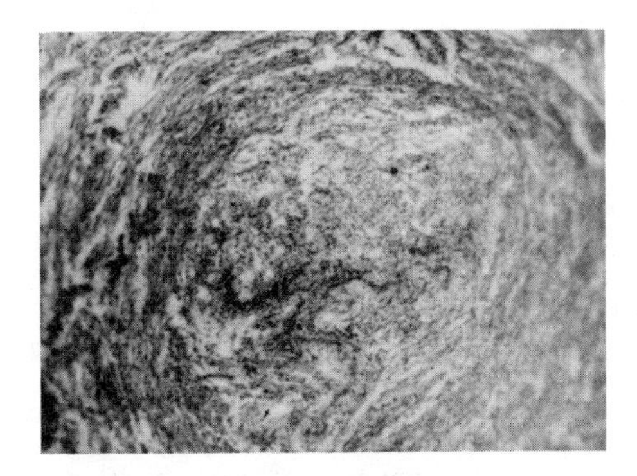

Fig. 6. Fibroid hyperplasia twisting with XKM-2 adhesive.

and shallow muscular stratum, the feasibility of sealing can be established to judge whether the sealing could be sustained.

EFFECT AND EVALUATION

We have performed sterilization operations with XKM-2 adhesive to nearly 10,000 women voluntarily being sterilized and conforming to the sterilization conditions. The success rate of sterilization has been up to 95%. The earliest operation was done 15 years ago and so far its effect has been satisfactory too.

The clinical practice was performed at first in the Department of Gynecology and Obstetrics of Xi'an No. 4 Hospital, and was then gradually popularized to others. It was found, based on the results obtained from long-term tracks, visits and check-ups of about 1,000 cases, that the success ratio of operations was influenced by several factors: (1) if the operated woman was in breast-feeding period, the rate was lower; (2) the uterus position (rear, central or front) had no clear influence on the success ratio and (3) the more the time after operation was prolonged, the firmer the sealing effect was and the higher the success ratio. We found that a few patients without achieving the desired results were mainly due to the XKM-2 adhesive not injected into the duct cavity, or due to the spasm of the uterus and oviduct, or the malformation of the uterus as well.

One case of sterilization with XKM-2 sealing could usually be completed within 5-10 minutes without surgery. It has been confirmed through many times of inquiry after sealing that most of the operated women did not have any side-reaction. During the operation, no case of uterus perforation, hemorrhage and infection was found. Most of the operated women did not feel any discomfort. Their menstrual cycles had no distinct variations and no abnormal phenomena was discovered after gynecological examination. It is truly an ideal sterilization method of safety, reliability, simplicity and easiness, and it has been well received by medical personnels and childbearing age women in China.

Once we adopted a reopening operation to a sterilized woman whose oviduct had been sealed with XKM-2 adhesive by removing XKM-2 polymer. Six months later that woman was pregnant normally and gave birth to a well-developed boy. That is a presage that this sealing method for sterilization with this kind of adhesive may develop to be a reversible operation. However, we have not gained enough experience in the reopening operation.

Adopting the XKM-2 adhesive to seal man's spermatic duct for sterilization has a similar effect as well. Thus, it is an equally effective method for man's birth control.

Besides, we performed the sealing of pancreatic duct to 12 patients whose pancreatic duodenum had been excised; through our continuous observation of 3 years, we found that the curative effect was very satisfactory. All the patients had restored to health and no pancreatic, intestines or biliary fistula arose (Fig. 7). After the excision of a pancreatic duodenum, the critical problem of reconstructing the alimentary canal is how to handle the pancreatic duct refraining from pancreatic fistula. The sealing method with XKM-2 adhesive is not only easy to operate, because it requires shorter time for the operation and is reliable, but it is also a good method to refrain from pancreatic fistula. In the past, the incidence of the ulcer after reconstructing the alimentary canal was relatively high,

Table 1. Effect of breast-feeding on success ratio.

Items	Success		Failure	
	Cases	Success ratio %	Cases	Failure ratio %
In breast-feeding-period	431	89.98	48	10.02
Not in breast-feeding period	408	95.55	19	4.45

but among all of patients sealed with XKM-2 adhesive no case of ulcer have occurred.

DISCUSSION

It was found, based on the results obtained from long-term tracks, visits and check-ups of about 1,000 cases, that the success ratio of operation was influenced by several factors:

1. Effect of breast-feeding: The results of 906 cases are shown in Table 1. It is thus clear that if sealing sterilization is operated not during the breast-feeding period, the success ratio can be higher.

2. Effect of uterine position. The results of 907 cases are shown in Table 2. Although for those the difference among the three positions is not clear, the success ratio for those operated on the front position is slightly lower.

3. Effect of the number of years after operation. According to the visitation of 906 cases, (Table 3), it is thus clear that the success ratio for those operated over three years is higher than that for those operated within 2 years.

4. Effect of spastic systole of oviduct. Under normal conditions, an oviduct usually opens and closes periodically on the uterine mouth in accordance with the regulation of opening for 7-10 seconds and then closing for 10-20 seconds. But there were few operated women who had the spastic systole of oviduct due to the stimulus of appliances and medicines or other psychological factors, so that the XKM-2 adhesive could not be injected into the oviduct cavity, but polymerized at the uterine corners, as a result the operation failed.

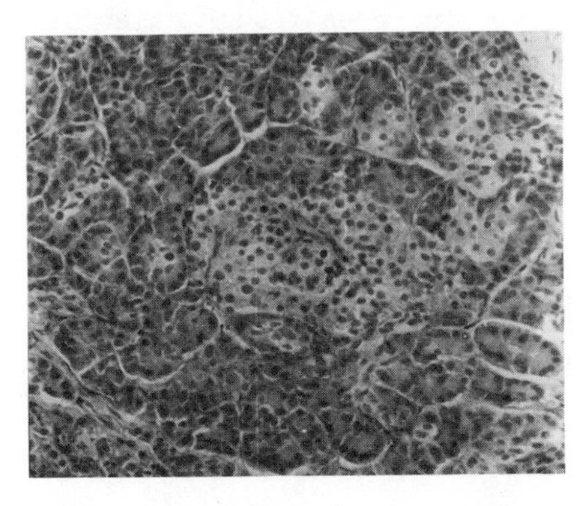

Fig. 7. A pancreatic duct-3 years after operation.

Table 2. Effect of uterine position on success ratio.

Position	Success		Failure	
	Cases	Success ratio %	Cases	Failure ratio %
Front	104	89.66	12	10.34
Central	250	92.59	20	7.41
Rear	486	93.28	35	6.72

Table 3. Effect of the number of years after operation on success ratio.

Number of years	Total Cases	Success		Failure	
		Cases	Success Ratio %	Cases	Failure ratio %
Within 1 year	302	279	92.38	23	7.63
Within 2 years	343	310	90.38	33	9.63
Within 3 years	105	100	95.24	5	4.76
Over 3 years	156	150	96.15	6	3.85

5. Effect of uterine deformity: Saddle-backed or double-cornered uterus has no great influence on the operation result, but double-uterus women could not be sterilized with success even after operated on for many times.

CONCLUSIONS

At present, the total population throughout the world has exceeded five billion. Faced with the important problem of rapid increase of population, every country is studying and adopting various measures of population control. However, even though numerous scientists have made great efforts for this purpose, there is no relatively ideal method for birth control. Our work has been carried out for 16 years with good achievements. A small drop of adhesive unexpectedly has such an ability to make a woman sterilized painlessly without surgery. So to say, this is a marvellous application of adhesive. Judging from the sterilization effect, we hope that a firm sealing has formed in the oviduct, but judging from the possible requirement of reopening, we also hope that the sealing should not be excessively firm with the aim of removing the polymer to recover the ability of child bearing. How to solve this contradiction is an important problem to be studied from now on. We believe that the possibility of solving the contradiction exists, and adhesives and high polymers will eventually play their important role with respect to human birth control.

REFERENCES

1. R.H. Davis, J.M. Donald, G.K. Facog and H.B. Facog, "Effect on Fertility of Introuterotubal Injection of Gelation," Obst and Gynec, 4(3) 446, 1973.
2. J.A. Zipper, Ennio Stachetti, and Mario Medel, "Human Fertility Control by Transvaginal Application of Quinacrine on the Fallopian Tube," Fertil. and Steril., 21(8), 1970.
3. R.M. Richard, "Female Sterilization by Electrocoagulation of Tubal Ostia Using Hysteroscopy," Am. J. Obst and Gynec, 117(6), 801, 1973.
4. J.F. Hulka and K.F. Omran, "Sterilization by Cryosurgery: A Report of Animal Fertility Studies," Fertil. and Steril., 22(2), 170, 1971.
5. Wang Ting-rui, Reproduction and Contraception, 1(3), 48, 1981.
6. Qin Zhao-yin and Ruan Chuan-liang," Occlusive Procedure of Pancreatic Duct for Pancreaticoduodenectomy," Chin. J. Sur. 23, 727, 1985.

Discussion

On the Paper by L.H. Lee

H.R. Bhattacharjee (Allied-Signal Inc., Morristown, N.J.): You mentioned the bioadhesive of Dr. J. Waite in your talk; my questions are:

(a) Can you tell us the future of such adhesives?

(b) Are you seeing any immediate applications of such adhesives in solving various problems you mentioned in your talk?

L.H. Lee (Xerox Corp.)

(a) There could be many future applications for such adhesives. Industrially, they may be useful for underwater applications. Medically, they can be used as biocompatible nonspecific adhesive for cell and tissue culture. One of these adhesives has been developed by Biopolymers, Inc. (Farmington, Conn.) Other promising applications include dental and ophthalmic treatments.

(b) One of the immediate problems is the supply of this type of adhesives. Some biotechnological concerns, e.g., Genex Corp., Gaithersburg, MD., have been contemplating to use the recombinant DNA technology to synthesize some of the peptides. If a sufficient quantity of this type of adhesives can be made available, many exploratory applications will appear.

On the Paper by S.J. Shaw

Y. Ersun-Hallsby (Dow Chemical Co., Midland, Mich.): Have you looked into mixing fluoroepoxy resins with epoxy resins?

S.J. Shaw (RARDE Walthem Abbey, U.K.): Although it has been realized that the incorporation of fluoroepoxies into conventional epoxy systems could bring about improvements in various desirable properties (e.g., reductions in water absorption and surface tension), no attempt has been made to verify these possibilities. A major problem which would be likely to restrict the success of such an approach concerns compatibility. The fluoroepoxies, particularly those containing high fluorine levels (a high value of n in the pendant C_nF_{2n+1} group) would probably exhibit incompatibility with most conventional epoxy resin systems. Incompatibility problems would probably also exist between conventional epoxies and silicone-amine curing agents.

Y. Ersun-Hallsby: To what extent have the fluoroepoxies been commercialized. What are the major contributions to cost? What if any difficulties are encountered in synthesis?

S.J. Shaw: Various intermediates for the manufacture of the fluoroepoxies together with the resins themselves are currently being manufactured by Allied Chemicals of Buffalo, New York. Although Allied are not manufacturing the silicone-amine and fluoroanhydride curing agents, some of these can be obtained commercially from alternative sources.

In common with most fluorinated polymer systems, the fluoroepoxies are currently expensive in comparison to their non-halogenated counterparts. An increase in volume production which could be expected to accompany commercial acceptance would be expected to reduce their cost significantly.

Although some difficulties in synthesis were encountered during the development of the fluoroepoxies (e.g., the means of incorporating the C_nF_{2n+1} group onto the benzene ring), they were all eventually overcome.

PART TWO:

STRESS AND INTERFACE

Introductory Remarks

Charles E. Rogers

Center for Adhesives, Sealants and Coatings
Case Western Reserve University
Cleveland, Ohio 44106

In order to obtain an adhesive bond which is stable in an application environment, it is necessary that several criteria be satisfied. One criterion is that the substrate surface must have a composition, structure and morphology that is conducive to wetting and spreading of the adhesive. That surface must have cohesive strength and not be susceptible to attack by environmental agents, especially water.

Another criterion is that the adhesive must be chosen in consideration of its interactions with the substrate surface so that it wets and spreads to obtain intimate molecular contact with the substrate. The cured adhesive must have adequate cohesive strength as well as an adequate adhesion bond to the substrate. Cure shrinkage must be minimal. The adhesive material also must not be susceptible to attack by environmental agents.

The stability of such a "proper" bond, in the terminology of J.J. Bikerman, then should not depend on the interfacial adhesion, per se, but rather on the cohesive properties of the adhesive and the substrate, including any oxide or other layers. Hence, it is necessary to understand the stress-strain distributions in the system and the mechanism of fracture processes in the components of the system. The effects of micro-defects, acting as stress concentrations to initiate crack or craze growth, is of particular concern. Perturbations in stress distributions due to added fillers or other causes can seriously alter the fracture threshold and other fracture characteristics.

Finally, the effects of environmental agents, especially water, on the above criteria must be established with a view to eliminate undesirable effects. This aspect has not been studied over the years as much as would be desired. This can be attributed primarily to the need to test many samples over considerable periods of time under a variety and range of exposure conditions. Few laboratories can afford the time, effort and money for truly exhaustive testing of many different formulations and substrates. Accelerated tests generally have many problems so that test results do not correlate with real-time tests or actual application performance. This constitutes a major problem for the development of better adhesives, sealants and coatings.

The papers in this session address aspects of most of these criteria. It is common practice now to modify substrate surfaces so as to optimize the adhesive bond strength. Very often organic coupling agents are used for this purpose. Schmidt and Bell describe the use of polymeric coupling agents to promote the interaction between a steel surface and an epoxy adhesive. These agents promise to be tougher and more hydrophobic than their low molecular weight counterparts. Yang, Rusch, Clough and Fateley have used Auger Electron Spectroscopy to study the bonding and hydrolysis of coupling agents on aluminum surfaces. Pike and Lamm review the use of inorganic primers for adhesive bonding. These seem to mask any differences in adhesion performance between aluminum samples with different preliminary surface treatments, raising the possibility of more simplified treatments. The inorganic primers are found to be hydrolytically stable and may be useful also for other substrates.

Studies of the stress distribution in polymeric composites have been made by Liu and Walker. Attention was focused on the stress distribution at the interface, the effect of void side on the distribution, and void growth due to local stresses. Parvin and Knauss address the issues of interfacial adhesive failure and the fracture mechanism and deformation of the interlayer in the zone at the crack front. Bonding improvement was obtained by elimination of air bubbles in the interlayer.

Most of the above papers considered the effects of humidity on the adhesion performance. Radiation, as an environmental agent, was considered by Dickinson. He has used electron bombardment, focused on the crack-tip, to study crack initiation and growth in several polymers. The enhanced crack growth is attributed to radiation-induced chain scissions, where fewer bonds can reform due to the applied stress.

These, and the other papers presented in the symposium, provide important information relating to the criteria governing adhesion. They especially direct attention at the effects of environmental exposure on adhesive system behavior in terms of fundamental considerations of adhesive and surface compositions and properties, interface/interphase modifications and failure mechanisms. Such information is essential for the development of successful adhesives, sealants, and coatings for extended performance applications.

ELECTRON BOMBARDMENT INDUCED CRACK INITIATION AND CRACK GROWTH IN POLYMERS AND POLYMER SURFACES

J.T. Dickinson
Department of Physics
Washington State University
Pullman, WA 99164-2814

ABSTRACT

When a notched, polymeric material is stressed, the notch opens into a wide crack-tip, exposing a region of high stress concentration. The consequences of electron bombardment of the tip of polyisoprene, butyl rubber, Kapton-H®, Teflon®, low density polyethylene, and linear high density polyethylene stressed under vacuum are under investigation. Evidence is presented for electron induced crack growth at stresses below that needed for crack growth due to stress alone. The electron current densities used in these experiments are sufficiently small that thermal heating of the zone near the crack tip tends not to dominate. To provide information on the physical phenomena involved, we present measurements of electron current, gas pressure, and sample load in response to both periodic and stationary electron bombardment of the sample. Video taping of the shape of the crack before and during bombardment are also presented. Experiments involving the bombardment of unnotched polymers under stress are also described, along with measurements of the influence of electron bombardment on the tear energy of an elastomer.

INTRODUCTION

High energy electrons bombarding a polymer interact with the polymer via inelastic collisions causing energy deposition in the material. These inelastic collisions may cause vibrational excitations of the molecules, ionization, and broken bonds. The chemical effects resulting from these interactions which tend to "strengthen" the material may include additional polymerization, crosslinking or branching of the polymer. Likewise, the chemical effects that tend to "weaken" the material may include bond scissions, molecular dissociation (via electronic excitations), electron-stimulated desorption of ions and neutral species (again, via electronic excitations), or thermal degradation and gas evolution (due to a temperature rise in the material being bombarded). All of these events can have mechanical consequences if the material is subjected to deformation.

In earlier work[1-5] and in this paper we will examine the consequences of simultaneously subjecting materials to mechanical stress and electron bombardment. The primary motivation for such an experiment is to examine the consequences of fracture in a high energy environment. Examples that are of interest may include stressed materials exposed to radiation, the space environment, advanced machining and cutting techniques, combustion of rocket propellents, and materials exposed to plasma environments. Secondly, the use of electron beam excitation of the crack-tip may further our understanding of the physics of fracture at a molecular level and could eventually lead to selective and controlled bond breaking and/or activated stress-dependent chemistry. An additional long-term goal is to perform various electron spectroscopies, e.g., electron energy loss spectroscopy, on stressed molecules to provide useful information about the state of these molecules. At this point, however, we concentrate on the mechanical and morphological response of polymers experiencing stress + radiation and consider the following materials: polyisoprene (PI), butyl rubber (IIR), Kapton-H®, Teflon® (PTFE), low density polyethylene (LDPE) and linear high density polyethylene (LHDPE). The radiation of interest here were beams of electrons with energies from 500 to 3000 eV.

EXPERIMENTAL

PI samples were cut from sheets provided by H.M. Leeper, Alza Corporation, which were replicate plaques of Goodyear Natsyn 2200 crosslinked with Hercules Di-Cup R. Butyl IIR samples were made from 100 ppw Butyl, 268 ppw of Vanfre AP, 5 ppw ZnO, 2 ppw stearic acid, 1.25 ppw of sulfur, 1 ppw Captax, and 1.224 ppw methyl tuads (tetramethylthiuram disulfide). The Kapton-H samples were supplied by E.I. Dupont de Nemours and Co. The Teflon (PTFE, polytetrafluoroethylene) samples were obtained from commercial rolls of Scotch 48, Thread Sealant and Lubricant produced by 3M. The low density polyethylene (LDPE) were from commercial sheets from U.S. Industrial Chemicals (PA 80037). The linear high density polyethylene (LHDPE) samples were also from U.S. Industrial Chemicals (LR 20175) with a density of 0.95 g/cm^3 and a characteristic melt index of 0.1.

Rectangular test specimens of the polymers were mounted in a vacuum system equipped for straining materials in tension. Two sample orientations that were studied were used. The usual test involved a notched sample mounted horizontally with the notch facing the electron beam. The materials chosen were deformed to yield an open, U-shaped crack, allowing convenient electron beam bombardment in the region of high stress. Side-on bombardment of both notched and unnotched specimens was also studied. Of interest was the occurrence of microcracking, and complete penetration of the electron beam through the specimen.

The samples were subjected to a constant strain rate, typically 0.01-3.0 %/s. A load cell (Sensotec, Model 11) was used to monitor the force applied to the sample. Most of the tests were carried out at a pressure of 10^{-5} Pa. A Varian Glancing Incidence Auger Electron gun with a 2-3-mm spot size was mounted so that the electron beam of 10-200 μA at kinetic energies of 0.5-3.0 keV would strike the sample at or near the focal point. In most experiments, the time that the beam was actually on the sample was minimized to avoid the build-up of surface charge which would tend to reduce the current density bombarding the polymer surface.

An electrometer connected to a metal collector mounted behind the sample was frequently used to measure the electron current in the beam and also to determine quite accurately the time the electron beam came on, off, or penetrated through the sample.

The rise in pressure in the vacuum system could be measured by means of a nude ionization gauge. The appropriate signals from the various transducers were all simultaneously digitized with 0.1-s time resolution using a LeCroy Data Acquisition System and stored on disk for later analysis. For some of the tests, a video camera was mounted such that a video recording of the crack and its propagation while under electron bombardment could be obtained.

RESULTS

Figure 1(b) shows a typical load vs time curve for a notched PI sample elongated to failure at a constant strain rate of 0.04 mm/s. In contrast, Fig. 1(a) shows results for a similar sample loaded in the same manner to which a 10-μA, 1.6-keV electron beam was swept across the notch. The vertical arrows represent the times when the electron beam passed over the notched crack-tip. The response of the sample to the passing electron beam consisted of dramatic jumps of the crack with accompanying drops in the force. The response is instantaneous on a time scale of sub-milliseconds. Similar tests on unstressed PI or PI stressed only into the elastic region showed no effect of cutting and no visible damage to the PI sample at these beam currents and exposure times. The last drop in load (Fig. (1a)) corresponds to the final

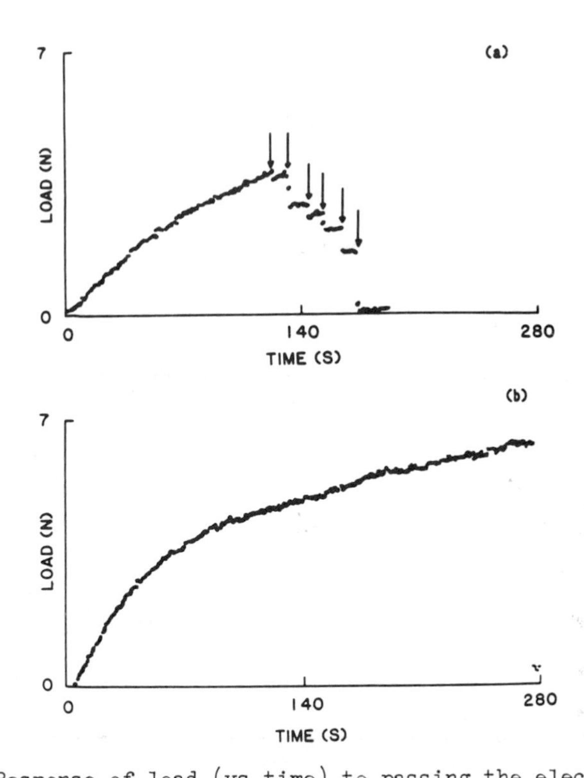

Fig. 1. (a) Response of load (vs time) to passing the electron beam across a stressed crack tip in polyisoprene. The sudden decrease in load is due to crack growth. Vertical arrows show when the electron beam was incident on the crack tip.
(b) Load vs time for a notched, unbombarded polyisoprene sample.

rupture of the specimen which was initiated with the elecron beam. At higher electron currents, the response was found to be larger jumps in crack length and therefore larger drops in the force for a given strain. Comparison of Figs. 1(a) and 1(b) show that the time and therefore the extension at failure are considerably smaller for the sample exposed to the electron beam.

Similar results have been found on a number of other polymers. For example, in Fig. 2(a) we show the response of notched low density polyethylene (LDPE) elongated at a constant strain rate of 0.6 %/s and periodically bombarded by an electron beam of 85 μA (current density 1×10^{-3} A/cm^2). The arrows indicate the times the electron beam first starts striking the notch. Again, the drop in load is due to crack growth and the last arrow on the right indicates where final rupture occurred. Figure 2(b) is the stress vs time for an identical notched sample strained at the same rate but with no electron bombardment applied. The unbombarded sample fails in a completely different fashion; namely, by continuous drawout as opposed to crack formation and crack propagation.

A number of calculations of the rise in temperature due to absorption of the electron beam were carried out. One of these followed a development by Jaeger[6] which assumes a semi-infinite material with heat supplied at a constant rate over the whole surface. Other calculations

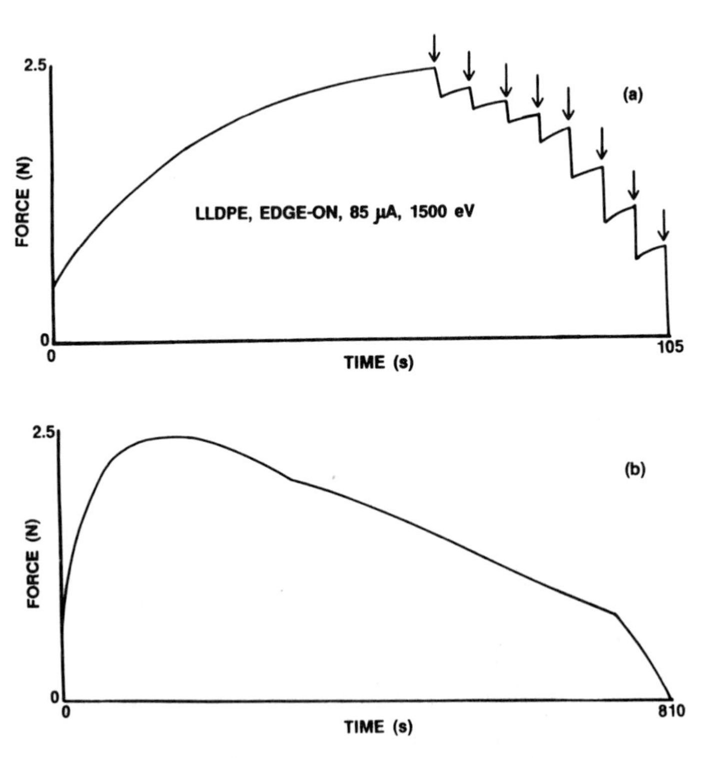

Fig. 2. Mechanical response of: a) notched LDPE exposed to an electron beam of 85 μA. The arrows show the times when the beam moves onto the stressed notch. The sudden drops in force indicate crack growth. b) an unbombarded sample.

employed finite difference methods applied to the exact geometric conditions of the sample and electron beam. All of these calculations showed temperature increases of only 5-20°C for the various materials, beam currents, and exposure times used in these experiments. We therefore conclude that the resulting crack growth is not dominated by the increase in the temperature of the polymer due to beam heating. An alternative mechanism involves direct bond breaking of molecules in tension due to inelastic electron scattering, resulting in irreversible chain scissions that lead to crack formation and growth.

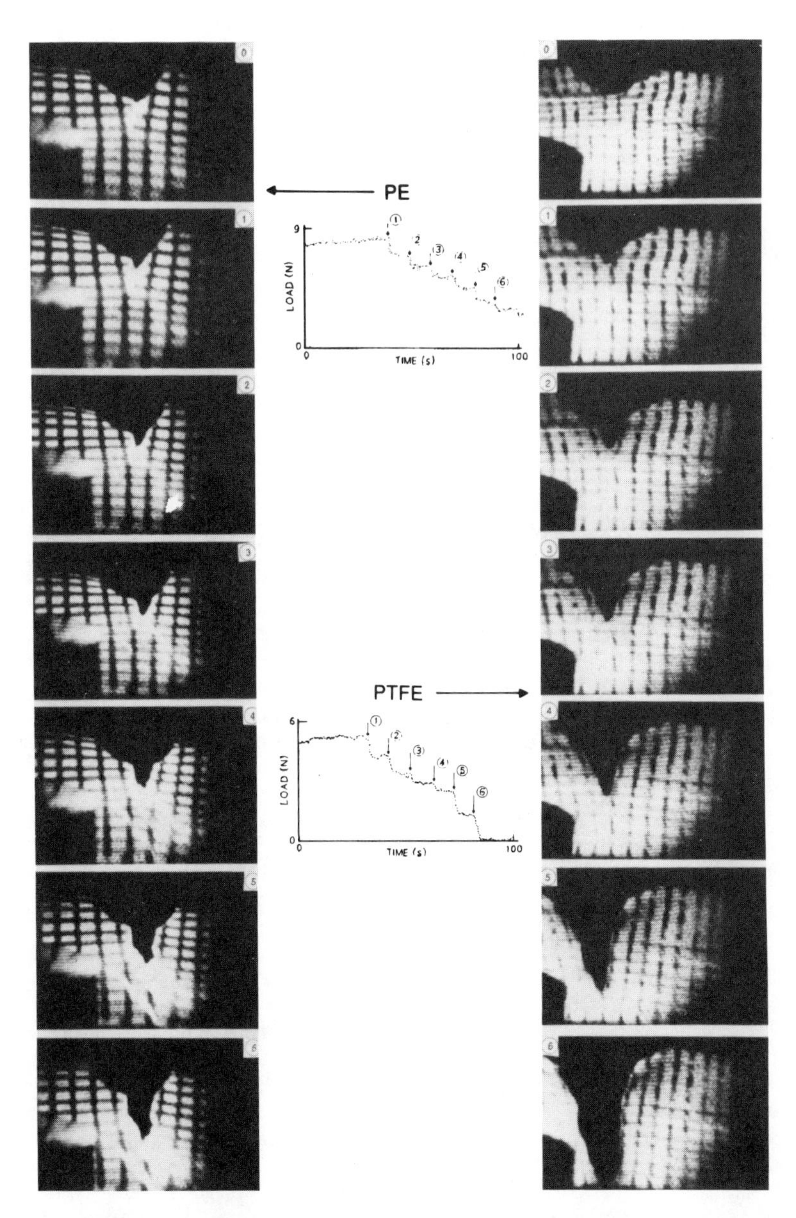

Fig. 3. Series of video images showing the propagation of the notch due to the application of an electron beam to two edge-on samples: on the left is LHDPE and on the right, Teflon. The load vs time curves, marked with numbered arrows corresponding to the numbered video frames, are shown in the center.

The consequences of combining stress and electron bombardment of notched samples can be further illustrated by viewing the crack with a video camera. Figure 3 shows the load vs time curves and corresponding frames obtained for electron beam bombarded notches in PTFE and LHDPE during constant strain rate loading. The electron beam was centered in the notch of both samples with a diameter of approximately 2/3 the notch width. With application of the beam, the crack propagated, the load dropped, and one observed in the video frame a pronounced sharpening of the crack. When the beam was removed, the crack was observed to arrest and would start to open. Repeated beam applications resulted in sharpening of the crack shape. In the case of Teflon, the unbombarded material tends to fail by chain slippage, resulting in the formation of fibrils parallel to the strain direction and no well defined crack. The formation of a relatively sharp crack in the bombarded case supports the concept of chain scissions broken by the incident radiation.

Unbombarded LHDPE samples failed by a crazing or thinning process where the regions of high stress concentration near the crack tip craze and drew out. In the LHDPE photographs, on the left hand side of Fig. 3, the bright spots correspond to the area of the sample where the thinning process is occurring. The electron induced damage is concentrated in a region that is much smaller than the width of the electron beam, within the thinned region indicating that local stress is a critical factor in the probability that an electron induced chain scission yields an irreversibly broken bond.

Another mode of electron beam induced fracture which we have examined involves the sample arrangement where an unnotched sample is stressed and the e-beam is allowed to strike the sample side-on. Of interest here is to determine if the electron beam could lead to sufficient damage of an unnotched sample to result in penetration of the beam through the sample. Thus, at time $t=0$ the electron beam is quickly applied to the center of an elongated sample, which results in an immediate drop in the electron current being collected behind the sample. If the electron beam penetrates through the sample, the collector current will suddenly increase. Unnotched samples of LHDPE, 50 μm in thickness and initially pre-stressed followed by some relaxation were hit with the electron beam.

Figure 4 shows the change in total gas pressure in the vacuum chamber, collector current (essentially the inverse of the current of the sample), and applied load for such a LHDPE sample with an electron beam current of 10 μA. The time to penetrate through this particular sample was 1.7 s. Observation of the sample during bombardment showed that the sample did not fail completely. Instead, only the center of the electron beam went through the sample, while the outer regions of the beam penetrated at a later time, opening up a "hole" and resulting in the slow increase in collector current and the slow decrease in pressure shown in Fig. 4. In terms of the time required to penetrate through the material, as one might expect, higher currents penetrated considerably faster. Application of the electron beam to unstressed samples of LHDPE for several hundred seconds of exposure caused no noticeable damage and no penetration. The mechanism for failure appears to be electron beam induced crack initiation. Under the stress and irradiation, microcracks are forming which quickly lead to a crack opening up through the material. If the stress in the specimen is sufficiently intense prior to bombardment, the newly formed crack usually leads to rapid failure.

Experiments involving the bombardment of a number of polymers including Kapton-H, PTFE, and LHDPE show, with varying sensitivities, that cracks can be initiated when the material is under stress. Figures 5 and 6 show SEM photos of the stressed regions for both bombarded and unbombarded Kapton-H and Teflon. In the case of Kapton-H, the microcracks are seen to be very sharp and perpendicular to the direction of stress. In Teflon, the fibrillation of the unbombarded surface is parallel to the direction of applied force, as expected. The bombarded surface shows a vast array of microcracks that are normal to the direction of stress and appear to consist of "cut" fibrils. The resulting fracture surface although oriented essentially normal to the direction of applied force, is "fractal" in appearance due to the random distribution of these short microcracks which link to form the main crack. The microcracks appear to span a number of fibrils which are "chopped" in the same region of the specimen.

Another experimental arrangement which we have investigated involves measuring the influence of electron bombardment on the tear energy of elastomers. A cleavage test piece, shown schematically in Fig. 7, has its tear region well exposed so that particles can be focused into the crack tip. For such an arrangement (no radiation), the tear energy, W, is given by:[7]

$$W = 2F/t, \tag{1}$$

where F is the force obtained for constant tear speed (unspecified), and t is the width of the narrow web connecting the upper and lower portions of the test specimen. In the usual test, F is translatable

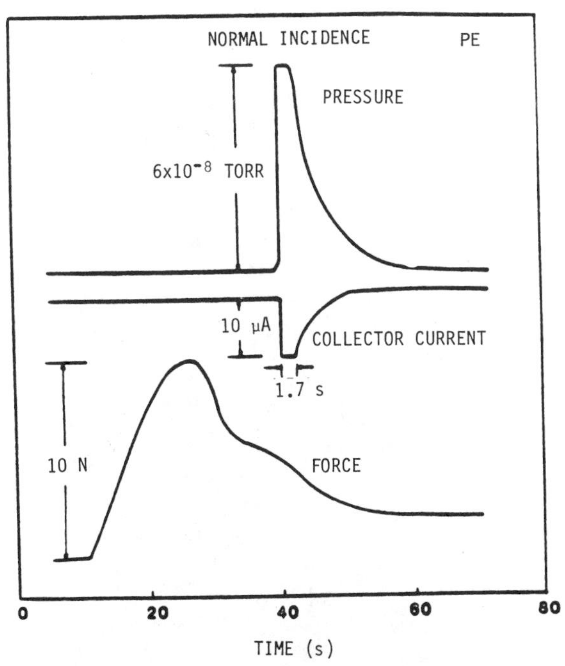

Fig. 4. Response of unnotched LHDPE under stress to electron bombardment normal to the surface. Shown are the change in total pressure vs time, collector current (the collector is immediately behind the sample) vs time, and sample load vs time.

into an average number of bonds being strained to failure in some zone near the crack tip. If we introduce a second mode of failure; namely, electron beam induced bond scissions, and if we maintain a constant tear speed, we expect the tear energy to decrease due to the breaking of fewer mechanical bonds.

In Figures 8(a) and 8(b) we show the force F during tear tests responding to the sudden application of electron beams of two magnitudes. When the beam was turned off, the force would eventually return to the original value. The drop in force vs current density, j, is shown for two different tear speeds in Fig. 8(c), which are seen to give essentially the same results. The solid curve is a simple model described below.

DISCUSSION AND CONCLUSION

First let us summarize the basic characteristics of electron beam induced fracture:

1) The polymers must be elongated beyond a certain stress state to observe crack growth under bombardment. However, no evidence of

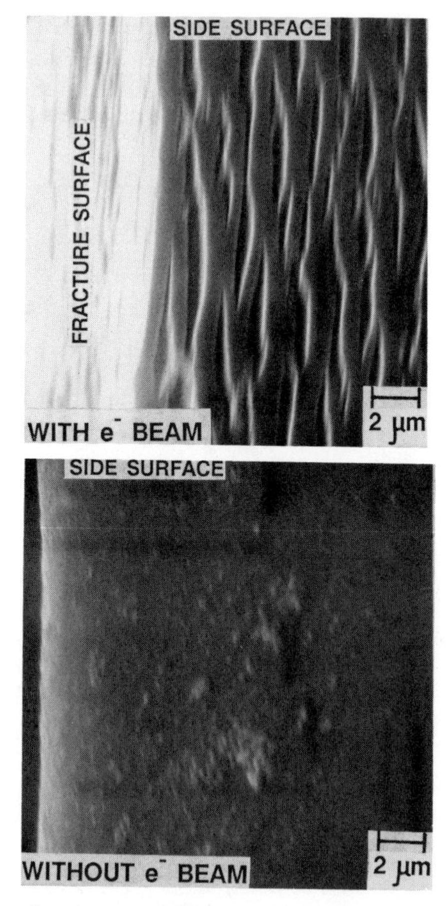

Fig. 5. Exposed and unexposed Kaptan-H surfaces of strained material. The bombarded surface shows considerable microcracking. The density of the microcracks is related to the stress and the electron beam intensity.

crack growth occurs before the electron beam is applied, i.e., it is still below the critical stress concentration.

2) The current densities necessary to obtain noticeable crack growth below critical stress levels were on the order of 10-100 μA/cm^2.

3) The higher the stress, the more evident the response to the electron beam.

4) The calculated heating effect of the electron beam on the time scale to see mechanical response is small. The change in gas pressure[1] due to evolution of gases from electron beam bombardment of the polymer is considerably smaller than that obtained at higher currents and follows precisely in time the electron beam current to the sample with no evidence of a "cooling curve."

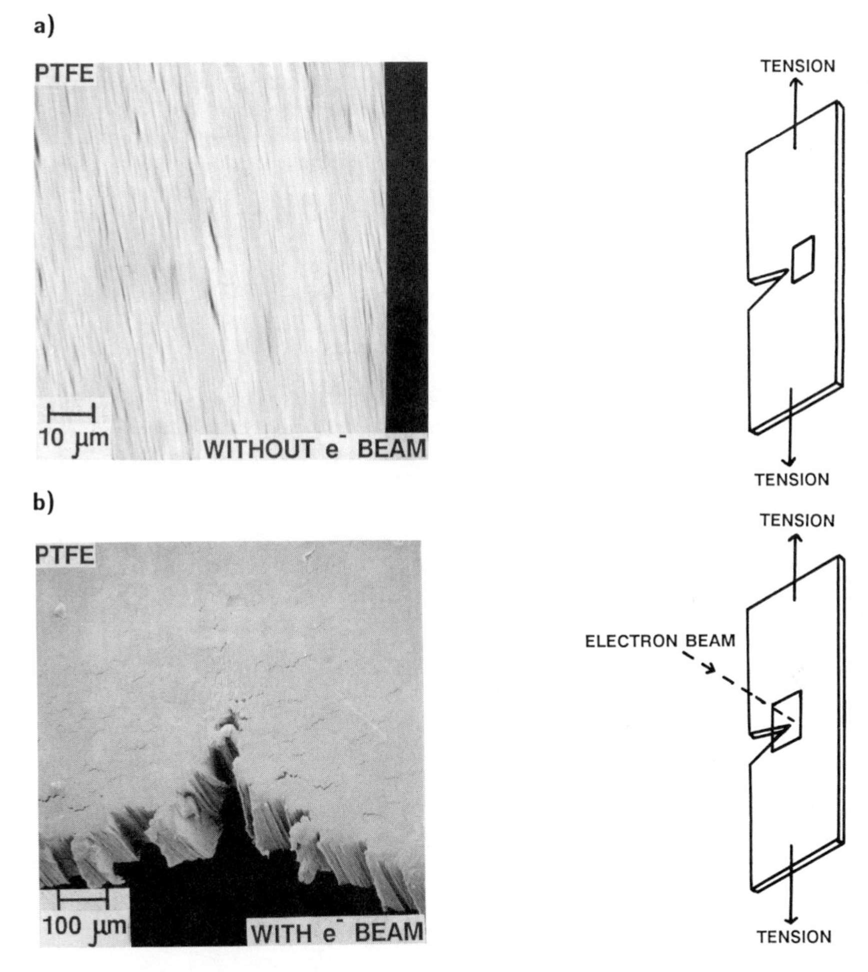

Fig. 6. The consequences of bombarding Teflon with 50-μA, 1500-eV electrons, side on.
a) SEM of the stressed material showing the resulting fibril formation parallel to the direction of force (vertical).
b) The resulting fracture surface and exposed surface full of microcracks. The fracture surface and the micro-cracks seem to be strongly related to the fibrils.

5) Samples of unnotched LHDPE responded to electron beam bombardment when stressed, resulting in penetration of the beam through the sample.

6) In the case of PTFE, the failure mechanism is completely changed from slipping type process to chain scissions due to electron beam interactions.

7) The tear energy of elastomers such as butyl rubber and polybutadiene are dynamically reduced when the crack tip was electron bombarded.

8) Striking evidence of highly localized microcracking of all of the polymers studied has been observed for combined exposure to stress and radiation.

These results suggest that the phenomenon of electron beam induced fracture is not dominated by thermal effects, but instead appears to be a direct consequence of electronic interactions; i.e., direct scissions of load bearing molecular chains by these electron collisions. The patterns seen in Kapton-H, Teflon, and several other polymers indicate that the microcracking is highly localized. Such instabilities are most likely due to variations in the morphology of the polymer, probably strain induced, making some regions much more susceptible to permanent bond scissions than others.

A simple argument for why higher stress regions are more sensitive is that bonds broken under stress are less likely to reform. Thus, a load carrying chain that undergoes a bond scission is far less likely to reform a load carrying bond and therefore remains permanently broken. In such regions of the polymer where greater numbers of irreversibly broken chains occur, crack initiation and/or crack propagation is enhanced. The linking of such damage into larger microcracks and/or main crack propagation transpires because of the resulting stress distributions from the array of damage in the bombarded region. In the elastic case, for example, Yokobori, et al., have shown that there are

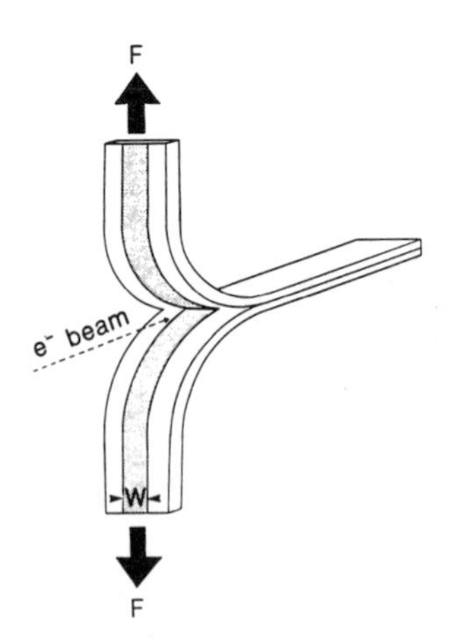

Fig. 7. Schematic diagram of cleavage specimen for elastomer tear test. Arrow indicates incident electron beam.

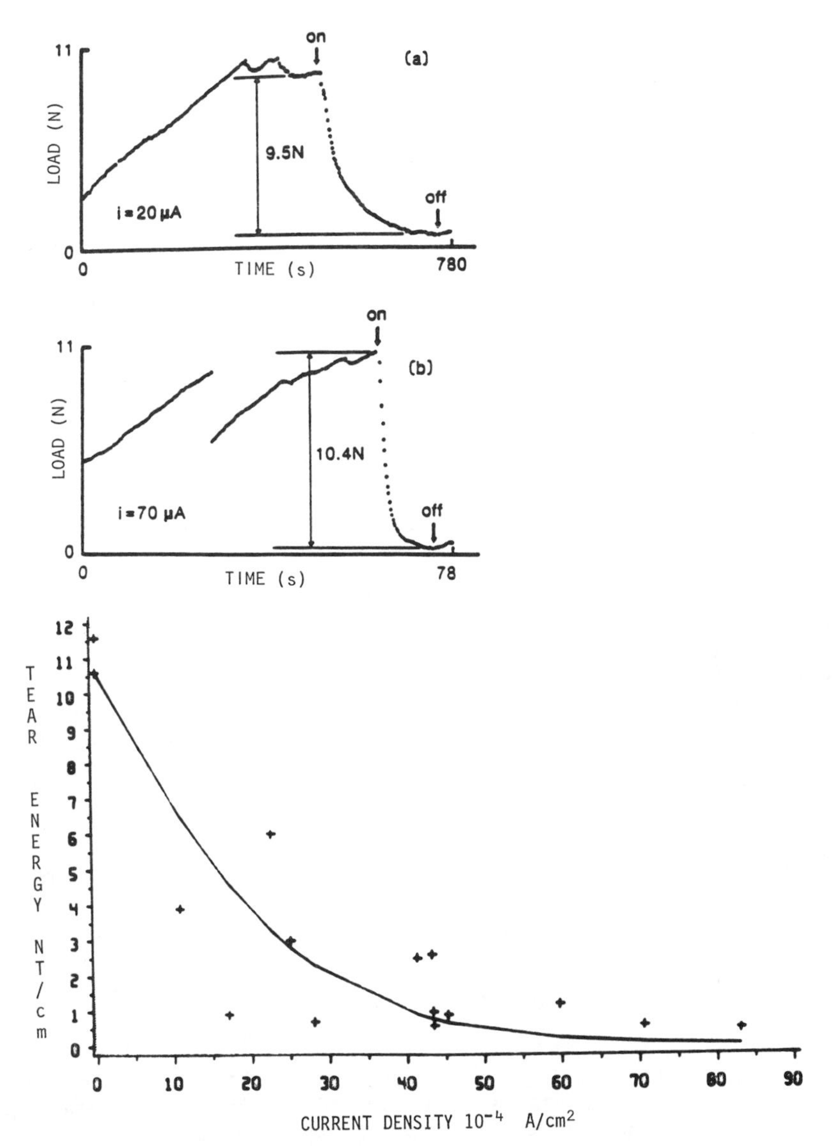

Fig. 8. Response in the force during a tear test when electron beam is applied to the tear region of butyl rubber.
a) Total electron current of 20 μA;
b) Total electron current of 70 μA;
c) The force (corresponding to tear energy measurements) vs current density for butyl rubber.
Data shown for two strain rates is essentially the same. The solid curve is a curve-fit from a model presented in the Discussion.

a variety of crack lengths, relative positions, and orientations where two adjacent cracks greatly increase the stress intensity factor due to strong interactions between the cracks.[8]

For the bombarded notch or the elastomer tear test, the highly strained region where the electrons cause fracture to occur is also inhomogeneous in terms of stress distributions. Nevertheless, in the case of the tear test, the measured drop in force vs current density dependence (e.g., Fig. 8(c)) is due to an average bond breaking rate induced by the beam. For the unbombarded tear test, Lake and Thomas[9] have argued that the magnitude of the tear energy, W_0, is directly proportional to the N, the number of load bearing chains per unit area of fracture surface produced by the tear, i.e.,

$$W_o = bN, \tag{2}$$

where b is a proportionality constant.

A simple model for the consequences of the bombarded tear would be as follows:

$$W(j) = W_o[1 + c(j) - f(j)g(s)], \tag{3}$$

where j is the current density in the electron beam, c(j) is a function representing electron induced crosslinking which strengthens the material, f(j) is a function representing the increasing amount of bond breaking with increasing current, and g is a function of stress, s, representing the loss of effectiveness of bond breaking with decreasing stress. As a beam is applied to the tear region, one observes the effect of a reduced stress at the crack tip (i.e., the crack opening decreases dramatically). Consequently, the effectiveness of the electrons is reduced. Since the stress level that results is a function of j we can represent the product of f*g by a single F(j). For a given material, physically reasonable functions for c and F are as follows:

$$c(j) = C_o j, \tag{4}$$

and

$$F(j) = tanh(j/j_o), \tag{5}$$

where C_o and j_o are scaling constants for a particular material. The tanh dependence represents a function which will increase nearly linearly, followed by a smooth asymptotic approach to unity. Thus with zero crosslinking (C_o=0), the tear energy approaches zero with increasing current. Thus substituting Eqs.4 and 5 into 3, W is given by:

$$W = W_o[1 + C_o j - tanh(j/j_o)]. \tag{6}$$

A nonlinear least squares fit to the butyl rubber tear energy data by this equation is the solid line shown with the data in Fig. 8(c), where the best fit corresponded to a C_o of essentially zero. Therefore the only significant parameter was j_o, which adjusts the manner in which the curve falls towards zero; the best fit was with a value of 1 x 10^{-3} A/cm^2. The model is currently being tested on data taken on polybutadiene, which is known to crosslink under electron beam irradiation and clearly requires a term like Eq. 4 in W(j).

It is likely that a more meaningful model for both the reduction in tear energy and the electron beam induced fracture described earlier will involve a) the response of the spatially inhomogeneous morphology of the stressed polymer, b) the fracture mechanics of an array of damage sites in the region of the crack tip, and possibly c) the dynamics of the interaction of radiation with the stress field in the crack tip region.

The prospect of performing controlled direct rupture of bonds under stress with external radiation sources appears promising and should lead to improved understanding of fracture in elastomers and polymeric materials. This work is being extended to other radiation sources such as UV photons and fast-atom bombardment as well as other types of materials. Clear indications of crack initiation and enhanced damage are again being found for the case where both stress and radiation interact simultaneously with the material.

ACKNOWLEDGEMENTS

The author would like to thank Les Jensen, Koksal Tonyali, and Mike Klakken of Washington State University and Alan Gent, University of Akron for their important contributions to this work. Also, I would like to thank Harold M. Leeper, Alza Corporation, for providing the PI samples and Clarence Wolf from McDonnell Douglas Corp. for providing the Kapton-H samples. This work was supported by the Office of Naval Research, Contract N00014-80-C-0213, NR 659-803, McDonnell Douglas Independent Research and Development Program, the Washington Technology Center, and the NASA-Johnson Space Center.

REFERENCES

1. J.T. Dickinson, M.L. Klakken, M.H. Miles and L.C. Jensen, J. Polym. Sci. Polym. Phys. Ed. 23,2273 (1985).
2. J.T. Dickinson, L.C. Jensen, M.L. Klakken, J. Vac. Sci. and Technol. 4A, 1501 (1986).
3. R. Michael, S. Frank, D. Stulik and J.T. Dickinson, to appear in Proceedings of 13th International Symposium on "Effects of Radiation on Materials", ASTM E-10, Seattle, WA, June, 1986.
4. J.T. Dickinson, M.L. Klakken and L.C. Jensen, in Proceedings of the 18th International SAMPE Technical Conference Materials For Space - The Gathering Momentum 18, 983-992 (1986).
5. J.T. Dickinson, K. Tonyali, M.L. Klakken and L.C. Jensen, J. Vac. Sci. and Technol. A 5(4) (1987).
6. J.C. Jaeger, Austral. J. of Sci. Res. 5, 1 (1952).
7. A.N. Gent and C.T.R. Pulford, in Developments in Polymer Fracture-1, E.H. Andrews, ed., Applied Science Publishers Ltd, London (1979), pp.159-160.
8. T. Yokobori, M. Uozumi and M. Ichikawa, Reports of Research Institute for Strength and Fracture of Materials, Tohoku University, 7(1), 25, (1971).
9. G.J. Lake and A.G. Thomas, Proc. Roy. Soc. London A300, 108 (1967).

AN EXPERIMENTAL ARRANGEMENT TO ESTIMATE THE FAILURE BEHAVIOR OF AN UNCROSSLINKED POLYMER UNDER HIGH SPATIAL CONSTRAINT

M. Parvin* and W.G. Knauss**

Graduate Aeronautical Laboratories
California Institute of Technology
Pasadena, CA 91125

ABSTRACT

In the search for the "constitutive behavior" of uncrosslinked and damaged polymers near the failure point we have identified a method for experimentally estimating such behavior. This work involves interferometrically measured displacements in a bonded cantilever-type specimen (DCB specimen). The method as outlined in references 1 and 2, consists basically of determining the displacement profile of the adherend in the region of the interlayer "crack tip", plus the determination of the relation between the force and displacement at the point of load application on the specimen. The interferometry measurements which allow precise evaluation of displacements are outlined here. In the process of developing a suitable specimen it has been demonstrated that increased pressure in the bond formation process generates superior bond durability including that in a wet environment.

INTRODUCTION

This study is motivated by a desire to better understand the requirements for and behavior of "tough" matrix materials for composites. Thermoplastic matrix materials are contemplated for the next generation of composites with the expectation that they are more forgiving and more tolerant to flaws under load. In attempts to synthesize materials for this purpose it is desirable to assess the effect of molecular structure on their mechanical failure behavior which controls the "tough" or "non-tough" characteristics. In an attempt to relate the molecular/microscopic structure of such a polymer to its macroscopic failure behavior it is appropriate, therefore, to develop methods that allow fairly detailed description of the material related to failure.

---

* Research Fellow.

** Professor of Aeronautics and Applied Mechanics

Failure behavior involves, grossly speaking, the development of a "yield-like" behavior occasioned by various microscopic processes depending on kinematic constraints, and the "strain softening phase" in which the material develops loss of cohesion. In a highly constrained environment such "strain softening" is usually associated with the development of voids, their growth and their coalescence. It is thus of interest to develop experimental techniques and analysis methods which allow estimation of these charateristics.

The failure of (thermoplastic) matrix materials occurs in a kinematically very constrained environment. This constraint is three dimensional due to the fiber geometry and it is difficult to simulate for testing purposes on a macroscopic scale. An alternate or simulating geometry is provided in a test procedure through which one can control the spacing between the stiffer phases. This may be achieved by embedding the matrix material between two relatively stiff plate or beam elements in the manner of an adhesive layer. The resulting geometry resembles a double cantilever beam (DCB) test geometry wherein one subjects a small amount of material to tension at the tip of a "crack." The geometric arrangement is shown in Fig. 1. By controlling the separation of the plate members, i.e., by varying the thickness of the matrix layer "d" it is possible to examine how the failure process of the assembly depends on the degree of constraint such as the layer thickness. In particular, it is of interest to examine the failure strength (yield, voiding ...) and the onset and their growth of voids as a function of kinematic constraint. Unfortunately we were, to-date, unable to generate test conditions with transparent (glass) beams that would allow us to observe void formation and their growth in situ and in a real time. We were thus forced to follow the procedure outlined here, to ultimately deduce this information from test results as described below. Two questions need to be answered first before such a proposed scheme can be accomplished: One needs to determine whether the matrix material can be made to adhere sufficiently well to the plates so that failure does not occur at the interface (bonding problem); the other question related to the precision with which macroscopic measurements can be made and then interpreted. The latter is the subject of a separate report.[1]

BONDING DEVELOPMENT

The bonded joint model, shown in Fig. 1, is a laminate of 2024 aluminum (adherend) bonded by polyvinylacetate as an initial model material. Some of the properties of polyvinylacetate (PVAc) that makes it a desirable model material to begin this study are: adhesion to polar surfaces such as aluminum, speed of bonding and apparently no chemical reaction required to achieve bond strength. While PVAc is an important ingredient of some pressure sensitive adhesives, we use it here not for its adhesives characteristics but because it is available in a narrow distribution of molecular weight (consistent and repetitive

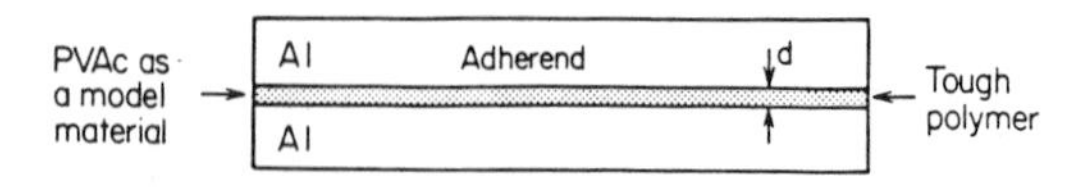

Fig. 1. Bonded joint model

supply) and because it has a glass transition temperature close to room temperature. Thus initial checkout of equipment and laboratory operation near the glass transition temperatures may be followed by experiments at increasingly lower temperatures to study the effects of temperatures on the failure mechanism. In this use of the model material the important feature is the relative purity and consistency of the material, and not a particularly high value of yield or other strength characteristics.

The selection of 2024-T3 aluminum is based on its extensive use, ease of surface preparation and its compatibility with PVAc. Aluminum has an elastic modulus of about 7×10^4MPa, which is higher by a factor of about 30 than that of PVAc, the latter being approximately 2200 MPa. This difference in material stiffness guarantees that the greater deformation will occur in the polymer, rather than in the aluminum without incurring plastic deformation in the latter.

To develop the technique for making specimens such as shown in Fig. 1, square aluminum samples coated with layers of polymer of different thicknesses (.025 mm-.25 mm) were brought together under different pressures of up to 350 kPa on a hot plate at 90-120°C for 1 hour. This plate was cut into 5 one-inch wide strips. A wedge was inserted between the aluminum adherends. The quality of the bond was then judged by the examination of the locus of failure. In a "good joint" the failure would be within the adhesive (cohesive failure), in contrast to the failure at the interface of aluminum strips and PVAc (adhesive failure). To determine whether the locus of failure was along an interface or through the polymer, the two parts of the separated specimen were soaked in an alkaline etching solution which attacks only the uncoated parts of the surface and stains it dark.

At low bonding pressure all bonds failed at the interface. The failure with this type of bonding is attributed to the existence of air bubbles. Increasing the applied pressure improved adhesion by producing some area of cohesive failure. However, at still higher pressures most of the adhesive squeezed out and no improvement was obtained. The development of bubbles can be eliminated (or at least reduced) by applying high pressure while simultaneously preventing the squeeze-out of the adhesive in some way. The details of this bonding process are discussed next.

SURFACE TREATMENT

To achieve the requisite bond strength, the following steps were taken in preparing the 2024-T3 bare aluminum for bonding with polyvinylacetate.

<u>Polishing</u>. The surfaces of the as-received aluminum sheets were polished with 600 grit sand paper to remove the spots of oxide, impurities and to reduce scratches from the surface. The samples were then washed and cleaned by acetone to remove any deposits left over from polishing. This step can be eliminated if the as-received contamination and oxide layer is only "moderate."

<u>Vapor degreasing</u>. The surfaces of the polished samples were exposed to the vapor of trichloroethane for 5-10 minutes to remove the oil and grease from the surface.

<u>Alkaline cleaning</u>. Alkaline solution was made by dissolving 22 gram Okaite No. 164 in one liter of water. For further degreasing,

the samples were soaked in this solution for 10 minutes at 82°C and were then immediately rinsed in warm water to prevent dry-on.

Alkaline etching The samples were next placed for 10 minutes in an alkaline solution containing:

Sodium hydroxide	:30 gram
Sodium phosphate	:1 gram
Distilled water to make	:1 liter at 68°C

During alkaline etching the aluminum surface containing copper, iron, manganese or silicon develops a black residual film (smut) that is insoluble in sodium hydroxide. In order to remove this film at the end of the etching, the samples were immersed in a solution of 300 gram Sulfuric acid, 50 gram sodium dichromate and distilled water to make one liter. The samples were then rinsed in cold water. A high etching temperature and long transfer time from the etching tank causes dry-on of the alkaline etch characterized by cloudy and stained areas.

Acid etching - Forest Product Laboratory process ("Optimized" FPL Etch)[3]. Following the alkaline etching treatment, the samples were immersed for 10 minutes in the optimized FPL etching solution containing:

Sulfuric acid (SO_4H_2) 66 BE	300 gram
Sodium dichromate $Na_2Cr_2O_7.2H_2O$	50 gram
2024-T3 aluminum file chips	1.5 gram
Distilled water to make	1 liter
at 68°C temperature.	

The FPL etching was followed by generous rinsing in cold running water. The samples were dried in an oven at 65°C or left in laboratory air to drip dry. During the entire process, care was taken that samples were not touched by hand or "contaminated" in any other way.

It should be noted that in the normal FPL process which has been the industry standard surface treatment for more than thirty years, there is no alkaline etching step. This step was found to improve the surface preparation by completely eliminating all the surface scratches still present after alkaline cleaning. Of course, the polishing process as the first step can be improved to eliminate the surface scratches, but this would be time consuming and, in fact, has the undesirable effect of removing the surface roughness. This step can be eliminated if the as-received aluminum surface is not badly oxidized or scratched. In the presence of high surface scratches or oxidation, it was found advantageous to increase the duration of the alkaline etching to 15 minutes. SEM micrographs at 50,000 magnification of the aluminum surfaces prepared by the method described above were typical of the FPL process.*

BONDING FORMATION AND EVALUATION

In order to be able to cause cohesive failure of the PVAc it is necessary to generate a durable bond between PVAc and aluminum. We,

---

* We are grateful to Dr. J. Venables of Martin Marietta for obtaining these records for us.

therefore, detail first the procedure involving high pressure bonding to obtain the requisite durability. To assess this durability we then test the bond strength in both "dry" and "wet" conditions to ascertain that during subsequent prolonged testing the bond does not fail before the adhesive gives way.

Bonding procedure In order to be able to produce significant pressure in the plates, we found it necessary to apply the following steps:

a) Square pieces (15 mm x 150 mm x 2.5 mm) of 2024-T3 aluminum were prepared by the surface treatment procedure described in the previous section.

b) The four sides of the faying surface of each sheet were covered with 12.5-mm wide masking tape. Then, the adhesive solution (20% PVAc in acetone) was very slowly poured over the exposed surfaces with full coverage. In contrast to dipping or brushing, very few visible bubbles are produced in this manner. The parts were then placed under a cover (to prevent contamination) to allow evaporation of the acetone from the adhesive solution and setting.

c) After 12 hours the masking tape was removed and was replaced by teflon strips of 0.075 mm thickness. The role of the teflon film was to prevent the adhesive from being squeezed out and also to control the thickness of the polymer film (Fig. 2).

d) The two aluminum plates with teflon strips in between were placed on top of each other and were placed in the press at 130°C under a load of 180 kN. This produced a pressure of about 350 kPa on the adhesive. (This is a crude estimate based on equal bulk modulus values for the PVAc and the teflon). When the two parts were brought together, only the PVAc coated areas of 125 mm x 125 mm were in contact (adhesive layer was thicker than the teflon spacer). When the pressure was increased, some PVAc was squeezed out along with bubbles until the adhesive was level with the teflon film. Since teflon possesses some compressibility, applying more pressure after this point spreads the adhesive more and forces out more bubbles. After 2 hours, the samples were allowed to cool under pressure and the bonded samples were removed from the press after 12 hours.

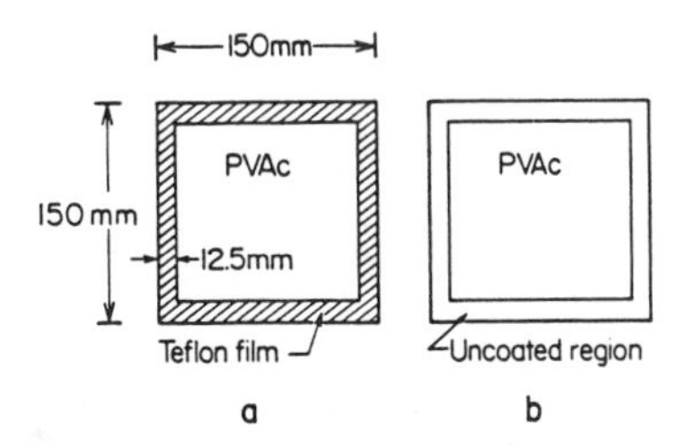

Fig. 2 (a) Aluminum plate covered with teflon strips along its circumference and PVAc in the central region. When the two plates are brought together and pressed, the teflon strips prevent the squeeze-out of the PVAc and also control the thickness of the interlayer.

Wedge Test Procedure: Dry To test the adhesive strength of PVAc, use was made of the wedge test[4] in the following manner.

The bonded panels were cut into four 25-mm wide test specimens with 25 mm from three sides discarded. The teflon spacer was removed from the fourth side to provide space for inserting a wedge. The wedge was then driven into the joint to produce a stress along the "bondline" (Fig.3). The wedge test simulates in a qualitative manner the effect of forces on adhesive joints. The deformation process at the opening point (crack tip) can be monitored by an optical microscope and the locus of failure can be studied after complete separation of the parts.

We emphasize at this point that these wedge tests are conducted only for the purpose of evaluating the durability of the bond, and they are not performed to derive material characterization data for the model adhesive. Tests relating to material characterization will be described in Item 4 of the next section. Here it is of interest to point out that in all durability wedge tests the failure occurred cohesively within the model adhesive. Thus the separated adherends were still covered totally with PVAc polymer: This observation is deduced from soaking them in an alkaline etching solution as mentioned before. For purposes of reference with respect to typical adhesive strengths it may be of interest to note that the "fracture energy" resulting from these "dry" tests is on the order of $G=4-10\ kJ/m^2$ (see Appendix A for the method of determination).

Wedge Test: Moist Environment Since we wish to ascertain the durability of the adhesive bond we attempt to accelerate the failure by exposing the test geometry to a "degrading" wet environment. To this end a wedge was forced into the bondline for a fixed distance as described above and the position of the crack tip noted. The crack stabilized a few seconds after inserting the wedge and remained stationary hereafter. The stressed specimen was then immersed in water at room temperature and removed periodically for the measurement of crack growth. While in all instances some time-dependent interfacial failure growth was observed, depending on the level of loading, one observed distinctly different failure behaviors as a function of bond preparation:

a) In "weak" bonds, produced at a pressure $p<0.35$ MPa, the crack started to grow immediately and continued to grow until complete separation occurred so quickly (a few minutes) that data recording was impractical.

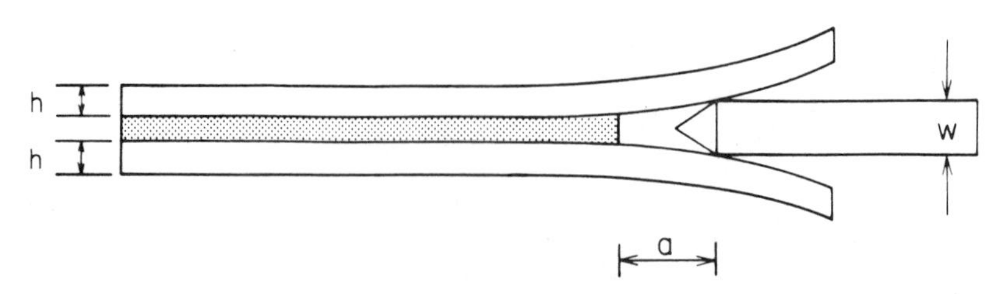

Fig. 3. Wedge test specimen.

b) When the bond was formed under higher, "intermediate" pressure (P~3.8 MPa), the rate of crack growth decreased significantly. Fig. 4 shows bond length histories resulting from end displacements held constant throughout the test. For comparison purposes we choose to characterize the end displacement of the "beam" in terms of an <u>initial</u> energy release rate G_o as computed in Appendix A. The crack propagated along the adhesive-adherend interface. It showed the tendency to move to the other interface but most of the time remained at one interface. In some cases, secondary cracks nucleated at the other interface; the main crack grew then toward the secondary crack and the failure path oscillated between the two interfaces.

c) In samples prepared at the highest pressures employed in this study (P~7.6 MPa), no immediate growth was observed upon insertion of the wedge. The crack remained essentially stationary and only after a long period of time (2-4 days) some crack propagation occurred close to the sample edge (Fig. 5); no noticeable growth occurred for four weeks after this initial extension.

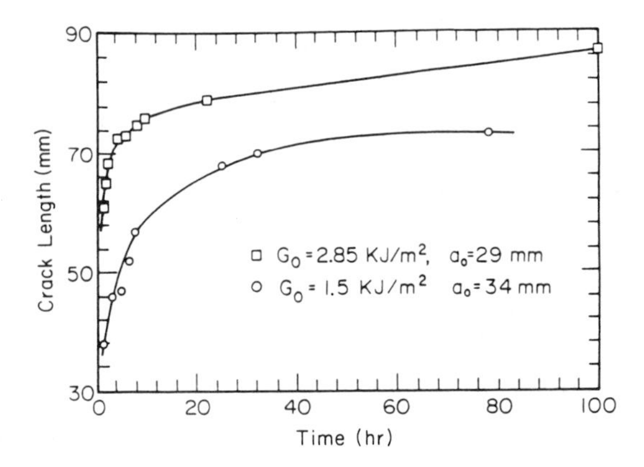

Fig. 4. Crack growth as a function of time in water at 20°C.

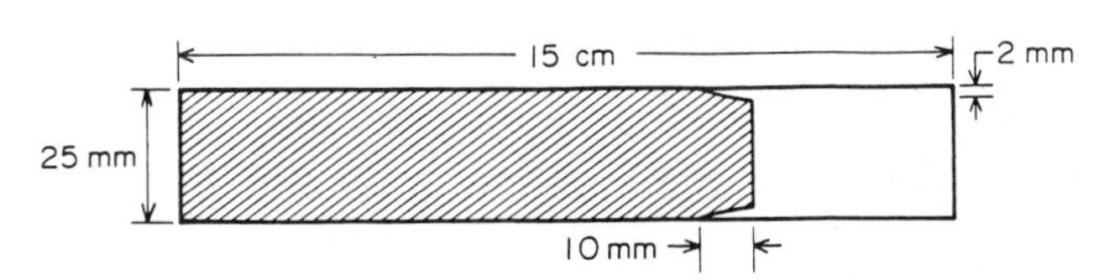

Fig. 5. Crack growth profile in a wedge test specimen in water. Crack opens slightly along the specimen edge over a short length.

There are at least three possible explanations for this marked effect of bond-pressure on bond performance. One reason would be that the high pressure forces a closer "mechanical interlock" than is possible without it. Second, the pressure may have eliminated any small flaws at the interface with a tightly adhering coating of the polymer on the etched surfaces so that water cannot get to the metal or the aluminum oxide interface. The fact that PVAc is a very compliant polymer at the temperatures used so far may be, at least in part, responsible for this behavior since it will flow easily in conformity with typical non-crosslinked polymers, and thus flow rather than break off from the irregularities on the adherend surface. For some aluminum epoxy joints a decrease in bond durability during exposure to water is reported.[5-7] The change of behavior from cohesive (in inert environment) to adhesive (in water) has been attributed to the change of oxide morphology (produced by etching) to hydroxide which does not have a strong adhesion to aluminum.[5] In the present study this effect was only observed in the bonds which were formed under relatively low pressure.

A third possibility exists which addresses the residual stresses associated with thermal cool-down during the manufacturing process: The high pressure compresses the material well into the time when the polymer solidifies and upon pressure removal in the solid state polymer expansion occurs. This expansion counteracts the thermal shrinkage associated with cool-down and thus acts to reduce residual stresses. That this mechanism is posible follows from the simple comparison of order of magnitude of volume strains associated with the two kinds of deformation sources. Although we do not know the bulk modulus of liquid PVAc precisely we infer from its glassy value of about 2,000 MPa (ν_g=0.35, E_g=1,800 MPa[8] a bulk modulus above the glass temperature of about 700 MPa. Thus the release of a bond pressure of 7 MPa gives rise to a volume increase of about 1%. By comparison a thermal change of ΔT=100°C results in a volume shrinkage of about 6.5% in the liquid state.*

Thus about 15% of the volume shrinkage is counteracted by the imposition of the pressure. While this comparison does not clearly argue for this mechanism of bond strengthening, the argument cannot be dismissed unequivocally that pressure bonding is effective through opposing the residual thermal manufacturing stresses.

EXPERIMENTAL RESULTS

We recall that the purpose of this study is to gain knowledge about failure behavior of thermoplastics under kinematic constraint: Yet, before proceeding to measurements in this regard, it is of interest to record some observations derived from the wedge tests. While the latter were performed primarily for purposes of verifying bond durability, it turned out that pertinent information could be assessed from these tests. These more qualitative results are described below and follow that exposition by initial measurements on the basis of optical interferometry.

Crack Tip Deformation Making use of specimens produced under a pressure of 7.6 MPa in a dry environment the deformation of the crack tip was carefully monitored by an optical microscope at different time

---

* Most of the temperature change occurs in the liquid state since room temperature is only about 10°C below the glass temperature.

intervals and for different amounts of beam separation (produced by wedges of varying wedge angles). The deformed zone is schematically shown in Fig. 6. The deformation at the edge of the interlayer (crack tip) results in highly stretched fibrils and voids which gives rise to what we choose to call a macro craze: (Fig. 7a, 7b). Such a macro-craze grows by the nucleation of pores ahead of the main crack (Fig. 7c). Formation of the voids was followed by void coalescence and rupture of the highly stretched fibrils (Fig. 7d). Macro crazes usually grew to about 3 mm in length and the voids stretched to about 0.075 mm before rupture of the ligaments occurred.

Crack Tip Deformation in Bulk PVAc The "craze zone" which was formed in the interlayer during the wedge tests was much longer than those normally observed in polymers. It was therefore of interest to study the crack tip deformation in bulk PVAc and to compare it with the behavior of the interlayer.

To this end single-edge-notch specimens (150 mm x 50 mm x 6 mm) were tested by a static loading apparatus. At relatively low stresses (2MPa) numerous crazes formed along curved lines normal to the contours of maximum principal stress (Fig. 8). This is the feature observed in many polymers, but the extent of the surface crazes in PVAc was much greater than in glassy polymers. For example, in polystyrene the crazes are formed in a band width of about 0.6 mm.[9] In PVAc as shown in Fig. 8 the craze band width is about 20 mm. We attribute this difference to the fact that at the test conditions the PVAc was considerably closer to its glass temperature than the polystyrene.

Gent [10] has suggested that the hydrostatic tension at the craze tip decreases the glass temperature T_g to room temperature and results in void nucleation. Lauterwasser and Kramer[11] have tested this idea experimentally and have concluded that the change in T_g which is on the order of about 3°C does not decrease it to the room temperature and may affect the yield stress only in a minor way. The low glass temperature of PVAc (28-30°C) is probably the main contributor to the high degree of fibrillation and formation of large voids. At our testing temperature of 22°C, any possible change in free volume associated with temperature and the dilatational stresses may facilitate the flow of the polymer and result in more void nucleation and expansion than would occur at lower temperatures.

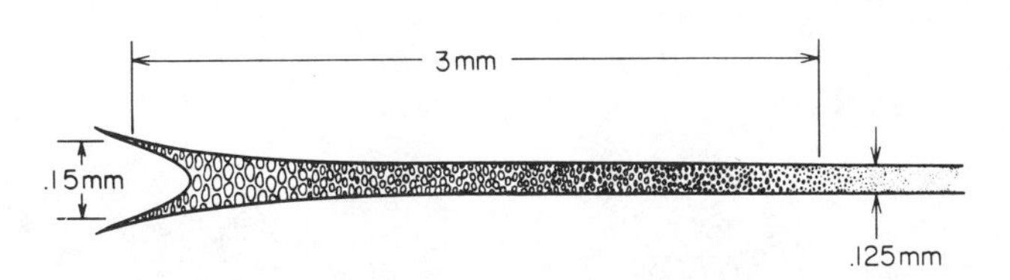

Fig. 6. Schematic representation of the deformed zone at the "crack tip."

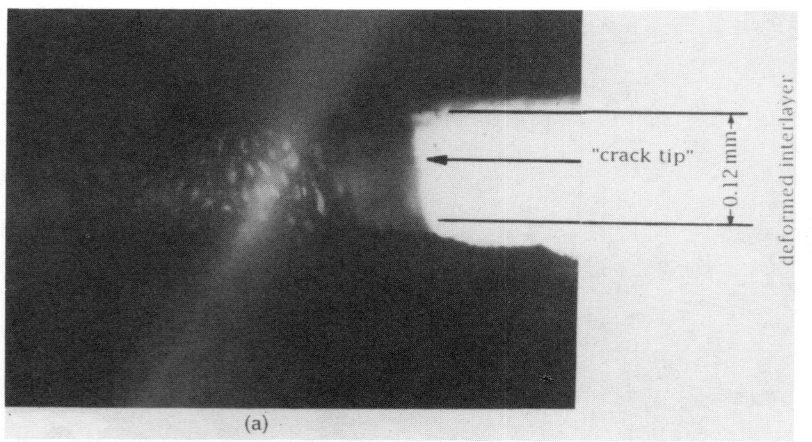
"crack tip"
0.12 mm
deformed interlayer

(a)

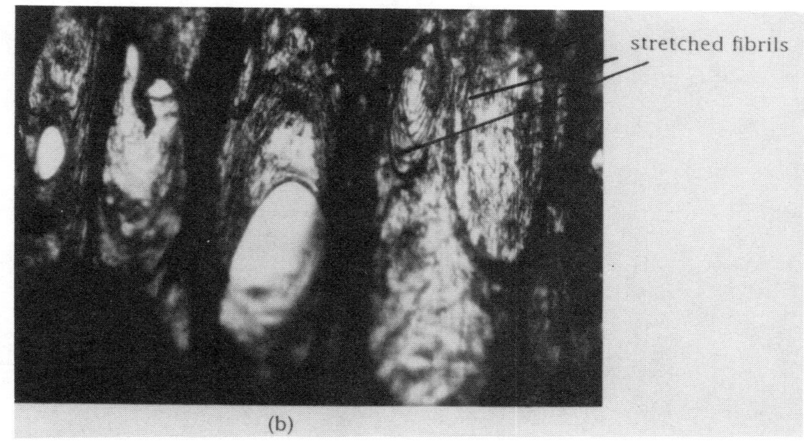
coalesced voids
.05mm

(c)

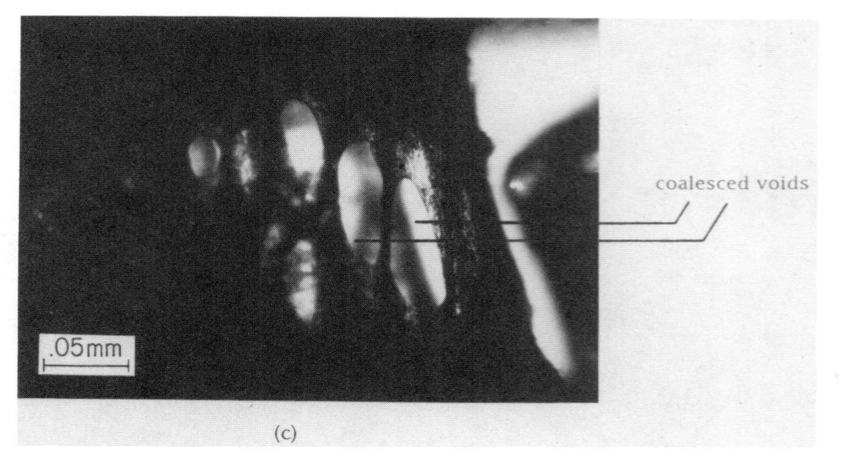
stretched fibrils

(b)

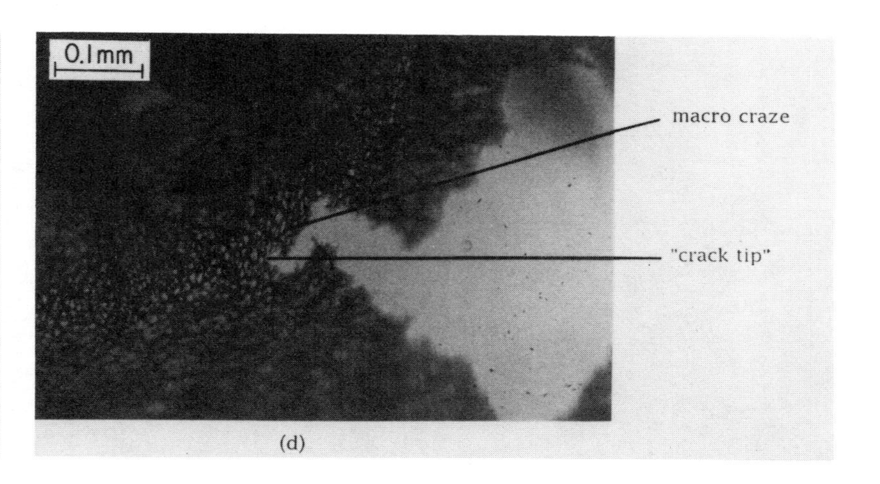
0.1mm
macro craze
"crack tip"

(d)

Fig. 7. The sequence of void formation and corresponding fibrillar structure. The deformation at the "crack tip" is accompanied by localized plastic deformation of highly stretched fibrils and voids (macro craze); Figs. (a) and (b). The "craze" zone grows by repeated nucleation of pores ahead of the main craze: Fig. (c). Formation of the voided structure is followed by void coalescence and rupture of the highly stretched fibrils, Fig. (d).

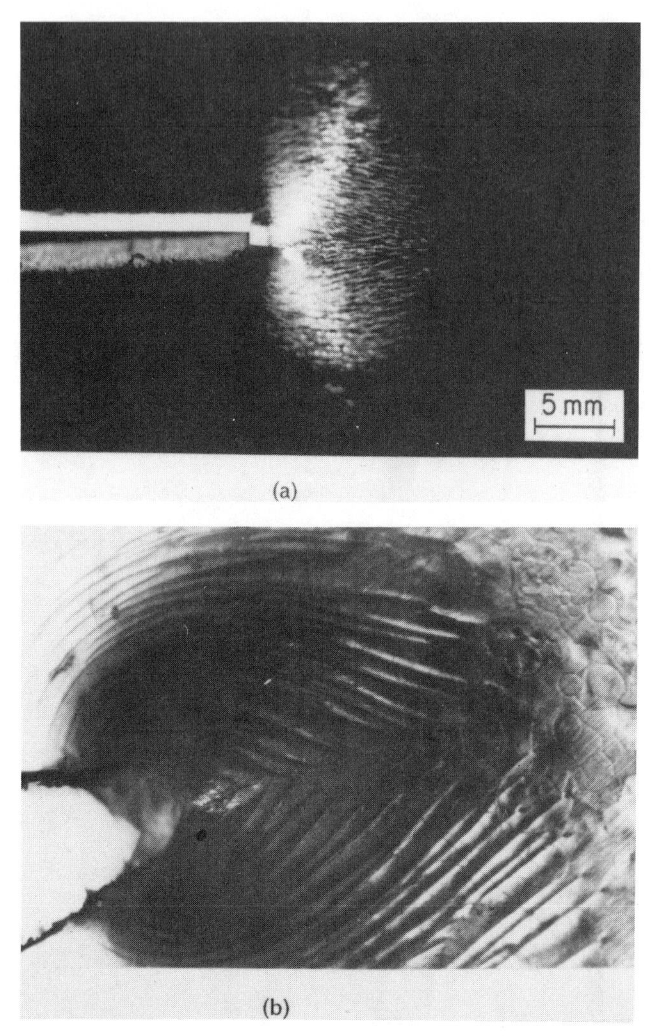

(a)

(b)

Fig. 8. (a) Craze at the tip of a crack in a SEN specimen PVAc at room temperature. (b) at 10x.

Effect of bond geometry The length of the macro craze showed a roughly inverse dependence on the thickness of the polymer layer. It was typically about 3 mm for an interlayer thickness of .125 mm and was reduced as the thickness increased. This result is related to the degree of confinement of the interlayer between the two substrates [12-14] and will be investigated in further studies.

The fracture surface of narrow specimens which were only 6-mm wide showed a tongue-shaped feature as shown in Fig. 9. When the failed parts were examined under an optical microscope, a step was observed in the PVAc layers covering the two plates. On one of the plates there was a "low step" in the central region surrounded by a "high step" region. The "tongue-shaped feature" was the result* of the intersection of these two regions. The width of the tongue increased with increasing specimen width until it covered the whole specimen width so that no tongue-shaped feature was observed in 25-mm wide specimens.

Experimental Work with Optical Interferometry The primary objective of this work is the determination of material behavior in confined spaces including voiding. To this end we employ optical interferometry. The method for determining the properties of the polymer as it approaches failure consists of deducing the cohesion forces from the measured displacements. This may be accomplished, in principle, by solving a mixed boundary value problem for the adherend which is represented as a linearly elastic body. In pursuing this road we expect that the boundary displacements are measured sufficiently precisely so that any rigid body motion can be identified and subtracted from the measurements.

The interference apparatus (Fig. 10) is basically a Michelson interferometer which consists of two mirrors M_1 and M_2, a beam splitter S, a light source L, and a camera C. A ray LA from the light source L (laser beam) is divided into two when it meets point A on the beam splitter S. The ray AB_1 is reflected to B_1 on the mirror M_1, crosses S and passes into the objective O_2. The second ray crosses S and is reflected by the mirror M_2 and then by the semireflecting face of S and finally superposed on the first ray in objective O_2.

Fig. 9. The fracture surface of a 6 mm wide sample, wedged apart; tongue-shaped feature represents a change in elevation, but fracture occurred cohesively at all points. Crack propagation from right to left.

---

* This phenomenon is probably related to the effect of adherend deformation as a function of "plate width" and associated with anticlastic curvature of the adherend.

In the DCB test the mirror M_2 is replaced by the DCB specimen one side of which is polished to produce a mirror surface. To allow observation of the fringes in the testing machine, the ray AB_2 passing through the splitter illuminates the top surface of the specimen via a mirror located above the specimen at a 45° angle relative to the top surface.

Destructive interference of the reflected waves (dark fringe) occurs when the path length difference (beam deflection) between the two flat faces of the specimen and the reference mirror are an odd number of half wavelengths. The laser operates at a wavelength of 632.8 nm. Therefore, each dark fringe represents an increment of 316.4 nm in the displacement of the polished surface of the specimen with respect to the plane of symmetry.

In the above setup, it is difficult to determine the zero fringe, so that all displacements are taken relative to some flat reference surface away from the crack tip. However, from a practical point of view this reference point cannot be taken too far from the crack tip because the fringe density in the crack tip region becomes high and because the optical beam width used is only 50 mm in diameter. Since we are, ultimately, interested not so much in the displacement itself as we are in its derivatives, we may subject the specimen to a small (<2°) rigid body rotation with respect to the axis of observation. Thus there will be a portion of the specimen surface which is located at right angles to the observation axis. Also, because the adherend plates undergo bending deformation involving anticlastic curvature, there occurs a saddle point in the elevation map. For measurements and reference purposes we choose thus a point on the specimen surface about

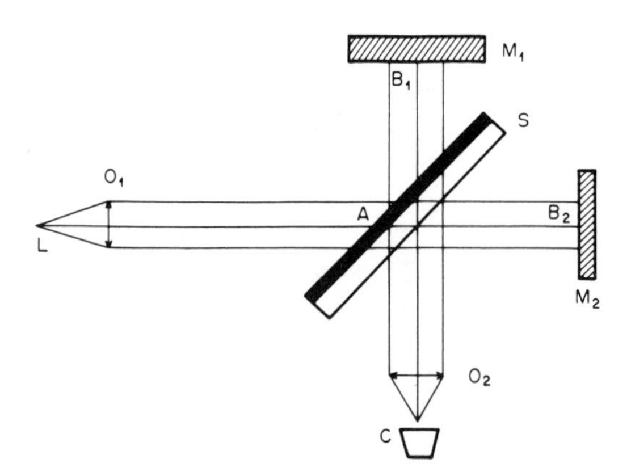

Fig. 10. Schematic drawing of optical arrangement in interferometry. M_1 and M_2 are flat mirrors, S is a semireflecting lens, L is the light source and C is a camera.

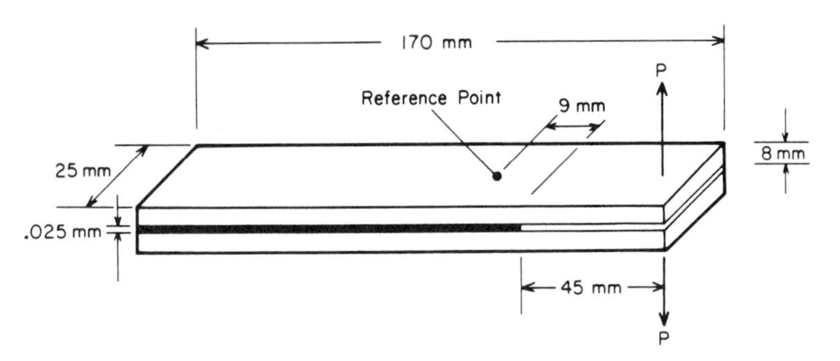

Fig. 11. Double cantilever beam used in interferometry measurements.

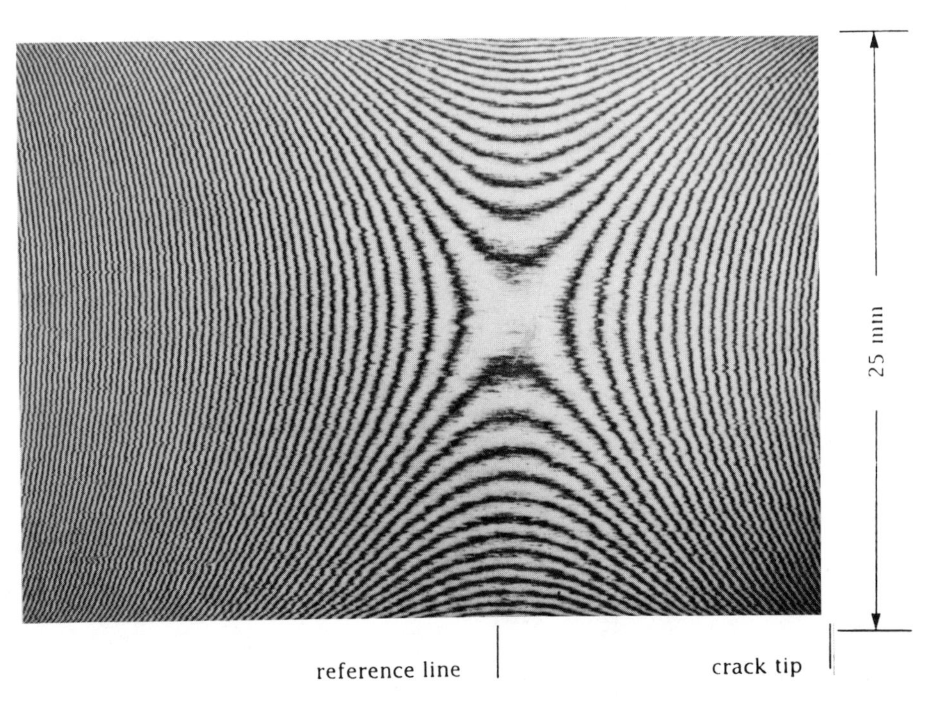

Fig. 12. Fringe pattern in a double cantilever specimen at P=222 N.

10 mm from the crack tip; by selecting an appropriate viewing direction (by a superposed rigid body rotation) through mirror adjustment, the saddle point can be brought into coincidence with the reference point after applying a load increment. This procedure is only a matter of convenience and does not affect the accuracy of the curvature of the adherends.

Results of Interferometry Measurements The DCB samples (Fig. 11) were loaded in an Instron testing machine at 0.5 mm/min cross head speed. At certain load levels the machine was stopped to capture the

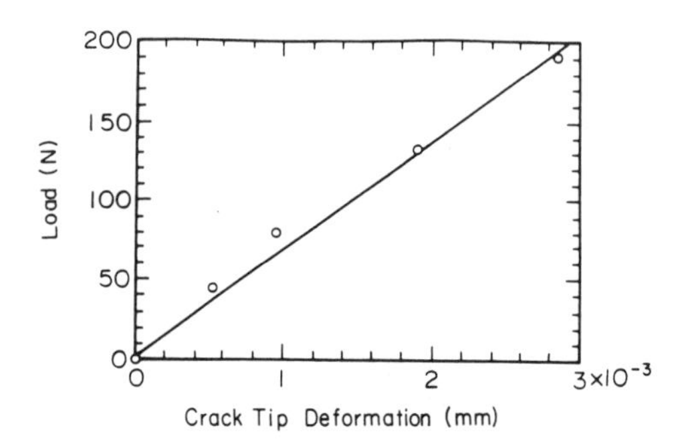

Fig. 13. Crack tip deformation as a function of load.

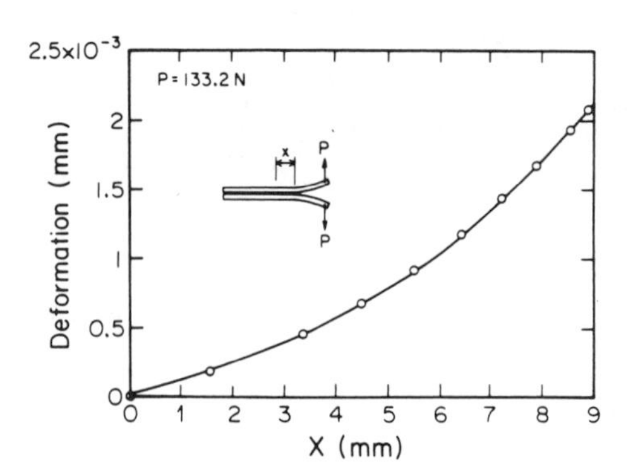

Fig. 14. Deformation of the bonded region.

fringe pattern with a camera (Nikon F3), a typical example of which is shown in Fig. 12. In order to take into account the fringe pattern that existed at zero load, these fringes were recorded and then deducted from the measurements at other load levels.

The fringes which are highly dense in the regions away from the "crack tip" and close to the loading point can be resolved under optical magnification ranging from 10 to 50. The crack tip deformation resulting from different loads as determined interferometrically is shown in Fig. 13 and a profile of the beam at a load level of 133.2 N is shown in Fig. 14.

ACKNOWLEDGEMENTS

We gratefully acknowledge the support of NASA through the grant NAF-1-474 under the program leadership of Dr. J.H. Crews, and of ONR through the grant N00014-84-K-0424, monitored by Dr. L.H. Peebles.

REFERENCES

1. M. Parvin and W.G. Knauss, "On the Failure Behavior of an Uncrosslinked Polymer under High Spatial Constraint," GALCIT SM Report 87-15 (1987).
2. T. Ungsuwarungsri and W.G. Knauss, GALCIT SM Report 85-4-1 (1986); to be published in Int. J. Fracture.
3. H.W. Eichner and W.E. Schowalter, Forest Product Research Laboratory Report No. 1813 (1950).
4. J.A. Marceau, Y. Moji and J.C. McMillian, Adhes. Age, 20, (10), 28 (1977).
5. J.D. Venables, J. Mater. Sci., 19, 2431 (1984).
6. T. Smith and R. Crane, 25th National SAMPE Symposium and Exhibition, May 25 (1980).
7. A.J. Kinloch, L.S. Welch and H.E. Bishop, J. Adhes. 16, 165 (1984).
8. I. Emri and W.G. Knauss, Polym. Eng. & Sci., 27, (1) 86 (1987).
9. A. Chudnowsky, Int. J. Eng. Sci., V 22, (8-10) 989 (1984).
10. A.N. Gent, J. Mat. Sci., 5, 925 (1970).
11. B.D. Lauterwasser and E.J. Kramer, Phil. Mag., 39, 469 (1979).
12. W.D. Bascom, R.L. Cottington and C.D. Timmons, J. Appl. Polym. Sci., 32, 165 (1977).
13. A.J. Kinloch and S.J. Shaw, J. Adhes., 12, 59 (1981).
14. S.S. Wang, J.F. Mandell and F.J. McGarry, Int. J. Fracture, 14, 39 (1978).

APPENDIX A

"Determination of the Strain Energy Release Rate "G" from Wedge Test"

Wedge test specimens for "adhesive joints" can be analyzed through the model of a double cantilever beam. Having combined simple beam theory and linearly elastic fracture mechanics, and considering the contribution of shear to elastic energy, the following expression may be derived for the strain energy release rate.[4]

$$G = \frac{Ew^2h^3}{16} \cdot \frac{[3(a+.6h)^2+h^2]}{[(a+.6h)^3+ah^2]^2},$$

where w is the wedge thickness; h is the adherend thickness; E is the modulus of elasticity of aluminum and a is the crack length. The terms h^2 and ah, are the results of shear contribution. The term, 0.6h, is an empirical correction factor for rotation of the beam ahead of the crack tip due to plastic deformation or crazing of the polymer interlayer.

FINITE ELEMENT MICROSCOPIC STRESS ANALYSIS OF FILLED POLYMERIC COMPOSITE SYSTEMS

Chi-Tsieh Liu, Russell Leighton and Greg Walker

Air Force Astronautics Laboratory
Edwards Air Force Base, CA 93523-5000

ABSTRACT

The elastic stress fields in a polymeric material containing two circular inclusions under uniaxial loading conditions are calculated using TEXGAP-3D finite element computer code. In the finite element model two boundary layers between the inclusion and the binder are included. In the analysis it is assumed that the moduli of the inclusion and the binder are different and they have three different distributions in the boundary layers. Particular attention is focused on the stress distribution at the interface between the inclusion and the boundary layer. In addition, the effects of the size and the location of the void, which is in the boundary layer, on the local stresses between the inclusion and the boundary layer, are studied and the role played by the local stresses on the growth of the void is discussed.

INTRODUCTION

It is well known that the material response and failure behavior of a particle-reinforced polymer material such as solid propellant are closely related to the damage state in the material. The damage may be in the form of microvoids and/or microcracks in the binder or in the form of the binder/particle separation known as dewetting. The presence of damage will redistribute the local stresses in the damage region which, in turn, may lead to additional damage in the material. Therefore, to obtain a fundamental understanding of the damage process, detailed knowledge of the local stress distribution around the filler particle is indispensible.

The elastic stress distribution around inclusions in an infinitely extended body subjected to uniaxial tension or shear at infinity were investigated by Yu and Sendecky,[1] Wong,[2] Goodier,[3] Shelly and Yu,[4] Mura and Cheng,[5] and Tandon and Weng.[6] A comprehensive experimental study of the role of the chemical bonding in adhesion was conducted by Gent.[7] The related problem of studying stress distribution around inclusions surrounded by boundary layers, with and without voids, has received little attention.

In this study, the elastic stress fields in a polymeric material containing two circular inclusions, surrounded by two boundary layers under uniaxial loading conditions, were determined using the TEXGAP-3D[8] finite element computer code. In the analysis, it was assumed that Young's moduli of the inclusion and the binder are different and there are three different distributions in the boundary layers. The focus of attention was on the stress distribution at the interface between the inclusion and the boundary layer. In addition, the effect of void size in the boundary layer on local stress distribution was studied and the role local stresses played in void growth was investigated.

FINITE ELEMENT ANALYSIS

One of the most powerful methods of numerical stress analysis, widely used in business and in practical design, is the finite element method. By use of this technique, the approximate solution with a reasonable accuracy can be obtained for the stress distribution in a wide range of structures with different material properties, complex geometries and loading conditions. In this study, the TEXGAP-3D elastic finite element computer code was used to determine the stress fields in the specimen. The specimen geometry, the applied load and the finite element model are shown in Fig. 1. Because the geometry of the specimen and the applied load is symmetrical, only half of the specimen was analyzed. Also, because of the symmetrical condition, shear stress and normal displacement along the horizontal centerline are zero. These conditions are satisfied by using the SLOPE boundary condition in the analysis. In the finite element analysis Young's moduli of the inclusion and the binder were equal to 6895 MPa and 0.6895 MPa, respectively. Three different types of distribution of Young's modulus, uniform, descent and ascent, in the boundary layer were considered. For the uniform distribution case, the moduli in the two boundary layers were equal to 0.6895 MPa. For the ascent H-S distribution type, the modulus in the

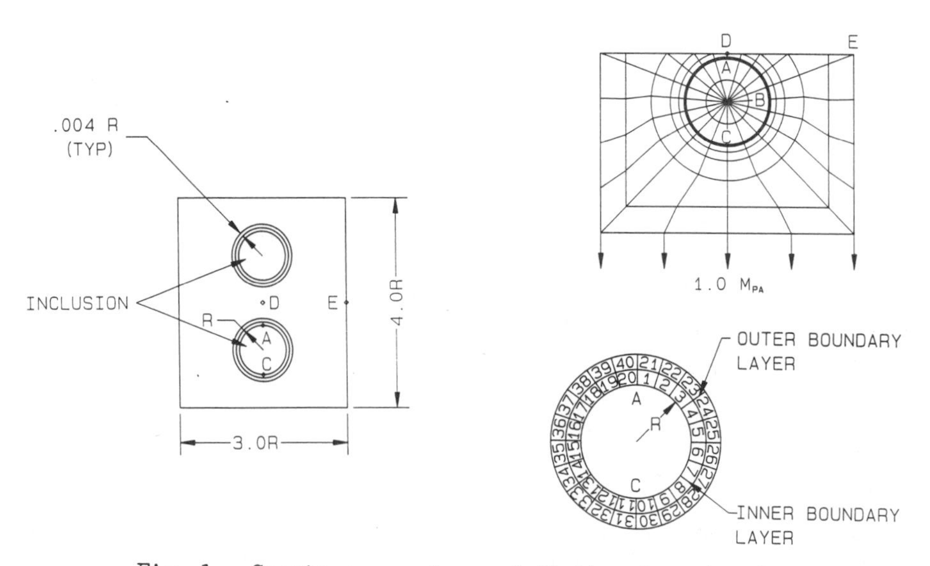

Fig. 1. Specimen geometry and finite element model

outer boundary layer was equal to 5.52 MPa and that in the inner boundary layer, which is adherent to the periphery of the inclusion, was equal to 13.79 MPa. When the order of the moduli in the boundary layers is reversed, the boundary layer is denoted as S-H type. In this study it was assumed that a void in the inner boundary layer was symmetric with respect to the specimen's vertical plane of symmetry. In the finite element analysis, the void was simulated by setting the modulus equal to 0.0001 MPa and the specimen was subjected to a uniform stress of 1.0 MPa. Hence, the stresses shown in the figures represent the stress concentration factor at a particular point. The results of the finite element analysis are discussed in the following paragraphs.

RESULTS AND DISCUSSION

Before giving a detailed discussion of the stress analysis results, I will briefly discuss the basic damage mechanism in highly filled polymeric materials such as solid propellants.

A composite solid propellant may be considered as a lightly crosslinked polymer, highly filled with coarse solid particles. When subjected to external loads, it behaves like a viscoelastic material. Since a highly filled composite solid propellant consists of a large number of fine particles, on the microscopic scale it can be considered nonhomogeneous. When this material is stretched, the different sizes and distribution of the filler particles, the different crosslink density of polymer chains, and the variation of the bond strength between the particles and the binder can produce highly nonhomogeneous local stress and strength fields. Because of the particle's high rigidity relative to the binder material, the magnitude of the local stress is significantly higher than that of the applied stress, especially when the particles are close to each other. Since local stress and strength vary in a random fashion, the failure site does not necessarily coincide with the maximum stress location and it also varies in a random fashion. In other words, the location and degree of damage will also vary randomly in the material. The damage may be in the form of a microcrack and microvoid in the binder, or in the form of the binder/particle separation known as dewetting. Experimental findings[9] reveal that for a densely filled solid propellant, the stresses tending to cause local failure are nearly equal at the binder/particle interface and in the binder itself; and that failure, either cohesive failure in the binder or adhesive failure at the interface, occurs when either the local binder rupture strength or the interfacial bond strength is lowest. The stresses tending to produce failure are greatest when the particles are close to each other when their line of center most nearly coincides with the direction of the external load. When a microcrack is generated in the binder between particles, it will grow toward the particles in the direction of the load. As the microcrack approaches the binder/particle interface, interfacial failure or dewetting may occur. When the particle is dewetted, the local stress will be redistributed. With time, additional binder/particle separation and vacuole formation takes place. This time-dependent process of dewetting nucleation, or damage nucleation, is due to the time-dependent processes of stress redistribution and binder/particle separation. Once dewetting occurs, the growth of the void around that particle may become larger with increasing strain, and the preferred direction of void growth seems to be in the direction of the applied load. Depending on the solid propellant formulation and testing conditions, damage growth may take place as successive nucleation and coalescence of the microvoids or as the material tears. As damage grows, it will eventually give rise to one or more dominant cracks that propagate rapidly to fracture. Therefore, to obtain a fundamental understanding of the damage process and fracture behavior in particle-filled polymeric materials such as solid propellants, detailed knowledge of the local stress distribution around the filler particle is indispensible.

Having discussed the damage mechanisms in particle-filled polymeric materials, I will now discuss the finite element stress analysis results.

For the geometry of the specimen considered in this study, and when there are no inclusions in the specimen, the stress state in the specimen is one dimensional and the nonvanishing stress is the normal stress σ_y which has a magnitude equal to the applied stress. However, the presence of inclusions in the specimen changes the stress state continuously from a one-dimensional state near the edges of the specimen to a three-dimensional state near either the center of the specimen or the surface of the inclusion as shown in Fig. 2. Figure 2 also reveals that maximum stresses occur at the center of the specimen. The magnitude and distribution of the stresses depend on the type of the boundary layer. For the uniform type of boundary layer, the maximum principal stress σ_1 is equal to 5.06 MPa whereas for the H-S type or the S-H type σ, is equal to 6.58 MPa.

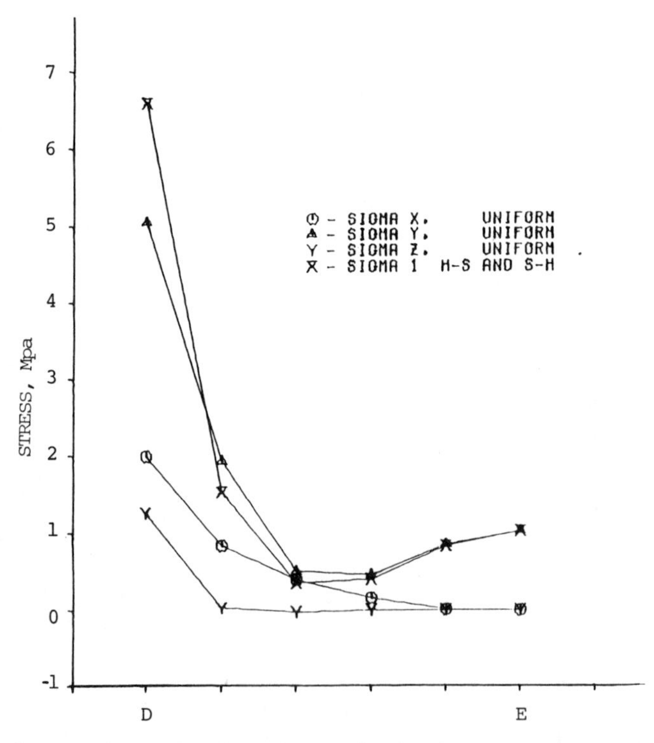

Fig. 2. Stress distribution along the horizontal plane of symmetry (without void).

The presence of a void in the boundary layer will redistribute the stress in the neighborhood of the void and the stress between the two particles. The magnitude and the distribution of the redistributed stress will depend on void length and the type of boundary layer. A detailed discussion on the stress distribution in the presence of a void is presented in the following paragraphs.

Figure 3 indicates that the peak stress of σ_1 along the horizontal plane of symmetry decreases as the void length is increased, and it shifts toward the edge of the specimen. It also indicates that the magnitude of σ_1 for the uniform type of boundary layer is smaller than that for the H-S and the S-H types of boundary layers and that the difference between the magnitudes of σ_1 for the two nonuniform boundary layers are negligible.

The distribution of σ_1 along the vertical plane of symmetry between the two inclusions is shown in Fig. 4. According to Fig. 4, when there is no void in the boundary layer, σ_1 has a maximum value at the midpoint between the two inclusions, and the value of σ_1 is decreased as the surface of the inclusion (point A) is approached. This figure also reveals that the value of σ_1 is increased when the inclusion is surrounded by either a H-S type or a S-H type boundary layer, and these two types of boundary layers produce essentially the same magnitude and distribution of σ_1 along the specimen's vertical plane of symmetry.

From the above discussion, one notes that the interaction between the two inclusions significantly increases the stress σ_1 at the midpoint between the two inclusions. If the stress is higher than the fracture strength of the binder and if the bond strength is strong enough, failure will occur at the high stress location, generating a microcrack in the binder. The growth behavior of the microcrack, i.e., whether it

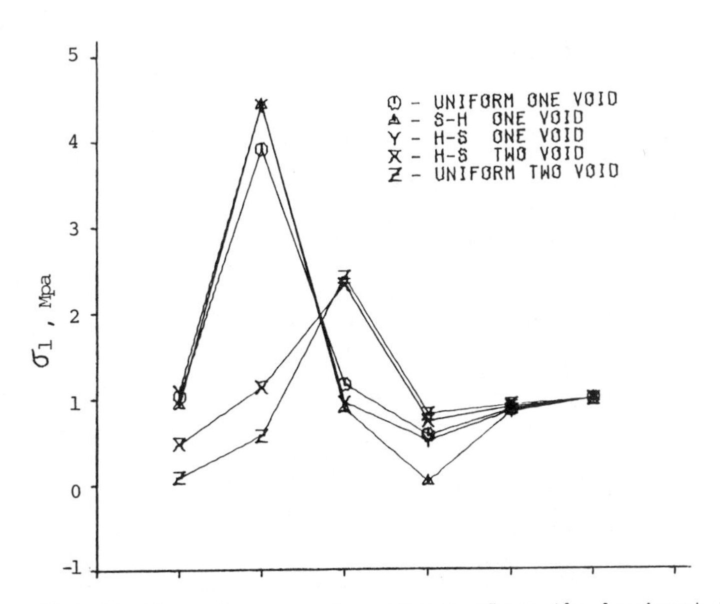

Fig. 3. Distribution of max. prin. stress along the horizontal plane of symmetry (with void).

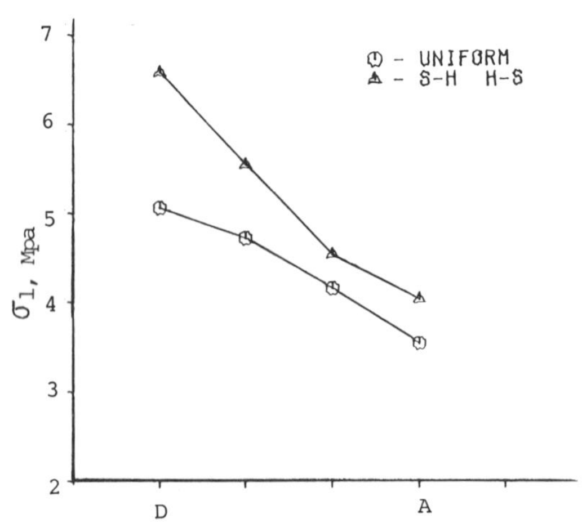

Fig. 4. Distribution of max. prin. stress along the vertical plane of symmetry (without void).

will grow parallel or perpendicular to the applied load, will depend on a number of factors such as the distance between the two inclusions, local stress and strength, and local microstructure of the binder and inclusion system. Experimental findings,[9,10] reveal that, depending on the system studied, microcracks may grow either toward the inclusion or parallel to the surface of the inclusion. When the microcrack grows toward the inclusion and if the bond strength at the interface between the binder and the inclusion is weak, dewetting will occur. The dewetting process will be controlled by the local stress state near the void tip and the local stress will be redistributed as the dewetting process continues. In order to see how the local stress distribution is affected by the growth of the void the distribution of stress along the periphery of the inclusion, with and without voids in the inner boundary layer, is determined and discussed in the following paragraph.

Typical plots of distribution of σ_1 in the inner boundary layer are shown in Figs. 5 and 6. From Fig. 5, it can be seen that, when there is no void present, the magnitude of σ_1 decreases as the void length is increased from point A, and reaches a minimum value at point B at which point it begins to increase again. From Fig. 6, it can be seen that, when there is a void present in the inner boundary layer, the distribution of σ_1 follows the same trend as that observed in Fig. 5, and that the peak stress of σ_1 moves with the tip of the void. Plots of the peak stress of σ_1 at the tip of the void in the inner boundary layer and near the tip of the void in the outer boundary layer (as void grows) are shown in Figs. 7 and 8, respectively. From Figures 5 and 6 it is seen that σ_1 at point A are significantly higher than those at point C due to the interaction between the two inclusions, regardless of whether or not there are boundary layers. From Figures 7 and 8, one notes that when there is no void, the value of σ_1 at point A is 3.55 MPa for the uniform boundary layer case, and 4.04 MPa for the nonuniform (H-S and S-H) boundary layer case. When a void exists in the inner boundary layer and when the boundary layer is the uniform type, the values of σ_1 at the tip of the void decreases as the void length is increased. However, when the boundary layer is the nonuniform type, the value of σ_1 at the void tip decreases initially, and then increases again as the void length is increased. In Fig. 8 it is seen that the values of σ_1 near the void tip in the outer boundary layer decrease initially, and then increase again with increasing void length. This trend has been observed for both the uniform and the nonuniform boundary layer cases, except that the increase in σ_1 value for the H-S and S-H boundary layer cases is significantly larger than that for the uniform boundary layer case, as shown in Fig. 8.

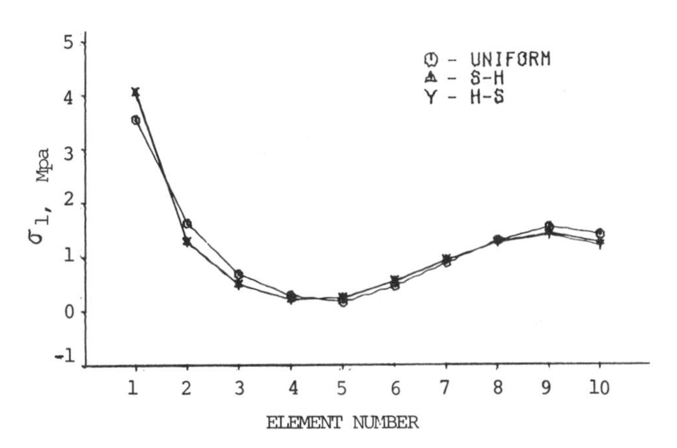

Fig. 5. Distribution of max. prin. stress along the periphery of the inclusion (without void).

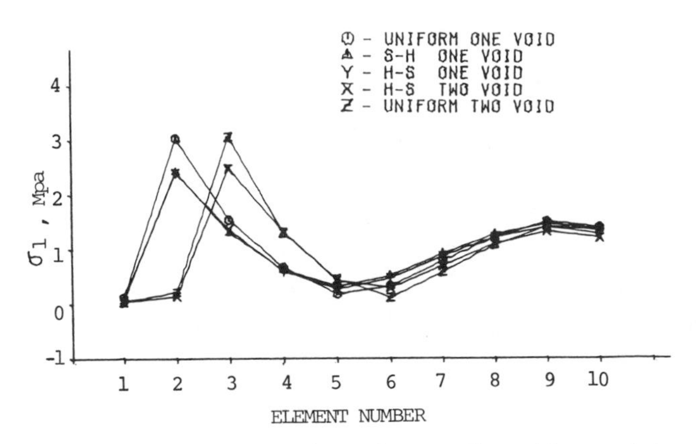

Fig. 6. Distribution of max. prin. stress along the periphery of the inclusion (with void).

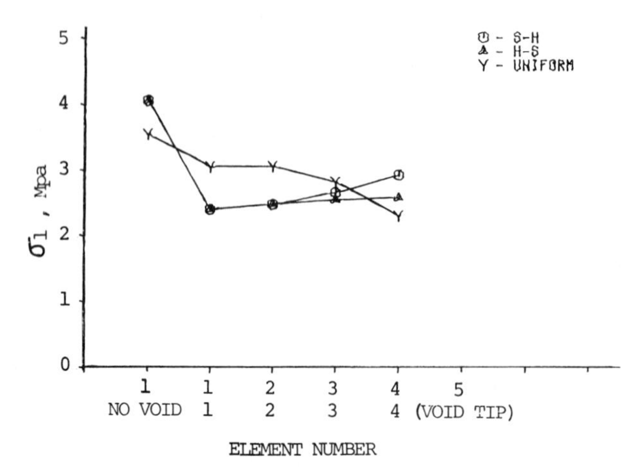

Fig. 7. Maximum prin. stress at void tip as void grows (inner boundary layer).

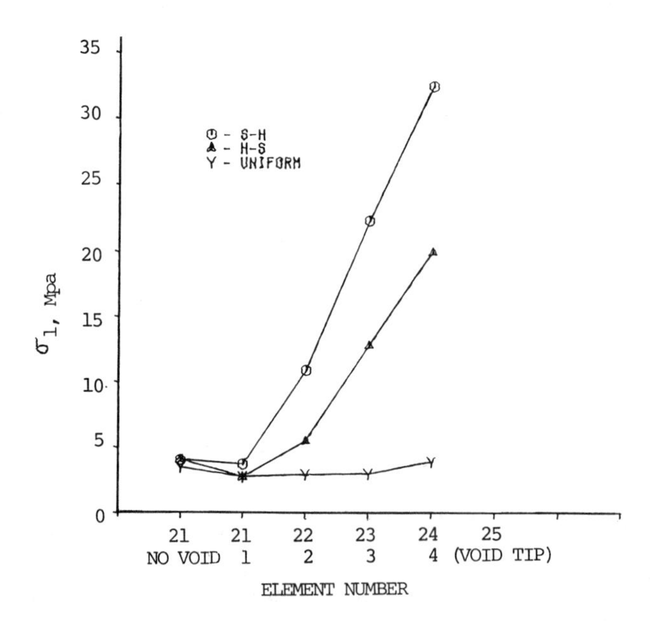

Fig. 8. Maximum prin. stress near void tip as void grows (outer boundary layer).

The above discussion indicates that when there is no void, a comparison of the uniform boundary layer with the nonuniform boundary layer shows that the nonuniform boundary layer has a higher probability of generating a void at point A. However, when a void is generated at point A, the tendency for the void to grow in the inner boundary layer is higher for the uniform-type boundary layer than for the H-S or S-H type boundary layer. Based on the results of the stress analysis, the stresses near the tip of the void are reduced when the voids grows sufficiently longer, which, in turn, implies that the tendency for the void to grow is reduced. The reduction of stresses near the void tip is probably due to a number of reasons such as the void growing into a lower stress region, the reduction of interaction between the two inclusions, and the reduction of interaction between the voids. In addition, when the length of the void increases, the magnitude of σ_1 near the void tip in the outer boundary layer also increases. Depending on void length and type of boundary layer, the magnitude of σ_1 in the outer boundary layer can be significantly higher than that in the inner boundary layer as shown in Figs. 7 and 8. In Fig. 8, for the uniform boundary layer case, the magnitude of σ_1 in the outer boundary layer becomes higher than that in the inner boundary layer when the void length reaches six elements in length. This implies that for the uniform boundary layer case, the void will grow in the inner boundary layer until the tips of the void approach tangent to the periphery of the inclusion and to the direction of the applied load. Under this condition, the tendency for the void to grow along the periphery of the inclusion can be greatly reduced, and the probability of the fracture

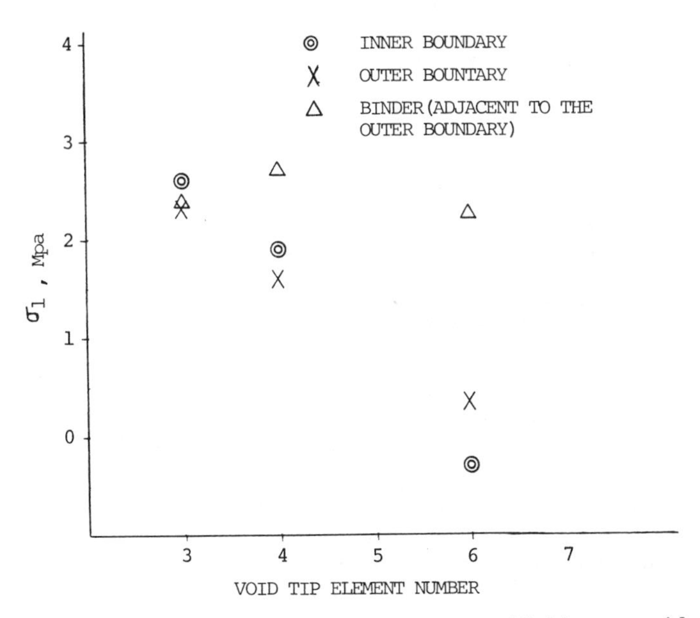

Fig. 9. Maximum prin. stress at and near void tip as void grows (special case).

initiation in the binder can be significantly increased. This type of damage process has been observed by Bill[9] in his study of cumulative damage in solid propellants. Referring back to Fig. 8, it is seen that a similar trend exists for the nonuniform boundary layer case, except that a void will be generated in the outer boundary layer when a short void two elements long is generated in the inner boundary layer. It is also shown in Figs. 7 and 8 that the magnitude of σ_1 near the void tips in the outer boundary layer is much larger than that at the void tips in the inner boundary layer when the void length is equal or greater than six elements. This indicates that the tendency of generating a void in the outer boundary layer is higher for the nonuniform boundary layer case than for the uniform boundary layer case.

It should be pointed out that the above discussion is centered on the stress distribution near the void tips, and the distribution is symmetrical with respect to the vertical center line of the specimen. In order to see how an unsymmetrical void will affect the stress distribution near the void tips, a stress analysis was conducted. In the stress analysis the unsymmetrical voids are located in the inner boundary layer of each inclusion in the center region of the specimen. They are located symmetrically with respect to the specimen's horizontal plane of symmetry, and unsymmetrically with respect to the specimen's vertical plane of symmetry, with one fourth of its length (two elements long) located to the left side of the vertical plane of symmetry. The results of stress analysis are discussed in the following paragraphs.

The results of stress analysis indicate that the maximum values of σ_1 at (near) the right side tip of the unsymmetrical void in the inner (outer) boundry layer are 0.103 MPa (2.58 MPa), 1.0 MPa (15.4 MPa), and 1.57 MPa (25.8 MPa) for the uniform, H-S, and S-H types of boundary layers, respectively. And they also indicate that the maximum values of σ_1 at (near) the left side tip of the unsymmetrical void in the inner (outer) boundary layer are 4.18 MPa (3.93 MPa), 3.04 MPa (4.02 MPa), and 2.56 MPa (6.21 MPa) for the uniform, H-S, and S-H types of boundary layers. These results indicate that, for the uniform boundary layer case, the left side void has a tendency to grow in the inner boundary layer, whereas the right side void has a tendency to grow into the outer boundary layer. However, for the nonuniform boundary layer cases, both the left and right side voids have a tendency to grow into the outer boundary layer.

In the above paragraph it is pointed out that, for the nonuniform boundary layer case, the magnitude of the σ_1 near the void tips in the outer boundary layer are much higher than that in the inner boundary layer. This indicates the effect of nonuniform distribution of modulus on the stress distribution near the tip of the void in the composite system. In order to determine how the moduli of the outer boundary layer and the binder affect the local stress near the void, a stress analysis was conducted. In the analysis three void lengths six elements long, eight elements long (symmetrical void), and eight elements long (unsymmetrical void) were considered, and the modulus of the binder is equal to 5.52 MPa with the moduli of the inner and outer boundary layers equal to 13.79 MPa and 0.6895 MPa, respectively. The results of the stress analysis are discussed in the following paragraph.

The results of the stress analyses, plotted the maximum value of σ_1 in the binder and in the boundary layers as a function of the right side length of the void, are shown in Fig. 9. Figure 9 indicates that the maximum value of σ_1 at the void tip in the inner boundary layer and that near the void in the outer boundary layer decreases with increasing void length. It also indicates that, for the symmetrical void case, the magnitude of σ_1 in the outer boundary layer is lower than that in the inner boundary layer, and the converse is true for the unsymmetrical void case. It is interesting to note that when the right side tip of the void reaches elements 4 (symmetrical void) and 6 (unsymmetrical void), the magnitude of σ_1 in the binder near the outer boundary layer is higher than that in the inner and the outer boundary layers, especially for the unsymmetrical void case. This implies that a void may be generated in the binder material. Referring back to Figs. 9 and 8, it is noted that the magnitudes of σ_1 in the outer boundary layer for the special case are significantly lower than those for the H-S and S-H type boundary layer cases. This indicates that the change of the moduli of the outer boundary layer and the binder material can have a significant effect on the local stress distribution near the void.

The results of the stress analysis indicate that there are two regions in the specimen where stress concentration can take place. Therefore, microcrack initiation and propagation will not, in general, be restricted to a single point. If binder failure occurs first at the midpoint between the two inclusions, the growth behavior of the microcrack will depend on the state of the redistributed stress field near the microcrack. When the microcrack approaches the inclusion, and if the interface bond strength is low, failure at the interface may occur before the microcrack reaches the surface of the inclusion. However, if the interface has a strong bond, or the boundary layer is stiff enough, the microcrack may grow away from the inclusion. On the other hand, if the interface failure occurs first and the interfacial crack grows in the interface between the binder and the inclusion, the stress at the interfacial crack tip will decrease as the interfacial crack grows longer. Under this condition, the higher stress between the two inclusions and near the interfacial crack tip may cause fracture in the binder. When this kind of fracture behavior occurs in the specimen, a complex phenomenon of crack interaction will take place, affecting the subsequent crack growth behavior.

CONCLUSIONS

The results of the stress analysis reveal that when there is either no void, or a shorter void, in the boundary layer, a three-dimensional stress state exists in the center region of the specimen and near the surface of the inclusion. However, when a longer void exists in the boundary layer, and for the cases considered in this study, the stress state in the center region approaches two-dimensional. It also reveals that the magnitude of the local stress near the void tips depends on the type of boundary layer and the void length. If the magnitude of the local stress near the void tips is higher than the cohesive strength of the binder, instead of growing in the interface, the void will grow into the binder.

Although the present analysis sheds some light on the relationship between the stress state and damage process in a filled polymeric composite system, additional work is needed to investigate the stress state and microcrack interaction in a more representative composite polymeric system under a complex loading condition.

ACKNOWLEDGEMENTS

This work was supported by the Air Force Office of Scientific Research. The authors would like to express their appreciation to Ms. K. Pace for her assistance and to Mr. L. Austin for typing and editing the manuscript.

REFERENCES

1. I.W. Yu and G.P. Sendecky, "Multiple Circular Inclusion Problems in Plane Elastostatics," ASME J. Appl. Mech., 41 215 (1974).
2. I.-Chih Wong, "A Contact Stress Problem for a Rigid Smooth Sphere in an Extended Elastic Solid," ASME J. Appl. Mech., 32 651 (1965).
3. J.N. Goodier, "Concentration of Stress Around Spherical and Cylindrical Inclusions and Flows," Trans ASME J. Appl. Mech., 55 39 (1933).
4. J.F. Shelly and Y.Y. Yu, "The Effect of Two Rigid Spherical Inclusions on the Stress in an Infinite Elastic Solid," ASME J. Appl. Mech., 33 68, (1966).
5. T. Mura and P.C. Cheng, "The Elastic Filed Outside on Ellipsoidal Inclusion," ASME J. Appl. Mech., 44 591 (1977).
6. G.P. Tandon and G.J. Weng, "Stress Distribution In and Around Spheroidal Inclusions and Voids at Finite Concentrations," December 7-12, 1986.
7. A.N. Gent, "The Role of Chemical Bonding in the Adhesion of Elastomers," Project Number 092-555, Technical Report No. 11, Institute of Polymer Science, The University of Akron.
8. K. Kathiresan and S.N. Atluri, "Three-Dimensional Homogeneous and Bi-material Fracture Analysis for Solid Rocket Motor Grains by a Hybrid Displacement Finite Element Method," AFRPL-TR-78-65. Georgia Institute of Technology, Atlanta, Georgia, 30332.
9. K.W. Bills, Jr., "Solid Propellant Cumulative Damage Program," AFRPL-TR-68-131, Aerojet General, Sacramento, CA, October 1968.
10. L.R. Cornwell and R.A. Schapery, "SEM Study of Microcracking in Strained Solid Propellant," Metallography, 8, pp. 445-452 (1975).

INORGANIC PRIMERS FOR ADHESIVE BONDING: A STATUS REPORT

R.A. Pike and F.P. Lamm

United Technologies Research Center
East Hartford, CT 06108

ABSTRACT

The use of inorganic primers in adhesively bonded joints offers an alternate approach to standard organic primers or coupling agents for improving the environmental resistance of bonded structures. Up to a tenfold improvement in resistance to crack growth, as measured by wedge crack tests, has been found in specific cases using the inorganic primer. In addition, application of inorganic primers on aluminum adherends has resulted in elimination of the differences normally associated with varying metal surface treatments when organic primers are employed. The inorganic primers investigated, formed by hydrolysis of metal alkoxides on a treated adherend surface, are equally effective after room or elevated temperature conversion to their oxide forms. As with organic primers, a thickness effect has been demonstrated. A review of the results obtained with inorganic primers in bonded joints at United Technologies Research Center (UTRC) and other laboratories is presented, and reasons are discussed for the demonstrated improvements in environmental stability.

INTRODUCTION

Benefits in weight saving and manufacturing costs have led to increased use of adhesively bonded structures in the automotive, aircraft and aerospace industries. To be a viable alternative to, for example, metal fasteners, these adhesive bonds should maintain the strength typical of conventional fastener systems. In many applications the bonds are put under a variety of environmental and mechanical stresses, such as exposure over long periods of time to wet environments and to temperature cycles which can result in a loss of bond strength. The loss of strength can result from the extension of interfacial disbonds and other deformations that occur in the adhesive and which are exacerbated by moisture. Because of this deficiency, extensive research and development efforts have been undertaken to define methods and identify materials which improve bonded joint performance in humid conditions. Thus, it is essential that before bonding, the adherend is cleaned and chemically pretreated to produce a surface which interacts with the adhesive to develop bond strengths to meet application requirements. A variety of pretreatments for aluminum has been developed to produce improved bondability, including acid etching (FPL), and anodized treatments with sulfuric (SA), chromic (CAA) and phosphoric acid (PAA). These pretreatments typically utilize corrosive conditions, anodization equipment or environmentally hazardous chemicals, such as chromium. The PAA treatment is generally accepted as the most effective surface treatment in terms of bond strength and durability for adhesives curing at 121°C. In-depth surface analysis using scanning transmission electron microscopy showed that PAA treatment produces fine oxide protrusions of greater length and width than other surface treatments. These whiskers are believed to account for the strength enhancement achieved with joints made using PAA-treated adherends. Thus, mechanical interlocking by whisker reinforcement of an adhesive appears to play a role in increasing adhesive bonding.[1] The probability that chemical interaction is of major importance, depending upon the polymer/metal combination, has also been demonstrated.[2]

The objectives of the investigation, involving the use of inorganic primers as substitutes for standard organic resin primers were (a) to find an effective primer system which would eliminate the use of environmentally undesirable chromium containing compounds (b) to improve the environmental resistance of bonded structures and (c) if possible to define the relative roles of mechanical interlocking, chemical bonding and physical forces (acid-base interaction) in adhesion and determine the factors which control these phenomena in bond formation.

The effect of substituting the normally used organic primer with an amorphous inorganic aluminum oxide primer on the tensile lap shear and T-peel strength of adhesively bonded aluminum was shown to result in equivalent wet and dry strengths for both primer systems.[3] This result could not be predicted from the relatively smooth topography of the inorganic-primed surface since the honeycomb-whisker protrusions related to enhanced bond strength in phosphoric acid anodized (PAA) systems compared to FPL etch treatments were not available to provide a high degree of mechanical interlocking.[3]

Bond integrity is associated with the interfacial adhesion in a bonded joint and related to hydrolytic stability of the oxide, which is generally believed to be the limiting factor in bond durability.[4] Thus, improvements in oxide stability usually improve durability, as measured by wedge crack tests. Such improvements which have been reported to date include coating Forest Products Laboratory (FPL) acid-treated surfaces with phosphorous nitrilo compounds,[4] and phosphoric acid dip treatments on chromic acid anodized (CAA)[5] and sulphuric acid anodized (SA)[6] surfaces. These postanodizing treatments do not alter the rough topography (in fact, they enhance the surface roughness)[6] of the anodized surfaces, in marked contrast to the generation of a relatively smooth oxide surface when the inorganic amorphous primer is employed. Surface treatments for titanium alloys have also been evaluated and ranked in terms of moisture resistance.[7] Those which form micro rough surfaces such as CAA or sodium hydroxide anodizing methods have been found to be superior to acid etching procedures such as phosphate fluoride.

The work reported in this paper briefly summarizes the previously published results with inorganic primers in bonded joints as well as new evidence as to the effectiveness of this approach to improve the environmental stability of bonded joints. In addition, preliminary data from other laboratories as to the effectiveness of such materials is presented.

GENERAL EXPERIMENTAL PROCEDURES

Material Definitions

Epoxy Film Adhesives

Hysol EA-9649 - A dicyandiamide-cured (177°C) aluminum powder-filled epoxy resin containing asbestos and rubber additive with scrim support.

FM-300 - A dicyandiamide-cured (177°C) titanium dioxide powder rubber modified epoxy resin containing a brominated epoxy with scrim support.

Hysol EA-9628 - An aluminum powder-filled epoxy resin with aromatic amine hardener (121°C) with scrim support.

Epoxy Primers

Hysol EA-9205 - Epoxy resin system in MEK containing strontium chromate.

BR-127 - Epoxy resin system in MEK containing strontium chromate.

The 2024-T3 or 6061 aluminum alloy was phosphoric acid-anodized at room temperature, using 12% phosphoric acid with either 8V for 25 min. or 10V for 20 min., or treated by FPL acid-etch using the standard sodium dichromate-sulphuric acid formulation. The inorganic primer was formed by applying a one percent toluene solution of E-8385 (sec-butyl) aluminum alkoxide from Stauffer Chemical Co. (or other inorganic alkoxide) to the treated surfaces.

Titanium 6-4 alloy was etched using commercially available Pasa Jel 107 from Smetco Company and was similarly coated with the aluminum alkoxide. The alkoxide was converted to oxide primer by solvent evaporation at RT or by heating at temperatures up to 325°C. Hysol EA9205 and American Cyanamid BR127 epoxy primers were used for the control samples.

Adhesive-bonded joints were prepared using commercial samples of Hysol EA9649 and FM-300 from American Cyanamid, 177°C curing scrim supported film adhesives and Hysol EA9628, a 121°C curing supported film. Wedge crack tests were carried out according to ASTM D-3762 using 95% relative humidity (RH) at 49°C and 71°C for the 121°C and 177°C curing adhesives respectively. Exposure times at these conditions ranged from 4 to over 250 hours. Crack lengths were measured under 20X magnification in millimeters.

RESULTS

A. Effect of Metal Surface Treatment The initial investigation of inorganic primers involved the bonding of 2024-T3 aluminum which was pretreated by PAA with EA-9649 and EA-9628, 177°C and 121°C heat-cured adhesives respectively. Two organic primers, BR-127 and EA-9205, both of which were spray applied and cured by heating to 121°C prior to bonding, were used as controls. The inorganic primer generated from sec-butyl aluminum alkoxide was found to provide a four (BR-127) to ten

(EA-9205) fold improvement in crack propagation resistance compared to the two organic primers.[8]

The response of the 177°C adhesive to primed surfaces which were FPL etched was similar. As with the PAA treatment, the inorganic-primed surface was again superior to the organic-primed samples, but the margin of improvement was less. The data also indicated that for organic-primed surfaces bonded with 177°C curing adhesives, the FPL surface treatment on 2024 aluminum provided better crack resistance than did PAA treatment. This result was in agreement with previously reported work using stress-rupture tests which showed that on primed surfaces, FPL-etch was superior to PAA for joints bonded with the higher temperature curing epoxy adhesives.[9]

The effect of primer on the crack resistance of PAA- and FPL-treated aluminum bonded with the 121°C curing adhesive EA9628 was also determined.[8] In both cases, the inorganic primer was found to produce improved crack resistance compared with organic-primed controls. Unexpectedly, it was found that with the inorganic primer generated from the aluminum alkoxide the crack propagation resistance of both the FPL-treated and PAA surfaces was essentially the same with both adhesives; the difference between the two surface treatments using the organic primers was negated by application of the amorphous alumina primer.

Wedge crack tests on 2024 aluminum using an acid-etch system, SmutGo, which reportedly leaves a very smooth oxide layer on the surface have been carried out by Martin-Marietta.[10] These workers found that there was a substantial improvement in the crack resistance of FM-300 bonded specimens using the inorganic alumina primer compared to BR-127 organic primer. For example, the organic primed specimens completely failed in 2 hours compared to the inorganic primer which showed a crack growth of 2.7 cm in 200 hours.

Similar results on 2024 aluminum have been obtained at UTRC using the Pasa Jel 101 acid-etch system which contains no chromic acid, i.e., the inorganic primer showed marked improvement over BR-127 primed adherends, particularly with EA-9649, as illustrated in Fig. 1. It was also demonstrated that a five-minute application of FPL-etch at RT was an effective surface treatment when combined with the inorganic primer as illustrated in Fig. 2. Thus, acid-etching carried out for short periods at room temperature, as is essential for field repair situations, can result in a high level of environmental resistance in bonded joints.

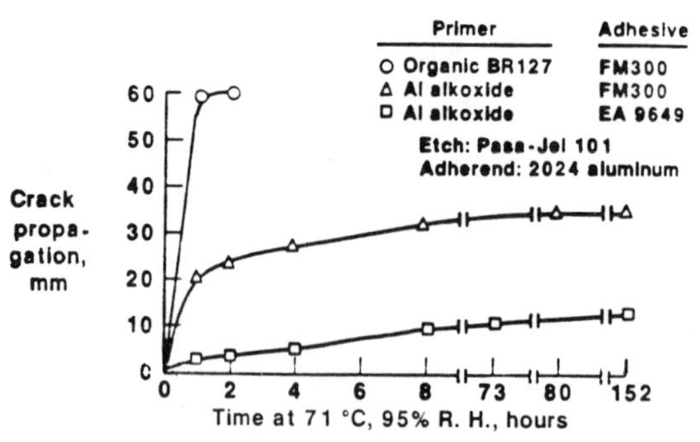

Fig. 1. Effect of non-chromate etch on crack growth.

B. Effect of Primer Conversion Temperature Thermal weight loss data (TGA) on the metal alkoxide compounds in terms of weight loss showed these materials (at 10°C/min heat rate) generally had organic loss temperatures in the 280°-290°C range. Such temperatures are prohibitive for structural aluminum adherends which have been treated to induce certain mechanical properties for specific applications. An alternate mechanism to thermal decomposition for organic loss and alumina formation with the alkoxide compound, would involve hydrolysis by the water of hydration on the oxide of the etched or anodized aluminum to produce an alcohol and the amorphous alumina. The generated alcohol would be lost by evaporation along with the carrier solvent at relatively low temperatures. Thus, the alkoxide would act as a desiccant or dehydrating agent to remove the water of hydration on the oxide surface.

It has also been observed that the interaction of the alkoxide with residual Al-hydroxyl groups on the oxide surface may play a prominent role in the chemical bonding of the primer to the substrate surface.[11]

To determine if there was a minimum temperature at which the alkoxide to amorphous oxide conversion could be carried out, alkoxide-coated, PAA aluminum samples were heated at different temperatures in the range RT to 325°C and wedge crack specimens prepared using EA9649 adhesive. The results are listed in Table 1. The total crack growth for 100-122 hours exposure to temperature and humidity was found to be essentially the same over the conversion temperature range investigated and in agreement with the original test results for the alkoxide primer. The ability to use the lower conversion temperatures minimizes the risk of mechanical property degradation of aluminum substrates. As a result, the use of the inorganic primer in field repair systems is an area of potential application.

C. Effect of Primer Thickness It is well known that organic primers are used at a preferred thickness generally on the order of 0.2-0.4 mil. To determine if there was a thickness effect with the inorganic primer related to crack propagation, a series of three wedge crack tests were run with PAA, FPL and CAA treated 2024 aluminum bonded with Hysol EA-9649 adhesive. It had been established that at the one percent solution concentration used, one coat of the aluminum alkoxide resulted in approximately 150 nm of primer thickness.[3] The results for the first two surface treatments are shown in Figs. 2 and 3 for specimens exposed to temperature and humidity for over 160 hours. With the PAA-treated surface bonded with EA-9649 it was apparent that coatings made with a thickness greater than 450 nm (3 coats) were

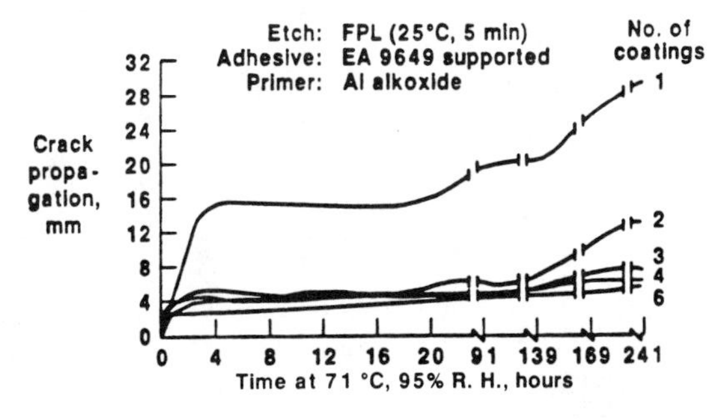

Fig. 2. Effect of primer thickness on crack growth.

superior to thinner coatings and that after 6-7 hours exposure very little additional cracking occurred; failure was cohesive within the bondline. The one and two primer coated specimens exhibited some degree of adhesive failure as determined by visual 20X magnification of the fracture surface. The FPL-treated surface bonded with the same adhesive indicated that two coats of the inorganic primer, ~300 nm, was sufficient to produce a low level of crack propagation. The CAA-surface treated aluminum exhibited no substantial difference between the organic-primed surface and one coating of the inorganic primer. Thus, it appears that the effective thickness of the inorganic primer is directly related to the degree of oxide porosity and the depth of the porous oxide layer resulting from each of the three surface treatments. As has been reported,[4] these factors decrease in terms of surface treatment PAA > FPL > CAA. The latter treatment produces a thick dense oxide (low porosity) having no oxide protrusions.

D. Effect of Aging on Inorganic Primer The ability of any primer system to be protective and usable over an extended period after application is one of the primary reasons for using a primer since it acts as a protective coating for the activated metal oxide surface. In actual practice, it is not unusual for treated and primed metal surfaces to be placed in storage for up to two years prior to bonding. To determine the effect of aging on inorganic primed PAA-treated 2024 aluminum, wedge crack tests were performed over a period of a year after initial primer application. The specimens were bonded initially, after 5 months and one year exposure to atmospheric conditions using EA-9649 adhesive. In each instance the same low level (~5 mm) of crack extention was obtained with failure within the adhesive bondline. Thus, the inorganic primer appears to have adequate stability under exposure to the laboratory atmosphere.

Table 1. Effect of Alkoxide Conversion Temperature on Crack Propagation

	Conversion Temperature (°C)				
	20	70	120	220	325
Initial crack length (mm)[a]	47.0	48.5	48.0	47.0	46.0
Increase in crack length (mm) after:[b]					
1 hr	1.9	1.2	1.4	1.85	1.5
4 hrs.	0.3	0.5	0.7	1.0	0.75
20 hrs.			1.2	1.24	1.0
34 hrs.	1.5	1.1			
100 hrs.			0.2	1.3	1.0
122 hrs.	1.45	2.6			
Total Δa(mm)	5.15	5.4	3.5	5.4	4.25

[a] Crack length a_0 resulting from wedge insertion.
[b] At 71°C/95% RH, 2024 aluminum adherends, PAA-treated, EA9649 adhesive.

Table 2. Aluminum Lap Shear Strengths of Inorganic and BR-127 Primers With Reliabond 398 Adhesive[a]

Primer	Test Temperature, Lap Shear Stg. MPa (psi)		
	Pre-Test Conditions	RT[d]	177°C[d]
Inorganic[b]	10 min. at test temp.	33.9 (4920)	17.9 (2600)
BR-127[c]	10 min. at test temp.	23.4 (3400)	primer failure
Inorganic[b]	50 hrs. at 177°C	26.6 (3860)	19.7 (2860)
BR-127[c]	50 hrs at 177°C	too weak to test	-----

a. 2024-T3, PAA anodized 20 min. at 10V (UTRC).

b. aluminum alkoxide applied from 1% toluene solution heated 85°C, 45 minutes (UTRC).

c. BR-127 primer spray applied (0.2 mils) at Northrop and cured at 121°C for 30 minutes.

d. each test an average of 4-8 specimens.

Samples of PAA-treated 2024 aluminum specimens primed with the inorganic primer were also shipped from UTRC to Northrop Corporation for bonding and testing. Northrop compared the UTRC inorganic primed specimens with Northrop primed BR-127 organic primer using Reliabond 398 film adhesive.[12] The results of the tensile lap shear tests are listed in Table 2.

As shown, the inorganic primer provided improved room temperature strength compared to the BR-127 primer and gave a high level of strength retention after a 50-hour soak at 177°C. In contrast, the organic-primed samples lost essentially all strength after the heat treatments.

These results reflect not only the stability at room temperature of the inorganic primed surfaces but indicate the level of thermal stability which can be achieved with inorganic primed bonded structures compared to organic primers.

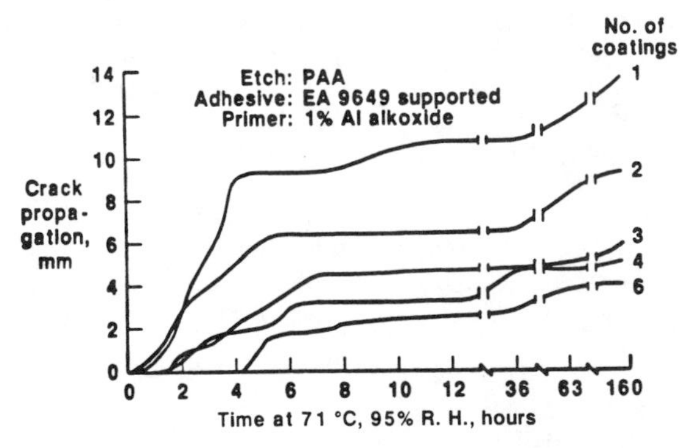

Fig. 3. Effect of primer thickness on crack growth.

Table 3. Metal Alkoxides as Inorganic Primers in PAA-treated Aluminum 2024-T3 Alloy

Alkoxide	Comparative Wedge Crack Results[a]	
	a_o[b], mm	Total Crack Propagation, mm/Hours
sec-Butyl aluminum, E-8385[c]	47.0	3.5/256
Precursor of spinel, E-8385.4[c]	50.0	3.0/172
Precursor of mullite, E-8385.3[c]	49.5	4.6/172
Aluminum, phosphorous, RS-1[c]	47.5	4.3/100
Zirconium n-propoxide	48.0	5.0/217
Tetraisopropyl titanate	47.0	6.0/217
Tetraethyl silicate	49.0	9.8/217
A-1100 Silane[d]	46.5	9.8/217

a. At 71°C/95% RH, EA 9649 adhesive.
b. Crack length resulting from wedge insertion.
c. Supplied by Stauffer Chemical Company.
d. Applied as a 1% solution in 95/5 water/isopropyl alcohol solution.

E. <u>Effect of Mixed Alkoxides on Adhesion</u> To determine if mixed or other metal alkoxides might enhance further the effectiveness of the aluminum alkoxide-based primer in the application of metal alkoxides to substrates other than aluminum, i.e., titanium, steel, ceramics, glass and carbon, etc., for improved environmental resistance in adhesively bonded joints, several alternative alkoxides, listed in Table 3, were evaluated using the same techniques developed for the aluminum alkoxide.

These results showed that the precursor alkoxide compositions based on spinel ($Al_2O_3 \cdot MgO$), mullite ($3Al_2O_3 \cdot 2SiO_2$) and phosphate modified aluminum alkoxide gave the same low cohesive crack extension 3-5 mm when applied as primers to PAA-treated aluminum and bonded with EA-9649 adhesive. The zirconate and titanate alkoxides were nearly as effective; the silicon-based alkoxide and silane providing the least protection.

With FPL-treated 2024 aluminum bonded with Hysol EA 9649, the spinel, mullite and cordierite ($2Al_2O_3 \cdot 2MgO \cdot 5SiO_2$) precursor alkoxide primers were, as indicated in Table 4, somewhat less effective than the aluminum alkoxide.

The results suggest that on FPL- and PAA-treated surfaces the alkoxides with the higher silicon content are less effective as inorganic primers.

F. The Effect of Inorganic Primer on Other Surfaces A preliminary hypothesis in the use of inorganic primers on metal oxide surfaces was that the most effective alkoxide would be the material which matched the adherend surface oxide, i.e., titanium alkoxides would be superior to aluminum alkoxide on titanium adherends. In some instances, this has not proven to be the case. Studies at Virginia Tech[13] have shown that the aluminum alkoxide is markedly more effective as an inorganic primer than a titanium alkoxide for bonding titanium 6-4 wedge specimens which were treated with phosphate fluoride etch. Unlike with aluminum, the surface treatment used to activate the titanium surface prior to alkoxide primer application has a more pronounced effect on the environmental resistance of the resulting bonded structure. For example, titanium surfaces treated with Pasa Jel 107 resulted in much lower crack growth rates (UTRC), 4-7 mm in 176 to 217 hours, when the titanium adherends were primed with either aluminum or titanium alkoxide primers. The rinse time after etching was found to be critical; short rinse times (less than three minutes) were most effective. This difference may be related to differences in surface oxide morphology or oxide type generated by the two treatments and the ability of the inorganic primer to adhere to the treated surfaces.

G. The Effect of Adhesive Type It was anticipated that since the inorganic primer surface probably consists of a high degree of Al-OH units compared to the alumina surface generated by FPL- or PAA-treatments that there may be a variation in response of adhesive systems in terms of wetting and spreading which would ultimately affect the level of adhesion. This appears to be the case in a comparison of EA9649 vs. FM300 using the inorganic primer and BR127 as the organic control with the Pasa Jel treatment a shown in Fig. 1. Preliminary data comparing the two adhesives is listed in Table 5.

As noted, FM300 with BR127 primer gave a lower crack growth with both treatments on 2024 alloy when the primer was cured at 65°C rather than 120°C. This has also been found to be true from work carried out at

Table 4. Metal Alkoxides as Inorganic Primers on FPL-treated Aluminum 2024 Alloy

Alkoxide	Comparative Wedge Crack Results[a]	
	a_o[b], mm	Total Crack Propagation, mm/Hours
sec-Butyl aluminum, E-8385[c]	48.0	6.0/241
Precursor of spinel, E-8385.4[c]	51.0	7.0/188
Precursor of mullite, E-8385.3[c]	48.0	9.6/188
Precursor of cordierite, E-8385.3[c]	49.5	18.6/188

a. At 71°C/95% RH, EA 9649 adhesive.
b. Crack length resulting from wedge insertion.
c. Supplied by Stauffer Chemical Company.

Martin-Marietta.[6] On 6061, the higher cure for BR127, 120°C, gave equal or improved performance. The EA9649/BR127 combination was superior to FM300 on 2024/FPL and equal to FM300 with the remaining combinations. The inorganic primer gave improved results, compared to BR-127, with the EA9649 adhesive with 6061/FPL and 2024/PAA systems and the two primers were equivalent with the other combinations. With possibly one exception, 2024/PAA/ FM300/BR127 cured at 65°C, the inorganic primer system was found to be equivalent or superior to the organic-primed FM300-bonded specimens. Clearly, additional investigation must be carried out to understand the effects of alloy, primer cure temperature, adhesive composition and surface treatment found in this preliminary study.

H. Characterization of the Aluminum Alkoxide Inorganic Primer The exact nature of the formation of the amorphous hydrated alumina primer from an aluminum alkoxide applied to a metal oxide surface is undetermined. The data listed in Table 1 suggests that the primary mechanism for primer formation may be hydrolysis of the aluminum alkoxide under conditions which involve limited amounts of moisture.

Under certain circumstances, i.e., limited amounts of water, aluminum alkoxides can produce a stable amorphous boehmite, AlO(OH).[14] Formation of a stable boehmite type oxide has also been found to occur on aluminum surfaces during exposure to steam in the presence of alkali.[15] Thus, if true, the structure and chemical properties of the resulting amorphous alumina primer will be controlled by the moisture content of the anodized or etched oxide layer to which the alkoxide is applied as well as the availability of atmospheric moisture. Studies are currently underway to test this hypothesis. Auger analysis of aluminum surfaces to which the aluminum alkoxide was applied at room temperature and heated to 50°C gave Al, O, OR ratios which matched the formula $Al_nO_{n-1}(OH)_{(n+2)-x}(OR)_x$; where n = 25, O = 24, OH = 23, OR = 4. After heat treatment at 325°C the ratios changed slightly to n = 29, O = 28, OH = 29, OR = 2. Thus, even after exposure to high temperature there appeared to be small amounts of residual alkoxy groups.

The possibility thus exists for the formation of a direct chemical linkage between the monohydroxy alumina, poly-AlO (OH), and the adhesive. Consideration of possible acid-base effects leading to enhanced adhesion must also be explored. An additional factor which may also be involved is related to stress gradients which could be present when a resin system is applied to a micro rough surface due to polymer component segregation. With the more brittle 177°C curing adhesive (EA9649), it was found that the initial crack length on PAA-treated surfaces was of the order of 60-65 mm for the organic-primed system compared with an average of 47 mm for the inorganic-primed specimens (see Table 1). There was very little difference between the two primers on the FPL-treated specimens with the inorganic-primed samples producing initial crack lengths averaging 3-5 mm less than those in the organic-primed samples. Further analysis of the effect of surface treatment and adhesive type on initial crack length of wedge crack specimens to elucidate the cause of such differences seems warranted.

Although the inorganic primer can be used as a replacement for organic primers, in most cases eliminating the use of chromate materials, tests have shown that when used concurrently, i.e., an organic primer applied over the inorganic primer, very acceptable joint properties are obtained.

Table 5. Wedge Crack Propagation with EA9649 and FM300 Adhesives

Aluminum Alloy	Surface Treatment	Adhesive	Crack Propagation (mm)[a]		Exposure[b] Time, Hrs.	Cure Temp[c] BR127, °C
			BR127	Al Alkoxide		
2024	FPL	EA9649	4.0	4.0	152	65
2024	FPL	FM300	26.0	20.0	152	65
2024	FPL	FM300	50.0	26.0	100	120
2024	PAA	EA9649	8.0	3.0	256	65
2024	PAA	FM300	10.0	15.5	256	65
2024	PAA	FM300	26.0	16.0	152	120
6061	FPL	EA9649	31.0	9.0	80	65
6061	FPL	FM300	45.0	32.0	80	65
6061	FPL	FM300	31.0	30.0	100	120
6061	PAA	EA9649	2.0	2.0	150	65
6061	PAA	FM300	1.0	1.0	150	65
6061	PAA	FM300	3.0	1.5	160	120

a. Values average of four tests; in general ± 1.5 mm spread.
b. Specimens exposed to 71°C, 95% RH.
c. 65°C for 0.5 hr. 120°C for 1 hour.

CONCLUSIONS

The use of an inorganic primer in contrast to organic primer has been shown to mask any differences in adhesive performance normally associated with two aluminum surface treatments, PAA and FPL. This finding may allow the elimination of the use of the chromium-containing organic primers for corrosion protection. The primer generated from an aluminum alkoxide over a range of temperatures appears equally effective independent of conversion temperature, and is hydrolytically stable under long-term exposure at room temperature. The structure of the primer suggests the possibility of direct chemical interaction with an applied adhesive. Preliminary studies have shown that the inorganic primer is effective on metal surfaces with short treatments at room temperature which are adaptable to field repair situations. In addition to aluminum alkoxides, mixed alkoxides containing silica, and magnesia as well as other metal alkoxides appear to be effective for specific situations. Initial results indicate that inorganic primers may also impart environmental resistant properties to adherends other than aluminum. To optimize the performance of such primers the mechanisms of formation and the structure must be further defined and the effect of adhesive composition on bonded joint properties determined.

ACKNOWLEDGEMENTS

The technical assitance of R.N. Girouard and J.P. Pinto of UTRC, the interest and encouragement of Professor J.P. Wightman and his staff at the Center of Adhesion Science of Virginia Polytechnic and State University and the willingness of H. Clearfield, G.D. Davis and J.S. Ahearn of Martin-Marietta Laboratories and G. Beckwith of Northrop Corporation to share their results pertaining to testing of the inorganic primer concept and the supply of metal alkoxides by W. Gentit of Stauffer Chemical Company is gratefully acknowledged.

REFERENCES

1. J.D. Venables, D.K. McNamara, J.M. Chen, T.S. Sun and R.L. Hopping, Appl. Surf. Sci., 3, 88 (1979).
2. W. Brockmann, O.D. Hennemann and H. Kollek, Int. J. Adhes. and Adhes., 2, 33 (1982).
3. R.A. Pike, Int. J. Adhes. and Adhes., 5, 3 (1985).
4. D.A. Hardwick, J.S. Ahearn and J.D. Venables, J. Mater. Sci., 19, 223 (1984).
5. P. Poole and J.F. Watts, Int. J. Adhes. and Adhes., 5, 33 (1985).
6. D.J. Arrowsmith and A.W. Clifford, Int. J. Adhes. and Adhes., 5, 40 (1985).
7. S.R. Brown, 27th National SAMPE Sym., May 4-6, 1982, p.363.
8. R.A. Pike, Int. J. Adhes. and Adhes., 6, 21 (1986).
9. H.S. Swartz, J. Polym. Sci. Appl. Polym. Symp., 32, 65 (1977).
10. H.M. Clearfield, Martin Marietta, private communication.
11. J.P. Wightman, Virginia Tech. Center for Adhesion Science, private communication.
12. G.T. Beckwith, Northrop Corp., private communication.
13. J.A. Filbey and J.P. Wightman, Abstracts 9th Annual Meeting the Adhesion Society, February 9-12, 1986, p.25a.
14. B.E. Yoldas, J. Am. Ceram. Soc., 65, 387 (1982).
15. D.G. Altenpohl, Corrosion, 18, 143 (1962).

THE BONDING AND HYDROLYSIS OF THE Cr (III) FUMARATO COUPLING AGENT ON ALUMINUM SURFACES STUDIED BY AUGER ELECTRON SPECTROSCOPY AND X-RAY PHOTOELECTRON SPECTROSCOPY

Charles Q. Yang[a*], Thomas W. Rusch[b], Stephen P. Clough[b], and William G. Fateley[c]

[a]Department of Chemistry
Marshall University
Huntington
West Virginia 25701-5401

[b]Perkin-Elmer Corporation
Physical Electronics Laboratory
6509 Flying Cloud Drive,
Eden Prairie, Minnesota 55344

[c]Department of Chemistry
Kansas State University
Manhattan, Kansas 66506

ABSTRACT

Cr(III) fumarato coordination compound is a successful coupling agent for aluminum-polyethylene adhesion. When an aluminum foil was pretreated with the Cr(III) fumarato coupling agent and coated with polyethylene, the aluminum-polyethylene composite films demonstrated extraordinary water-resistance and stability in chemically aggressive environments. The formation of chemical bonding between Cr(III) fumarato coordination compound and aluminum oxide was studied by Auger electron spectroscopy (AES) and X-ray photoelectron spectroscopy (XPS). The AES studies also demonstrated that the bonding between Cr(III) fumarato coordination compound and aluminum oxide was stable in water at room temperature, but underwent hydrolysis at elevated temperature.

INTRODUCTION

Reinforced plastics and polymer composites have found many applications in modern industries as material scientists modify their properties by introducing coupling agents to treat the reinforcing inorganic substrates. Generally, a coupling agent acts as a "molecular bridge" between two incompatible materials: a hydrophilic mineral and a hydrophobic organic polymer. Without the use of coupling agents to

*Author to whom correspondence should be sent.

treat inorganic substrates, the bonding between an organic polymer and an inorganic substrate can easily be attacked by water molecules diffusing through the polymer into the interface. At the polymer-mineral interface, water molecules compete with the polymer for mineral surface and cluster into films on that surface. As a result, interfacial adhesion failure and metal corrosion occur. The physical and chemical properties of the polymer composite deteriorate rapidly when the composite is used in wet conditions or under other harsh environments. When an inorganic substrate is pretreated with coupling agents, however, the interfacial bonding is significantly enhanced. Coupling agents function to compete with water molecules for the mineral surface and to prevent water from clustering into films at the interface. With the application of coupling agents, the composite materials usually demonstrate superior water resistance and stability in an aqueous environment.[1]

The aluminum-polyethylene composite film is a very important and widely used packaging material. However, good adhesion between aluminum and polyethylene is difficult to obtain due to the low polarity of polyethylene. Traditionally, improvements in aluminum-polyethylene adhesion were obtained by introducing polar groups into the polymer chains by various means, such as oxidation during high temperature extrusion coating and copolymerization of ethylene with unsaturated carboxylic acids. These chemically modified polyethylenes were not very successful in providing water-resistant adhesion to aluminum, because the polar groups introduced offered sites of attack by water, which reduced bond strength rapidly when the composite film was in contact with moisture or other chemically aggressive materials.[2]

In the early 1970's DuPont developed a new coupling agent, an aqueous solution of Cr(III) fumarato coordination compound (known as Volan 82), which has proven to be a very effective bonding agent for polyethylene and oxide-bearing metals.[3] When an aluminum foil was first treated with Volan 82 and then coated with low density polyethylene by extrusion coating to form a polyethylene-aluminum-polyethylene composite film, the commposite film demonstrated not only high bonding strength and good durability but also extraordinary water-resistance. In our previous research, the composite films still possessed initial bonding strength even after being soaked in sodium hydroxide solution, acetic acid solution, sea water and tap water for three years (Fig. 1).[4] Without the application of Volan 82, however, adhesion failure occurred when the composite films were immersed in water for only a few days. Because of the superior water-resistant bonding between aluminum and polyethylene with the application of Volan 82, the composite film has been widely used as shielding on underground communication cables in the People's Republic of China since the mid-1970s.[5]

Cr(III) fumarato coordination compound in aqueous solution is an inorganic polymer with a complex, pH-dependent structure, in which Cr^{3+} ions are linked by fumaric or hydroxyl bridges (Scheme 1).[6-9] When Cr(III) fumarato coordination compound was precipitated from its aqueous solution, an infrared spectroscopic study of the precipitate demonstrated that free carboxylic groups existed in its molecule.[10] This probably indicates that some fumaric acid forms side groups in a Cr(III) fumarato coordination compound molecule.

Due to the unique performance of coupling agents in the modifications and improvements of polymer composites, studies of the bonding mechanism of coupling agents and the interface chemistry of polymer composites have attracted considerable interests among material scientists.[11]

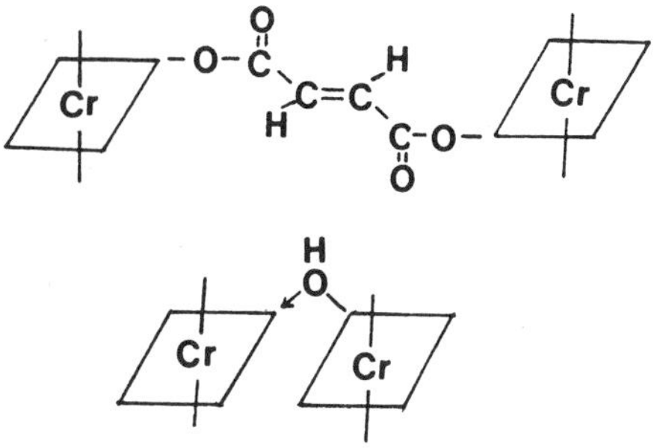

Scheme 1

In our previous research, it was demonstrated that the diffusion of the Cr(III) fumarato coupling agent into the interfacial polyethylene layer and the formation of interfacial crosslinking of the coupling agent under extrusion conditions contribute to the excellent aluminum-polyethylene adhesion of the composite films.[12,13] In this research, AES and XPS have been used to study the nature of the bonding between the Cr(III) fumarato coupling agent and the surface aluminum oxide, and the hydrolysis of the coupling agent on aluminum surfaces.

EXPERIMENTAL

1. A PHI TFA 4000 Auger electron spectrometer in the Physical Electronics Laboratory of Perkin-Elmer Corporation was used to carry out all the AES measurements except the AES line scan, which was made in the Research Laboratory of Nanjing Chemical Industries, Nanjing, People's Republic of China with a PHI 550 scanning Auger microprobe. The beam voltage and current in all the experiments were 5.00 kV and approximately 1 μA, respectively.

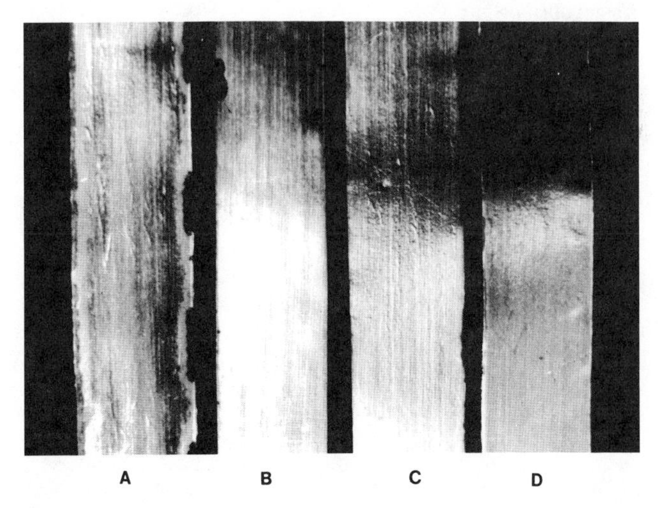

Fig. 1. The aluminum-polyethylene composite films soaked in different mediums for three years: (A) in 0.6% aqueous NaOH solution; (B) in 0.5% aqueous acetic acid solution; (C) in sea water; (D) in tap water.

2. A PHI 5400 X-ray Photoelectron Spectrometer with a 400-watt magnesium X-ray source in the Physical Electronics Laboratory was used to measure the Al2p binding energies (BEs) of the untreated and treated aluminum samples. A PHI 550 X-ray Photoelectron Spectrometer in the Research Laboratory of Nanjing Chemical Industries was used to study the Ag3d BEs on the aluminum surface.

3. The Cr(III) fumarato coupling agent was prepared by the reaction of $Cr(NO_3)_3 \cdot 9H_2O$ and an excess of fumaric acid under reflux. The chromium concentration in the final product was ~0.24% by weight.

RESULTS AND DISCUSSION

An aluminum foil was first washed with a 5% NaOH solution, thoroughly rinsed with water and dried at 100°C for 5 min. A part of the aluminum foil was kept for the AES experiment, while the rest was dipped in a dilute aqueous Cr(III) fumarato coupling agent solution (chromium concentration 0.05%) at 50°C for 2 minutes, thoroughly rinsed with water and dried at 100°C for 5 min. Both the untreated and the treated aluminum foils were measured with AES (Figs. 2 and 3). Chromium was not detected on the untreated aluminum surface by AES, while carbon, calcium and nitrogen peaks appeared in Fig. 2 were possibly due to surface contamination. Iron and silicon detected on this surface (Fig. 2) were probably the trace elements contained in the aluminum foil that have segregated to the surface. In the spectrum of the treated aluminum surface (Fig. 3), the peaks at 527 eV and 572 eV are associated with the chromium LMM Auger electrons and are obviously due to the Cr(III) fumarato coupling agent bonded on the aluminum surface. An intense carbon KLL Auger peak at 269 eV observed in Fig. 3 is due to the fumarato groups in the coupling agent molecules. The retention of the Cr(III) fumarato coupling agent on the treated aluminum surface was also confirmed by a secondary ion mass spectroscopic (SIMS) study in our previous research.[4] Chromium peaks and peaks of the fragments of the fumarato group, such as $CH{=}CHCO^+$ and $C{=}CHCO^+$ were found in the SIMS spectrum of the treated aluminum surface.[4] The retention of the coupling agent molecules on the aluminum surface can be considered as an indication of the possible chemical bonding of the coupling agent to the surface aluminum oxide.

For comparison, an aluminum foil was treated with an aqueous chromium nitrate solution (chromium concentration 0.05%) in the same way as it was treated with the coupling agent, and then measured with AES. No chromium was detected on this surface by AES. It is important to note that only when a Cr^{3+} ion is coordinated by fumaric acid, can it become so chemically reactive as to be successfully bonded to the aluminum surface.

The relative atomic concentrations of aluminum, oxygen, carbon and chromium of the Volan 82-treated aluminum surface measured by AES are presented in Table 1. After Cr(III) fumarato coordination compound was precipitated from an aqueous solution, the elemental analysis of the precipitate showed that the ratio of carbon atoms to chromium atoms was 5.2:1.0.[10] It should be noted that the atomic ratio of carbon to chromium on the treated aluminum surface is 5.1:1.0 as demonstrated in Table 1, which is very close to the same atomic ratio in a Cr(III) fumarato coupling agent molecule.

The possible chemical bonding between the Cr(III) fumarato coupling agent and the surface aluminum oxide was also studied with the AES line scan. An aluminum foil was first treated with an aqueous solution of silver nitrate and nitric acid (0.1 N:1N) in a condition that permitted a partial deposition of silver on the aluminum surface. This aluminum was treated with the Cr(III) fumarato coupling agent by the same method previously described. Both silver and chromium were detected on this surface by AES. When the surface was measured by X-ray photoelectron spectroscopy, the Ag $3d_{3/2}$ and $3d_{5/2}$ BEs were found to 368.0 eV and

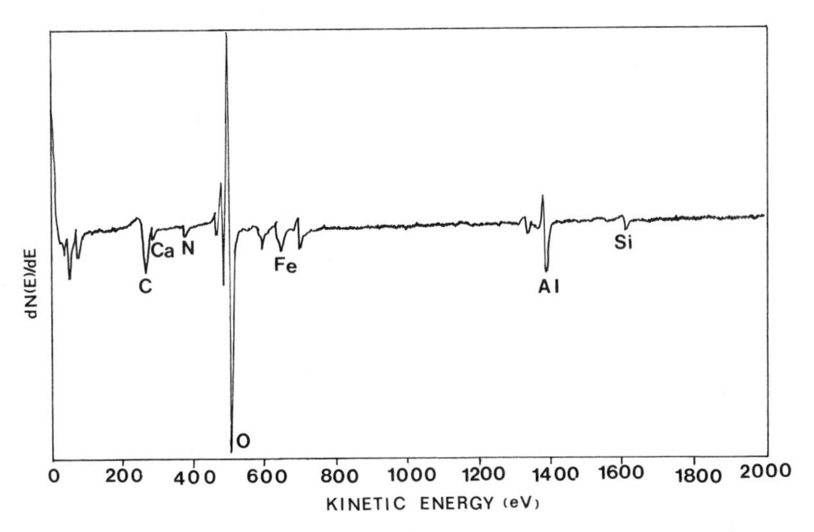

Fig. 2. AES spectrum of an untreated aluminum foil.

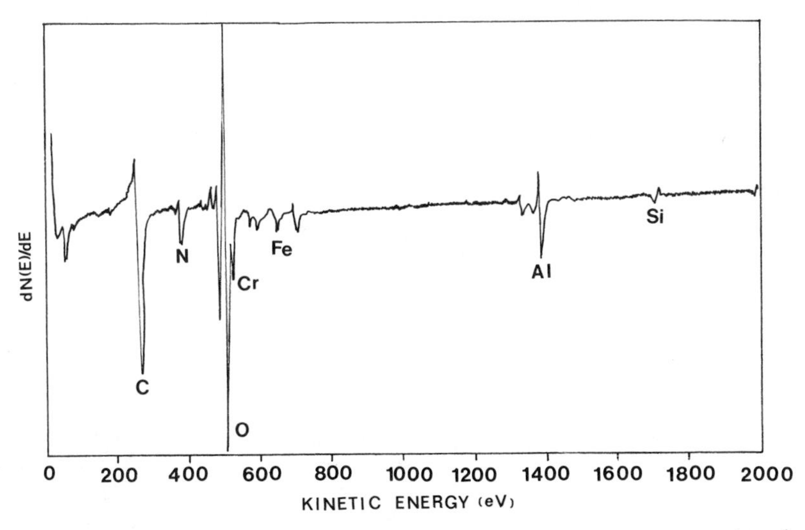

Fig. 3. AES spectrum of the aluminum foil treated with the Cr(III) fumarato coupling agent.

Table 1. Relative Atomic Concentration of Aluminum, Oxygen, Carbon and Chromium (%) Provided by AES.

Volan 82-treated aluminum	Al	O	C	Cr
Before immersion	17	35	40	7.8
After 4 hrs. immersion in water	13	40	40	6.7
After 40 hrs. immersion in water	14	38	42	6.7
After 8 hrs. extraction with hot water	29	70	0.6	0.5*

* The 0.5% Cr on this aluminum surface was due to O KLL Auger peak interference. No Cr signal was detected in the 518-538 eV energy region by AES multiplex measurements.

Fig. 4. AES line scan of the aluminum foil partially deposited by silver and treated with the Cr(III) fumarato coupling agent.

374.0 eV respectively, of which C1s BE (284.6 eV) was used as a standard to calibrate the changing effect. This indicated that silver deposited on this aluminum surface was metallic silver.[14] The signal intensities of aluminum, chromium and oxygen along a line on the aluminum surface was measured (Fig. 4). The AES line scan presented in Fig. 4 shows that aluminum, chromium and oxygen have the same relative atomic concentration distribution along the line, i.e., the chromium atomic concentration increases where the aluminum atomic concentration increases due to a decrease in silver deposition. This result suggests that the Cr(III) fumarato coupling agent is not bonded with metallic silver on the surface, but only with the aluminum oxide. This should be considered as another evidence of the chemical bonding between the coupling agent and the aluminum oxide on the surface.

X-ray photoelectron spectroscopy was also used to investigate the bonding mechanism of the Cr(III) fumarato coupling agent on the aluminum surface. Shown in Fig. 5 are the Al2p photoelectron spectra of an untreated aluminum foil, and the aluminum foils treated with the Cr(III) fumarato coupling agent at two different temperatures (50°C and 35°C, respectively). The BE of metallic aluminum at 72.70 eV was used as a calibration standard.[15] The differences in Al2p BES between aluminum oxide and metallic aluminum for three aluminum foils are summarized in Table 2. On the untreated aluminum surface, Al2p BE of aluminum oxide is 75.85 eV, with BE difference between aluminum oxide and metallic aluminum as 3.15 eV. However, the Al2p BE of aluminum oxide was shifted to 75.30 eV and 75.25 eV when the aluminum foils were treated with the coupling agent at 50°C and 35°C, respectively (Table 2). The experimental results demonstrated above were repeated by using a series of treated and untreated aluminum samples. All the XPS data demonstrated a chemical shift of ~0.6 eV for the Al2p BE of aluminum oxide when the aluminum foils were treated with the coupling agent. The Al2p electron BE shift indicates that an aluminum oxide with a new chemical environment probably existed on the aluminum surface as a result of the treatment of the surface with the coupling agent. This observation can be interpreted as another evidence of the chemical bonding between the surface aluminum oxide and the coupling agent, i.e., the chemical shift (~0.6 eV) of Al2p BE was caused by the chemical bonding between the surface aluminum oxide and the coupling agent.

Since Cr(III) fumarato coordination compound has a complex and pH-dependent structure, it is almost impossible to separate the compound from its aqueous solution without changing its structure. This was the reason why we were unable to compare the Cr2p BE of the pure coupling agent with that on the treated aluminum surface.

Table 2. Al2p Binding Energy Difference.

Aluminum Sample	Al2p binding energy difference between metallic aluminum and aluminum oxide
untreated	3.15 eV
treated with Volan 82 at 50°C	2.60 eV
treated with Volan 82 at 35°C	2.55 eV

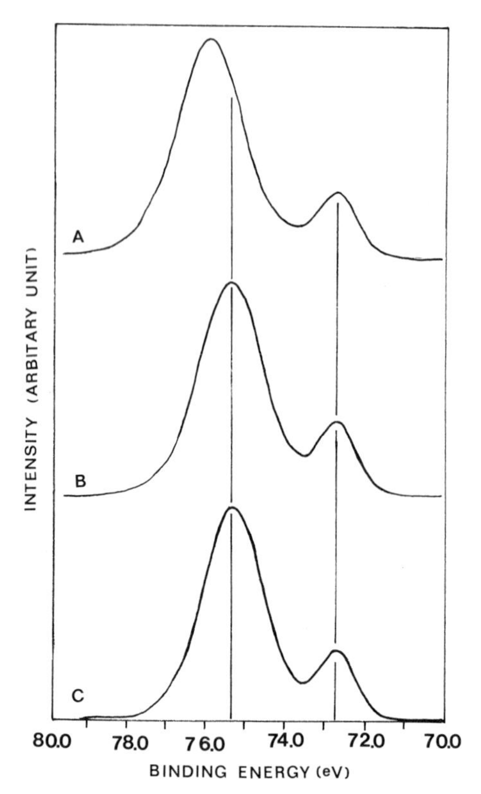

Fig. 5. Al2p photoelectron spectra of (A) the untreated aluminum foil; (B) the aluminum foil treated with the Cr(III) fumarato coupling agent at 50°C; (C) the aluminum foil treated with the Cr(III) fumarato coupling agent at 35°C.

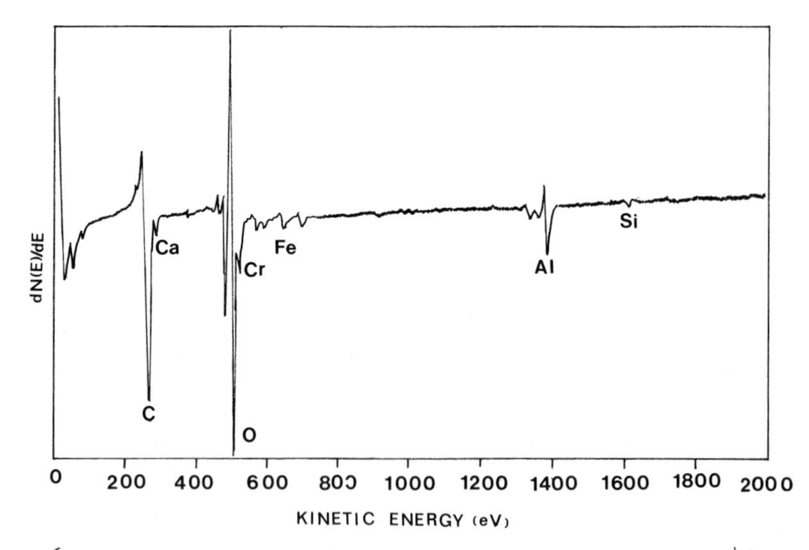

Fig. 6. AES spectrum of the treated aluminum foil after 40 hours water extraction at room temperature.

The hydrolysis of the Cr(III) fumarato coupling agent on the aluminum surface was also studied by AES. The aluminum foil treated with the coupling agent was immersed in water at room temperature, while the water was changed every 30 minutes. After 4-hour immersion, the aluminum foil was dried and measured by AES. Another treated aluminum sample was immersed in water at room temperature for 40 hours while the water was changed every eight hours, then dried and measured by AES (Fig. 6). Calcium and nitrogen detected on these surfaces were probably due to contamination. The relative atomic concentrations of aluminum, oxygen, carbon and chromium measured by AES for both samples are summarized in Table 1. It can be seen that the bonding between the coupling agent and aluminum oxide is relatively stable in water at room temperature, because the atomic concentration of chromium was only slightly decreased during the 40 hours of immersion in water. The chromium atomic concentration decreased from 7.8% to 6.7% during the first 4-hour immersion. This is probably because some coupling agent molecules were physically adsorbed on the aluminum surface. It has been reported that the coupling agent molecules physically adsorbed on a substrate surface can be easily removed from the surface by water extraction at room temperature.[16] After the first 4-hour extraction, a prolonged immersion of the treated aluminum foil in water did not further reduce the chromium atomic concentration on the aluminum surface (Table 1). The coupling agent molecules, which were not removed during the 40-hour water extraction at room temperature, were probably those chemically bonded to the surface. The slight increase in oxygen atomic concentration during this period was probably due to the hydrolysis of surface aluminum oxide. The retention of the coupling agent on the aluminum surface after the 40 hours of immersion is considered to be additional evidence of the chemical bonding between the coupling agent and the surface aluminum oxide.

The treated aluminum foil was extracted in a Soxhlet extractor with hot water for 8 hours, and then examined by AES (Fig. 7). The relative atomic concentrations on this surface are also summarized in Table 1.

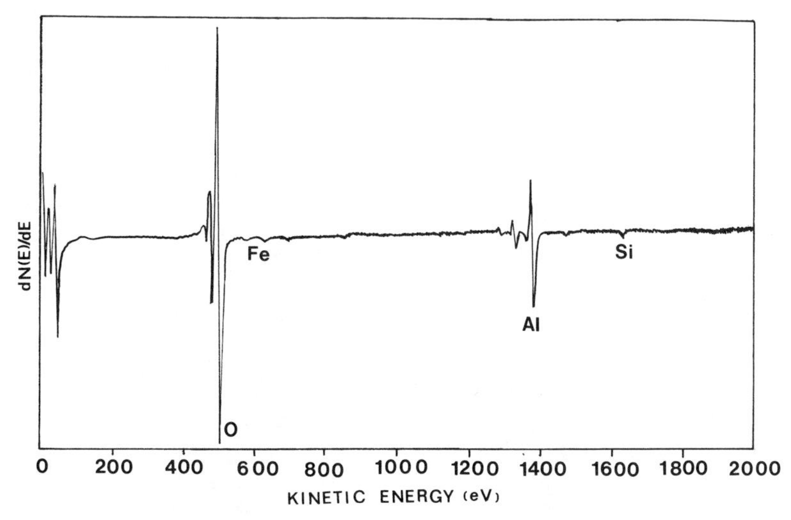

Fig. 7. AES spectrum of the treated aluminum foil after 8 hours Soxhlet extraction with hot water.

The extracted aluminum surface was found to be clean aluminum oxide, since only aluminum, oxygen, and a small amount of silicone and iron, which were trace elements contained in the aluminum, were detected on this surface. The 0.5% chromium atoms detected on this surface (Table 1) was due to oxygen KLL Auger peak interference, because no chromium signal was observed in the 518-538 eV energy region by AES multiplex measurements. The complete removal of the coupling agent from the aluminum surface with hot water extraction demonstrates that the bonding between the Cr(III) fumarato coupling agent and the surface aluminum oxide is a reversible process and is able to undergo hydrolysis under attack of water molecules at elevated temperature. This observation reveals that the chemical bonding between the coupling agent and aluminum oxide is formed possibly through a dehydration process as demonstrated in Scheme 2. The aluminum oxide formed at room temperature on aluminum surface has been proved to be amorphous aluminum oxide[17], which contains a great number of hydroxyl groups. When a clean aluminum surface was treated with the coupling agent solution, the coupling agent molecules can be adsorbed on the surface aluminum oxide by hydrogen bonding with a replacement of water molecules normally adsorbed onto the surface. The same mechanism has also been reported for adsorption of some hydroxyl-containing compound on aluminum surfaces.[18] Covalent bonds are probably formed between the aluminum oxide and the coupling agent molecule through a dehydration process (Scheme 2) as was demonstrated by the data presented above. Since the Cr(III) fumarato coupling agent is an inorganic polymer with multiple bonding sites, multiple bonds can be formed between a coupling agent molecule and the aluminum oxide, which stabilize the bonding of the coupling agent on the surface. Even though hydrolysis of the

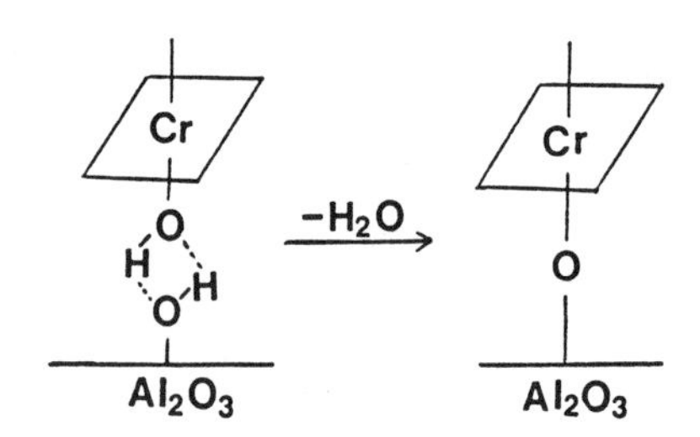

Scheme 2

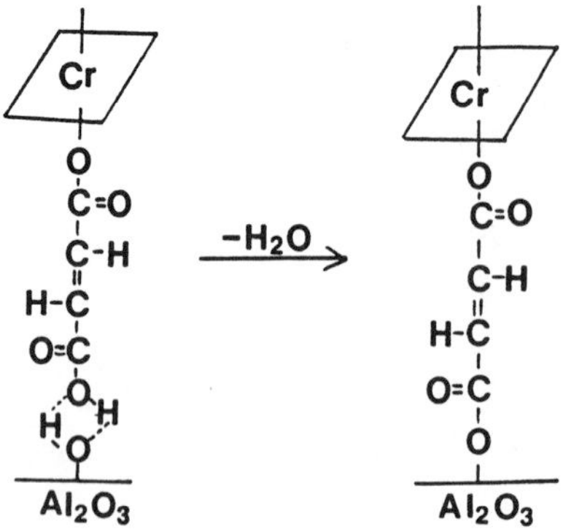

Scheme 3

chemical bonding between the coupling agent and the aluminum oxide may occur in water at room temperature, the formation of multiple bonds renders the coupling agent difficult to be removed by water extraction at room temperature. The hydrolysis process appeared to be accelerated by an elevated temperature. Under such conditions, the hydrolysis product could be removed by the water extraction as was illustrated in Fig. 7. Another possible chemical binding mechanism is demonstrated in Scheme 3. An infrared spectroscopic study of the coupling agent[10] and a study of the atomic ratio (chromium/oxygen) in the coupling agent molecule[10] in our previous research demonstrated that the Cr(III) fumarato coupling agent contains free carboxylic groups as side groups in their molecules. Therefore, a fumarato bridge can be formed between a coupling molecule and the aluminum oxide as is demonstrated in Scheme 3.

Macroscopically, the excellent water resistance of the aluminum-polyethylene bond demonstrated by the composite films also indicates the formation of the chemical bonding between the Cr(III) fumarato coupling agent and the aluminum oxide on the aluminum surface.

It should be noted that since the specific surface area of aluminum foil is very small and the concentration of coupling agent solutions used to treat aluminums is usually 0.2% or less, the amount of the coupling agent bonded to an aluminum surface is exceedingly small. It is for this reason that studying the chemistry of the coupling agent on a metal surface is extremely difficult. In this research, however, AES and XPS appeared to be sensitive techniques suitable for the studies of the surface chemistry of the treated aluminum foils.

CONCLUSIONS

All the AES and XPS data presented in this paper indicate that chemical bonding was probably formed between the Cr(III) fumarato compound and the surface aluminum oxide when aluminum was treated with the coupling agent. The bonding between the coupling agent and the aluminum oxide was stable during the 40-hour immersion in water at room temperature, but underwent hydrolysis during the hot water extraction.

ACKNOWLEDGEMENTS

The authors wish to acknowledge the Physical Electronics Division of Perkin-Elmer Corporation for providing the opportunity for conducting this research in its Eden Prairie Laboratory. This research was also partially supported by U.S. Department of Energy, Grant #De-FG02-85ER13347.

REFERENCES

1. E.P. Plueddemann, in: Interfaces in Polymer Matrix Composites, E.P. Plueddemann, ed., Academic Press, New York, pp.199-201 (1974).
2. J.A. Robertson and J.W. Trobilcock, TAPPI 58:106 (1975).
3. A.J. Deyrup, US 3,787,326 (Assigned to DuPont Company), January 22, 1974.
4. Q.X. Yang and Q.L. Zhou, "The ESCA and AES Studies of the Interfacial Chemical Bonding Between Aluminum and Chromium (III) Fumarato Coordination Compound," Adhesive Chemistry-Developments and Trends, L.H. Lee, ed., Plenum, New York (1984).
5. 753 Adhesive Shielding Research Group, in: "Scientific and Technological Information of Post and Telecommunication, Suppl.,"

Information Institute of Post and Telecommunication, Beijing, China pp.1-17 (1978).

6. H.B. West, US 3,725,448 (Assigned to DuPont Company), April 3, 1973.
7. H.B. West, US 3,843, 699 (Assigned to DuPont Company), October 22, 1974.
8. C.C. Cumbo, US 3,775,455 (Assigned to DuPont Company), November 27, 1973.
9. C.C. Cumbo, US 3,950,506 (Assigned to Dupont Company), April 13, 1976.
10. C.Q. Yang, Master's Degree Thesis, Nanjing University, Nanjing, China (1981).
11. E.P. Plueddemann, Silane Coupling Agents, Plenum, New York, (1982), Chapter 1.
12. C.Q. Yang and Q.L. Zhou, Appl. Surf. Sci. 23:213 (1985).
13. C.Q. Yang, Mater. Chem. Phys. 15:401 (1986).
14. C.D. Wagner, W.M. Riggs, L.E. Davis, J.F. Moulder and G.E. Muilenberg, Handbook of X-ray Photoelectron Spectroscopy, Perkin-Elmer Corporation, Physical Electronics Division, Eden Prairie, Minnesota pp.110-111 (1979).
15. C.D. Wagner, W.M. Riggs, L.E. Davis, J.F. Moulder and G.E. Muilenberg, Handbook of X-ray Photoelectron Spectroscopy, Perkin-Elmer Corporation, Physical Electronics Division, Eden Prairie, Minnesota pp.50 (1979).
16. M.E. Schrader, I. Lerner and F.J. D'Oria, Mod. Plast. 45:195 (1967).
17. F. Keller and J.D. Edwards, "The Behavior of Oxide Film on Aluminum," in: "Proceedings of Intern. Conf. Surf. React.," Pittsburgh, Pennsylvania (USA) (1975).
18. G.D. Davis, J.S. Ahearn, L.J. Matienzo and J.D. Venables, J. Mater. Sci. 20:975 (1985).

INVESTIGATION OF STEEL/EPOXY ADHESION DURABILITY USING POLYMERIC COUPLING AGENTS I SYNTHESIS AND CHARACTERIZATION OF EME COUPLING AGENTS

R.G. Schmidt* and J.P. Bell

*Research and Development
Dow Corning Corporation
Midland, MI 46868

*Dept. of Chemical Engineering &
Institute of Materials Sciences
University of Connecticut
Storrs, Connecticut 06268

ABSTRACT

Ethylene mercaptoester (EME) polymeric coupling agents containing 23-90 wt% mercaptoester units have been synthesized starting from ethylene-vinyl acetate (EVA) random copolymers. The EME copolymers are semicrystalline, exhibit good thermal stability and have been shown to have the ability to interact chemically with both epoxy resins and iron ions through the mercaptoester functional group. Both the initial adhesion strength (EME 90) and the corrosion protection (EME 23) were increased substantially beyond those of the controls when the EME copolymers were applied as coupling agents in steel/epoxy adhesion systems. The wet adhesion durability results were analyzed in terms of the potential number of interfacial bonds formed, hydrophobicity and surface energy of the interfacial region.

INTRODUCTION

All adhesion scientists believe that water is a very destructive environment for metal/polymer adhesion systems. Several reviews have been written which address this problem.[1-4] Diffusion of water into the interfacial region can reduce the strength of metal/polymer adhesion systems by a number of different mechanisms.[4-8] Since water molecules are very strong hydrogen bonding agents, they can readily break non-covalent bonds between the metal and polymer and form new hydrogen bonds with the oxide surface of the metal. The result is a weak water layer which can reduce the strength of the entire adhesion system.[5] Water can also weaken the interfacial region by initiating corrosion and/or hydration reactions with the base metal, oxides or the polymer itself.[6-8] Metal/polymer systems which exhibit strong bonding and are also immune to the above mentioned strength-loss mechanisms have yet to be developed.

In an attempt to improve adhesion durability, a number of researchers[9-15] have employed various low molecular weight coupling agents which presumably can form chemical bonds "across" the metal/polymer interface. These ventures have met some success, but further improvements are desirable.

In our laboratories, ethylene mercaptoester (EME) copolymers with various mercaptoester unit concentrations (23-90 wt%) have been synthesized and employed as coupling agents in steel/epoxy thick film adhesion systems. These polymeric coupling agents promise to be tougher, more hydrophobic and less sensitive to thickness than their low molecular weight counterparts. It is believed that these features will result in a coupling agent region and hence an adhesion system with increased strength and durability.

The EME copolymer coupling agents are also attractive since they can be used as a tool to study the effect of various interfacial properties on adhesion durability. Varying the concentration of functional groups on the EME coupling agents will not only have an effect upon the number of successful bonds formed across the steel/epoxy interface, but will also influence the hydrophobicity and surface energy of the coupling agent region. Any of these factors could ultimately govern the durability of the adhesion region. The ultimate goal of this research is to gain an understanding of the effect of each of these factors on the steel/coupling agent/epoxy interphase. Although a specific steel/epoxy system is being used, it is believed that the information obtained will be applicable to metal/thermoset adhesion systems in general.

The present study deals first with the synthesis of the EME copolymers and the characterization of their physical properties. Special attention is directed toward relating the physical properties to adhesion concerns. In addition, the ability of the EME copolymers to interact chemically with both the steel and epoxy phases has been investigated. Finally, wet environment peel adhesion results from samples treated with various EME coupling agents covering a wide range of functional group concentrations have been used to evaluate which factors actually control adhesion durability and corrosion protection in these systems.

EXPERIMENTAL

Materials High molecular weight (60,000-90,000) ethylene-vinyl acetate (EVA) (Elvax®, DuPont) and ethylene-vinyl alcohol (Evalc) (Eval®, Kuryray, Japan) random copolymers were used as starting materials in the synthesis of EME copolymers. Mercaptoacetic acid and p-toluene sulfonic acid were obtained from Aldrich in their highest reagent grade purities and used following purification by vacuum distillation and recrystallization respectively. Sodium methoxide (Aldrich) was used without further purification. All solvents used in the EME synthesis were distilled and stored over molecular seives (4Å) prior to use. Trimethylolpropane trithioglycolate, 95+% (TTTG) (Evans Chemetics) was used as received.

Ferric chloride hexahydrate (Aldrich), 1-octanol (Fisher), citric acid (Fisher) and ammonium hydroxide (Baker) were obtained in reagent grades and used without further purification.

Peel adherends (1"x4") were cut from 4"x10" 1010 SAE 20-mil thick carbon steel plates (Q Panel) using a squaring sheet metal shear blade. The plates were wiped with a damp cloth and acetone degreased before undergoing the specified pretreatments. Epon 1001® (Shell Development Company) was dissolved (40 wt% solids) in an equal weight solvent mixture of xylenes, cellosolve and MIBK prior to mixing with Versamid 115® (Miller Stephenson) (80 phr) polyamide curing agent. Pressure sensitive polyethylene tape was obtained from 3M.

Methylene iodide, formamide and glycerol were purchased from Aldrich Chemical Company in purities of 99% or above for use in contact angle measurements. Tri-m-tolyl phosphate (97%) was obtained from Pflatz & Bauer, Inc.

EME Synthesis EME copolymers were prepared staring from EVA copolymers in a two-step reaction (Fig. 1). The following scheme is for the preparation of EME 47 from EVA 40. In the first step, EVA was saponified using sodium methoxide in a refluxing methanol/xylenes solution to produce an ethylene-vinyl alcohol (EVAlc) intermediate: To a 500 ml 3-neck flask equipped with a magnetic stirrer, Dean-Stark trap and condenser was added 125 ml xylenes, 125 ml methanol, 1 g sodium methoxide and 10 g EVA copolymer. The mixture was refluxed with slow, steady mixing for 20 hours in order to obtain a completely saponified copolymer. As the reaction proceeded, the mixture was replenished with methanol in amounts equivalent to that collected in the distillate. Following reaction, the mixture was cooled to room temperature and then neutralized with aqueous hydrochloric acid. The mixture was filtered and the white powder intermediate was washed repeatedly to remove neutralization products.

Once dried, the EVAlc copolymer was reacted with mercaptoacetic acid in the presence of p-toluene sulfonic acid catalyst to yield the EME product. The second step of the reaction was driven to the desired product by removing the water byproduct using a Dean-Stark trap: Prior to reaction, 6 ml mercaptoacetic acid was added to 20 ml xylenes and 0.5 g zinc granules in a 100 ml reaction flask. One drop of hydrochloric acid (37%) was added and the mixture was heated under nitrogen to 45°C for ten minutes. Hydrogen gas, which reduces any disulfides that may be present to thiols, is produced at the zinc surface.[16,17] After cooling to room temperature the zinc granules were removed by filtration and the filtrate was introduced into a clean 100 ml reaction flask equipped with a constant nitrogen purge, magnetic stirrer, Dean-Stark trap and condenser. Toluene sulfonic acid (0.3 g) and 2 g EValc copolymer were added to the mixture and heated to 110°C for 1 hour. The hot reaction mixture was quenched dropwise in a swirling, nitrogen

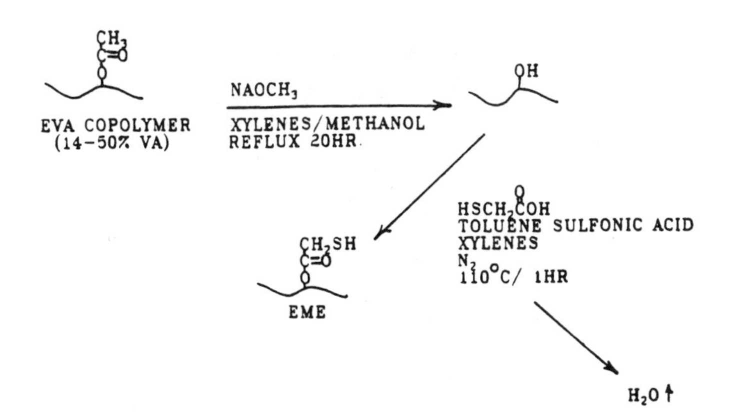

Fig. 1. EME Synthesis route

purged, 50 ml xylenes bath. Under these conditions the polymer remains in solution. The residual reactants were removed by extensive extraction with distilled and deionized water in a continuous extraction apparatus. Gas Chromatography was used to confirm the effectiveness of the water extraction.

Thermal Properties Thermal stability of the EME copolymers was analyzed using a DuPont 950 Thermal Analyzer equipped for thermogravimetric analysis. Samples were heated from room temperature to 800°C at a heating rate of 25°C/min with constant air flow (50 cm^3/min). DSC scans were performed on a Perkin-Elmer DSC-2 with heating and cooling rates of 10°C/min. All DSC samples were annealed for five minutes at 140°C before cooling to the desired starting temperatures to eliminate any prior thermal history effects.

IR Spectroscopy A Nicolet 60SX Fourier Transform Infrared (FTIR) Spectrometer was employed to obtain spectra of polymer samples prepared either by solvent casting on a NaCl plate or through pellet formation with KBr. The signal-to-noise ratio was enhanced by the averaging of 32 scans.

SEM/EDX A 1200B Amray Scanning Electron Microscope equipped with an EDAX 1900 detector and data station was used to analyze polymer samples for the presence of sulfur and iron. Prior to analysis the samples were mounted on aluminum stages and carbon coated in a high vacuum chamber.

Peel Sample Preparation To provide easy handling and insure identical treatments to every sample, the steel plates were placed in glass racks (capacity: 30 samples) prior to the pretreatment procedures. The 1"x4" steel plates were prepared for bonding using acetone degreasing and ammoniated citric acid[18] pretreatments in a nitrogen-purged glove box. Following distilled water rinsing and xylenes bath, the EME coupling agents were applied to the steel plates from solution (0.25 wt% in xylenes/methanol) at 60°C under nitrogen.

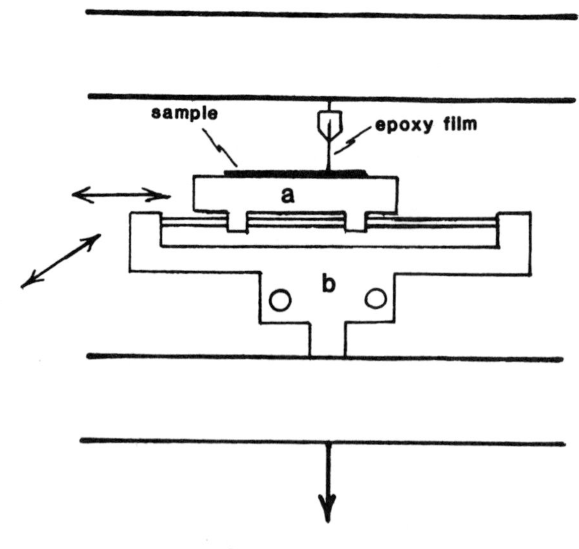

Fig. 2. 90° peel test apparatus.

A 1 minute immersion in xylenes/methanol bath completed the treatment. Ellipsometry and X-ray photoelectron spectroscopy results indicated that the coupling agent thicknesses were approximately 50Å. Epon 1001/Versamid 115 (5-mil dry thickness) films were applied to the pretreated samples one hour after mixing at room temperature using a thin film applicator (Gardner Labs). The films were cured for 7 days in air at room temperature. Post curing for 9 hours at 80°C was found to be necessary to remove the residual solvent and complete the cross-linking reactions.[19] The back and sides of the samples were masked with polyethylene tape prior to 57°C distilled water bath exposures.

90° Peel Test Following specified water exposures, samples were scribed to a width of 0.7 inches with a razor blade and immediately tested for adhesion strength using a 90° peel test apparatus designed in our laboratory (Fig. 2). The 90° peel test fixture attached to the crosshead of an Instron® tensile test machine. The sample stage slid freely on low friction bearings, enabling the peel angle to remain at 90° as the crosshead moves downward. A TM-S Instron tensile tester equipped with either a 500 g, 2000 g or 50 pound load cell and a chart recorder was used to perform the measurements. The peel rate for all tests was 0.4 in./min. The average peel force was taken from the chart recorder output of peel force vs. debonded length. All of the peel test variables (i.e., crosshead speed, peel angle, epoxy thickness, epoxy composition) were chosen so as to minimize as many extraneous contributions to the peel force (such as plastic deformation of the peel strip) as possible. Therefore, one should be careful when comparing other peel test results with the ones reported here.

Surface Energy Measurements An IMASS® contact angle analyzer was used to obtain the advancing and receding contact angles of four test liquids on polymeric surfaces which were dip-coated from solution onto steel plates. The test liquids consisted of tritolyl phosphate, methylene iodide, formamide and glycerol. These liquids were chosen due to their relatively low vapor pressures and their wide range of polar and dispersive components of surface tension. The test liquids and samples were stored, and the measurements were completed, at the equilibrium temperature of the contact angle analyzer (38°C).

From the measured values, the equilibrium contact angle from each of the four test liquids was calculated (assuming a roughness factor of unity) and paired in all possible combinations (i.e., 1-2, 1-3, 1-4, 2-3, 2-4, 3-4).[20] The polar, dispersive and total surface energies of the polymer surfaces were calculated for each pair using the harmonic mean equation by means of a simple computer program.[20] The mean of the resulting values is reported.

RESULTS

EME Synthesis EME copolymers ranging from 23-47 wt% mercaptoester units were prepared, starting from EVA random copolymers with 18-40 wt% vinyl acetate units (Fig. 1). Also, a higher mercaptoester unit copolymer (90 wt%) was prepared starting from an EVAlc copolymer. Fig. 3 shows the IR spectra of the starting material, intermediate, and final product in the synthesis of EME 47 from EVA 40%. As the acetate groups were saponified to hydroxyl groups, the disappearance of carbonyl (1739 cm^{-1}) and ester (1240 cm^{-1}) absorptions of the EVA, accompanied with the large build-up of hydroxyl group absorbance (3400 cm^{-1}) in the EVA1c spectra, indicate essentially complete conversion to the intermediate. When the EVAlc is reacted with mercaptoacetic acid, strong new cabonyl and ester absorptions appear respectively at

1731cm^{-1} and 1160cm^{-1} in the product spectra. The very weak absorption at 2569cm^{-1} is characteristic of thiol groups. On the other hand, the strong absorption centering at 1287cm^{-1} can be assigned to the CH_2-SH wag motion of the mercaptoester group. The essentially complete disappearance of the OH band at 3400cm^{-1} indicates very high conversion of the EVAlc to the EME copolymer.

Thermal Analysis At a heating rate of 25°C/min. the first sign of significant weight loss (>2%, by TGA) for the EME copolymers ranged from 350°C for EME 23 to 230°C for EME 90. It has been shown[21] that this first weight loss can be attributed to the cleavage of mercaptoester groups from the backbone. The EME copolymers are semi-crystalline. The crystalline regions present are apparently due to the ordering of ethylene segments within the copolymers; the percent crystallinity ranges from 28% for EME 23 to approximately 0% for EME 90. For the same EME copolymers (EME 23-90), the melting, glass transition and decomposition temperatures vary from 83-59°C, -110--108°C and 350-230°C respectively.

Chemical Interactions The attractiveness of the EME copolymers is that the mercaptoester side groups have the potential to form chemical bonds with both the steel substrate and the epoxy adhesive or film. The chemical reaction between the mercaptoester functionality of the copolymer and epoxide rings of the epoxy resin was monitored using FTIR spectroscopy as described below. The proposed reaction route is shown in Fig. 4.[22]

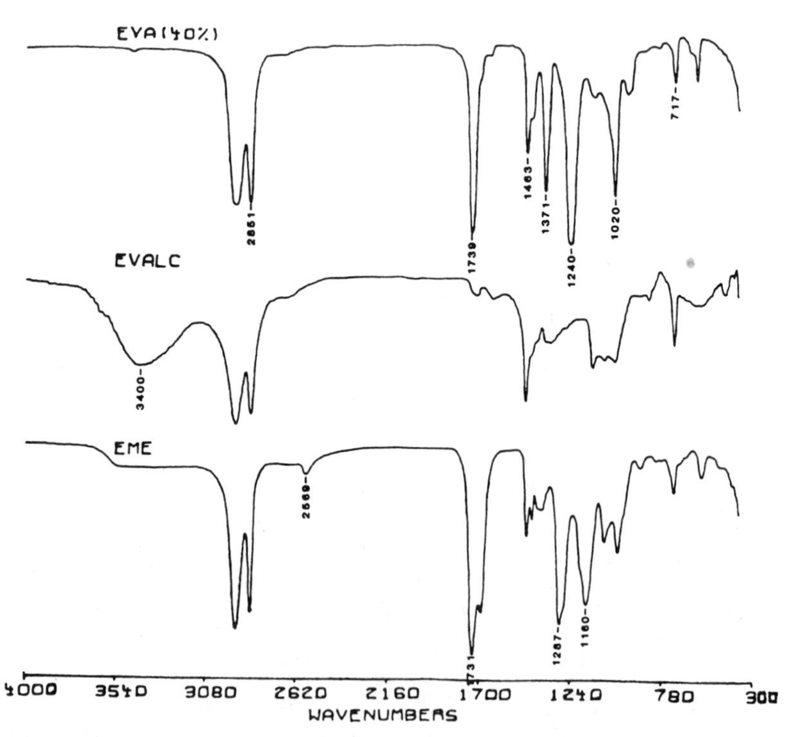

Fig. 3. IR spectra of starting material (EVA),reaction intermediate (EVAlc) and product (EME) obtained starting from EVA wt.%.

As the reaction proceeds, hydroxyl groups are created at the expense of thiol and epoxide ring functionalities. Both the thiol group (S-H stretch) and the epoxide ring appear as relatively weak bands in the IR spectra. Therefore, TTTG (trifunctional low MW mercaptoester compound) and Epon 828 model compounds were used in this study to increase the concentration and hence the absorption of infrared radiation by the reactive groups.

A TTTG/Epon 828 mixture was cast on a NaCl plate and exposed to 80°C for up to 9 hours (identical to post-cure conditions). Figure 5 shows the IR spectra after 0, 2 and 9 hour exposures to 80°C in the (a) hydroxyl, (b) thiol and (c) epoxide ring absorption regions. The steady build-up of hydroxyl group absorption along with the subsequent decline in thiol and epoxide ring absorptions indicates that the reaction proceeds as proposed.

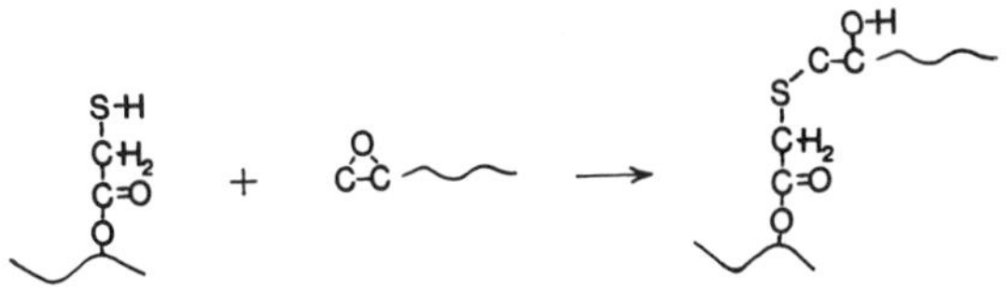

Fig. 4: EME coupling agent/epoxy reaction route

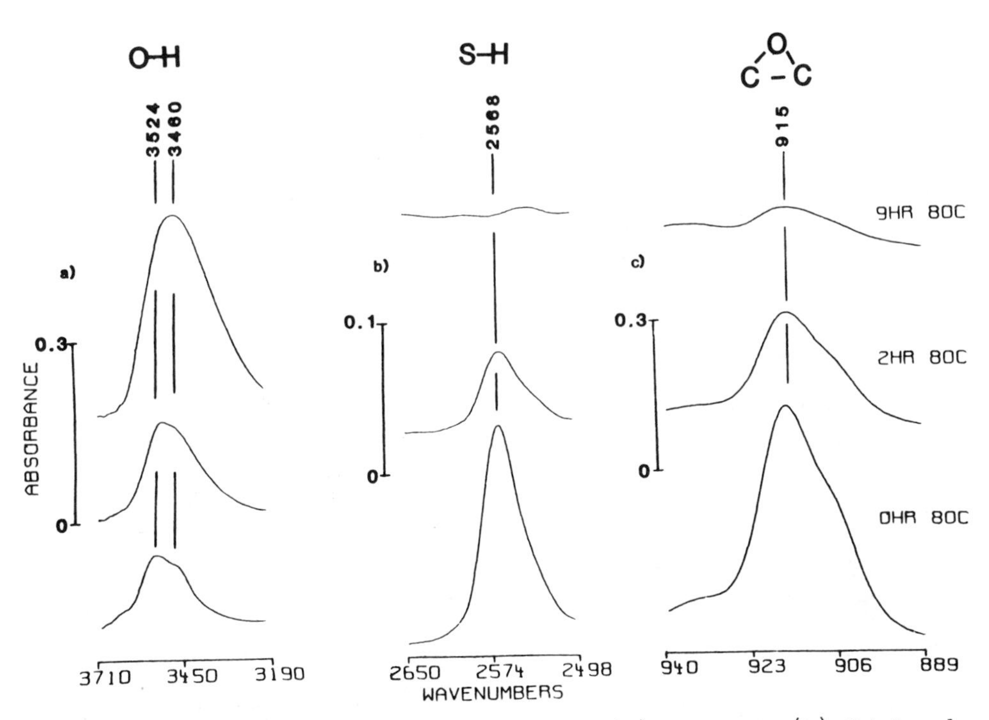

Fig. 5. IR spectra of TTTG/Epon 828 in the (a) hydroxyl, (b) thiol and (c) epoxide ring absorption regions as a function of 80°C exposure time.

A similar study was completed using samples prepared by casting a thin film of Epon 1001 on an EME 47-covered NaCl plate. Although the reactive group absorptions were very weak, the IR spectra indicated that the reaction proceeded as in the model system. The IR spectra of Epon 828 alone was also monitored during 80°C exposure. The spectra showed that Epon 828 had only a very slight tendency to self-polymerize under these conditions. This supports the previous conclusion that the reduction of the epoxide ring absorption primarily occurred due to reaction with the coupling agent thiol group.

The EME copolymers are believed to have the ability to form coordinate bonds with a properly prepared steel surface. Coordinate bonds are the strong localized bonds that are formed between metal ions (in this case iron ions) with incompletely filled d-orbitals and electronegative atoms possessing one or more pairs of unshared electrons in their outermost valence shell.[23,24] The thiol group, through losing its proton, and also the carbonyl oxygen can donate a pair of electrons and form coordination bonds with metal ions.[25] If both of these bonding mechanisms occur, a five-membered chelate ring is formed with free ions on the steel surface. If the iron oxide surface were that of perfect crystal face, all the iron valence and coordinate bonds would be satisfied. However, oxide films on most metals are believed to be disorganized with incompletely coordinated metal ions present especially at protrusions and discontinuities on the surface. It is these vacant coordination sites that can potentially be filled by the electronegative sulfur and oxygen atoms of the EME coupling agents.[24]

To confirm that the EME copolymers do indeed have the ability to form complexes with iron ions, they were exposed to a 1.1×10^{-3}M solution of $FeCl_3 \cdot 6H_2O$ in octanol. When 10 ml of a 1.0 wt% EME 31 in xylenes solution was mixed with 10 ml of the yellow-green ion containing solution a drastic color change was observed. Immediately the solution color turned dark pink and then slowly faded to a clear, colorless solution within one hour after the addition of the copolymer. Apparently some type of chemical interaction took place between the iron ions and the EME copolymers. Following the color changes and extensive washing with ethanol, the precipitated polymer was subjected to an Energy Dispersive X-ray Analysis (EDX).

Figure 6 shows the EDX spectra of bulk EME copolymer before and after iron ion exposure. The significant peak (Fig. 6b) at the characteristic iron binding energy of 6.40 keV indicates that indeed the polymer successfully scavenged iron ions from solution. Similar experiments using a LDPE solution showed neither color changes nor significant iron concentrations in the EDX spectra. The sources of the Al and Si peaks present in these spectra were believed to be the aluminum sample mounts and laboratory contamination respectively.

To date, IR spectra of iron oxide powder treated with the EME coupling agents suggest the successful formation of Fe-S bonds. However, conclusive evidence of bonding via the carbonyl group of the ester and hence the formation of a chelate has yet to be obtained.

Adhesion Durability Table 1 lists the 90° peel adhesion values for wet and redried samples prepared with acetone degreasing (controls) three different EME coupling agent/citric acid and TTTG/citric acid (low MW coupling agent) pretreatments. The samples have been soaked in 57°C water for the times indicated. Also, the average amount of time

Table 1. 90° Peel Strengths (g/in) of Wet and (Redried) Steel/EME/Epoxy Systems Following Specified Exposures to 57°C Water Baths. Peel rate: 0.4in/min.

Treatment	Peel Strength (G/in) Immersion Time (Hrs. 57C Water)						Corrosion Protection Hrs.
	0	1	3	5	11	24	
Control	78	69 (57)	17 (30)	9 (24)	4 (16)	2 (5)	14
EME 23	47	26 (40)	16 (31)	11 (26)	10 (23)	9 (19)	46
EME 47	284	139 (196)	47 (83)	36 (71)	16 (57)	13 (44)	32
EME 90	729	319 (543)	36 (311)	24 (187)	16 (66)	11 (30)	20
TTTG	684	106 (650)	28 (180)	11 (17)	8 (19)	3 (9)	16

Samples dried 1 hour under vacuum at 50°C.

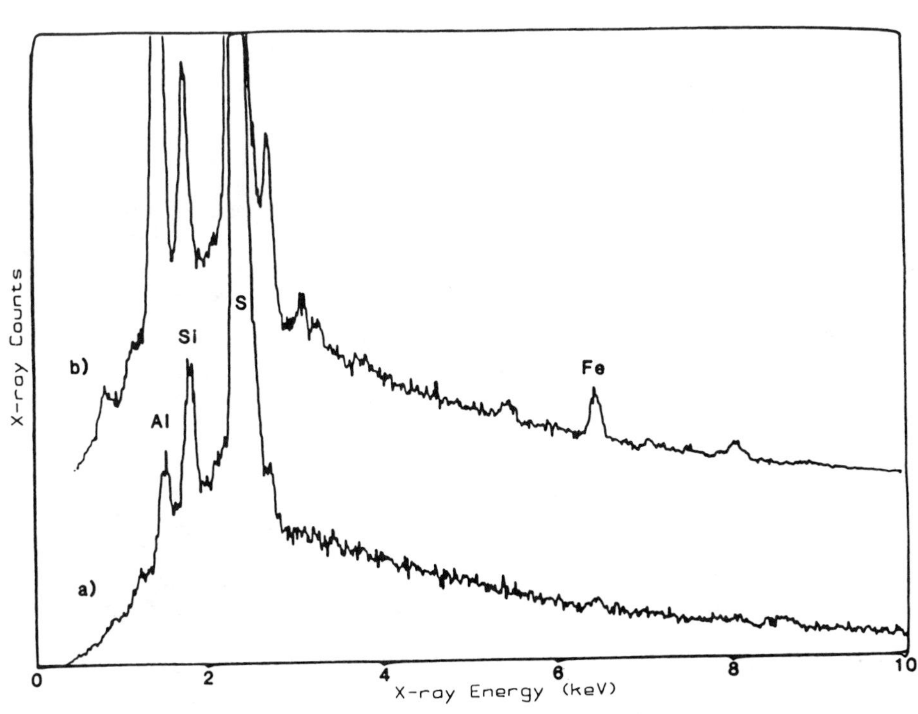

Fig. 6. EDX spectra of EME 31 (a) before and (b) after exposure to $FeCl_3$, $6H_2O$/Octanol solution.

Table 2. Surface Tensions and Fractional Polarities of EME Copolymer and Epoxy Surfaces.

Polymer	γ^{Ps}	γ^{Ss}	X^{Pss}
Epoxy (1001/V115)	9.9	40.8	0.24
EME (23)	3.0	37.2	0.08
EME (47)	8.4	43.1	0.19
EME (90)	14.8	46.2	0.32

s* Surface energy in dynes/cm.
** $X^P = \gamma^P / \gamma^S$

that the coupling agent/opoxy resin systems protect the steel adherends from corrosion under these conditions is reported. The appearance of an average of three or more pits per sample was used as the criterion for the presence of significant corrosion.

The following conclusions may be drawn from these peel data:

1. Presence of the coupling agent/citric acid pretreatments enhanced both the adhesion durability and corrosion protection beyond those of the controls.
2. Both the initial and redried adhesion strengths increased significantly with an increase in coupling agent functionality.
3 The ability of the coupling agents to protect the steel substrates from corrosion decreased with an increase in coupling agent functionality.
4. Within 11-24 hours of exposure to 57°C water the wet adhesion strengths of all the coupling agent samples dropped to essentially equal values.

Possible explanations will be given in the Discussion section.

Surface Energy When an EME copolymer coupling agent is applied to a metal surface before bonding, it greatly reduces the surface energy of the metal substrate. How well an epoxy adhesive wets the substrate will be governed by the surface energetics of the two contacting phases. The surface energy of the EME copolymers have been determined using contact angle measurements and the harmonic-mean equation.[20,26] Table 2 lists the calculated values of the polar components (γ^p) and total (γ^s) surface energies along with the resulting fractional polarities (X^p where $X^p = \gamma^p / \gamma s$) of these polymers.

It is apparent that the concentration of polar reactive groups on the EME coupling agents has a significant effect on the fractional polarity and hence the wettability of the adherend surface. Therefore, along with the potential number of chemical reaction sites and hydrophobicity, the surface energy of the EME coupling agents should also be considered when analyzing the steel/EME/epoxy adhesion systems.

DISCUSSIONS

The polymeric nature of the EME coupling agents should provide significant advantages over the more common low molecular weight coupling agents. Unlike low molecular compounds, polymers are capable of bearing a load which should reduce the sensitivity of the effectiveness of the coupling agents to the thickness of this layer.

In addition, low molecular weight coupling agents are normally quite permeable to water. Therefore, their presence can actually promote the infiltration of destructive water molecules into the interfacial region. Since the EME copolymers consist of a relatively few reactive hydrophilic groups on a hydrophobic hydrocarbon backbone, their use should reduce the water permeability of the interphase.

Unlike the low molecular weight compounds, the viscoelastic properties of the polymeric coupling agents should enable them to help relieve internal stresses in metal/thermoset adhesion systems. This is advantageous since significant stresses can develop during the post-cure cooling of the thermoset, primarily due to the thermal expansion coefficient mismatch between the metal and the polymer.[27-30] The crosslinked structure and the elevated glass temperatures of most thermosets limit their stress relaxation ability. Epoxy resins have thermal expansion coefficients exceeding that of steel by an order of magnitude.[28] Hence, a large degree of stress can develop. This interfacial stress has been cited as the cause of failure in a number of adhesion systems.[27,28,31,32]

The thermal properties of the EME copolymers indicate that they will remain stable under normal epoxy curing and use conditions. The relatively low crystallinity contents and glass temperatures suggest that the EME copolymers should have significant mobility at room temperature to help relieve stresses developed in the interphase due to the steel/epoxy thermal expansion coefficient mismatch.

Table 1 shows that the initial adhesion strength was enhanced by almost an order of magnitude over the controls by employing the EME 90 coupling agent. On the other hand, the more hydrophobic EME 23 coupling agent was the most successful at protecting the steel adherend from corrosion reactions. Corrosion protection was extended over three times as long as that achieved in the control systems.

The peel strength results in Table 1 suggest that maximum initial (dry) adhesion is obtained by incorporating a large concentration of polar reactive moieties along the polymer coupling agent backbone. However, such practice is not without its consequences. The increased hydrophilicity associated with an increase in polar group concentration reduces the ability of the coupling agent/resin system to protect the steel adherend from corrosion. Apparently water molecules can reach the steel surface in sufficient concentration to initiate corrosion reactions more readily when a coupling agent that is less resistant to water permeation is present in the interphase.

Therefore, the results suggest that the properties required of the interfacial region for maximum dry adhesion strength differ from those required for maximum wet adhesion durability and corrosion protection. Consequently, the optimum coupling agent/resin adhesion system for a structural joint will most likely differ from that for a protective coating.

The increase in initial adhesion strength that is observed with an increase in coupling agent functionality is presumably due to an increase in the number of successful bonds formed across the steel/epoxy interface. However, the effect that the mercaptoester unit concentration has on the surface energy of the adherend is believed to also be important, especially for low mercaptoester concentrations.

The interfacial defect and fracture energy adhesion models both predict that in order to maximize adhesion (minimize interfacial defects, maximize work of adhesion) the fractional polarities of the adhering phases should be equal.[33] That is, $X^p{}_1=X^p{}_2$ where $X^p=\gamma^p/\gamma^s$ and $\gamma^s=\gamma^d+\gamma^p$. The values in Table 2 indicate that the optimum wetting condition of the epoxy on the EME coupling agent should occur with an EME composition somewhere between 47 and 90 wt% mercaptoester units. Also, the large difference in fractional polarities between the epoxy and EME 23 suggets that the wetting between these two phases will be significantly less than optimum. Indeed the relatively poor initial peel strength values obtained when the EME 23 coupling agent is employed (Table 1) support this conclusion.

TTTG has been used previously as a coupling agent in steel/epoxy structural joint systems.[18] Good adhesion durability was obtained in these systems for up to 750 hours immersion in 57°C water. However, it is very difficult to compare joint and peel systems since the geometry of the joint system delays the diffusion of water into the interfacial region. The TTTG coupling agent contains a significantly higher concentration of mercaptoester groups by weight than EME 90. As shown in Table 1, peel specimens constructed with TTTG exhibit a greater rate of strength loss and a lesser degree of corrosion protection in the presence of hot water than those prepared using EME 90. Based on the previous discussion this is not surprising since the TTTG is expected to be more water permeable. On the other hand, the initial adhesion strength of the EME 90 samples exceeded that of the TTTG samples. This contradicts the previous observations indicating that a greater concentration of reactive groups leads to greater initial adhesion strength. The ability of the EME 90 to relieve, at least in part, interfacial stresses due to its viscoelastic nature is believed to be the major reason for this contradiction. However, the possibility that the EME 90 coupling agent layer is simply stronger than the TTTG coupling agent region also exists.

CONCLUSIONS

EME coupling agents containing 23-90% mercaptoester units by weight have been synthesized starting from high molecular weight EVA random copolymers. The EME copolymers are semicrystalline and exhibit good thermal stability. In addition, it was shown that the coupling agents have the ability to interact chemically with both an epoxy resin and free iron ions via the mercaptoester functional group.

As coupling agent functionality was increased the initial adhesion strength of the steel/EME/epoxy peel (90 degree) systems increased, approaching a ten-fold improvement over the controls with the EME 90 coupling agent. These results have been attributed to the increase in both the number of interfacial chemical bonds formed and the adherend surface energy with an increase in EME functionality.

All of the coupling agent samples exhibited an increase in corrosion protection beyond that of the controls. The best corrosion protection was obtained with the EME 23/epoxy resin system which protected the steel adherend for an average of 48 hours in 57°C water baths. However, the degree of protection decreased significantly with an increase in the concentration of mercaptoester units. This suggested that the corrosion protection obtained was governed by the hydrophobicity of the interfacial region.

Although the TTTG low molecular weight coupling agent contains a higher concentration of functional groups than EME 90, samples prepared with EME 90 exhibited greater initial adhesion strength. It has been hypothesized that the ability of the viscoelastic polymeric coupling agent to relieve interfacial stresses could be responsible for this discrepancy.

ACKNOWLEDGEMENTS

The Adhesives and Sealants Council is gratefully acknowledged for providing the funds for this research. The authors are also appreciative of E.I. DuPont Company and National Starch and Chemical Corporation for supplying the EVA copolymers and helpful technical information, respectively.

REFERENCES

1. A.J. Kinloch, J. Adhes. 10, 193 (1979).
2. D.M. Brewis, J. Comyn and J.L. Tegg, Int. J. Adhes. and Adhes., 1, 35 (1980).
3. J.D. Venables, J. Mater. Sci., 19, 2431 (1984).
4. R.G. Schmidt and J.P. Bell, in Adv. Polym. Sci., Vol. 75, (K. Dusek, Ed.) Springer Verlag, Berlin, p.33 (1986).
5. R.A. Gledhill and A.J. Kinloch, J. Adhes., 6, 315 (1974).
6. J.D. Venables, D.K. McNamara, J.M. Chen and T.S. Sun. Appl. Surf. Sci., 3, 88 (1979).
7. H. Leidheiser, Jr. and M.W. Kendig, Ind. Eng. Chem. Prod. Res. Dev., 17, 42 (1978).
8. J.W. Holubka, J.E. deVries and R.A. Dickie, Ind. Eng. Chem. Prod. Res. Dev., 23, No. 1, 63 (1984).
9. E.P. Plueddemann, Silane Coupling Agents, Plenum Press, New York (1982).
10. J.M. Park and J.P. Bell, in Adhesion Aspects of Polymeric Coatings, (K.L. Mittal, Ed.) Plenum Press, New York, p.205 (1982).
11. A.J. DeNicola, Jr. and J.P. Bell, in Adhesion Aspects of Polymeric Coatings, (K.L. Mittal, Ed.) Plenum Press, New York p. 443 (1982).
12. Cleveland Society for Coatings Technology, J. Coat. Technol., 51, No. 655, 38 (1979).
13. T.T. Kam and R.K. Hon, J. Coat. Technol., 55, No. 697, 39 (1983).
14. S.J. Monte and G. Sugarman, in Additives for Plastics, Vol. 1 (R.B. Seymour, ed.) Academic Press, p.169 (1978).
15. P.E. Cassidy and B.J. Yager, Rev. Polym. Technol., 1, 1 (1972).
16. M. Sato, T. Hirano and T. Kan, J. Agr. Chem. Soc., (Japan), 15, 783 (1939).
17. M.R.F. Ashworth, The Determination of Sulphur-containing Groups, Vol. 3, Chapt. 10, Academic Press, New York (1977).
18. A.J. DeNicola Jr., Ph.D. Thesis, U. of Connecticut (1981).
19. R.G. Schmidt and J.P. Bell, Report to Adhesive and Sealant Council, October (1985).
20. S. Wu, Polymer Interface and Adhesion, Marcel Dekker Inc., New York, Chapts. 1-5 (1982).
21. T. Bunning, R.G. Schmidt and J.P. Bell, Report to Adhesive and Sealant Council, August (1986).
22. H. Lee and K. Neville, Handbook of Epoxy Resins, McGraw-Hill Company, New York, p.5-38 (1967).
23. F. Basolo and R.C. Johnson, Coordination Chemistry, W.A. Benjamin, Inc., New York (1964).

24. J.C. Bolger and A.S. Michaels, in Interface Conversion for Polymer Coatings, P. Weiss and G.D. Cheever, Eds., Elsevier, New York (1968).
25. F.P. Dwyer and D.P. Moeller, Chelating Agents and Metal Chelates, Academic Press, New York, p.124 (1964).
26. S. Wu, J. Polym. Sci., 73, 590 (1980).
27. S.G. Croll, in Adhesion Aspects of Polymeric Coatings (K.L. Mittal, ed.) Plenum Press, New York, p.107 (1982).
28. M. Shimbo, M. Ochi and K. Arai, J. Coat. Technol., 56, No. 713, 45 (1984).
29. N.J. Delollis and O. Montoya, Appl. Polym. Symp., 19, 417 (1972).
30. A.J. Durelli, V.J. Parks and C.J. del Rio, Acta. Mech., 3, 352 (1967).
31. A. Saarnak, E. Nilsson and L.O. Kornum, J. Oil. Col. Chem. Assoc., 59, (12), 427 (1976).
32. J.L. Prosser, Mod. Paint Coat., 67, (7), 47 (1977).
33. S. Wu, J. Adhes., 5, 39 (1973).

Discussion

On the Paper by J.T. Dickinson

Ed Kresge, (Exxon Chemical Co.): Is there a low critical energy for crack propagation as there is for typical chemical degradation such as O_3 attack?

J.T. Dickinson, (Washington State University): In the cases we have investigated there are thresholds in both the applied load (which translates to a G_c) and the current density, j, necessary to induce both crack growth and also microcracking. This is a form of environmental degradation and certainly has similarities to chemical attack. Our work with excimer laser bombardment of stressed polymers appears also to show similar effects regarding thresholds.

Lyle R. Kallenback (Phillips Petroleum Co.): Filler particles and fibers in polymers changes crack propagation. What happens in your systems?

J.T. Dickinson: We have not as yet examined the consequences of electron beam irradiation of stressed, filled polymers. I have an experiment designed which takes a macroscopic T-peel adhesive test where the crack tip is bombarded. I predict that we can make the adhesive failure shift to cohesive failure due to induced bond scissions in the polymer adhesive. I think it is an excellent idea to pursue the fracture experiments in filled systems, e.g., particulate filled elastomers.

On the Paper by M. Parvin and W.G. Knauss

R.A. Pike (United Technologies Research Center): How did you know that all acetone was out of the PVAc film prior to pressure?

M. Parvin, (California Institute of Technology): We don't know.

R.A. Pike: Were bond thickness the same in both low and high pressure bonds?

M. Parvin: Nearly the same.

H. Bhattacharjee (Allied-Signal, Inc., Morristown, N.J.):

(a) The title refers to composites, however, you only talked about PVAc, why so?

(b) Why was it necessary to use masking tape first and then Teflon? You could have used Teflon all along.

Parvin: PVAc is used as a material model. The experimental procedure and measurement techniques developed may be applied to any composite material where a small volume is under high stresses.

PVAc is applied to the aluminum surface in a fluid form. Teflon cannot prevent the flow of the PVAc solution unless it is tightly adhered to the Al surface. Using masking tape and then replacing it by teflon was found to be more convenient.

On the Paper by C.T. Liu

B.H. Edwards (3M Company, St. Paul, Minn.): Have you considered the case of a soft particle in a glassy matrix e.g., a rubber filled epoxy?

C.T. Liu (Air Force Rocket Propulsion Laboratory): No, I didn't consider the case of a soft particle in a glassy matrix. This is because I am interested in the stress distribution in composite solid propellants, which are filled elastomers containing solid particles. The rigidity of the solid particle in approximately 5 orders of magnitude higher than that of the matrix material.

David Kirby (Whirlpool Corp.): Have you considered superimposing the residual stresses induced from processing onto the calculated stresses caused by mechanical loading?

C.T. Liu: In this study, I didn't consider the effect of residual stress on the stress distribution near the inclusions in the composite system. The typical residual stress in solid propellant grain induced from processing is the stress due to thermal shrinkage when the solid propellant is cooled from the cure temperature. In stress analysis, the residual stress can be considered as the initial stress.

On the Paper by R.A. Pike and F.P. Lamm

E.P. Plueddemann (Dow Corning Corp., Midland, Mich.): Have you tried inorganic primers on polished aluminum surface without pre-etching?

R.A. Pike (United Technologies, Research Center): No, but we have tested the inorganic primer on as-received solvent-cleaned 2024 aluminum which was not etched. The inorganic primer in wedge crack tests showed a substantial improvement in crack growth resistance compared to an organic primed control. However, the crack resistance was inferior to that obtained on an etched aluminum adherend.

G. Surendra (University of Missouri-Rolla, MO):

(a) What kind of interaction are taking place between primer and adhesive?

(b) Any peel strength measurement?

R.A. Pike:

(a) Not known, probably chemical interaction.

(b) Yes, but no data in the presentation of the adhesion result was better.

R. Evans (Case Western Reserve University): Can it be used on glass?

R.A. Pike: Yes

On the Paper by R.G. Schmidt and J.P. Bell

Tinh Nguyen (National Bureau of Standards, Gaithersburg, MD): What is the orientation of the polymeric coupling agents on the steel surface?

R. Schmidt (University of Connecticut): I don't know. I think that would be interesting to examine.

Vinod Gupta (3M Company, St. Paul, Minn.): What is the maximum service temperature of these polymeric coupling agents (i.e., thermal stability)?

R. Schmidt: Thermogravimetric analyses of the EME copolymers in air show that their decomposition temperatures vary from 350°C for EME 23 to 230°C for EME 90. Therefore, they should be stable under most epoxy cure and use conditions.

John Fitzgerald (Eastman Kodak, Rochester, N.Y.): Have you tried mixing the coupling agents in the resin formulation?

R. Schmidt: No, we have only applied the coupling agent directly to the substrate surface from dilute solution. However, I expect that the limited solubility of the EME copolymers in common solvents would make it difficult to incorporate a significant quantity into a high solids resin formulation.

R.A. Pike (United Technologies Technical Center): What criteria were used to identify corrosion?

R. Schmidt: Visual observation of corrosion pit formation.

PART THREE:

ADHESIVES FOR SPACE AND HARSH ENVIRONMENTS

Introductory Remarks

Willard D. Bascom

Department of Materials Science and Engineering
University of Utah
Salt Lake City, Utah 84112

When we hear the phrase "harsh environments" the first thing we are likely to think of is military aircraft operating at high temperatures and humidities that limit the use of polymer adhesives. Certainly, the bulk of the research on high performance polymer adhesives has been directed at these applications. Actually there is a broad spectrum of harsh environments for polymer adhesives includes, seashore and deep ocean operations, the oil patch industries, in vivo materials, strong electric fields and fatigue loads. Less obvious but just as demanding are applications that require long-term durability; 20 years or more under relatively mild environmental conditions.

The polymers that are used in these various applications differ widely in chemical constitution, of course. Also, each has operational limits. In his excellent article in this Session, Dr. Pat Cassidy reviews the state of the science of polymers for aerospace, geothermal and undersea environments. He reminds us that the upper use temperature for any reasonable time is 400-600°C with short time excursions to 700–800°C and that these upper limits have been with us for a long time (since 1960) with only minor improvements. Perhaps as Cassidy suggests, these are in fact the upper limits of organic backbone polymers. Organometallics may push these limits further and two of the Session papers discuss these materials.

Another "harsh environment" into which polymers, especially adhesives, are being pushed is low cost, high production rate applications. The automotive industry cannot afford the extensive and expensive surface cleaning and conditioning that is used in the airframe industries. The car manufacturers want to rapidly bond to oily metal surfaces. Clearly this is a problem in surface chemistry (liquid-liquid displacement) that challenges the organic chemist more severely than the hot-wet problem. The high production rate that these same industries demand can presently be met with the cyanoacrylates and hot melt adhesives but it is not difficult to predict that these materials will be pushed to higher mechanical properties and longer durability. Dr. Holubka addresses this point in his discussion of the testing for adhesive bond durability at the Ford Motor Co.

As the organic chemist struggles for new adhesive polymers he is aware of three factors that complicate his job. The first is that his ultimate customer, the engineer, is a very conservative chap who designs to some property, e.g., upper use temperature, with an often arbitrary safety factor. The operating conditions may be 250°C but the engineer wants 350°C assurance. Perhaps with more field experience with polymer materials these design allowables can be relaxed. Processing is another humbling factor. A newly developed high performance polymer may have superb temperature or moisture resistance but its processability is awful. Enter the formulator who dopes the base polymer, so carefully crafted by the synthetic polymer chemist, with additives for better melt-flow properties. Additives which invariably reduce the very properties that were designed into the polymer molecule. This is a problem that begs research. Polymer formulation is an art form that is seldom mentioned in university polymer courses and the available texts are essentially cookbooks.

Finally, the synthesis of novel polymers is a costly business with yields that are best described as miniscule. Scale up to levels that can be tested for mechanical properties is even more expensive. This problem is certainly not new but will not go away until micro-mechanical test methods are developed. The polymer industry lauds the synthetic polymer chemist but is slow in helping him to bring the fruits of this labor to commercial viability.

Editor's Note: Dr. Sylvie Boileau's paper on "New Fluorinated Polysiloxanes" was published elsewhere. Instead, Dr. I. Yilgör's paper on "Synthesis and Characterization of Atomic Oxygen Resistant Poly(siloxane-imide) Coatings," originally presented to the Symposium on High Performance Polymers for Harsh Environments, is included in this Part.

AN OVERVIEW OF POLYMERS FOR HARSH ENVIRONMENTS;

AEROSPACE, GEOTHERMAL AND UNDERSEA

Patrick E. Cassidy

Department of Chemistry
Southwest Texas State University
San Marcos, Texas 78666

ABSTRACT

Three applications in which polymers have been subjected to extreme stresses are aerospace, geothermal and undersea exposure. All of these environments require extraordinary behaviour of coatings, seals or adhesives. The aerospace uses have received the most attention, and therefore, research efforts in the past twenty-five years have concentrated on the development of polymer backbones that are resistant to high temperatures and unusual chemical (atomic oxygen) environments. Several successful aromatic and inorganic backbones have been discovered, such as polybenzimidazoles and polyphosphazenes. Geothermal energy production also has extreme temperature requirements, but in an environment quite different from aerospace. Here hydrolytic and reductive stresses are seen rather than oxidative, and therefore, less exotic materials are used including rubber blends and polymer-concrete composites. Finally, for undersea electronic applications time is the greatest enemy of a seal meant to provide a barrier to water intrusion. Achieving a fifteen-year lifetime of a rubber-sealed device is quite difficult using economical materials. Here additives in rubber can be critical as are processing and adhesion technology. The most challenging task is to develop valid accelerating aging techniques to estimate useful life.

INTRODUCTION

The past twenty-five years have seen the science of polymers elevate from common to specialized uses. As more advantages become obvious for the application of polymers to high stress situations, more research and development took place toward specialty materials and more well-developed testing techniques. These stresses have encompassed high temperature and oxidation, toward which organic materials have inherent weaknesses, hydrolytic thermo-reduction, severe heavy abrasion and very long-term exposure, for which few materials have been tested or their lifetimes predicted.

Table 1. Applications, Properties and High Stress Factors

	Aerospace	Geothermal	Undersea
Properties Required	Strength-to-weight ratio Adhesion Modulus Ablation Thermal Stability	Hydrolytic and Thermo-reductive Resistance Elasticity	Adhesion Impermeability Marine Growth Resistance Corrosion Prevention
"High Stress" Factor	Temperature Oxidation Stress Water	Temperature Hydrolysis Reduction H_2S Abrasion	Hydrolysis Time

Three applications which will be discussed in this paper and which demonstrate the wide range of stresses to which modern polymers are being subjected are: aerospace devices, geothermal energy recovery and undersea exposure. Although the applications are only three in number, they vary greatly in the destructive forces which act on the polymer. Applications, required properties and high stress (factors) situations are listed in Table 1.

One of the first areas encountered by the specialty polymer chemist was the aerospace use of polymers. This application requires resistance to extreme oxidation effects and temperatures. Thermo-oxidative stability is one of the mot difficult properties to retain while maintaining processability. Still other requirements for this application are strengh-to-weight ratio (several polymers are available which demonstrate high strength and modulus and low density), adhesion and ablation (where a polymeric coating is consumed in a high-energy reaction thereby removing thermal energy from the surface of a spacecraft). All of these, of course, are generally intended for a short-term use so that aging phenomena usually are not a consideration. The types of detrimental or degrading forces experienced by polymers in aerospace are temperature, oxidation, environmental water (which has a long-term effect particularly in composite and adhesive joints) and high mechanical stress situations, which tend to accelerate degradation.

The next application for polymers in extreme service is perhaps the newest of these and that is in the area of recovery and use of geothermal energy. This application exposes the polymer not only to high temperatures, but also to hydrolytic thermo-reduction and abrasion. Oxidative degradation is not considered. Also, this application is generally a short-term one, or at least it can be, in many of the downhole uses. This use differs from aerospace use also in that these materials are commonly elastomeric and used as seals in either static or dynamic applications. The types of detrimental forces imposed on polymers in geothermal applications are high temperatures, hydrolysis, chemical reduction, salt, hydrogen sulfide and abrasion.

The third use, undersea, seems mild; but a fifteen-year anticipated lifetimes undersea makes time as ominous an enemy as temperature to polymeric barriers. In this application, the intensity of the degrading forces is not so extreme; however, those that are present must be withstood for very long periods of time; fifteen years is not uncommon. This application puts a whole new type of stress on the polymer and is one where aging becomes significant and, of course, where accelerated aging techniques can be applied. Properties that are sought for this application are adhesion, low permeability, corrosion prevention and marine growth resistance.

Aerospace Uses Since the genesis of the discovery of thermally-stable or high temperature polymes in the early 1960's, much research in this area has been done.[1] The property of thermal stability was first recognized in the properties of aramids and polybenzimidazoles.

The needs of aerospace applications for such high temperature polymers became the initial thrust of research in this area. Such applications include: ablative coatings, fire-resistant paints and cloth, adhesives and wire coverings. Several other uses, however, have been discovered which were not anticipated, for example: reverse osmosis membranes, high modulus fibers and liquid crystals. New types of problems and materials to solve them can be expected for developing needs such as solar energy recovery.

Many thermally stable, yet intractible, materials were realized by early work on aromatic heterocyclic- and amide-containing backbones. With processability in mind, further synthetic work sought, successfully, tractible materials with good stability. Backbones with heterocyclic, amide, ether, ester, ketone, trifluoromethyl and/or phosphazene units now appear to be the most promising processable materials. Furthermore, inorganic systems have not been as thoroughly investigated as have organic systems.

Since early high-temperature polymers were in the brick dust category of being insoluble and non-moldable, logically, syntheses turned from new backbones to methods of preparing tractible, known compositions. Moreover, several techniques emerged to permit processing: the use of solvent combinations such as N,N-dimethylformamide (DMF) with lithium chloride, staging of backbone cyclization reactions, incorporation of bulky groups and hexafluoroisopropylidene moieties pendant to polymer backbones and incorporation of crosslinking sites within oligomers or as end groups. More recently, a variety of commercially-available monomers (diacids, dianhydrides, tetra- and diamines, bisphenols, etc.) containing either trifluoromethyl groups or the hexafluoroisopropylidene moiety have allowed the synthesis of soluble, thermally stable, film-forming materials.

When one investigates a plot of relative thermal stabilities vs. the year of discovery (Figure 1), an interesting result emerges. The maximum stabilities attained each year have not increased significantly in over more than twenty years. All polymers show a heat resistance of 400-600°C in air and nitrogen with a few excursions to 700 or 800°C. For good reason, then, common opinion is that the maximum stability has been reached for organic backbones.

Thermal stability is often quite difficult to define; it also varies significantly with test environments and the definition is not standard to most published studies. Perhaps the easiest test method is thermogravimetric analysis (TGA); however, it is also the least revealing for determining reductive-type degradation or long-term use. Perhaps a more realistic test method for determining thermal stabilities of polymers is isothermal gravimetric analysis (IGA).

Labortory chemists generating data on new polymers typically report TGA values, where a 10% weight loss temperature constitutes the failure point or arbitrary limit of stability.

Polybenzimidazole (PBI) was first reported by Marvel and Vogel in 1961.[2] This heterocyclic system was the first of many efforts towards developing thermally stable systems. In 1958 and 1959, patents on aliphatic PBI systems established the way to a cascade of papers dealing with aromatic polymers. These were reviewed by Jones[3] in 1968 and in 1969 by both Levine[4] and Korshak.[5] A two-step route via the condensation of a tetra-amine with a dicarboxylic acid (or acid derivative) to give a soluble poly(amic-amine) intermediate in the first step:

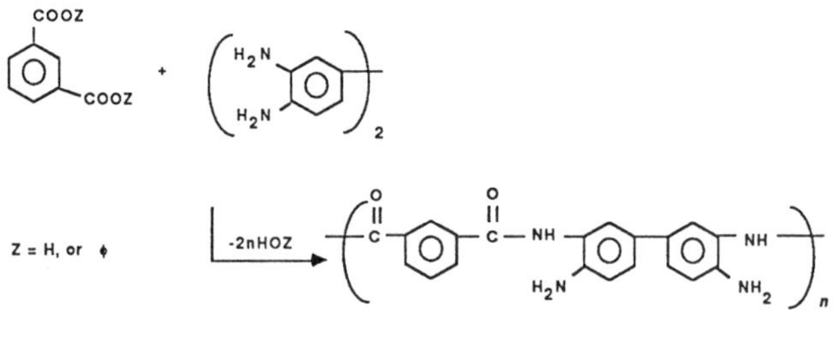

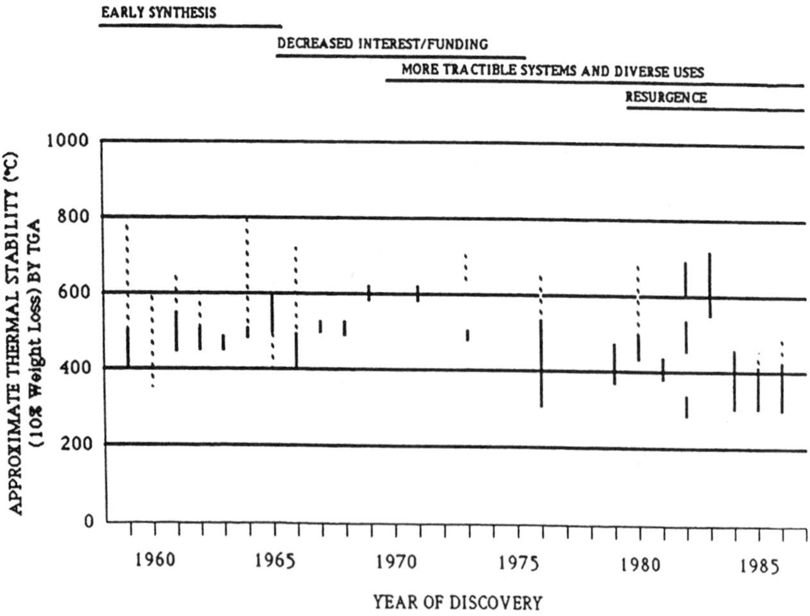

COOZ COOZ + (H_2N, H_2N)$_2$

Z = H, or ϕ

-2nHOZ

poly(amic-amine)

EARLY SYNTHESIS
DECREASED INTEREST/FUNDING
MORE TRACTIBLE SYSTEMS AND DIVERSE USES
RESURGENCE

APPROXIMATE THERMAL STABILITY (°C) (10% Weight Loss) BY TGA
1000
800
600
400
200
0
1960
1965
1970
1975
1980
1985
YEAR OF DISCOVERY

Fig. 1. Thermal stability v year of discovery.

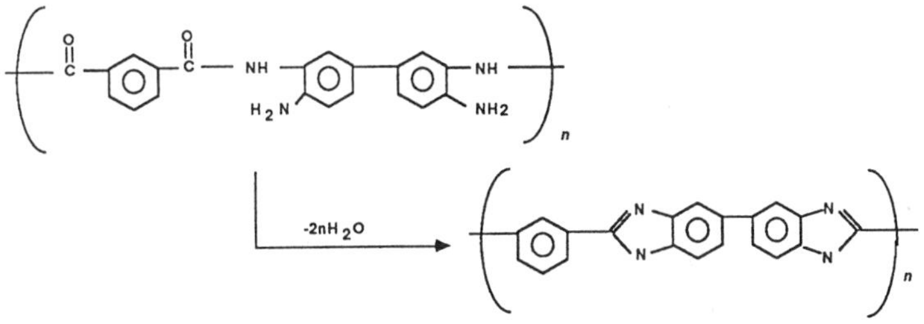

polybenzimidazole

The second step is the complete cyclization to the benzimidazole function.

A better route, suggested by Higgins and Marvel, employs condensation of the tetra-amines with the bis-bisulfite adduct of isophthalaldehyde.[6] Applications of PBI were reviewed by Leal in 1975.[7]

Following the success of the nylons, aromatic analogs were investigated with similar results. These materials were pursued commercially to yield Nomex, Fiber B or Kevlar (du Pont), X-500 (Monsanto) and Exten (Goodyear). Recently, Morgan[8] investigated aramids as liquid crystals and documented the structure/property relationship. Only interfacial and solution polymerizations were used to prepare the aramids. A recent innovation in aramid chemistry involved diphenylphosphite or phosphrous trichloride in the polymerization.[9] These phosphorylation methods appear to be facile routes to products with good mechanical properties.

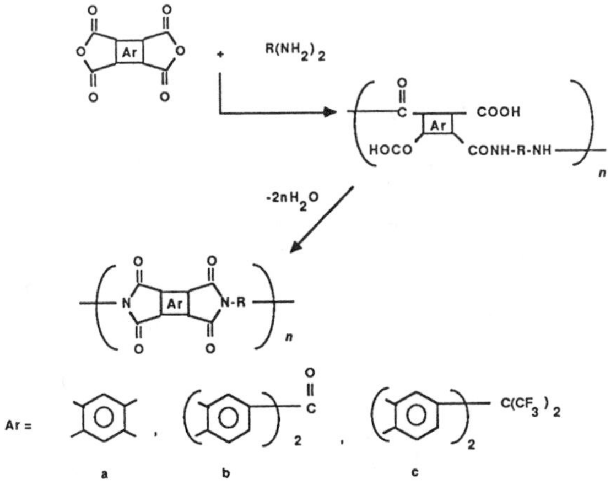

R = a variety of aromatic residues

Polyimides, one of the earliest thermally stable polymers, comprise a widely studied and used class of thermally stable polymers. There are over twenty commercial products which contain the imide function and which are used as coatings, films, fibers molding compounds, composites, adhesives and gas separation membranes. They can exist at 200°C for extremely long periods (nearly 20,000 hours). Polyimides are synthesized in two steps by the condensation of dianhydrides with diamines followed by dehydration. In most cases, because of intractibility of the final product, the intermediate poly(amic-acid) is processed and then cyclized.

However, in the case where Ar=c (a dianhydride containing the hexafluoroisopropylidene moiety), the final fully-imidized material is soluble and can be solution cast into films. Also, materials containing the hexafluoroisopropylidene group show promising results as semipermeable membranes.[10] Other versions contain reactive (unsaturated) end groups which allow processing of the polyimide oligomer.

The development of polybenzoxazoles was similar to that of its nitrogen analog, polybenzimidazole. A large number of these polymers were produced in the early 1960's.[11] The synthesis of polybenzoxazole is realized by the two-step condensation of a bis(o-aminophenol) with a dicarboxylic acid, done in both solution and melt processes. Siloxane-benzoxazole copolymers have also been made with improved solubility and retention of stability.[12] The thio analogue, polybenzothiazole, shows even more promise than the oxo derivative, and is presently being investigated.

In 1970 Economy, Nowak and Cottis[13] revealed an aromatic polyester with outstanding thermal properties (Ekonol, Carborundum Company). Two routes, from p-acetoxybenzoic acid or phenyl p-hydroxybenzoate, yielded the insoluble, high-temperature thermoplastic.[14] It displayed good mechanical properties. Work on Ekonol has continued and is focused on polymer conformation and processing.

Polyquinoxalines have proved to be one of the better high-temperature polymers with regard to both stability and potential application. Interest in these polymers peaked in the late 1960's and centered on their ability to act as metal adhesives. The aromatic backbones are derived from the condensation of a tetra-amine with a bis-glyoxal. They were first synthesized in 1964 by de Gaudemaris and Sillion in France[15] and by Stille and Williamson.[16] In 1967, a soluble phenylated version of this polymer was produced by Hergenrother and Levine.[17] Recently, attempts have been made to improve the final properties of polyquinoxalines; these have met with limited success.

Polypyrrones are representative of ladder polymers. These polymers were first synthesized in 1965 from the solution condensation of tetra-acids, or their derivatives, with tetra-amines, or their salts, via two intermediates which are soluble. A review appeared in 1974, with very little work apparent after that time.[18]

Work by Allcock at Pennsylvania State University[19] and by others at Horizon and Firestone[20] resulted in the development of broad temperature range, fire-resistant elastomers of polyphosphazenes. Several good reviews are available[21], and research on this inorganic rubber is still being done. The polymer is produced by a ring-opening polymerization of the cyclic dichlorotrimer $(Cl_2PN)_3$ in bulk with heat, with or without several catalysts. Several industrial and medical uses of this polymer have been proposed and attempted.[22]

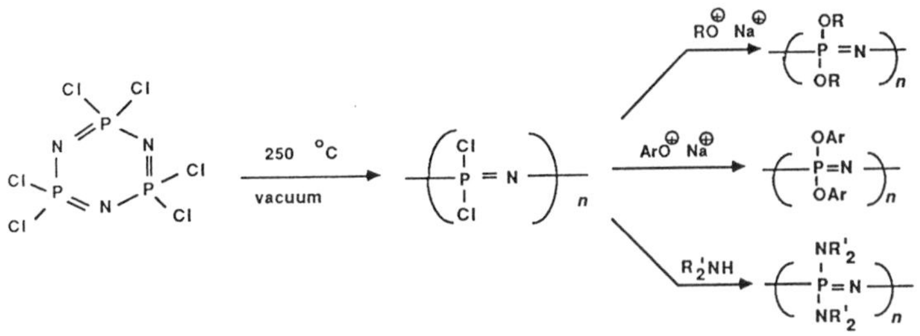

A class of polymers receiving more attention and speculation is organometallic polymers (or metal-containing backbones). New applications being sought for polymers (conductivity, transparency and selective adsorption or permeability) are primarily due to the apparently limited uses and thermal stabilities of known organic backbones. Some of these are polymers containing metal phthalocyanines, phosphinated-bridged metals and poly-Schiff base-metal complexes. Much more work remains to be done in this area. Representative backbones for the polymers listed above are contained in Table 2.

Geothermal Elastomers In the late 1970's and early 1980's the Department of Energy, in response to high oil prices and fears of petroleum resource depletion, intensified its effort in the production of energy from geothermal sources. A key part of this effort was the development and testing of elastomers to provide seals such as packers and O-rings for sealed drill bits.

The recovery of energy from geothermal sources is a process which seems relatively simple and straightforward on initial inspections, but to the polymer scientist poses some very serious problems. Geothermal energy production appears to be much like oil recovery except that the materials are subjected to much higher mechanical, chemical and thermal stresses when one deals with geothermal conditions as compared to petroleum recovery.

The development of polymeric material for geothermal energy has been equated to that which has been seen in the past for aerospace technology. That is to say some special polymers are necessary with properties heretofore not imagined. Furthermore, the quantity of these materials used will be very small; and, therefore, the costs may be very high. In addition, again owing to the low commercial value, development and testing of polymers for this purpose has not been done at the level of effort necessary to alleviate these research shortcomings.

It is very important to note that cost of materials takes on a new appearance when one is dealing with drilling technology. Because of the small amount of polymeric material used, the critical applications in which they are used and the high cost of drilling, the actual costs of materials is truly inconsequential. There are four general areas into which geothermal wells can be divided. The first, and one of the most common, is called hydrothermal convective which can be either vapor-dominated or liquid-dominated. Both of these types have been and

Table 2. Representative Backbones For Thermally Stable Polymers

Name	Structure	Comments
Polybenzimidazole(PBI)	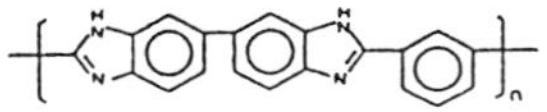	Tm>400 °C, m. w. to 60K, TGA(N_2) 500 °C, sol. in polar aprotics, flame resistant golden fiber
Aramids	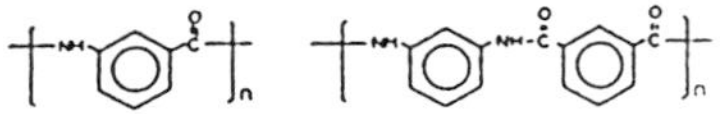	High Tm, TGA(N_2) 450 °C, Use temp ~370 °C (short term) flame resistant fabric
Polyimide	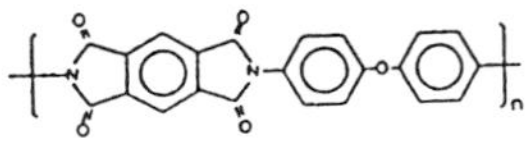	High Tm, TGA(N_2) up to 580 °C insol., sol. when $-CF_3$ appears in backbone, films and foams
Polybenzoxazole	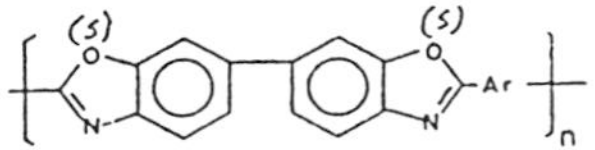	o-intractible, TGA(N_2) 500 °C S-TGA(air, 1% wt. loss) 600 °C, both o and s used as adhesives
Poly(phenyl quinoxaline)PPQ	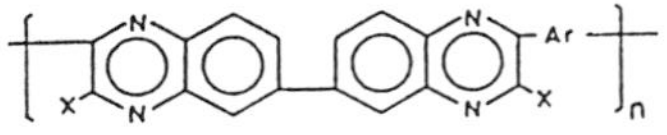	Tg~370 °C, TGA(N_2) 500 °C, m. w. to 300K, SST adhesive, yellow films from m-cresol
Polyester	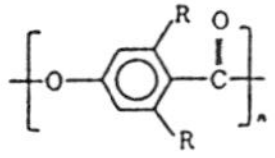	Use temperature 325 °C, fabrication by compression sintering, insoluble high flexural modulus, thermoplastic, machninable
Polypyrrone	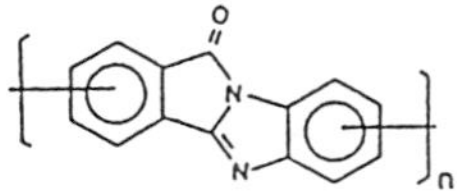	TGA(air) 600 °C, film former, high modulus, sol. in polar aprotics
Polyphosphazene	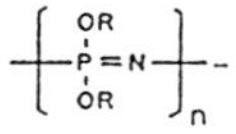	Elastic, soluble, Tg as low as -84 °C, Tm up to 390 °C, TGA(Ar) 410 °C

are now used and, in fact, are the most widely used of any type of geothermal area. The temperatures range up to 390°C and the pressures to 3,000 psi. Salinity can be either quite low (1000 ppm) or astoundingly high (250,000 ppm), although 95% of the geothermal sources have less than 2% salinity. A further complication is the presence of carbon dioxide and hydrogen sulfide. From these data it is obvious why polymeric materials, especially elastomers, have a rather short life expectancy in such environments. One of the more well known commercial elastomers has been said to have an operating temperature of 260°C in air. In fact, 260° is the lower end of the temperature for some of these wells.

The second kind of geothermal resource area is termed geopressured. This type involves highly pressurized water with dissolved solids and gases, primarily methane. They have been found as a result of oil well drilling and have occurred at depths of 8500 to 16,000 feet, at temperatures of 90-105°C and at pressures of 5,000 to 13,000 psi. Generally the pressure increases 1 psi for each foot of depth and the temperatures increase 3-4°C for each 100 feet depth.

The third type of potential energy source for geothermal recovery is hot dry rocks. This type has been explored only recently due to the difficulty of drilling and the short lifetime experienced for the drill bit (12 hours average for air drilling). It is proposed that temperatures can go to as high as 650°C in this type of resource.

The last type of heat source for geothermal energy is magma. Essentially nothing is known about this extremely high temperature energy source (above 650°C) and the development of this source in the far future is merely conjecture at this point. Perhaps the most important factor which has not been considered in depth is that nearly all of the research to date has sought thermally stable polymers for an oxidative medium. However, in the geothermal application the chemical stresses are not oxidative but reductive and hydrolytic. That means that all previous test data for thermal stability of polymers are not applicable to geothermal conditions. Evidence of this well-known thermally stable polymer in air or nitrogen behave very poorly in a simulated geothermal environment.

The reasons for using polymeric materials in place of metals are four-fold: corrosion prevention, erosion resistance, less scaling or mineral deposit and sealing of sliding surfaces. The first three problems occur most often in surface equipment and the last is encountered both downhole and on the surface.

Even though it might appear that some elastomers will exist at some of the lower geothermal temperatures and chemical environments, one must consider what happens to the mechanical properties under these conditions. When elastomers are placed under stress in a geothermal environment they degrade and flow or embrittle quite rapidly and fail to hole a seal, the reason for which elastomers are used.

One use of elastomers is a packer seal which is used to seal between the drill pipe and the wall of the well. Significant mechanical stresses of course are introduced when the packer is compressed in its actual application. Some packers fail by embrittlement and others by viscous flow at the pressures and temperatures realized.

Another use of elastomers is as O-rings or Reed ovals which seal the lubricant in the bearing of the drill bit cutter cone. This particular

O-ring must withstand the wobble of the cone while being subjected to 2500 psi confining pressure and up to 100 psi differential pressure. It further must withstand the mechanical abrasion of the particles which are generated in the drilling process and the chemical degradation forces exerted by both the cooling fluid and the lubricant which it is retaining.

Another well-known area for the use of elastomers is in cable coverings and sealants for well logging equipment. Generally, this equipment does not experience the extremes of temperature or mechanical pressure or long-term use as the previous items, but these elastomers are subjected to a stress which the others do not see, gas and liquid permeation with subsequent debonding. When logging equipment experiences, even for relatively short periods, high pressure, temperature and water exposure, the components of the well can diffuse through a cable jacket, cause debonding and when the cable is removed from the well the gases expand and develop a condition on the cable which has been termed "the bends." Therefore, in this case, both adhesion and resistance to permeability become additional requirements over simple hydrolytic and thermal stability.[23]

Four types of commercial elastomers were tested in various formulations, a nitrile rubber (Buna-N), ethylene-propylene-diene terpolymer (EPDM), fluorinated ethylene-propylene rubber (Viton and AFLAS, a TFE-propylene copolymer), and an epichlorohydrin. In general, the nitrile, EPDM and fluorinated rubbers of various formulations show the most promise. It is important to note that the additives which are used in these applications can make significant differences in the lifetimes of the pieces.

Several commercial materials considered for geothermal applications are given in Table 3.

Another technique which showed promise for use in downhole applications was that of using a high-Tg polymer. In this case, the part would need to be fabricated, heated to use temperature and placed in position on the metal part and then placed in the geothermal well. For this type of application graft and block copolymers of styrene-phenylquinoxaline were investigated. To produce such a copolymer, the synthesis of a polystyrene backbone with PPQ grafts was attempted with limited success.

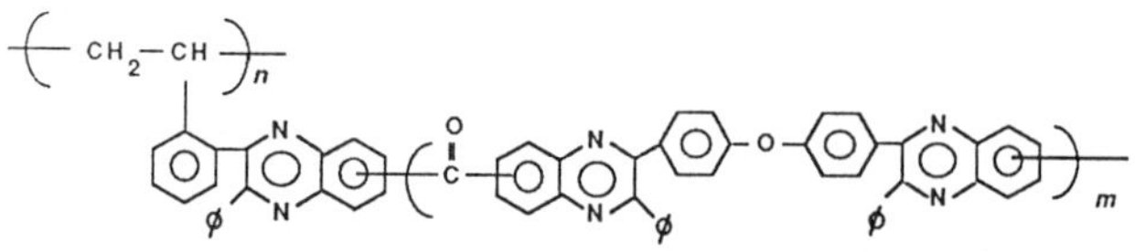

The investigation of elastomers for O-rings required materials with good abrasion resistance, high tear strength, resistance to attack from lubricants, drilling muds and H_2S as well as retention of properties at high temperatures. The role of the O-rings or Reed ovals is to seal out abrasive particles from the bearing surface of the cutter cone and to seal in the petroleum-based lubricant. A Buna-N material performed

Table 3. High Temperature Elastomers

Trade Name	Type	Decomposition Temperature	Cost $/lb.	Comments
Buna N	Nitrile	150 °C	6	Good oil resistance Poor H_2S and steam resistance
Viton E-40-C	Fluoro Elastomer	290 °C	35	Fair steam resistance Poor H_2S resistance
Viton G (peroxide cure)	Fluoro Elastomer	290 °C	40	Improved steam resistance Poor H_2S resistance
Kalrez (fully fluorinate)	Fluoro Elastomer	400 °C	200	Best for H_2S Poor for steam
EPDM	Polyolefin	250 °C	8	Excellent for steam H_2S unknown
Silicone	Siloxane	300 °C	17	Poor steam and H_2S resistance

satisfactorily up to 225°C. When the seal was pre-heated slightly before testing the life span was increased noticeably. This situation would be more realistic than simple cold testing. Ethylene-propylene elastomers gave poor performance at 200°C, and deterioration was accelerated by lubricants. Both Viton (duPont) and Florel (3M) showed use temperatures of 235°C either with or without abrasives. Fluorosilicone rubbers exhibited severe breakdown below 200°C. Similar results were obtained for a fluoroinated polyphosphazene (originally Firestone PNF-200 now Ethyl EYPEL^{TM}.). Although Kalrez showed no apparent degradation at 200°C, compression set was a problem in failure for this material to maintain a seal.

Another concept which is being tested in order to allow marginally stable materials to be used as O-rings is to apply a surface coating of a fluorinated polymer of Parylene. This would accomplish two goals, one is to make the surface a low friction one and the second is to decrease the access of chemiclly damaging species to the elastomer. A 400 Å coating of fluorinated polyethylene has been applied to samples of Kalrez (du Pont), Buna-N and Viton. Some small improvement in chemical resistance was recognized.

Undersea Exposure The last unusual condition to which polymers have been subjected is not one which imposes extremes in temperatures or chemical attack but instead one of the most severe stresses-time. Long-term maintenance of properties such as adhesion and low water permeability are critical. Elastomers are used to isolate undersea electronic devices, such as sonar transducers and hydrophones, which must maintain their integrity for a long period of time, perhaps as much as 15 years. The forces acting on these materials are of course hydrolysis, water leaching, corrosion at the adhesive interface, marine growth and mechanical degradation by sonic agitation.

The materials which are commonly used in systems such as these are W- or G-type Neoprene, butyl rubber and recently, polyurethane. Neoprenes have the advantage of easier processing and butyl is superior in permeation resistance. Urethane liquid resins are being found to provide excellent stability and, when needed, the ability to be used as repair materials in field applications.

The ability of rubber to isolate a system form water is, of course, the key property. Water will permeate commercial Neoprene formulations, even those up to 1/4" thick at an appreciable rate considering a 15-year use period. Rates on the order of 1-2 mg/cm^2/day have been measured for some formulations. The permeation rate and water absorption (leading to swelling, another problem) vary considerably with the components of the rubber. For example, some additives form soluble chlorides which promote water association. A further complication is that permeability is enhanced with aging and sonic excitation. Figure 2 shows the effect of long-term aging on permeability of Neoprene. Although the permeation rate of salt water is constant over nearly a year at 60°C, the rate at which pure water permeates increases significantly.

Permeation is not the only mechanism by which water can enter. Another is failure of the metal-rubber bond. This problem area involves not only elastomer composition, processing, and cure but also primer and adhesive used on the metal. If all factors are not carefully controlled, debonding and corrosion can occur to bring about rapid electronic failure.

The one advantage that polymer and material scientists have in dealing with undersea uses is that acceleration methods can be used to predict lifetimes. Accelerated life testing (ALT) becomes important and can compress 15-year of lifetime use into a few months. Of course the acceleration is accomplished by raising the temperature, to as much as 75°C for Neoprene, as long as the higher temperature is a valid one. For example, at 90-95°C Neoprene degrades by mechanisms which are not present at 20°C so the acceleration is not realistic.

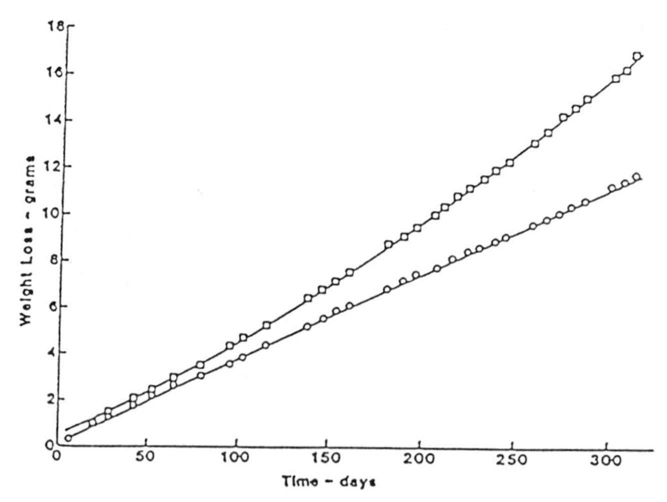

Fig. 2. Permeation of deionized water (□) and 3.5% salt water (•) through neoprene-G. Temperature: 60▪°C.

Great advances are now being made in development of commercial materials for undersea applications and in accelerated test methods to predict long-term life expectancy.

ACKNOWLEDGEMENTS

The author thanks the Robert A. Welch Foundation (Grant No. AI-524) for support during the preparation of this article. Also data were supplied by the Department of Energy, Naval Research Laboratory/Underwater Sound Detachment, Orlando and Texas Research Institute.

REFERENCES

1. (a) P.E. Cassidy, Thermally Stable Polymers: Synthesis and Properties;, Marcel Dekker, Inc., N.Y. (1980).
(b) P.E. Cassidy and N.C. Fawcett, Polyimides in Encyclopedia of Chemical Technology, R.E. Kirk and D.F. Othmer Eds., Vol. 18, 3rd ed., John Wiley and Sons, Inc., New York, pp. 704-718 (1982).
2. H. Vogel and C.S. Marvel, J. Polym. Sci., 50, 511 (1961).
3. J.L. Jones, J. Macromol. Sci., C2, 303 (1968).
4. H.H. Levine, in: Encyclopedia of Polymer Science and Technology, 11, p. 188 (1969).
5. V.V. Korshak, Heat Resistance Polymers, Moscow, p.248 (1969) in: Israli translation (1971).
6. J. Higgins and C.S. Marvel, J. Polym. Sci., Polym. Chem. Ed.., 8, 171 (1970).
7. J.R. Leal, Mod. Plas., p.60, August (1975).
8. P.W. Morgan, J. Polym. Sci., Polym. Symp., 65, 1 (1978).
9. N. Ogata and M. Harada, J. Polym. Sci., Polym. Lett. Ed., 15, 551 (1977).
10. G.R. Husk, P.E. Cassidy and K.L. Burns, Macromolecules, in press.
11. Y. Iwakura, K. Uno and Y. Imai, Makromol. Chem., 77, 33 (1964).
12. L.W. Breed and J.C. Wiley, J. Polym. Sci., Polym. Chem. Ed., 14, 183 (1976).
13. J. Economy, B.E. Nowak and S.G. Cottis, SAMPE J., 6, 6 (1970).
14. J. Economy, R.S. Storm, V.I. Matkovich, S.G. Corris and B.E. Nowak, J. Polym. Sci., Polym. Chem. Ed., 17, 2207 (1979).
15. G.P. deGaudemaris and B.J. Sillion, J. Polym. Sci., B2, 203 (1964).
16. J.K. Stille and J.R. Williamson, J. Polym. Sci., A2, 3867 (1964); ibid. B2, 209 (1964).
17. P.M. Hergenrother and H.H. Levine, J. Polym. Sci., A-1, d5, 1453 (1967).
18. B. Nartisissov, J. Macromol. Sci. Rev. Macromol. Chem., C11;(1), 143 (1974).
19. R.J. Ritchie, P.J. Harris and H.R. Allcock, Macromolecules, 12, 1014 (1979).
20. (a) K.A. Reynard, A.H. Gerber and S.H. Rose, Synthesis of Phosphonitrilic Elastomers and Plastics for Marine Applications, Horizons, Cleveland, Ohio. Naval Ship Engineering Center, AMMRC CTR 72-29, (AD 755188) (1972).
(b) K.A. Reynard, R.W. Sicka, J.E. Thompson and S.H Rose, Poly(acryloxyphosphazene) Foams and Wire Coverings, Horizons, Cleveland, Ohio, Naval Ship Engineering Center Contract No., N00024-73-5474, (1975).
21. (a) H.R. Allcock, Chem. Rev., 72, 315 (1972).
(b) H.R. Allcock, Science, 193, 1214 (1976).
22. (a) R.E. Goldsberry, M.J. Adamson and R.F. Reinisch, J. Polym. Sci., Polym. Chem. Ed., 11, 2401 (1973).

(b) J.K Valaitis and G.S. Kyker, J. Appl. Polym. Sci., **23**, 765 (1979).

23. P.E. Cassidy and G.D. Rolls, Polymeric Materials in Geothermal REcovery, Department of Energy, Material Sciences. Workshop on Polymers, Case Western Reserve University, Cleveland, Ohio, (1978).

ULTRAVIOLET CURABLE SILOXANE POLYMERS

Y. Okamoto, D. Crossan, K. Ferrigno and S. Nakos

Loctite Corporation
705 North Mountain Road
Newington, CT 06111

ABSTRACT

It has been demonstrated that UV curable silicone polymers can be prepared by the condensation reaction of an acryloxyalkylchlorodimethylsilane and silanol-terminated polydimethylsiloxanes. These acrylate-siloxanes are readily cured by UV irradiation in a few minutes compared to the overnight cure of RTV silicones. It is also demonstrated that the physical, thermal, and electrical properties of the acrylate-silicones are comparable to those of RTV silicones, and a wide variety of physical properties can be readily obtained by adjusting the polymer molecular weight, types and levels of silica reinforcing fillers.

INTRODUCTION

Siloxane polymers are widely used in coatings, sealants and adhesives, because of their excellent physical, thermal, and electrical properties. One drawback of conventional RTV siloxane polymers is their slow curing. Recently, UV curable siloxane polymers have attracted much interest, and research efforts to develop such polymers are increasing due to a combination of their excellent siloxane polymer properties with fast UV cure response.

It has been reported that UV curable siloxane polymers are produced by (1) UV radical polymerization of acrylate-siloxanes, (2) hydrosilation reaction of vinylsiloxanes and hydrogensiloxanes, (3) polyaddition reaction of vinylsiloxanes and mercapto-siloxanes, and (4) UV cationic polymerization of epoxy siloxanes.[1-3] It seems that acrylate-siloxanes have advantages of better shelf stability and less side reactions over other types of UV curable siloxanes.

Various methods have been reported to develop such UV curable acrylate-terminated siloxane polymers, for example:[4-9]

1. reaction of glycidylalkyl acrylates with amine-terminated siloxanes.

2. reaction of propargyl acrylates with hydrogensilane-terminated siloxanes.
3. reaction of isocyanatoethyl acrylate with silanol-terminated siloxanes.
4. reaction of hydroxyalkylacrylates with chlorosilane-terminated siloxanes.
5. ring-opening polymerization of octamethylcyclotetrasiloxane in the presence of diacrylatedisiloxane.
6. anionic ring-opening polymerization of octamethylcyclotetrasiloxane terminated with acryloxyalkylchlorodimethylsilanes.
7. reaction of hydrogensilane-terminated siloxanes with allyl chloride, followed by reaction with acrylic acid.

The method we have examined is the incorporation of acrylate groups by the condensation of acryloxyalkylchlorodimethylsilanes and silanol-terminated polydimethylsiloxane (PDMS). This paper will discuss the polymer preparation, characterization, and evaluation of properties of several UV curable telechelic acrylate and methacrylate-siloxane polymers.

EXPERIMENTAL

Typical procedures for preparing telechelic acrylate and methacrylate-siloxanes are as follows:

Preparation of Acryloxypropenylchlorodimethylsilane: Into a 12 L flask fitted with a condenser, a mechanical stirrer, a funnel and a thermometer were charged 2200 g (20.0 mole) of propargyl acrylate, 1.52 g of phenothiazine, and 50 g of 2% chloroplatinic acid in butyl acetate. The reaction mixture was heated to 65-70°C and 1990 g (21.0 mole) of chlorodimethylsilane was added dropwise to the reaction mixture while maintaining the temperature of 65-70°C. After the addition was completed, the reaction mixture was stirred at 65°C until all Si-H absorption in the IR spectrum (2200 cm^{-1}) disappeared. The mixture was then distilled at 82°C/13.3 Pa to give 3.24 kg of the product, which was stabilized by adding 200 ppm of hydroquinone monomethyl ether.

Preparation of Acrylated-Siloxanes: Into a 12 L flask fitted with a mechanical stirrer, a funnel and a thermometer was charged silanol-terminated polydimethylsiloxane, 9325 g (0.77 meq. of SiOH) and 72 g of pyridine. To this reaction mixture, acryloxypropenylchlorodimethylsilane, 176 g (0.82 mole) was added dropwise at 65°C. The reaction mixture was stirred at 65° for an additional 3 hours and then small amounts of methanol were added to destroy the unreacted acryloxypropenylchlorodimethylsilane. The precipitated pyridinium salt was removed by filtration to obtain colorless, transparent acrylate-silicone, PDMS 28A, which contains less than 1 ppm of chloride ion.

Formulation and Curing: Acrylate or methacrylate-siloxanes were formulated with 5-35% of fumed silica, Wacker HDK 2000, and 2% of diethoxyacetophenone (DEAP) photoinitiator in a Myers mixer with a steam heated jacket for 2 hours under vacuum. Then the formulated (meth)acrylate-siloxanes were cured in a mold, 15 cm x 15 cm x 0.15 cm, under UV irradiation (365nm), 70 mW/cm^2, for 1 minute and 2 minutes per side for acrylate and methacrylate-siloxanes, respectively.

Flammability Testing: Fiberglass cloth, 12.5 cm x 15 cm was coated by dipping it in 30% PDMS 28A in a methylene chloride solution. After removing the solvent in air, the PDMS 28A coated on fiberglass cloth was cured by UV irradiation. Flammability was evaluated by measuring the burning distance and time on a vertically positioned 15 cm x 2.5 cm fiber cloth strip after igniting by propane torch for 5 seconds.

RESULTS AND DISCUSSION

Polymer Preparation

Acrylate-siloxanes were prepared by the following sequence of reactions:

$$CH_2{=}CH\text{-}COO\text{-}CH_2\text{-}C\equiv CH \quad + \quad H\text{-}Si(CH_3)_2Cl$$

$$\downarrow$$

$$CH_2{=}CH\text{-}COO\text{-}CH_2\text{-}\overset{\overset{CH_2}{\|}}{C}\text{-}Si(CH_3)_2Cl \quad + \quad CH_2{=}CH\text{-}COO\text{-}CH_2\text{-}CH{=}CH\text{-}Si(CH_3)_2Cl$$

$$\underline{1} \qquad\qquad\qquad\qquad \underline{2}$$

Hydrosilation of propargyl acrylate with chlorodimethylsilane produced the acrylate capping agent which contained approximately 80% internal, 1, and 20% terminal addition products, 2.

The acrylate-capping agents were characterized by GC/MS, IR, and ^1H, ^{13}C and ^{29}Si NMR, and assignments of ^{13}C and ^{29}Si NMR are as follows:

```
      131
  CH3 CH2         O                     CH3                    O
   |  ||  67      || 128                 |     143             || 128
Cl-Si- C-CH2-O-C-CH = CH2,          Cl-Si-CH = CH-CH2-O-C-CH = CH2
   |  140        165     129             |  131      66    165     129
  CH3                                   CH3
  1.8                                   1.9
  18.9 (Si)                             18.1 (Si)
```

Then acrylate-siloxanes were prepared by the condensation of 1 and 2 with PDMS silanol in the presence of base.

$$\underline{1} \;\&\; \underline{2} \quad + \quad H\text{-}[O\text{-}Si(CH_3)_2]_n\text{-}OH$$

$$\downarrow$$

$$CH_2{=}CH\text{-}COO\text{-}C_3H_4\text{-}Si(CH_3)_2\text{-}[O\text{-}Si(CH_3)_2]_n\text{-}O\text{-}Si(CH_3)_2\text{-}C_3H_4\text{-}OOC\text{-}HC{=}CH_2$$

The methacrylate-siloxanes were also prepared using allyl methacrylate, instead of propargyl acrylate, by procedures identical to those mentioned above.

Acrylate-siloxanes were also characterized by ^{29}Si NMR. The assignment of spectra is as follows:

-19 to -24 ppm	polydimethylsiloxane backbone.
-3.1 ppm	terminal silicon in PDMS 28A and diacryloxypropenyltetramethyldisiloxane.
7.8 ppm	terminal silicon in PDMS 28M and dimethacryloxypropyltetramethyldisiloxane.

Table 1: Acrylate and Methacrylate-Siloxanes

Polymer	Molecular Weight	Viscosity (mPa s)
PDMS 12A & 12M	12,000	800
PDMS 28A & 28M	28,000	3,500
PDMS 50A & 50M	50,000	18,000

A&M: acrylate and methacrylate-siloxane respectively.

In the case of presence of unreacted silanol, the ^{29}Si NMR should reveal the silanol peak at -11 ppm.

Table 1 shows the approximate polymer molecular weights and their viscosities.

Physical Properties

Physical properties of acrylate-siloxanes formulated with 35% silica reinforcing filler, Wacker HDK 2000, were summarized in Table 2.

The physical properties of methacrylate-siloxane were comparable to those of acrylate-siloxane polymers. When the siloxane polymer molecular weight increased, tensile strength and elongation increased and tensile modulus and durometer decreased, which is an indication of siloxane polymer becoming more elastomeric.

The physical properties of acrylate-siloxane polymers also varied according to the level of silica reinforcing fillers as shown in Table 3.

Table 2: Acrylate-Siloxane Polymer Physical Properties

(Formulated with Wacker HDK 2000)

Physical Properties	PDMS 12 A	PDMS 12 M	PDMS 28 A	PDMS 28 M	PDMS 50 A
Tensile Strength (MPa)	4.49	6.87	6.77	8.28	6.51
Elongation (%)	220	313	569	684	777
Tensile Modulus					
@ 100% Elong. (MPa)	2.01	1.93	0.94	0.79	0.37
@ 200% Elong. (MPa)	4.12	4.40	1.77	1.66	0.7
Tear Strength (kN/m)	13.6	24.5	22.1	42.0	25.6
Durometer, Shore A	53	51	38	40	29

Table 3: Effect of Filler on the PDMS 28A Properties

(Filler: Wacker HDK 2000)

Physical Properties	Filler Levels, Wt %			
	0	15	25	35
Tensile Strength (MPa)	0.33	1.41	4.04	6.77
Elongation (%)	142	255	437	569
Tensile Modulus @ 100% Elong. (MPa) @ 200% Elong. (MPa)	0.27 --	0.48 0.99	0.70 1.53	0.95 1.77
Tear Strength (kN/m)	0.88	3.68	12.25	22.05
Durometer, Shore A	14	30	36	38

As expected, all physical properties were improved with the formulation of silica reinforcing filler.

The data in Tables 2 and 3 indicate that a wide variety of polymer physical properties can be readily obtained by adjusting polymer molecular weight and/or level of silica reinforcing filler.

Physical Properties at High and Low Temperatures

It is well known that siloxane polymers possess excellent physical properties at low temperatures. The acrylate-siloxane polymer, as expected, also showed excellent physical properties at -60°C, and all properties were almost identical to those at room temperature. As temperature increased to 150°C, the acrylate-siloxane started gradually losing its properties, especially on elongation, as shown in Table 4.

Compression Set

Compression set is one of the important properties for applications such as gasketing. All PDMS (meth)acrylate-siloxane polymers showed excellent compression set at room temperature. At elevated temperatures such as 150°C, the compression set of acrylate-siloxane polymers was poor, probably due to the presence of residual C=C unsaturation in the capping agent. On the other hand, the compression set of methacrylate-siloxane polymers which contained no C=C unsaturation was far superior to that of acrylate-siloxane polymers, and the compression set of 20-40% at 150°C was readily obtained with the proper formulations.

Curing Properties

Cure speeds of (meth)acrylate-siloxanes were examined employing a specimen with 0.15 cm thickness under 70 mW/cm^2 UV irradiation and the results are shown in Tables 5 and 6.

As expected, cure speed of PDMS 28A was slightly faster than that of PDMS 28M. PDMS 28A was cured and all properties reached a plateau at 45 sec/side. On the other hand, although PDMS 28M was cured for 45 sec/side, the tensile strength increased with further curing (Table 7).

Table 4: Acrylate-Siloxane Physical Properties at Various Temperatures

(PDMS 28A with 35% Wacker HDK 2000)

Properties	Temperature (°C)		
	-60°	25°	150°
Tensile Strength (MPa)	7.69	6.76	6.36
Elongation (%)	572	569	429
Tensile Modulus @ 100% Elong. (MPa) @ 200% Elong. (MPa)	 0.94 2.09	 0.94 1.76	 1.03 2.41
Tear Strength (kN/m)	30.4	22.1	20.0

Table 5: Acrylate-Siloxane Compression Set

(PDMS 28A with 25% Wacker HDK 2000)
Time = 70 hr.

Temp (°C)	Compression Set (%)
RT	0
38	0
43	5
49	13
54	16
65	21
150	100
150*	<40

*Proprietary Formulation

Table 6: PDMS 28A Cure Properties

(With 35% Wacker HDK 2000, 2% DEAP)

Properties	Cure Time (Sec/Side)					
	15	30	45	60	90	120
Tensile Strength (MPa)	N	5.40	7.10	6.76	6.72	---
Elongation (%)	O	640	566	569	519	---
Modulus @ 100% Elong. (MPa) @ 200% Elong. (MPa)	 C U	 0.58 1.42	 0.88 1.97	 0.94 1.76	 0.96 2.14	 --- ---
Tear Strength (kN/m)	R	29.2	25.9	22.1	2.36	---
Durometer Shore A	E	30	41	38	41	---

Table 7: PDMS 28M Cure Properties

(With 35% Wacker HDK 2000, 2% DEAP)

Properties	Cure Time (Sec/Side)					
	15	30	45	60	90	120
Tensile Strength (MPa)	N	1.02	6.48	6.74	8.12	8.28
Elongation (%)	O	272	591	633	605	684
Modulus @ 100% Elong. (MPa) @ 200% Elong. (MPa)	 C U	 0.47 0.81	 0.25 1.82	 0.27 1.62	 0.28 1.99	 0.28 1.66
Tear Strength (kN/m)	R	3.1	39.0	38.7	40.8	42.0
Durometer Shore A	E	23	37	39	40	40

Thus, we recommend a curing time of 60 sec/side for acrylate-siloxanes, and 120 sec/side for methacrylate-siloxanes.

Next, two types of photoinitiators, DEAP and 1-(4-dodecylphenyl)-2-hydroxy-2-methylpropane-1-one, Darocur^R 953, in various concentrations were examined in PDMS 28A with 25% HDK 2000.

As shown in Tables 8 and 9, 1.5% DEAP (Table 8) and Darocur^R 953 (Table 9) seem to be optimum concentrations. The slight deterioration in properties when the concentration of Darocur^R 953 is above 1.5% is probably due to plasticization caused by the long alkyl chain fragment of the photolysis products.

Electrical Properties

Electrical properties of PDMS 28A with 25% Wacker HDK 2000 were measured and the results are as listed in Table 10.

The dielectric constants are below 3.0, (i.e., low electrical loss) which are excellent for meeting modern electronic packaging requirements including printed wiring board applications.[10]

Thermal Properties

Thermal stabilities of PDMS (meth)acrylate-siloxane polymers were investigated employing thermogravimetric analysis (TGA). As shown in Table 11, thermal stability of acrylate-siloxane polymers was almost identical to that of methacrylate-siloxane polymer, and was comparable to or slightly superior to that of conventional RTV siloxane polymer. Acrylate-siloxane polymers started losing their weight gradually at approximately 250°C, and severe weight loss took place when temperature reached approximately 400°C. As expected, acrylate-siloxane polymers with silica reinforcing fillers showed higher char residue than unfilled acrylate-siloxane polymers.

Flammability

Besides thermal stability, the flammability of acrylate-siloxane polymers was also examined by burning the acrylate-siloxane coated on fiber glass cloth. PDMS 28A was completely burned in 25 sec., but PDMS 28A containing 10ppm of platinum self-extinguished in 8 sec., showing good fire resistance. These preliminary results strongly indicate the possibility of developing flame-retardant acrylate-siloxane polymers by proper formulation and compounding.

Table 8: Effect of DEAP Concentration

(PDMS 28A with 25% Wacker HDK 2000)

Properties	Concentration, %				
	0.5	1.0	1.5	2.0	2.5
Tensile Strength (MPa)	3.93	2.70	4.12	3.84	3.90
Elongation (%)	467	378	474	465	461
Modulus @ 100% Elong. (MPa) @ 200% Elong. (MPa)	 0.54 1.27	 0.48 1.09	 0.55 1.27	 0.61 1.33	 0.54 1.22
Tear Strength (kN/m)	23.1	13.5	26.4	25.5	23.6
Durometer Shore A	29	30	33	30	31

Table 9: Effect of Darocur[R]953 Concentration

(PDMS 28A with 25% Wacker HDK 2000)

Properties	Concentration, %				
	0.5	1.0	1.5	2.0	2.5
Tensile Strength (MPa)	2.86	2.96	3.68	2.89	3.42
Elongation (%)	399	367	427	380	422
Modulus @ 100% Elong. (MPa) @ 200% Elong. (MPa)	 0.48 1.17	 0.56 1.34	 0.66 1.46	 0.51 1.19	 0.50 1.19
Tear Strength (kN/m)	8.9	15.4	20.1	16.8	13.3
Durometer Shore A	27	33	33	---	33

Table 10: PDMS 28A Electrical Properties

(With 25% Wacker HDK 2000)

Dielectric strength, volts/mil	516
Dissipation factor, 100 Hz 1000 Hz 1 M Hz	.00040 .00103 .00309
Dielectric const., 100 Hz 1000 Hz 1 M Hz	2.73 2.72 2.71
Volume Resist., Ohm-cm	2.0×10^{13}
Surface Resist., Ohm/sq.	$> 2 \times 10^{17}$

Table 11: TGA of Acrylate-Siloxane Polymers

(With 35% Acker HDK 2000)

Temp(°C)	Weight Loss (%)		
	PDMS 28A	PDMS 28M	RTV
50	0	0	0
100	0.1	0.3	0
150	0.6	0.6	0.2
200	1.2	1.0	0.7
250	1.8	1.4	1.7
300	2.4	2.0	2.9
350	3.3	2.9	4.3
400	5.8	5.2	7.9
450	8.2	7.8	11.6
500	13.7	14.8	23.5
550	34.8	36.7	51.0
600	46.9	47.0	56.2

CONCLUSION

UV curable siloxane polymers were prepared by the condensation of polydimethylsiloxane-silanol with (meth)acryloxyalkylchlorodimethylsilane capping agents. Acrylate- and methacrylate-siloxanes were readily formulated with silica reinforcing filler and then cured completely in a few minutes under UV irradiation.

A wide variety of physical properties was readily obtained by adjusting polymer molecular weights, types and/or levels of the silica reinforcing fillers. Overall (meth)acrylate-siloxane polymes are comparable to conventional RTV siloxane polymers with much faster cure response.

ACKNOWLEDGEMENTS

The authors would like to thank Loctite Corporation for the permission to publish this paper, R. Trottier and R. Bernard for GC/MS, TGA, and NMR analyses, and Louis Baccei for his helpful suggestions and encouragement.

REFERENCES

1. A. Kurita, N. Hayashi, and N. Kida, Jpn. Kokai Tokyo Koko, JP(. 60-190456 (1985).
2. J.N. Clark, C.L. Lee, and M.A. Lutz, Polym. Mat. Sci. Eng., 52, 442 (1985).
3. M. Hatanaka and A. Kurita, U.S. Pat. 4,451,634 (1984).
4. Y. Sata and H. Inomata, U.S. Pat. 4,293,397 (1981).
5. S. Lin and S. Nakos, U.S. Pat. 4,503,208 (1985).
6. T.R. Williams and M.D. Nave, Polym. Mat. Sci. Eng., 52, 279 (1985).
7. R.C. Chromecek, W.G. Deichert, J.J. Falcetta, and M.F. VanBuren, U.S. Pat. 4,276,402 (1981).
8. Y. Kayakami, Y. Miki, and Y. Yamashita, Jpn Kokai Tokyo Koho JP84-78236 (1984).
9. R.P. Eckberg, U.S. Pat. 4,348,454 (1982).
10. C.A. Garper, Electronic Packaging & Production,52, Nov. 1985.

EVALUATION OF ADHESIVELY BONDED COMPOSITE/METAL BONDS IN SIMULATED AUTOMOTIVE SERVICE ENVIRONMENTS

J.W. Holubka and W. Chun

Ford Motor Company
P.O. Box 2053
Dearborn, MI 48121

ABSTRACT

The effects of a simulated service environment on the locus and bond failure mechanism of composite/metal adhesive bonds have been studied using a cyclic corrosion test. The effects of metal adherend, metal primer, and adhesive polymer chemistry on bond durability have been evaluated. Two adhesives were examined in this study, a two component urethane adhesive, and a two-component epoxy polyamide adhesive. For composite/composite bonds, composite fiber tear-out was the primary mode of bond failure for both adhesives. The highest retention in strength (85%), after cyclic corrosion testing, was observed with the urethane adhesive. The epoxy-polyamide adhesive showed a 62% retention in bond strength. Metal pretreatments and primers that improve corrosion resistance of painted steel were found to improve bond durability, when applied to the substrate prior to adhesive application and cure. The highest retention in bond strength after cyclic corrosion exposure was observed with bonds prepared using phosphated steel substrates primed with a conventional automotive cathodic electrocoat primer.

INTRODUCTION

Durability is one of the most important requirements of a structural adhesive for automotive applications. Water, humidity, corrosion and cyclic exposure are key factors affecting the durability of adhesive bonds. Environmental effects on adhesive bonds have been discussed by a number of researchers.[1-7] Kinloch has reviewed the science of adhesion relating mechanics, mechanism of bond failure, as well as surface and interfacial chemistry to bond durability.[1-3] The durability of urethane and epoxy adhesive in wet environments has also been described.[4-6] Pocius has related adhesive composition and properties to bond durability.[7] We have previously reported on the effects of polymer composition on the durability of adhesive bonds in salt spray and water immersion tests[8], and have reported on the effects of polymer composition on the mechanical behavior of adhesives.[9] These studies, however, did not subject the adhesive bond to the cyclic environmental conditions that would be expected to be observed in service.

In this paper, we describe the results of our durability study of adhesive bonds in a cyclic corrosion test. The effects of adhesive composition, type of substrate, and substrate pretreatment on the durability of adhesive bonds are described and compared.

EXPERIMENTAL

Materials The adhesives and primers used in this study were materials that were used according to procedures defined by the adhesive manufacturer. Only a generic description of the adhesive composition and cure chemistry is presented. The epoxy adhesive is an epoxy-polyamide two-component material. The urethane adhesive is a two-component material based upon a polyester polyol that is crosslinked with a MDI based crosslinker. The primers used to pretreat the metal substrate prior to bonding including two solvent-based urethane primers, a solvent-based epoxy ester spray primer, and a conventional automotive, epoxy-based cathodic electrocoat (ecoat) primer. The steel substrates used were cold rolled, unpolished bare steel (CRS) and Bonderite 40 phosphated steel, both obtained from the Parker Company. The composite material was sheet molding compound (SMC) from Rockwell.

Adhseive Bonding Technique. Standard procedures for preparing adhesive bond specimens were used. The composite was initially sanded with 240 grit emery paper and then thoroughly rinsed with methylene chloride. Phosphated steel substrates were heated at 150°C for 5 minutes prior to bonding to remove adsorbed water. Adhesive bonds using the two-component epoxy and urethane adhesive were prepared as one-inch overlap shear specimens. Bond thickness was defined using 0.16-cm long wires of 0.076-cm diameter.

Adhesive Testing and Evaluation. The cyclic corrosion test that was used for this study was based on a modification of the Ford Arizona Proving Ground Test.[10] The scheme below summarizes the test. Fifty bonded lap shear coupons were prepared for evaluation in the cyclic corrosion experiment. The bond strengths of five samples were determined after every five cycles (five days + weekend), through 50 cycles of this test.

RESULTS AND DISCUSSION

Adhesive Chemistry. The adhesives chosen for this study are representative of the materials that are currently being considered for assembly of automotive structures involving composites. The urethane and epoxy adhesives are two-component materials. The two-component epoxy adhesive comprises Epon 828, a bisphenol A--epichlorohydrin epoxy resin, and a polyamide resin. The crosslinking reaction results in the formation of stoichiometric amounts of hydroxy groups (through an epoxy/amine NH reaction). Amide O=C-NH-linkages also are present in the crosslinked network. This is a polar crosslinked network and previous work has shown that adhesives of this type have a high water uptake.[11] The two-component urethane adhesive comprises a polyisocyanate and a polyester polyol. Curing of this material results in a crosslinked network consisting of ester and urethane linkages and essentially no free hydroxy groups. This adhesive has a significantly lower water uptake than the epoxy adhesive. The lower water uptake is probably attributable to the isocyanate/polyol cure reaction that leaves a

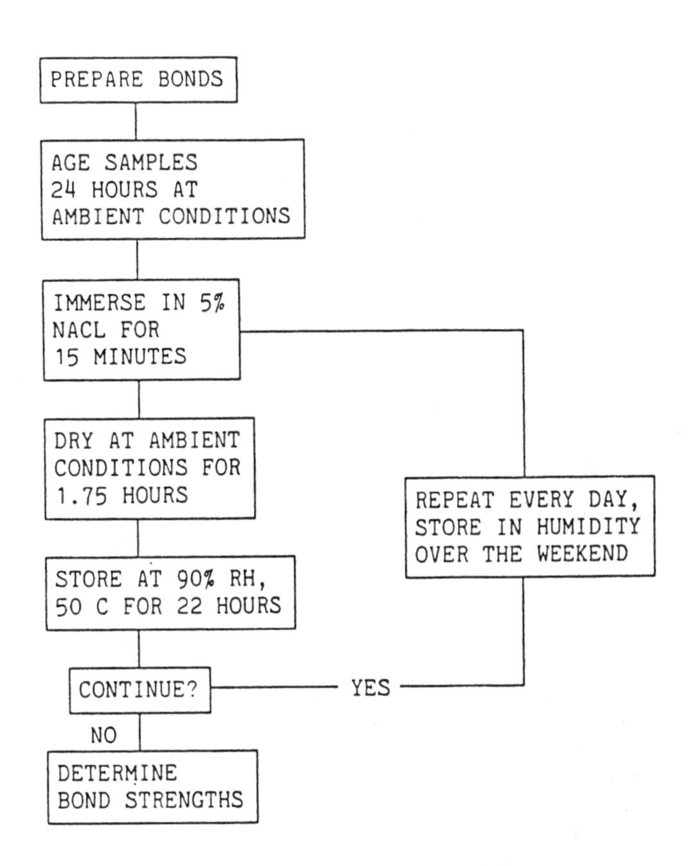

minimal amount of free polar functionality left in the cured adhesive to associate with water.

Adhesive Primer Chemistry. Adhesive primers are often used to prepare the surfaces of metals prior to bonding. Table 1 compares the corrosion resistance of two commercial adhesive primers that are formulated with urethane resins, with conventional epoxy cathodic electrocoat and epoxy ester spray primers. Table 1 also compares the corrosion resistance of the primer with retention of bond strength for composite/primed cold rolled steel bonds after 14 days of salt spray testing. The durability of the bond appears to be related to the corrosion resistance of the primer used to prepare the steel surface prior to bonding. The cathodic electrocoat primer, showing the best corrosion resistance, afforded bonds having 91% retention of bond strength after salt spray, whereas the urethane primer afforded bonds having only 43% retention of bond strength. A similar trend in corrosion resistance has been observed previously and is related to the hydrolysis resistance of the coating polymer.[11] The locus of bond failure is different for the cathodic electrocoat primer and the urethane primer. For the cathodic electrocoat primer, the locus of bond failure remains entirely in the composite (fiber tear) after salt spray. The locus of bond failure for bonds prepared with the urethane primer changes from fiber tear in the untested bond to primer adhesion loss for bonds exposed to salt spray.

Fractography. Figures 1-8 show the type of failures that a variety of epoxy and urethane adhesively bonded systems experience in the cyclic corrosion test. Figures 1-4 show the bond failure that was observed in 50 cycles of testing for the epoxy adhesive. The test matrix for the epoxy adhesive included bonds joining composite/cold rolled steel, composite/urethane primed cold rolled steel, composite/cathodic electrocoated cold rolled steel, and cathodic electrocoated phosphated steel. Figures 5-8 show the results of an identical series of bonds incorporating the urethane adhesive. Bond strengths for the urethane and epoxy bonds described above are reported as a function of corrosion cycle in Tables 2 and 3.

The locus of bond failure for composite/composite bonds was generally in the bulk composite for both adhesives; fiber tear-out was the predominant mode of failure. The reductions in bond strengths that were observed for composite/composite bonds in cyclic corrosion were comparable for both adhesives. The reductions in bond strengths, therefore, reflect the loss of composite properties in the cyclic corrosion test in the absence of any corrosion processes.

Differences in locus of bond failure were noted between urethane and epoxy adhesives for composite/cold rolled steel bonds. The predominant mode of failure experienced by the urethane adhesive in both unexposed and cyclic corrosion exposed bonds was adhesion loss at the adhesive-metal interface. Complete reduction in initial bond strength for bonds prepared with the urethane adhesive was observed within 5 cycles of testing. The poor performance of the urethane adhesive over cold rolled steel reflects poor initial adhesion of the adhesive esulting from the lack of polar functionality (i.e., free hydroxy groups) in the cured adhesive that are available to interact with the steel surface, as well as hydrolysis of the urethane adhesive chemistry at the steel interface through the action of corrosion generated hydroxide. Both of these mechanisms of adhesion failure have been observed previously for urethane coating adhesion loss in corrosion.

The performance of the epoxy adhesive was slightly better than that of the urethane adhesive for composite/cold rolled steel bonds. The locus of failure was associated with the composite (fiber tear) for both the unexposed bond and the bonds exposed to initial cycles of testing (5-15 cycles). After 5 cycles of the cyclic corrosion test, 45% of the initial bond strength was retained (compared to 0% for the urethane). After prolonged testing the locus of failure changed from fiber tear to adhesion loss at the adhesive/metal interface, and the strength of the bond dropped to zero. The progression of adhesion loss during the corrosion test can readily be seen by observing the encroachment of corrosion products under the adhesive as the exposure of the bond to corrosion increases to fifty cycles.

Primers applied to cold rolled steel prior to bonding significantly improve bond durability in the cyclic corrosion test and change the locus of bond failure. Two primers were studied to determine the effect of adhesive primers on bond durability to cold rolled steel. These primers included a highly corrosion resistant cathodic electrocoat primer and a corrosion sensitive urethane primer. The results of this study are summarized in Tables 2 and 3, and Figs. 2-3, 6-7. Application of primers to cold rolled steel prior to bonding was found to improve bond durability in the cyclic corrosion tests. The improvement in bond durability was related to the corrosion resistance of the primer. Pretreatment of cold rolled steel with the urethane primer resulted in a significant improvement in adhesive bond

Table 1. Effect of Primer on Composite/Steel Bond Strength

Primer Chemistry	Adhesion Loss in Salt Spray (mm)	Initial Bond Strength (MPa)	Bond Strength After Salt Spray (MPa)
Epoxy Electrocoat	0	6.27	5.92
Epoxy Ester	2-3	6.55	5.28
Urethane	4-5	6.36	2.73
Epoxy Urethane	0.5-1	6.19	4.65

Table 2. Cyclic Corrosion Test Results: Epoxy Adhesive

APGE Cycle	Bond Strength (MPa)				
	Composite/ Composite	Composite/ CRS	Composite/ Urethane-CRS	Composite/ Ecoat-CRS	Composite/ Ecoat-Phos. Steel
0	4.1	5.3	5.7	6.7	5.2
5	4.2	2.4	4.6	5.0	4.4
10	2.9	0.2	3.0	4.1	4.1
25	3.2	0.2	1.4	3.2	3.8
35	2.9	0.2	1.3	2.7	3.3
50	2.6	0.2	1.1	1.8	3.7

a. Abbreviations used: APGE is laboratory Arizona Proving Ground cyclic corrosion test; CRS is cold rolled steel; Ecoat is cathodic electrocoat primer; Ecoat-Phos. Steel is cathodic electrocoat primer applied to phosphated steel.

Table 3. Cyclic Corrosion Test Results: Urethane Adhesive

APGE Cycle	Bond Strength (MPa)				
	Composite/ Composite	Composite/ CRS	Composite/ Urethane-CRS	Composite/ Ecoat-CRS	Composite/ Ecoat-Phos. Steel
0	3.3	2.3	4.2	4.0	4.3
5	2.9	0.2	3.8	4.2	3.7
15	2.4	0.0	1.5	3.0	2.8
25	2.4	0.0	0.5	2.3	2.6
35	2.7	0.0	0.3	1.3	2.6
50	2.8	0.0	0.5	0.2	2.2

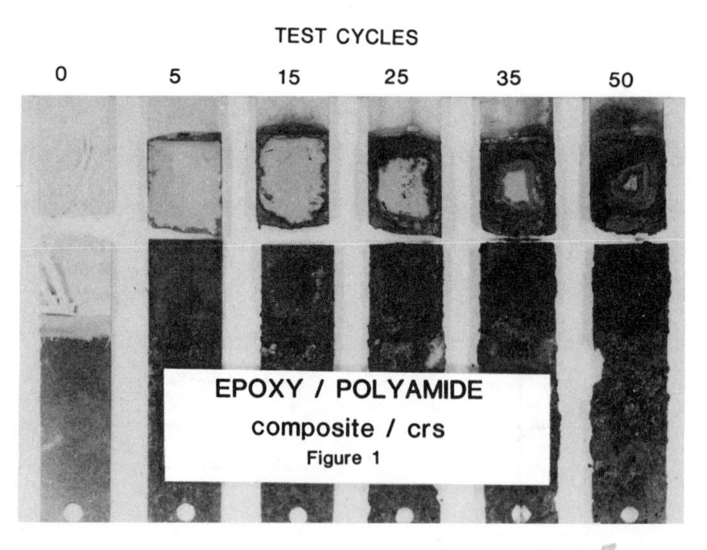

Fig. 1. Bond fractures for epoxy adhesive composite/CRS bond in APGE.

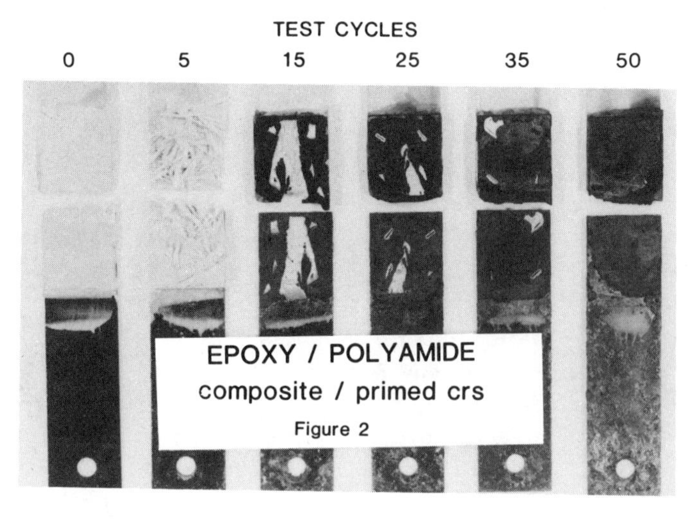

Fig. 2. Bond fractures for epoxy adhesive composite/urethane primed CRS bond in APGE.

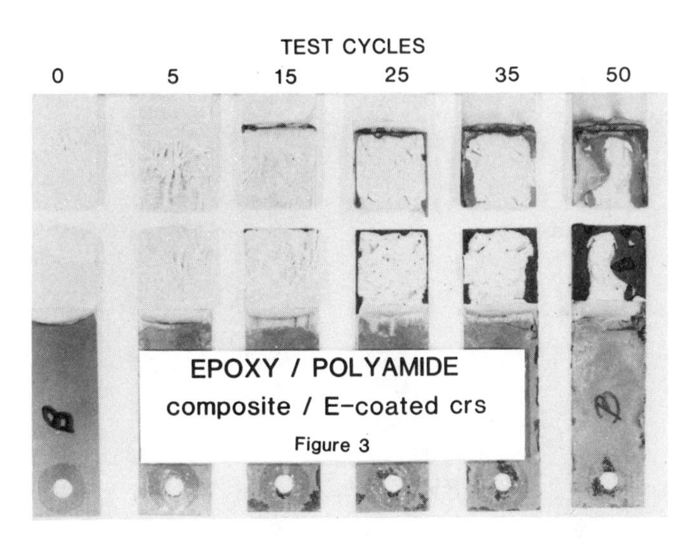

Fig. 3. Bond fractures for epoxy adhesive composite/ecoat primed CRS in APGE.

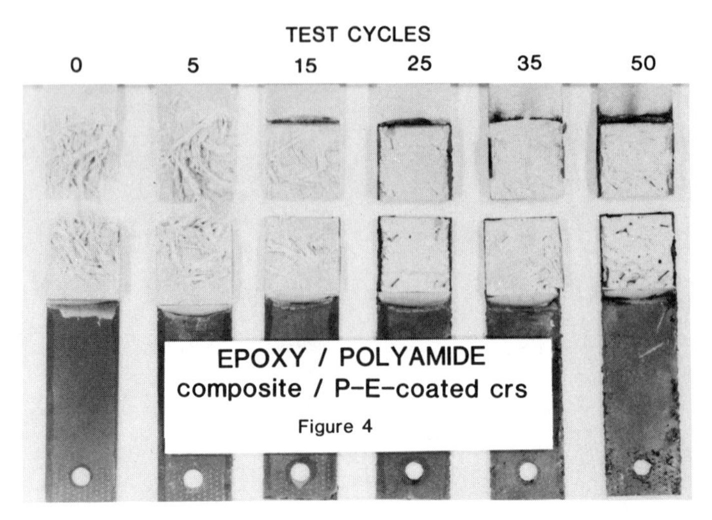

Fig. 4. Bond fractures for epoxy adhesive composite/ecoat primed phosphated steel in APGE.

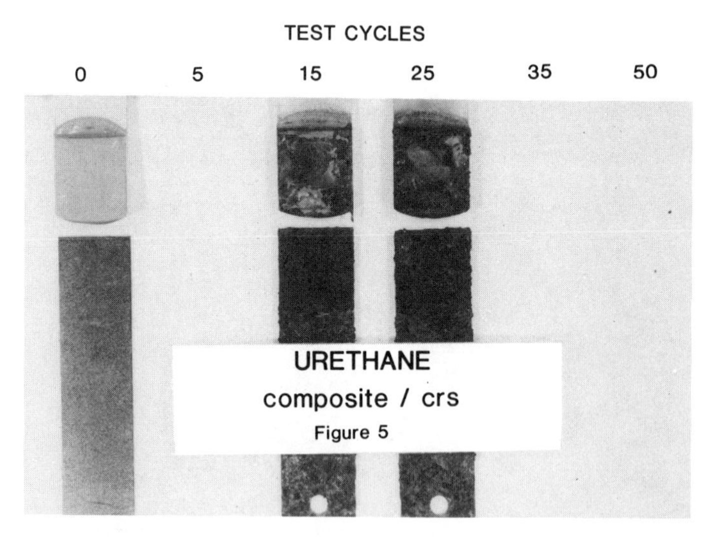

Fig. 5. Bond fractures for urethane adhesive composite/CRS bond in APGE.

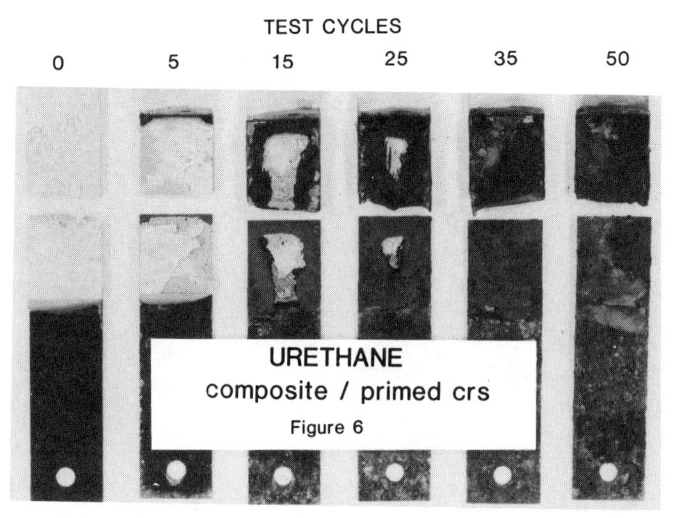

Fig. 6. Bond fractures for urethane adhesive composite/urethane primed CRS bond in APGE.

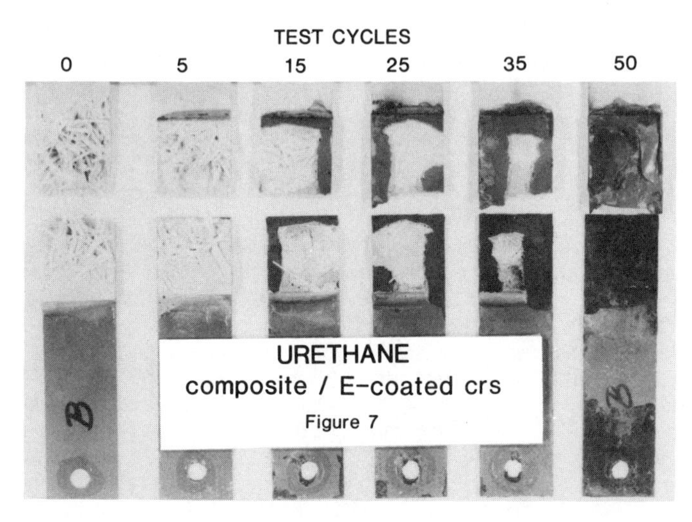

Fig. 7. Bond fractures for urethane adhesive composite/ecoat primed CRS in APGE.

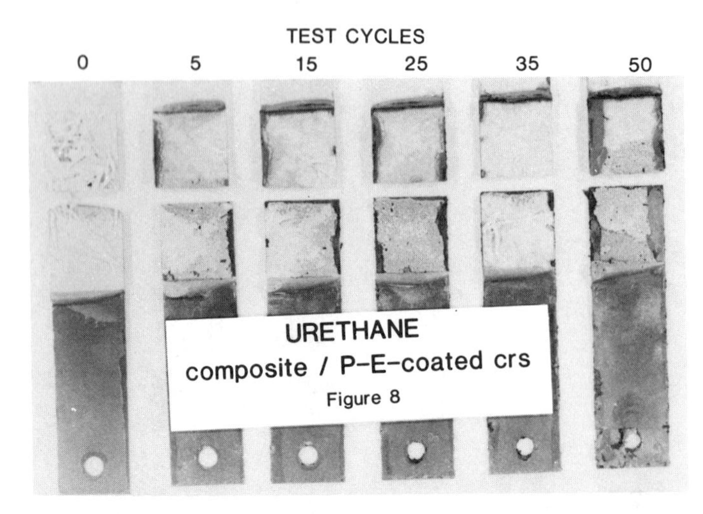

Fig. 8. Bond fractures for urethane adhesive composite/ecoat primed phosphated steel in APGE.

durability for both adhesives, 19% and 11% of the initial bond strength was retained, after 50 cycles of testing, for the epoxy and urethane adhesives respectively. Associated with this improvement in adhesive bond durability was a gradual change in the locus of failure as a function of corrosion cycle. As the number of corrosion cycles increased, the failure progressively changed from fiber tear to adhesion loss at the primer/metal interface for bonds prepared with both adhesives. Eventually, nearly complete primer/cold rolled steel adhesion failure occurred.

More complicated bond failures occur when a highly corrosion resistant primer, like the cathodic electrocoat primer, was used to prepare the metal surface prior to bonding. For bonds to the cathodic electrocoat primed cold rolled steel, the locus of failure was in the composite (fiber tear) for the first fifteen cycles for both adhesives. After longer exposure, the locus of failure occurs cohesively within the adhesive, changing to progressive electrocoat primer/cold rolled steel interfacial adhesion failure toward the completion of the cyclic corrosion exposure (40-50 cycles).

The effect of a metal conversion coating on bond durability is illustrated by comparing Figs. 3-4 and 7-8. Application of the corrosion resistant electrocoat primer to phosphated steel results in improvements in primer-metal adhesion. Bonds prepared with this particular substrate fail either by composite failure (fiber tear) early in the corrosion cycle or by cohesive failure in the adhesive at longer exposures. Little or no failure in electrocoat primer was observed over phosphated steel. Preparation of the substrate to maximize corrosion resistance afforded bonds having the best durability in the cyclic corrosion test, 50-70% of the initial bond strength was retained for bonds prepared with the epoxy and urethane adhesives after completion of the 50-cycle corrosion test. The effects of bulk adhesive composition appear to be less important in determining bond durability than is the chemical stability at the metal interface.

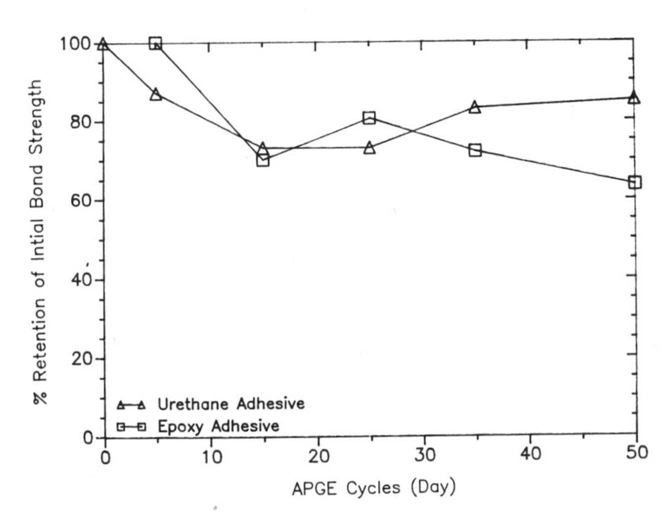

Fig. 9. Percent retention of urethane and epoxy adhesive bond strength as a function of corrosion cycle.

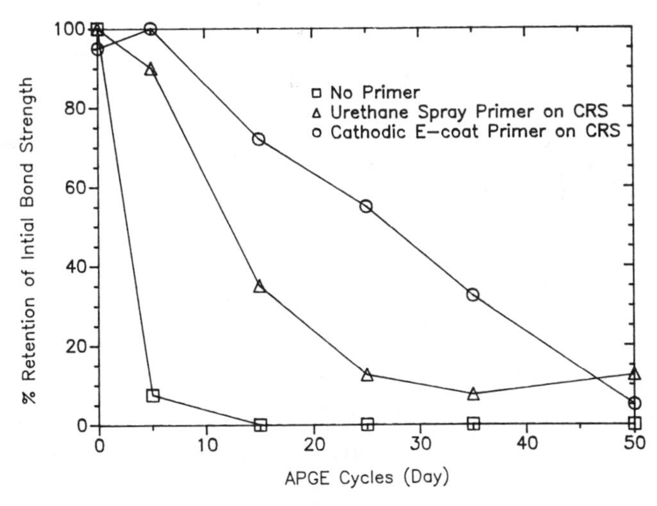

Fig. 10. Effect of steel pretreatment on percent bond retention of urethane adhesive bond strength as a function of corrosion cycle.

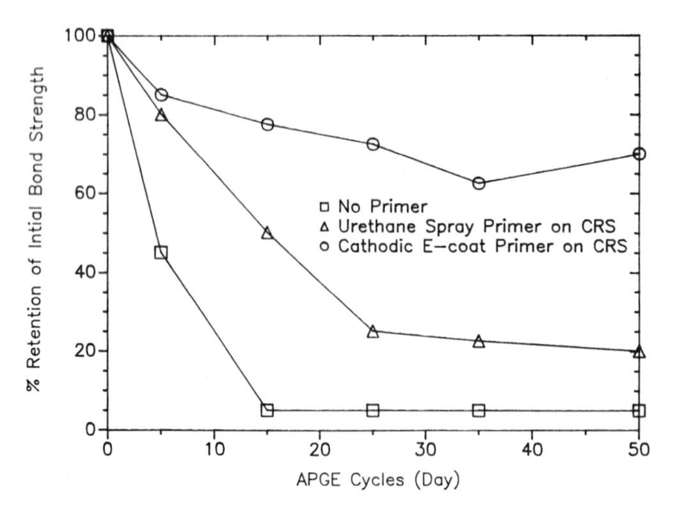

Fig. 11. Percent retention of epoxy adhesive bond strength as a function of corrosion cycle and steel pretreatment.

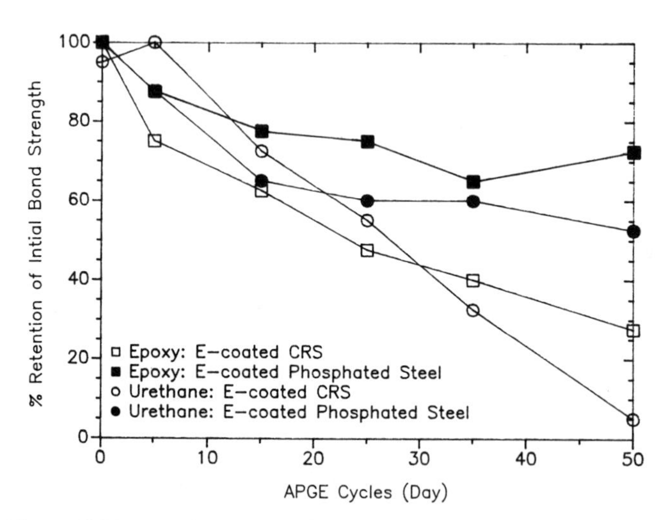

Fig. 12. Effect of conversion coatings on epoxy and urethane adhesive bond durability in cyclic corrosion.

Adhesive Bond Strength After APGE Testing. Figures 9-12 compare percent retention of bond strengths for both the urethane and epoxy adhesives in composite/composite, composite/cold rolled steel, composite/primed CRS, composite/electrocoated CRS, and composite/ electrocated phosphated steel as a function of corrosion test cycle. The results show that for composite to composite bonds (Fig. 9), the effects of adhesive composition are not observed. The bond strengths are shown to drop off at a comparable rate for both the urethane and epoxy adhesives. The structural strength of the composite, rather than the performance of the adhesive, appears to control the durability of the adhesive bond in the corrosion test. For composite to cold rolled steel adhesive bonds, the effects of adhesive primer on bond durability are readily apparent (Figs. 10-11). For both adhesives, bonding to untreated cold rolled steel results in a rapid drop off in bond strength as a function of corrosion test cycle. For composite to pretreated steel surfaces, the effects of the pretreatment rather than the effects of adhesive chemistry appear to control the rate of bond strength reduction in the cyclic corrosion test. The performance of composite bonds to the urethane primed cold rolled steel is improved relative to the corresponding bonds to cold rolled steel. Further improvements in bond durability are shown for steel surfaces prepared with an electrocoat primer, and primers applied over conversion coated steel (Fig. 12).

REFERENCES

1. A.J. Kinloch, J. Mater. Sci., 15, 2141 (1980).
2. A.J. Kinloch, J. Mater. Sci., 17, 617 (1982).
3. A.J. Kinloch, "Durability of Structural Adhesive Joints," Applied Sciences Publishers, New York, 1983.
4. E.H. Andrews and A. Stevenson, J. Adhes., 11, 17, 1980.
5. H. Dodiuk, L. Drori and J. Miller, J. Adhes., 17, 33, 1984.
6. S.G. Abbot and N. Brumpton, J. Adhes., 13, 41, 1981.
7. A.V. Pocius, D.A. Wangsness, C.J. Almer and A.G. McKown, "Adhesive Chemistry," Lieng-Huang Lee, ed., Plenum Publishing Corporation, New York, 1984.
8. J.W. Holubka, W. Chun, A.R. Krause and J. Shyu, Polym. Mater. Sci. and Eng., 53, 574 (1986).
9. K.A. Mazich, J.W. Holubka and P.C. Killgoar, Ind. Eng. Chem. Prod. Res. Dev., 25, 273 (1986).
10. Ford Laboratory Test Method, BI-23-1, April 30, 1981.
11. J.W. Holubka, A.R. Krause and W. Chun, "The Effects of Adhesive Chemistry on the Fatigue Resistance of Adhesive Bonds in Service Environment," Proceedings of the Second Conference on Advanced Composites, November 18-20, 1986.

HIGH TEMPERATURE ORGANIC ADHESIVES--A REVIEW

Chang Chih-Ching

Beijing Institute of Aeronautics and Astronatics
Beijing, People's Republic of China

ABSTRACT

Research on high temperature polymers began in the 1960's. Since then, many high temperature polymers have been studied as adhesives. In this paper, I briefly review the syntheses, structures and properties of PBI, PI, PQ, PPQ, ATPQ, PT, PPT, PPS and pyrrones. At the same time, I discuss some of the high temperature adhesives.

INTRODUCTION

Because of the rapid development of space navigation, military aircraft, missiles, electronics and mechanical processing during the past two decades, the needs of high temperature adhesives[1] have been developed. For example, supersonic aircraft flying in upper atmosphere with 2.5-3.0 Mach number, the surface temperature of aircraft can reach 260-316°C. Since the beginning of the 60's, heterocyclic-aromatic polymers have undergone evaluation as adhesives.[2] The polymers such as PBI, PI, PQ, PPQ, ATPQ, PT, PPT, PPS and pyrrone have unique high temperature properties.

From the viewpoints of synthesis, structure and property, I shall discuss the needs and developments of these high temperature adhesives. The evaluation will be given at the end of this paper.

RELATIONSHIP BETWEEN HIGH TEMPERATURE PROPERTY AND POLYMERIC STRUCTURE

High temperature materials require the maintenance of their desirable mechanical properties in certain surroundings during the service time. The thermal resistance of materials is usually described according to the relationship between temperature and time.

Generally speaking, high temperature adhesives are defined as being capable of functioning under the following conditions[3] as shown in Table 1.

According to the above classification, polybenzimidazoles (PBI's) have demonstrated useful bond strength within Range I. High temperature polymers, such as PBI's, polyquinoxalines (PQ's), polyimides (PI's) and polyphenylquinoxalines (PPQ's) have been evaluated for Range II applications. Acetylene-terminated polyphenylquinoxalines (ATPQ's) have shown good strength within Range III.

The glass temperature of polymers are closely related to the polymer main chain, the size of side groups, the symmetry of polymeric structure, the molecular weight and the rigidity of polymeric chain. Of the above factors, the last is the most important. Obviously, by increasing the aromatic ring and the heterocyclic ring, the rigidity of the polymeric chain is effectively increased resulting from the increasing difficulty of the internal molecular rotation. Since the structural arrangement of the rigid chain is almost symmetrical, the glass temperatures of heterocyclic-aromatic polymers are higher than those of normal aliphatic polymers (Table 3).[5]

Table 1. Classification of high temperature adhesives-Method I

Range	Use Temperature, °C(°F)	Use Time
I.	538-816(1000-1500)	2-10 min.
II.	482-538(900-1000)	1 hr.
III.	371-427(700-800)	24-200 hr.
IV.	260-371(500-700)	200-1000 Hr.
V.	204-232(400-450) 121-176(250-350)	30,000 hr. 1-5 years

Table 2. Classification of high temperature adhesives-Method II

Range	Use Temperature, °C(°F)	Use Time
I.	538-760(1000-1400)	Seconds of minutes.
II.	288-371(550-700)	Hundreds of hours.
III.	177-232(550-450)	Thousands of hours.

Table 3. Glass temperatures of some polymers

Polymers	T_g, °C
PI's	260-380,400
PBI's	310,430
PPQ's	317,406

Most of the heterocyclic-aromatic polymers belong to the fused-ring or ladder-type structure. Because of the increased heterocyclic rings and aromatic rings, the defect in a unit volume of polymeric structure is substantially decreased. Thus, the density and relative strength are simultaneously increased. Moreover, those C-H bonds and N-H bonds attached to the condensed polynuclear polymers render them difficult to oxidize. Thus, the factors, mentioned above, tend to raise the thermal resistance of heterocyclic-aromatic polymers.

In summarizing the experimental results of their predecessors, three essentials have been put forward for a thermally stable polymer: (1) the polymer should have high thermal stability and possesses a high softening point (T_s), a high melting point (T_m) or a high glass temperature (T_g), and (2) the thermally stable polymer should have a high decomposition temperature (T_d), and (3) the thermally stable polymer should be oxidation resistant. Obviously, the heterocyclic-aromatic polymers possess nearly all the above three properties.

POLYBENZIMIDAZOLES (PBI's)

Polybenzimidazole (PBI) was the first all-aromatic heterocyclic polymer to succeed as a high temperature adhesive in 1961.[6] It is formed from the melt polycondensation of 3, 3′, 4, 4′-tetraaminobiphenyl (TAB) and diphenyl isophthalate. These reactions include two steps as follows:

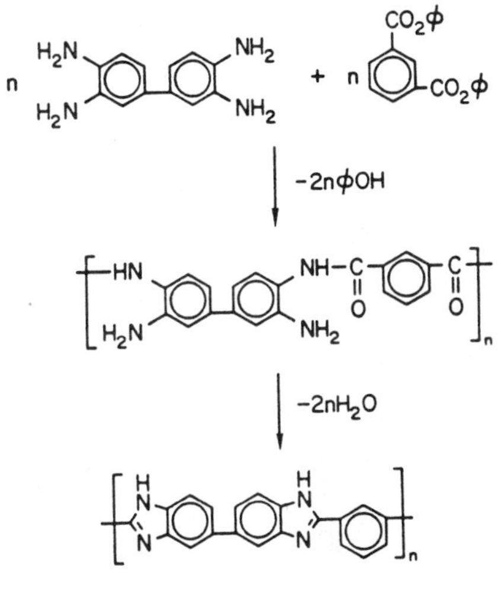

PBI exhibits good short-time higher temperature behavior at 538°C. Comparing with polyimides, some PI's decompose at 482°C or higher. In application, it involves the preparation of a low molecular weight oligomer (dimer to trimer). This oligomer can be dissolved in polar solvents such as: N, N-dimethyformamide, pyridine, dimethyl sulfoxide, formic acid, conc. sulfuric acid and N-methyl pyrrolidone. To produce the insoluble and infusible properties of a crosslinked material, PBI must be postcured at 399°C under 200 psi. In the above polycondensation, the evolution of a large amount of volatiles (phenol or water) takes place. Thus, lots of pinholes result. It should be mentioned that postcuring temperature of PBI is related directly to its shear strength. Postcuring provides the cross-linkage for raising the shear strength at higher temperatures.

Research has indicated that the N-H bond in PBI can easily be oxidized. Using the phenyl group to replace the hydrogen atom, the thermo-oxidative stability can be improved,[7] as illustrated in the following structure:

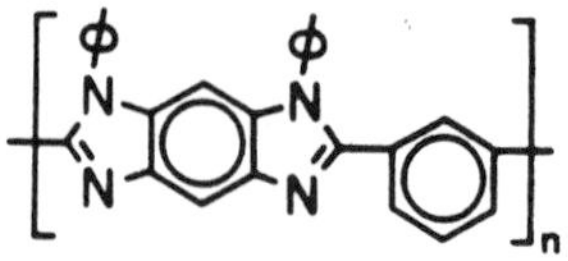

By introducing the ether bond to PBI, both solubility and thermal resistance can be improved.[8] The PBI, shown as follows, loses 20.1% weight at 720°C in the air.

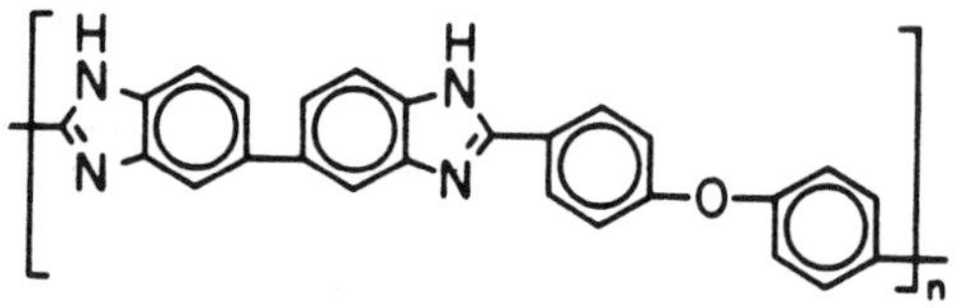

POLYIMIDES (PI's)

Polyimide (PI) is the most important heterocyclic-aromatic polymer. Initially, it was formed from the stoichiometric reaction of 4,4'-diaminodiphenyl ether with sym-tetracarboxylic acid dianhydride. The polyamic acid form was first obtained. N,N-dimethylformamide, dimethyl sulfoxide and N-methyl pyrrolidone have normally been used as solvents. In adhesive work, the polyamic acid was cyclodehydrated thermally,[9,10] PI's are stable at 260°C for a long time. Although the adhesive film is heated to 275-300°C for several months in air, the tenacity can be maintained. PI's may be used at 500°C for a short-term exposure. Hydrolytic instability, evolution of volatiles and processing difficulties are the three major shortcomings of common PI's.

By using the reaction of 3,3',4,4'--benzophenone tetracarboxylic acid dianhydride (BTDA) and metals the product can be processed and has good adhesion to metals.[11]

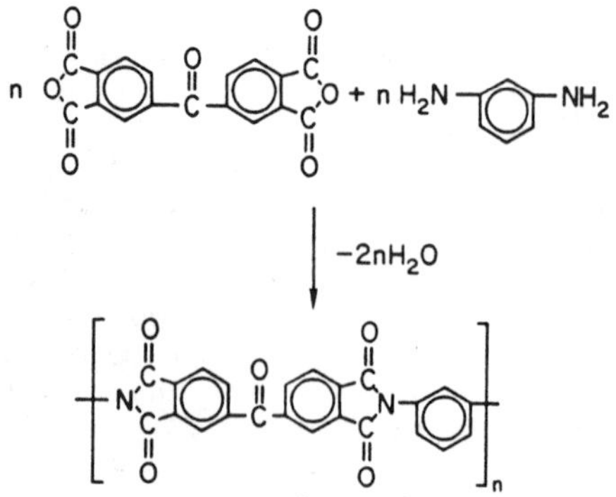

$-2nH_2O$

The polymer, from the reaction of BTDA and 3,3'-diaminobenzophenone (DABP), has surprisingly good thermal stability and processability.[12]

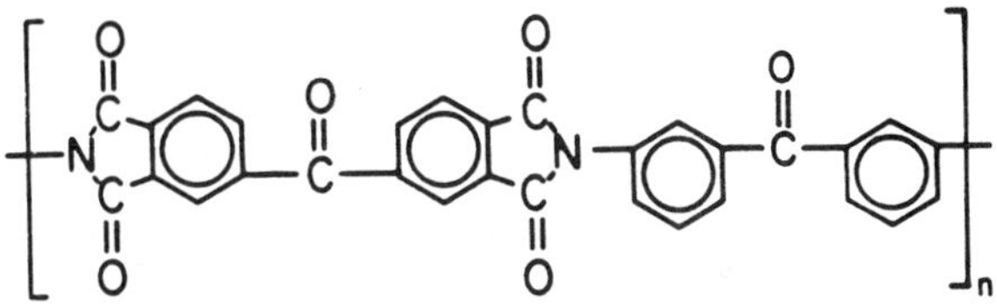

A fluorine-containing PI is made from the reaction of 2,2'-bis(3,4-dicarboxyphenyl)-hexafluoropropane (6F) and aromatic diamine.[13]

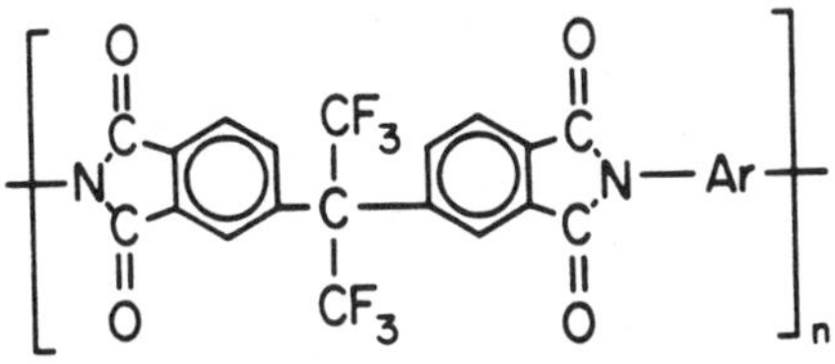

By varying the structure of the aromatic rings, their T_g's can be varied as shown in Table 4.

It is worthy to note that the acetylene-terminated imide (ATI) and nadic-terminated imide (NTI) can be used above 260°C.[14,15] The polymerized products have fewer pinholes because no volatile is evolved upon further polymerization. Their structures are as follows:

Table 4. Glass Temperatures of 6F(PI)'s

Ar	T_g (°C)
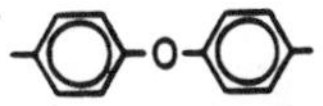 O	285
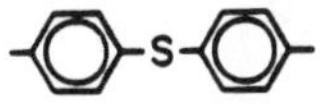 S	283
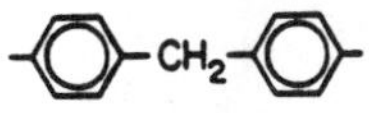 CH_2	291
	297
	326
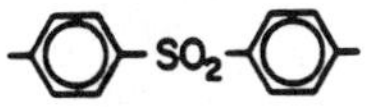 SO_2	336
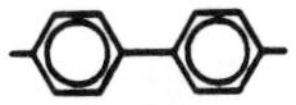	337
	365

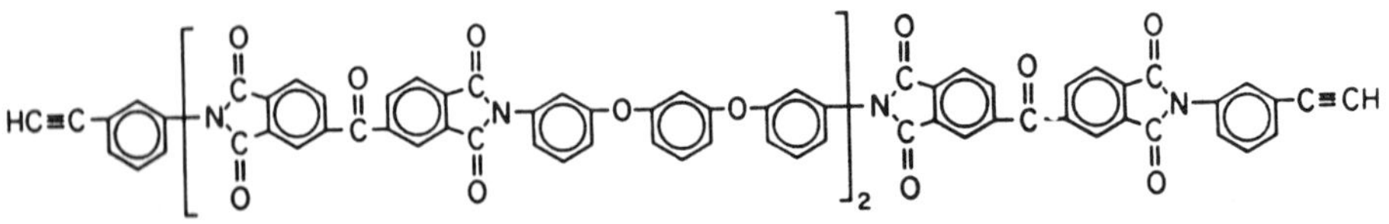

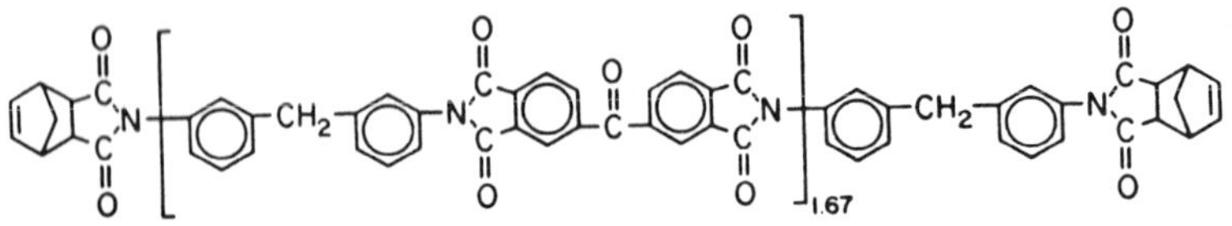

HC≡C– [–N(CO)₂C₆H₂(CO)₂–C(=O)–C₆H₂(CO)₂N–C₆H₄–O–C₆H₄–O–C₆H₄–]$_2$ –N(CO)₂C₆H₂(CO)₂–C(=O)–C₆H₂(CO)₂N–C₆H₄–C≡CH

[–C₆H₄–CH_2–C₆H₄–N(CO)₂C₆H₂–C(=O)–C₆H₂(CO)₂N–]$_{1.67}$ –C₆H₄–CH_2–C₆H₄–N(CO)₂

POLYQUINOXALINES (PQ's)

Research on polyquinoxaline (PQ) was first reported in 1964.[16] One type of PQ from the reaction of TAB with 4,4'-oxybis (phenylglyoxal hydrate) produced a monoether PQ.

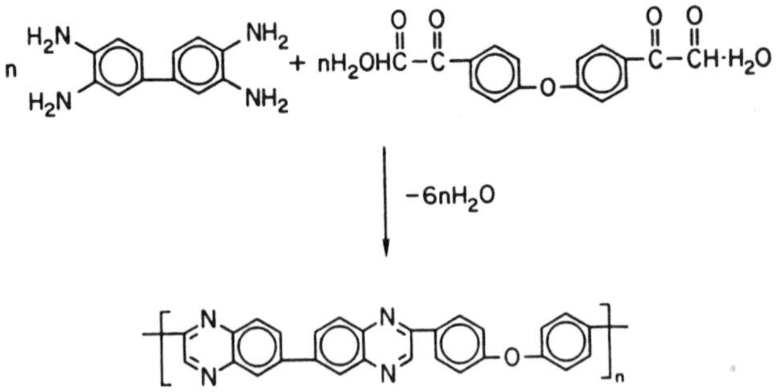

$$n\ (H_2N)_2C_6H_3–C_6H_3(NH_2)_2 + nH_2OHC–CO–C_6H_4–O–C_6H_4–CO–CH{\cdot}H_2O \xrightarrow{-6nH_2O} [\text{PQ}]_n$$

Research has indicated that there are three isomers.[17]

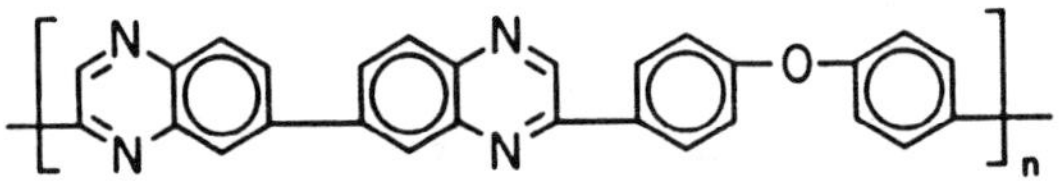

3,3' – ISOMER (A)

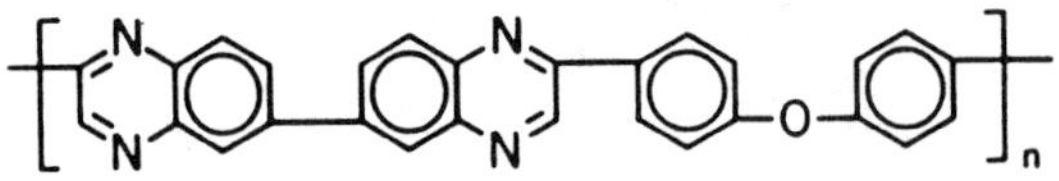

2,2' – ISOMER (B)

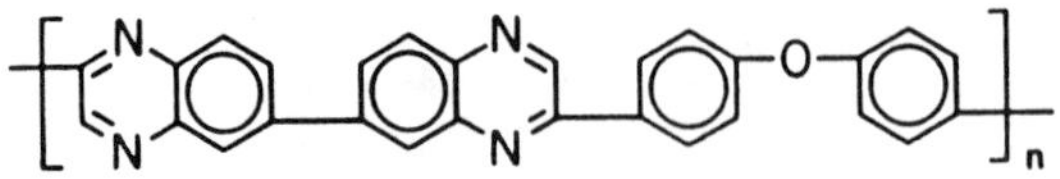

2,3' – ISOMER (C)

Among these isomers, isomer (B) is the main component. PQ is easily dissolved in m-cresol. Under the processing conditions of a temperature at 371-399°C and a pressure of 200 PSi, only a small amount of volatiles is evolved.

Monoether PQ and diether PQ were compared with respect to their physical properties in Table 5.[18]

During thermal analysis, the decomposition usually occurs at weak C-H bonds. In order to improve the thermal resistance of PQ, the phenyl group is employed to replace the active hydrogen atom.[19] From the reaction of TAB and 4,4'-oxydibenzil (ODB), polyphenylquinoxaline (PPQ) is formed as follows:

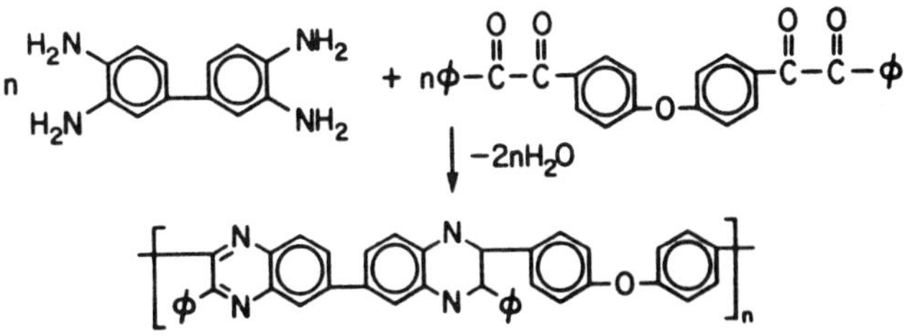

$$-2nH_2O$$

PQ and PPQ can be used to 288-371°C. But the properties of PPQ, such as solubility, oxidation resistance and melt flow, are better than those of PQ.

In addition, the thermal resistance of PQ may be improved with the end-capping.

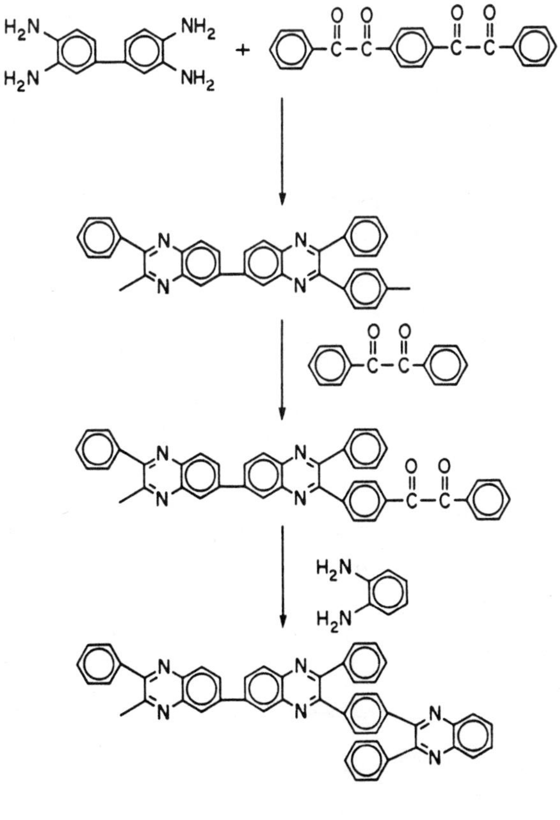

(I)

(II)

(III)

Table 5. Physical properties of monoether PQ and diether PQ

PQ	T_g(°C)	T_d(°C)	400°C/200hr Loss of weight (%)
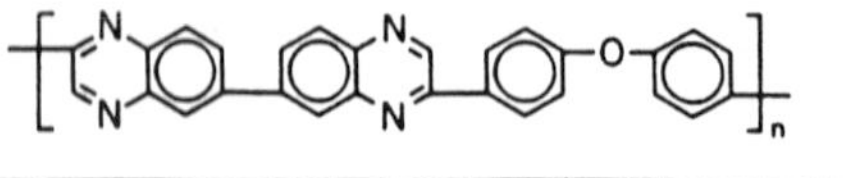	280	495	85
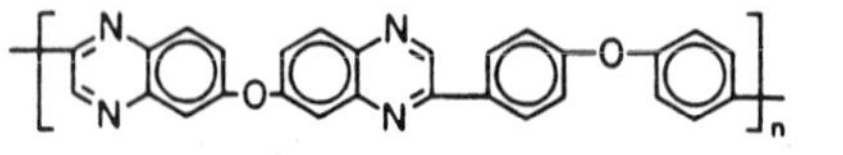	270	475	96

The weight loss indicates that thermal stability of the above series is in the order of (I)<(II)<(III). Obviously, different end-capping groups produce different effects.[20]

It has been mentioned that the linear PPQ has good thermal resistance. But owing to the thermoplastic nature of PPQ, it can not be used beyond T_g. In order to solve this problem, the methods of initiator addition, radiation, heat treatment are usually employed and a crosslinking agent is used to enhance the crosslinking reaction.

When PQ's are endcapped with acetylene, acetylene-terminated phenylquinoxalines (ATPQ) are obtained.[21]

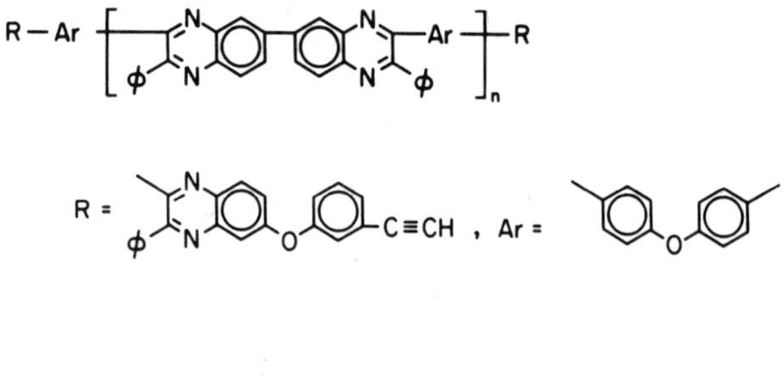

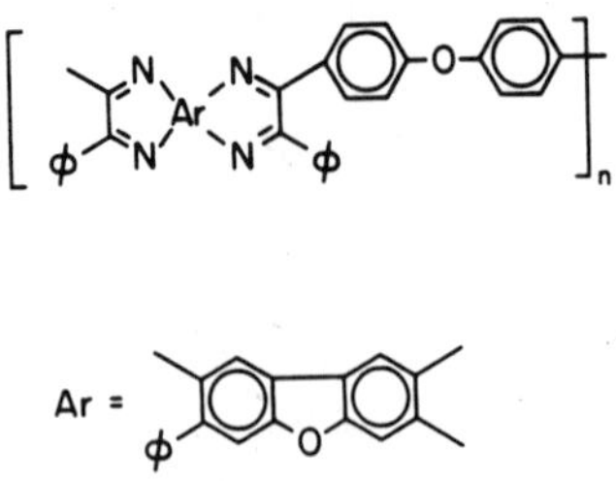

R—Ar —Ar— R

R = —C≡CH , Ar =

Ar =

After curing for 8 hours at 371°C, the T_g of the above ATPQ approaches 344°C

The T_g of the following polymer at 406°C[22] is derived by fused phenyl rings.

Although a good deal of research has been done on PQ, PPQ and ATPQ, the thermoplastic nature at high temperatures still remains unsolved.

OTHER HIGH TEMPERATURE POLYMERS

1. Polybenzimidazoquinazolines (Pyrrones) These polymers are called pyrrones for short. Pyrrones normally are prepared from the reaction between aromatic tetracarboxylic derivatives and aromatic tetra-amines. The oligomer is soluble in polar solvents. After undergoing high temperature cyclization, a family of ladder-type heterocyclic-aromatic polymers can be obtained.[23,24]

The following pyrrone has been used at 316°C for long-term and 427°C for some short-term applications.[25]

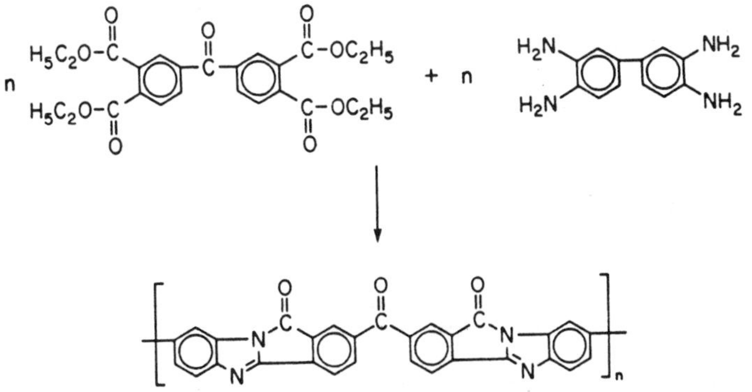

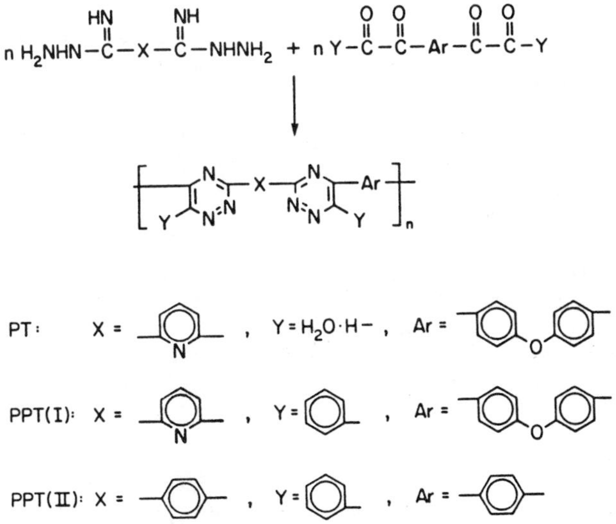

2. Poly-as-triazines (PT's) and polyphenyl-as-triazines (PPT's) Poly-as-triazine (PT) and polyphenyl-as-triazine (PPT) have been prepared from the polycondensation of 4,4'-oxydibenzil and aromatic bisamidrazone.[26,27]

PPT can be used at 260°C for some long-term applications. The thermal stability of PPT is better than that of PT. The PPT with the backbone of 1,4-phenylene is more thermally and oxidatively resistant.

3. Polyphenylene sulfide (PPS). The synthesis of polyphenylene sulfide (PPS) is shown as follows:[28,31]

$$n\ X-C_6H_4-SM \longrightarrow \left[C_6H_4-S \right]_n + nMX$$

$$n\ Cl-C_6H_4-Cl + nNa_2S \longrightarrow \left[C_6H_4-S \right]_n + 2nNaCl$$

PPS is thermally stable at 260°C for some long-term application.[32] PPS has good adhesion to glass, ceramics, steel and aluminum

SUMMARY

In this paper, I have reviewed several high temperature adhesives. The shortcomings of these adhesives are their costs and their higher processing temperature.[33] But owing to their unique high temperature nature, heterocyclic-aromatic polymers still receive much attention from chemists.

Among these high temperature polymers, PBI's are unique. Because of the processing temperature and volatiles, PBI's have a limited scope for production. Due to the advantages of PI's such as simple technology, synthetic properties and abundant resources, PI's have reached the industrial scale. The advantages of PQ's and PI's are their solubility characteristics and hydrolytic resistance. Pyrrones are ladder-type polymers. So far, pyrrones are the best radiation resistance materials. The last but not the least, PPS possesses good adhesive strength.

For future research on heterocyclic-aromatic polymers, we should focus at: (1) raising the capability to cyclize, (2) enhancing the thermal oxidative stability, (3) improving the processing property, and (4) reducing the cost.

In brief, heterocyclic-aromatic polymers are the best thermally resistant materials in synthetic polymers. Much attention has been paid to the application in the field of space application, high-performance aircraft and military vehicles. With the continuing research efforts of chemists, we believe that the future of heterocyclic-aromatic adhesives should be rather bright.

REFERENCES

1. L.H. Lee, "Developments in and Limitations of Adhesive Materials for Severe Environments and a Long Service Life," in Adhesive

Chemistry-Developments and Trends. Editor, L.H. Lee, p.675, Plenum, New York (1984).
2. Yang Yu-Kun et al., Synthetic Adhesives, (In Chinese) p.574 Scientific Publishing Agency, PRC (1983).
3. R.L. Patrick, Treatise on Adhesion and Adhesives, Vol.2, Materials, Marcel Dekker, Inc., New York, 408 (1969).
4. L.H. Lee, "Adhesives and Sealants for Severe Environments," Report presented to the Beijing Adhesion Society (April, 1986). Also, Inter. J. Adhes. and Adhes. 7, (2) 81 (1987.
5. Zhang Zhi-Dong, et al., Adhesion Principle and Useful Formulas 600, p.264, Hei Long Jiang Publishing Agency, PRC (1984).
6. H. Vogel and C.S. Marvel, J. Polym. Sci., 50, 511 (1961).
7. H. Vogel and C.S. Marvel, J. Polym. Sci., A1, (5), 1531 (1963).
8. R.T. Foster and Marvel, J. Polym. Sci., A3, 417 (1965).
9. W.M. Edward, U.S. Pat. 3,179,614 (1965).
10. A.L. Endrey, U.S. Pat. 3,179,631 and 3,179,633 (1965).
11. H.A. Burgman, et al., J. Appl. Polym. Sci., 12, (4), 805 (1968).
12. V.L. Bell, B.L. Stump and H. Gater, J. Polym. Sci., (14), 2275 (1976).
13. Fang E-Sheng, High-temperature Polymers, (In Chinese) p.4, Shanghai Scientific Reference Publishing Agency, PRC, (1980).
14. N. Bilow, A.L. Landis and L.J. Miller, U.S. Pat. 3,845,018 (1974).
15. H.R. Lubowitz, U.S. Pat. 3,528,950 (1970).
16. J.K. Stille and J.R. Williams, J. Polym. Sci., B.2, (2), 209 (1964).
17. J.K. Stille, et al., J. Polym. Sci., A,3, (3), 1013 (1965).
18. W.J. Wrasidlo, J. Polym. Sci., A-1, 8 (5), 1107 (1970).
19. W. Wrasidlo and J.M. Augl, J. Polym. Sci., B.7 (4), 281 (1969); A-1, 7 (12) 3393 (1969).
20. Lu Feng-Cai, et al., Polym. News, (3), 146 (1980).
21. R.F. Kovar, Polym. Preprints, 16, (2), 246 (1974).
22. F.L. Hedberg and F.E. Arnold, J. Polym. Sci., 12 (9) 1925 (1974).
23. V.L. Bell and G.E. Pezdirtz, J. Polym. Sci., B3, 977 (1965).
24. J.G. Colson, et al., J. Polym. Sci., A-1, 4(1), 59 (1966).
25. U.S. Pat. 3,632,441 (1972).
26. P.M. Hergenrother, J. Polym. Sci., A-1, 7(3), 945 (1969).
27. P.M. Hergenrother and D.E. Kiyohara, J. Macromol. Sci-Chem., A5 (2), 365 (1971).
28. R.W. Lenz, et al., J. Polym. Sci., 50, 351 (1962).
29. S. Tsunawaki and C.C. Price, J. Polym. Sci., A2 (3), 1511 (1964).
30. U.S. Pat. 3,380,951 (1968).
31. U.S. Pat. 3,408,342 (1968).
32. R.V. Jones, Hydrocarbon Processing, 51 (11), 89 (1972).
33. P.M. Hergenrother, Status of High Temperature Adhesives, in Adhesive Chemistry-Developments and Trends, Editor L.H. Lee, p.462, Plenum, New York (1984).

ORGANOTIN AND ORGANOTITANIUM-CONTAINING POLYDYES FOR COLOR PERMANENCE, REDUCTION OF LASER DAMAGE AND BIOLOGICAL RESISTANCE TO ROT AND MILDEW

Charles E. Carraher, Jr.,[a] Van R. Foster,[b] Raymond J. Linville,[c] Donald F. Stevison[c] and R.S. Venkatachalam[d]

[a]Florida Atlantic Univ.
Dept. of Chemistry
Boca Raton, FL. 33431

[b]DAP Home Improvement Products
Dayton, Ohio 45431

[c]Air Force Wright Aeronautical Labs
AFWAL/MLPJ
Wright-Patterson AFB
OH 45433

[d]Wright State University
Department of Chemistry
Dayton, Ohio 45435

ABSTRACT

Metal-containing polymeric dyes (polydyes) have been prepared by condensing difunctional organic dyes such as fluorescein and sulfonephthalein with Cp_2Ti, Cl_2 or R_2SnCl_2. These dyes show greatly enhanced color stability in solution as compared to the simple dye molecules and great resistance to bleeding of the polydye within solid plastics and films. In cases where the dye acts as a biocide, organotin moieties exert a synergistic effect by stabilizing the polydye and by enhancing its biotoxicity. Polydyes containing Cp_2Ti and or organostannane prevent burning of Lucite samples when they are cut by high energy laser beams and impart a moderate resistance to laser damage to latex films. A wide variety of other potential applications of polydyes are also discussed.

INTRODUCTION

The synthesis of titanium[1-6] and tin-containing[7-13] polyesters and polyethers has been accomplished (for instance).[1-4]

```
                                   Cp   O    O
                                   |    ||   ||
1   Cp2TiCl2+ -O2CRCO2-  →     -(-Ti-O-C-R-C-O-)-
                                   |
                                   Cp
```

2 $\quad Cp_2TiCl_2 + HO\text{-}R\text{-}OH \rightarrow \left(Ti(Cp)_2\text{-}O\text{-}R\text{-}O \right)$

$\left(Sn(R)_2\text{-}O\text{-}C(=O)\text{-}R'\text{-}C(=O)\text{-}O \right) \xleftarrow{^{-}O_2CR'CO_2^{-}} R_2SnCl_2 \xrightarrow{HOR'OH} \left(Sn(R)_2\text{-}O\text{-}R'\text{-}O \right)$

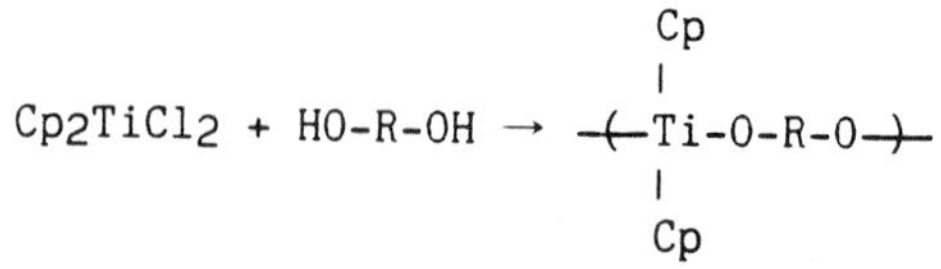

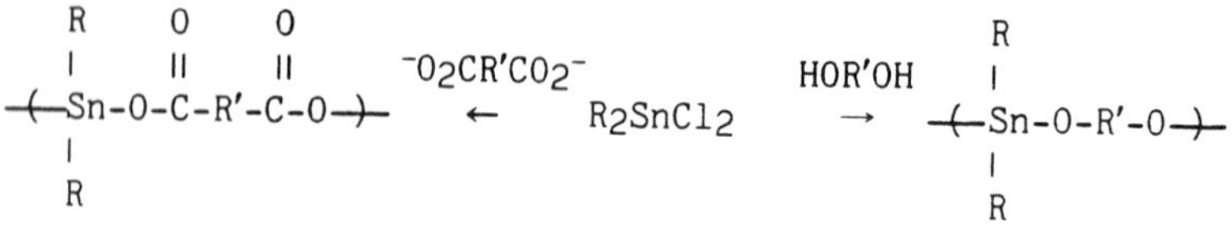

3 $\qquad\qquad$ 4

The structures of indicator dyes varies with pH. For dyes with structures similar to fluorescein and sulphonephthalein, two reactive sites are present through much of the pH range as shown below. Reactions of these dyes with diacid chlorides under mild conditions does not give polymeric materials.[14] However reaction with selected organometallic dihalides does give polymers (for instance [15-19]). Illustrative repeat units are given as 5 and 6.

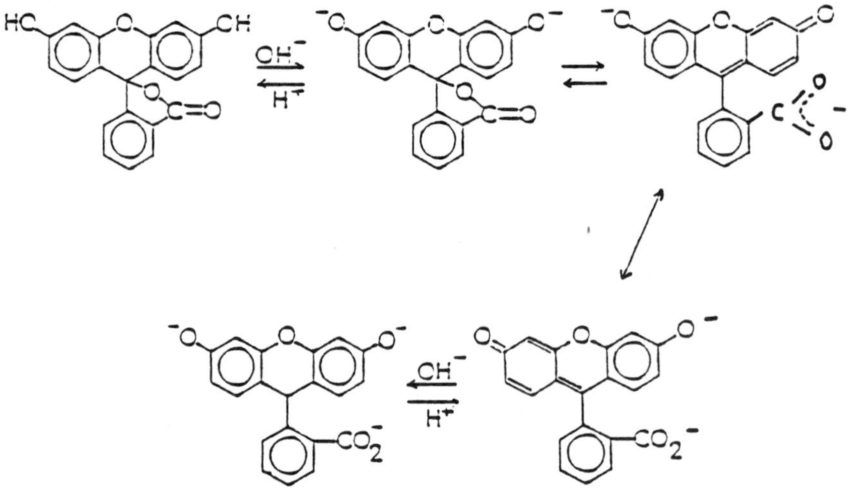

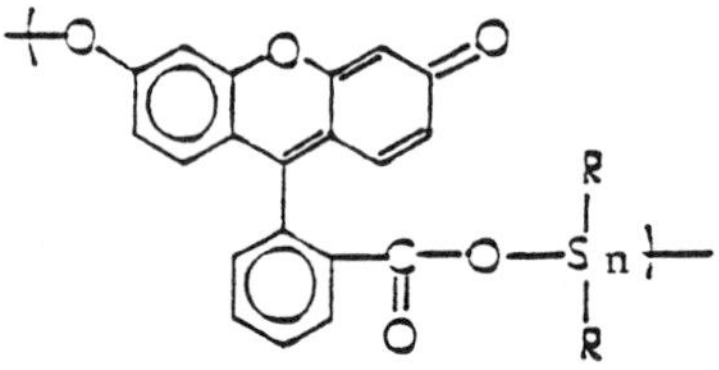

5

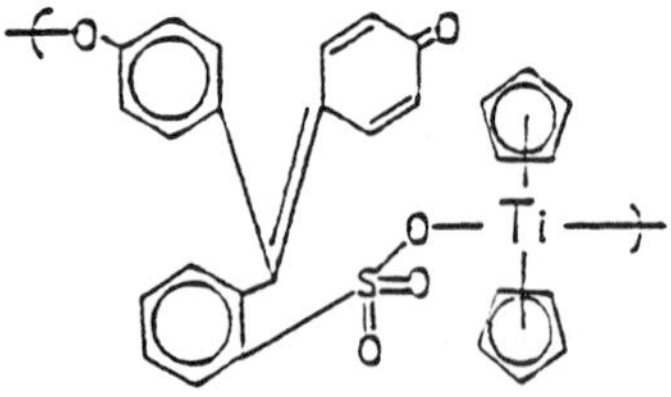

6

Fluorescein dyes have various uses. As dyes, they are employed in the textile field, as laser dyes, as food, drug and cosmetic colors and in coloring paper.[20-22] They are used in lacquers and in the printing industry.[20] Their fluorescent properties make them useful in specialty fields such as in producing theatrical effects in textiles, in fluorescent sign printing inks, to impart fluorescence to mineral oils and to detect water seepage and locate flaws in metal castings.[20] They also have been employed as biological stains, fungicides and photographic sensitizers.[20] Sulphonephthalein dyes are also widely employed and exhibit a relatively high color intensity.[20-22]

Other dyes have also been included in the backbone of analogous metal-containing polymers such as Congo Red, Eriochrome Black T and Indigo Carmine.[19] As a group these polymers are referred to as metal-containing polydyes or simply polydyes.

The advantage of including dyes as part of the backbone of polymer chains is the (potential) permanent, nonleaching nature of such polydyes compared with the monomeric dyes themselves where movement, even within apparently solid lattices (such as within plastics), occurs over extended time. This is especially important where environmentally acceptable biological activity is needed. An example is marine coatings, where the major commercial antifouling agents are monomeric and currently produce unacceptable levels of leaching. The polydyes may not only offer desirable coloring, but desirable ultraviolet "sink" activity due to the presence of the Cp_2Ti moiety which can absorb UV radiation without being degraded. This may lead to better weatherability of exterior coatings through retardation of ultraviolet-associated degradations. The ability to absorp energy in a nondestructive manner may also permit curing of coatings to greater depths than is normally possible. Tin-containing polydyes may act to inhibit mildew and rot-causing organisms.

This paper briefly describes the results of preliminary biological studies, several aging studies and finally preliminary studies involving high energy laser abatement of films and plastic materials containing small amounts (ppm) of selected polydyes.

EXPERIMENTAL

Synthesis of Materials The synthesis and structural characterization of the polydyes is described elsewhere (for instance Refs. [15-19]). Synthesis occurs by addition of aqueous solutions containing dye and any added base to rapidly stirred aqueous (for aqueous solution systems, Cp_2MCl_2 only) or organic (for interfacial systems) solutions containing the dihalo organometallic reactant.

The stannane-containing polydyes were generally oligomeric or low molecular weight polymers (DP=2-50), while the titanium-containing polydyes were generally high polymers (DP>100).[15-19]

Polydyes as Additives for Property Enhancement A number of the dyes have been employed as additives to plastics, sealants, caulks, paper materials, textiles and coatings. The following briefly describes general procedures employed. For paper, one to three drops of DMSO solution containing several ppm of dissolved polydye was applied and permitted to air dry. Several drops of DMSO solution were added to latex coating (exterior) materials. The resultant coatings containing polydye were applied to soft pine wood and allowed to dry. Flexible films were formed by pouring a thin layer onto a teflon-coated muffin pan and allowing it to dry. The resulting film was peeled from the pan and cut into appropriate sized samples. Some of these films were exposed to high energy radiation for specified times. The DMSO solutions containing polydye were added, again only two to four drops, to numerous plastics heated above their melt (or glass) temperatures. These plastics included polyethylene, polycarbonate (Lexan), polypropylene, high density polyethylene, polystyrene, poly(vinyl chloride), a number of copolymers, nylon 6,6 and PET (polyester). The polydyes were added as solids to a variety of sealants and caulks with hand mixing attempting to achieve an even mixing of the polydye material. A variety of fabrics have also been impregnated with DMSO solutions of the polydyes. These include cotton, denim, 50-50% cotton-polyester fabric, polyester, wool and nylon. The concentrations of dye and polydye (and where appropriate, film thickness) are as follows: solutions - 10^{-4} and 10^{-5} molar; impregnation of cloth, plastic and paper - $2x10^{-6}$ grams per gram of material; coatings and films - 200ppm and 0.2 mm thickness. Additions to molten plastics, coatings and caulks all appeared to be compatible. No effort was made to remove DMSO.

Biological Analyses A number of general biological analysis techniques were employed. Here is reported results of simple petri dish culture studies. For paper disk assays, the test organism is suspended in sterile water. Appropriate dilutions are made in Sabourand's dextrose agar and the solutions poured into petri plates. The samples (for solutions - paper disks are soaked in solutions containing the tested material or for solids - are "sprinkled" onto the plates) are added to the plates. The plates are incubated and results obtained.

Radiation For stability to high energy radiation, a Sylvania, Model 971 CO_2 laser was used (1.06 nanometers). Power was measured by a Model 213 Optical Power Meter and time of application was controlled employing a Systron/Donner Model 6150A Counter/timer. Burn through or radiation penetration was evaluated employing a specially designed apparatus described in reference 23.

RESULTS AND DISCUSSION

Stability to High Energy Radiation Brief laser stability tests were carried out using Lucite films impregnated with about 200 ppm tin-containing polydyes. The tested stannane polydyes were derived from diphenyltin dichloride condensed with Rose Bengal, Phloxine B, Eosin Y and Mercurochrome. In all cases, the tin-containing dyes lessened laser damage, allowing the polymeric material to be penetrated without excessive burning or melting. However, under the same conditions, Lucite films not containing the polydyes showed excessive burning. Figure 1 contains photographs of some of the polydye-containing films (thickness about 0.2 mm) "shot" with 5, 10, 25 and 50 watts of laser power. The polydyes derived from Rose Bengal and Mercurochrome showed no flame on exposure to the laser beam. The ability of the polydye impregnated materials to moderate the laser-energy is not expected since these metal-containing moieties possess the ability to "attract" radiation and to expel this radiation prior to subsequent molecular damage to the impregnated material.[28] It is surprising that such dramatic improvements in the stabilities of the materials occurs at such low polymer concentrations.

Plastic "plugs" impregnated with three to five drops of DMSO solution, $2x10^{-6}$ grams polydye per gram of plastic, were exposed to 500 watts per cm^2 of laser power to effect burn through (generally less

Fig. 1. Latex films containing 200 ppm of polydyes derived from diphenyltin dichloride and Phloxine B, Rose Bengal and Eosin Y exposed to laser energies of 5-50 Watts/cm^2 power.

a. Phloxine B - Φ_2SnCl_2
b. Rose Bengel- Φ_2SnCl_2
c. Eosin Y + Φ_2SnCl_2

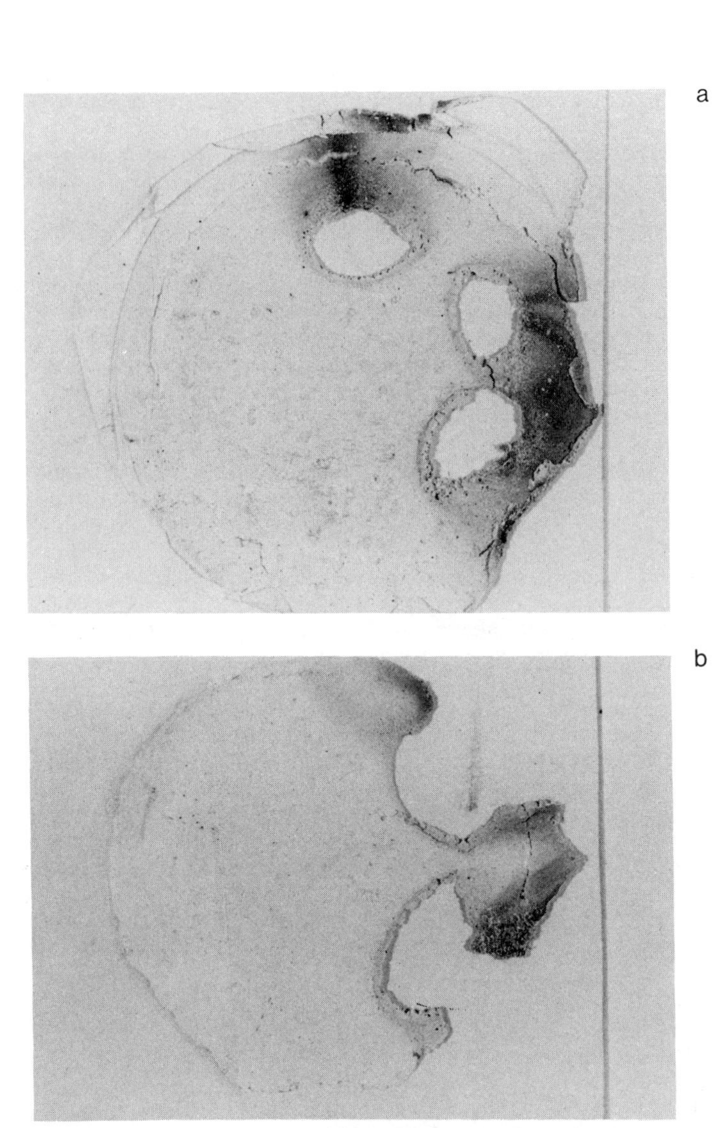
a
b

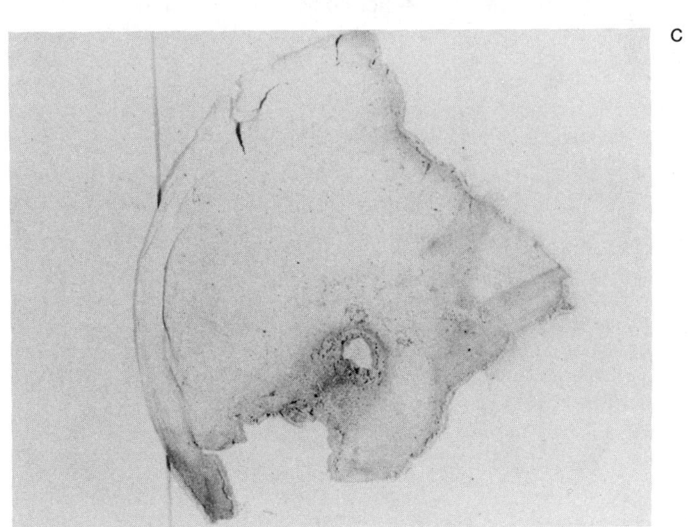
c

than 10 secs.). The solid polydye-impregnated plugs again did not burn. Burning did occur if voids were present for some samples.

Coatings (containing about 200 ppm polydye) on wood were also exposed to laser "blasts." The majority of the polydyes showed equal or greater minimum laser-burn-through-energies compared to the latex itself. For instance, the product from diphenylstannane dichloride and Phloxine B burns through at a laser power of 10 watts/cm^2 whereas an equivalent thickness (about 0.2 mm) of coating without polydye has a minimum burn through energy of 8 watts/cm^2. In another experiment, laser burn through times were determined as a function of power. Again, the coated portions containing added polydyes showed equal to greater burn-through times. For instance at 2 watts/cm^2 power, the coating containing diphenylstannane dichloride and Phloxine B polymer (100 ppm) has a burn-through (penetration) time of about 2.5 seconds whereas the latex itself has a burn-through time of about 1 second. Again, less burning and/or charring occurs with coatings containing added polydye.

Addition of such polydyes to plastics and coatings clearly may allow added flexibility and control of fabrication of such materials, particularly when laser beams are employed, allowing added use in such areas as microelectronics and biomedical materials.

Aging Tests with titanium and tin-containing polydyes derived from fluorescein dyes have been ongoing for about ten years. Tests with tin-containing polydyes derived from sulphonephthalein have been ongoing for about five years. Tests using materials derived from other dyes have been ongoing for times up to ten years. Test media include plastic plugs, solutions (10^{-4} and 10^{-5} molar), paper, textiles, coatings and films. In all cases the polydye impregnated materials and solutions have retained their coloration over the test period. Of interest, is the relatively short color-life of solutions containing the monomeric dyes. Noticeable color deterioration occurs after several months for such solutions, while the analogous polydyes have shown (apparently) full color retention over the test periods of up to ten years. Tested polymers include the following derived from Cp_2TiCl_2 - Bromothymol Blue, Bromophenol Blue, Eosin Y, Erythrosin B, Phloxine B, Bengal Red B and Fluorescein; dibutyltin dichloride - Bromocresol Purple, Thymol Blue, Phenolsulfonephthalein, Orange II, Mercurochrome; diphenyltin dichloride - Mercurochrome, Eosin Y, Erythrosin B, Fluorescein, Rose Bengal and Phloxine B; dimethyltin dichloride - Mercurochrome. Long term biological testing has just begun.

It is possible that reaction with the metal-containing moiety "locks" the dye into a stable form that resists deterioration. The presence of the metal-containing moiety may also contribute to the exceptional stability of the polydyes.

Biological Results Most tin-containing materials exhibit some inhibition to selected organisms. Many tin-containing nonpolymeric materials are employed as additives to topical medications and as additives to coatings for mildew and rot control. Polymeric tin-containing coatings have been successfully employed in the treatment of wood and ship hulls for reduction in weathering and barnacle prevention (for instance Refs. 24-26).

A two-step program was undertaken to briefly evaluate the biological activities of selected polydyes. A number of text organisms were employed including E. coli, Ps. aeruginosa, Ps. M-2 and S. aureus. For the first part of the study, simple culture plate biological assays were employed with the polydyes simply sprinkled onto the plate. In general, the polydyes successfully inhibited the test organisms while the corresponding monomeric dyes did not (for instance Table 1).

A second set of studies were undertaken where the polydyes were mixed in with caulks, etc. Simple compounding was not sufficient to achieve needed dispersion of the solid polydyes. Thorough mixing did produce samples that resisted attack by the tested microorganisms. Thus 0.2% by weight of a number of fluorescein-derived tin-containing dyes were compounded as solids with general acrylic latex paste and tested against A. niger. After one day no growth was found. After three days inhibition was about 60-75%, about the same as the monomeric organostannane dichloride itself. The tested polydyes included those derived from dimethyltin dichloride and Mercurochrome, diethyltin dichloride and Mercurochrome, dibutyltin dichloride and Mercurochrome, dimethyltin dichloride and Eosin Y and diphenyltin dichloride and Phloxine B. Thus the tin-containing polydes may act as both antibacterial and coloring agents, offering a multipurposed additive.

In another study, tin-containing polydyes were dissolved in DMSO. The DMSO-polydye solution was added to latex paste and mixed to give a homogeneous mixture. Films were then cast. Sections of the films were tested against A. niger. Results appear in Table 2. With the exception of Mercurochrome, none of the monomer-containing dyes inhibited bacterial growth. Conversely, a number of the polydyes showed good inhibition, in particular those derived from Mercurochrome. Such polydyes may prove useful in the long-term inhibition of rot- and mildew-causing organisms compounded into caulks, sealants and coatings.

Table 1. Inhibition as a Function of Organostannane Polyxanthene[a]

Dye	Organisms Tested Pseudomonas Aeruqinosa		Organisms Tested E. Coli	
	Weight Compound (mq)	Inhibition Zone (mm,diam.)	Weight Compound (mg)	Inhibition Zone (mm, diam.)
Eosin Y	10	0	10	1
Erythrosin B	10	0.8	1	0.7
Rose Bengal	4	1.3	10	1.6
Phloxine B	6	0	6	0
Mercurochrome	10	2	10	2.5

[a] All tested as solids with approximately one mm diameter delivery zones; all derived from diphenyltin dichloride.

Table 2. Results for materials (0.1% by weight) incorporated into acrylic latex pastes (films) tested against A. niger.

	Materials	Zone of Inhibition (mm)	Growth on Film
Controls	Me_2SnCl_2	0	+
	Et_2SnCl_2	0	+
	Bu_2SnCl_2	0	+
	$(C_6H_{11})_2SnCl_2$	0	+
	Φ_2SnCl_2	0	+
	Eosin B	0	+
	Mercurochrome	0	-
	Fluorescein	0	+
	Phloxine B	0	+
	Eosin Y	0	+
	Rose Bengal	0	+
Polymers	Eosin Y, Me_2SnCl_2	0	-
	Mercurochrome, Me_2SnCl_2	7	-
	Mercurochrome, Et_2SnCl_2	2	-
	Mercurochrome, Bu_2SnCl_2	10	-
	Phloxine B, Φ_2SnCl_2	0	-
	Eosin B, Φ_2SnCl_2	0	+
	Fluorescein, Φ_2SnCl_2	0	+
	Eosin Y, Φ_2SnCl_2	0	+
	Rose Bengal, Φ_2SnCl_2	0	+
	Mercurochrome, $(C_6H_{11})_2SnCl_2$	0	-

Note: "+' = growth on film sample and '-' = no growth on film sample.

REFERENCES

1. C. Carraher, Jr., J. Polymer Sci., A-1, 9, 366 (1971).
2. C. Carraher, Jr., CHEMTECH, 744 (1972).
3. C. Carraher, Jr., Interfacial Synthesis, Vol II (Edited by F. Millich and C. Carraher), Marcel Dekker, NY, Chapter 21, 1977.
4. C. Carraher, Jr. and S. Bajah, Polymer (Br.), 15, 9 (1974) and 14, 42 (1973).
5. C. Carraher, Jr. and P. Lessek, Eup. Polym. J., 8, 1339 (1972).
6. C. Carraher, Jr., J. Chem. Ed., 46, 314 (1969).
7. C. Carraher, Jr. and Dammeier, J. Polym. Sci., A-1, 10, 413 (1972).
8. C. Carraher, Jr., S. Jorgensen and P. Lessek, J. Appl. Polym. Sci., 20, 2255 (1976).
9. C. Carraher, Jr. and P. Lessek, Angew, Makrmol. Chem., 38, 57 (1974).
10. C. Carraher, Jr. and R. Dammeier, Makromol. Chem. 135, 107 (1970).
11. C. Carraher, Jr., G. Peterson and J. Sheats, Organic Coatings and Plastics Chemistry, 33 (2), 427 (1973).
12. C. Carraher, Jr. and G. Scherubel, Makrmol Chem., 160, 259 (1972) and 152, 61 (1972).
13. C. Carraher, Jr. and G. Scherubel, J. Polym. Sci., A-1, 9, 983 (1971).
14. C. Carraher, Jr., unpublished results.
15. C. Carraher, Jr., R.S. Venkatachalam, T. Tiernen and M. Taylor, Organic Coatings and Applied Polymer Science, 47, 119 (1982).
16. C. Carraher, Jr., D. Giron, D. Cerutis, W. Bert, R. Venkatachalam, T. Gehrke, S. Tsuji and H. Blaxall, Biological Activities of Polymers, (C. Carraher and C. Gebelein, Editors), Washington, DC, (1982).
17. C. Carraher, Jr., R. Schwarz, J. Schroeder, M. Schwartz and H.M. Molloy, Organic Coatings and Plastic Chemistry, 43, 798 (1980).
18. C. Carraher, Jr., L. Tisinger, G. Solimine, M. Williams, S. Carraher and R. Strother, Polym. Mater. Sci. & Eng., 55, 469 (1986).
19. C. Carraher, Jr., R. Schwarz, J. Schroeder and M. Schwarz, J. Macromol. Sci.-Chem., AIS 5, 773 (1981).
20. R. Kirk and D. Othmer (Editors), Encyclopedia of Chemical Technology, Volumes 8 and 15, Intersciences, NY, (1979).
21. D. Marmion, Handbook of U.S. Colourants for Food, Drugs and Cosmetics, Wiley-Interscience, NY, (1979).
22. E. Bishop (Editor), Indicators, Pergamon Press, NY, (1972).
23. C.M. Phillippi, D.F. Stevison and W.J. Ekman, U.S. Patent 3,926,034, issued December 16, 1975.
24. C. Gebelein and C. Carraher, Jr., (Editors) Bioactive Polymeric Systems, NY, (1985).
25. C. Gebelein and C. Carraher, Jr., (Editors) Polymeric Materials in Medication, Plenum, NY, (1985).
26. C. Carraher, Jr. and C. Gebelein, (Editors) Biological Activities of Polymers, ACS, Washington, DC, (1982).
27. R.V. Subramanian and K. Somasekharan, Advances in Organometallic and Inorganic Polymer Science, (C. Carraher, J. Sheats and C. Pittman, Editors), Chapter 2, Dekker, NY, (1982).
28. R. Field and P. Cowe, The Organic Chemistry of Titanium, Butterworths, Washington, DC, (1965).

SYNTHESIS AND CHARATERIZATION OF ATOMIC OXYGEN RESISTANT POLY(SILOXANE-IMIDE) COATINGS*

Iskender Yilgör

MERCOR Incorporated
2448 Sixth Street
Berkeley, California 94710

ABSTRACT

Synthesis of fully imidized, high molecular weight and soluble poly(siloxane-imide) segmented copolymers based on benzophenonetetra-carboxylic acid dianhydride (BTDA), 3,3'-diaminodiphenylsulfone (DDS) and α,ω-aminopropyl-terminated polydimethylsiloxane oligomers (PSX) have been demonstrated. During these studies, three different siloxane oligomers with number average molecular weights of 910, 2500 and 5000 g/mole were utilized. Level of siloxane incorporation was varied between 10 and 50 percent by weight. Copolymers obtained were soluble in polar solvents such as DMAC and NMP and showed fairly high molecular weights as indicated by their intrinsic viscosity values. They also displayed very good mechanical properties. DSC studies showed the formation of two phase morphologies with low and high temperature T_g values around -120 and +250°C corresponding to siloxane and polyimide transitions respectively. TGA analysis indicated fairly good thermal stability. Water contact angle measurements showed the formation of siloxane rich surfaces, which resulted in dramatic reductions in the equilibrium water absorption levels of these siloxane-imide copolymers compared with control BTDA-DDS polyimide. Incorporation of siloxanes also resulted in substantial improvements in the oxygen plasma resistance of these materials. When spray coated, siloxane-imide copolymers showed excellent adhesion to Kapton®. Oxygen plasma resistance of coated Kapton films showed substantial improvement over the uncoated control.

INTRODUCTION

High performance polymeric materials such as aromatic polyimides, graphite reinforced epoxy composites and fully fluorinated polymers

* This paper was presented to the ACS Symposium on "High Performance Polymers for Harsh Environments," April, 1987.

have been used in aerospace systems for various applications.[1-3] Polyimides are the most widely used polymers of this type as adhesives and also as a component of thermal control blankets due to their high temperature and UV stabilities and toughness. Polymer composites have potential for use as lightweight, high stiffness, low thermal expansion structural materials on spacecraft. Although these polymeric systems perform extremely well under fairly extreme conditions, their long-term stability in the Low Earth Orbit (LEO) environment is a major concern. Data obtained from earlier Space Shuttle flights (STS-1 through STS-5) indicate that substantial surface erosion and mass loss have occurred in many surface protective materials commonly used in aerospace applications such as Kapton®, Mylar®, Kevlar® and various polyurethanes.[4-6] It has been suggested that the degradation in these coating materials is primarily due to the reaction of atomic oxygen, predominant species in the LEO environment, with the polymer surface.[7] Due to extremely high velocities of the atoms (8 km/s) impinging on the shuttle surface, reactive oxygen atoms with energy levels of approximately 5 eV cause severe degradation on many polymers over fairly short periods of exposure. As a result of atomic oxygen degradation, the surface morphology, physical strength and thermal and optical properties of the polymers show significant changes. Similar results have also been obtained by various groups using oxygen plasma generators which simulate the effects associated with LEO in the laboratory.[8,9]

In order to obtain quantitative information on the interaction of some widely used spacecraft materials with conditions encountered in LEO environments, NASA researchers have evaluated various polymeric systems in a Space Shuttle flight (STS-8) experiment.[10,11] The materials evaluated included a series of aromatic polyimide films with varying backbone structures, both polymer- and metal-matrix composites, pigmented polyurethanes, a polydimethylsiloxane, a potassium silicate, an FEP Teflon® covered with a silver coating and two polysiloxane-polyimide films. The materials were mounted in the payload bay of the Shuttle and exposed to the LEO environment at an altitude of 225 km for 41.75 hours. This exposure provided a fluence of 3.5×10^{20} oxygen atoms/cm^2 perpendicular to the surface of the samples. The results of atomic oxygen degradation were determined by computing reaction efficiencies and characterizing surface and optical property changes. Analysis of the results indicated that composites protected by experimental thin-film metallic coatings were not eroded.[10] However, the polymeric matrix (graphite/epoxy) composites and all polymeric films and coatings with the exception of polydimethylsiloxane and polysiloxane-polyimide copolymers showed significant degradation. The polysiloxane-polyimide copolymers based on benzophenone tetracarboxylicacid dianhydride (BTDA), oxydianiline (ODA) and aminopropyl-terminated polydimethylsiloxane oligomers, containing 7 and 25 percent by weight of siloxane in their backbone compositions, showed reaction efficiencies of about an order of magnitude less than those of the aromatic polyimides. Polydimethylsiloxane and potassium silicate (inorganic) based paints were also very stable to atomic oxygen exposure. Interestingly, among nine polyimide samples with different backbone compositions, the Kapton® which is based on pyromellitic dianhydride (PMDA) and ODA showed the highest reaction efficiency.[10,11]

Laboratory experiments[8] using an oxygen plasma reactor at relatively high fluence atomic oxygen environment of 1.85×10^{22} atoms/cm^2, on four different polymeric materials including Kapton® and siloxane coated Kapton® films have also shown the same type of behavior. It was determined that the time needed to completely remove the Kapton was equivalent to 5 to 7 days in the LEO environment. On the other hand, siloxane coated Kapton film showed that the coating was resistant to oxidation after 125 equivalent days in LEO. The proposed degradation mechanism indicated that the coating was still in the induction phase of a diffusion-controlled oxidation.

Polysiloxane containing copolymers have several other advantages which make them attractive for applications as protective coatings. The preparation of linear block copolymers containing polydimethylsiloxanes and a variety of other organic blocks have been demonstrated.[12-15] These copolymers can often be easily processed either thermoplastically or from solution.[12] The transparent films obtained show mechanical properties ranging from elastomeric to quite rigid depending on the type, nature, weight percent and block length of the hard segments in the copolymers.[12,13] The covalent incorporation of a polysiloxane oligomer into the block copolymer backbone yields a microphase separated morphology due to the large differences in the solubility parameters of polysiloxane and the rigid blocks. It has further been established that in these copolymers there is a driving force for the siloxane, the lower energy component, to migrate to the air or vacuum interface. Thus contact angle measurements and photoelectron spectroscopy (ESCA) have consistently shown that the surface of the siloxane modified copolymers is predominantly the low energy polydimethylsiloxane.[15] Even at very low levels (2-5% by weight) of polysiloxane incorporation one can obtain materials with low-energy (siloxane dominated) surfaces, wherein the bulk properties of the material, such as strength, thermal and mechanical properties, are controlled by the hard block. Hence, as a coating applied over a substrate material, the tendency for the polysiloxane to migrate to the polymer-air interface may make the level of coating required extremely small.

Various research groups have been investigating the synthesis of siloxane-imide copolymers over the last decade.[16-23] The major interest in these type of copolymers was due to their hydrophobicity and superior performance as protective coatings in electronic circuits. However, most of the earlier investigators have utilized low molecular weight disiloxanes in order to obtain solubility and processability in the resulting copolymers.[16-18] It has later been shown that disiloxanes do not lead to the phase separated morphologies in the copolymers produced[12,24] and so are not as effective as higher molecular weight (M_n>600g/mole) polysiloxanes in providing the desired surface properties. When high molecular weight siloxane oligomers were utilized, the resulting polysiloxane-polyimide copolymers were no longer soluble or processable due to the increased block lengths of the rigid polyimide segments. This was a major problem in the application of these type of copolymer systems as thin film coatings. Recently, several groups have claimed the synthesis of soluble polysiloxane-polyimide copolymers.[25,26] However, in most of these systems, the level of the siloxane was very high and the backbone structures of the siloxane oligomers utilized were fairly complex.[27]

Therefore, in this study our major goal was to investigate the feasibility of the synthesis of soluble polysiloxane-polyimide block copolymers which could be cast onto a substrate from solution in the fully cured state. The conditions required for coating would be those necessary for solvent removal only, thereby making the process easily adaptable to existing coating facilities. It is also well known that high-molecular weight polysiloxane-polyimide copolymers with controlled structures are good adhesives and also display good strength and high-temperature stability.[28] This will eliminate the need for using an adhesion promoter in order to obtain good adhesion.

EXPERIMENTAL

Materials and Purification High purity, polymerization grade benzophenonetetracarboxylicacid dianhydride (BTDA), was obtained from Chriskev Company (Leawood, KS) and used as received. It was stored in a desiccator when not in use.

3,3'-Diaminodiphenylsulfone (DDS) was purchased from FIC Corporation (San Francisco, California). It was recrystallized from deoxygenated 60/40 (v/v) methanol/distilled water solution. After recrystallization DDS was dried in a vacuum oven at 80°C, overnight. (Yield 90%, m.p. 172-173°C).

Bis(3-aminopropyl)tetramethyldisiloxane (DSX) was obtained from Petrarch Systems, Inc. Octamethylcyclotetrasiloxane (D_4) was a product of Dow Corning. Both materials were used as received.

Various molecular weight, α,ω-aminopropyl-terminated polydimethylsiloxane oligomers (PSX) were synthesized in our laboratories by the base catalyzed equilibration of DSX and D_4 in bulk, at 80°C.[29,30] Average molecular weight of these oligomers were controlled by the initial stoichiometric ratio of DSX to D_4 in the reaction system. Table 1 provides the data on the synthesis and characteristics of the siloxane oligomers synthesized.

Reaction solvents, tetrahydrofuran (THF) and dimethylacetamide (DMAC) were carefully dried over calcium hydride and barium oxide respectively and freshly distilled (DMAC under vacuum) before use. Water contents of solvents after distillations were determined by Karl-Fisher titration and found to be less than 50 ppm in every case.

Reaction Procedure

Synthesis of Poly(amic acid-siloxane) Intermediates Poly(amic acid-siloxane) intermediates were synthesized in 250 ml, 3-neck round bottom reaction flasks fitted with a drying tube, nitrogen inlet, thermometer and a mechanical stirrer. All glasswares were dried overnight in an oven at 100°C before use. Reactions were conducted at room temperature, in two steps. At the end of these reactions, clear, viscous poly(amic acid-siloxane) solutions were obtained. A typical procedure for the two-step synthesis of a poly(amic acid-siloxane) intermediate based on PSX-910, BTDA and DDS was as follows:

In the first step 4.95 g (15.36 mmole) of BTDA was introduced into the reaction flask and dissolved in a mixture of 30 ml DMAC and 5 ml THF. Then 6.90 g (7.58 mmole) of PSX-910 was dissolved in 30 ml of THF, transferred into an addition funnel and added dropwise into BTDA solution. This process resulted in a clear solution of BTDA-PSX prepolymer. In the second step, a solution of 1.94 g (7.81 mmole) of DDS in 10 ml of DMAC and 5 ml of THF was added into the prepolymer solution dropwise, at room temperature and the reaction mixture was stirred overnight. The result was a clear, viscous solution of poly(amic acid-siloxane) intermediate, which was stored in a desiccator, at room temperature, until imidization.

Imidization Procedure A film of the poly(amic acid-siloxane) intermediate was cast on a Teflon plate and placed into a vacuum oven at 80-90°C for 4-6 hours where the solvents and water of imidization were removed. At the end of this process, IR spectroscopy and TGA studies indicated about 65% imide formation. The film was then placed in a furnace and the imidization was completed by curing the films at 200 and 250°C, one hour at each temperature. The films obtained by this method were clear and light brown in color.

Characterization Techniques The formation of poly(amic acid-siloxane) intermediates and the completion of imidization reactions were routinely followed by IR spectroscopy, using a Perkin Elmer Model 229B Infrared Spectrophotometer and a Nicolet MX-1 FT-IR Spectrometer.

Intrinsic viscosity measurements were performed at 25°C, in dilute NMP solutions using Ubbelohde viscometers.

Water-uptake of control polyimides and siloxane-imide copolymers were conducted at 23°C. Samples of polymer films (2.0x2.0x0.015 cm) were immersed into a distilled water bath at constant temperature and the weight gains were determined gravimetrically using an analytical balance. Samples were removed from the water periodically, surfaces were dried well by blotting and then weighed to determine the water uptake. Percent water uptake was plotted against time.

Stress-strain behavior of the polymers were obtained by using an Instron Model 1122 Tensile Tester. Dog-bone shaped test specimen were punched out of thin (0.015 cm) polymer films using standard ASTM dies and the tests were conducted at room temperature with a cross-head speed of 1.0 cm/min.

Glass temperatures of the copolymers were determined on a Perkin Elmer DSC-2, under helium or nitrogen atmosphere. The temperature range covered was from -150 to +300°C with a heating rate of 10°C/min. Each sample was scanned at least twice and the T_g values were usually determined from the second scan. Thermogravimetric analyses of the products were studied by using a Perkin Elmer TGS-2, under both air and nitrogen atmosphere. Heating rate was 10°C/min. TGA was also used to determine the percent imidization of the samples by measuring the amount of water released at various stages of the curing process.

Contact angle measurements were done on a Kernco Model G-1 Goniometer. Contact angles were measured at room temperature using distilled water. The volume of the water droplet was 2 µl. The average of 5 measurements was taken as the contact angle for each sample.

A Plasmod (Tegal Corporation, Richmond, California) operating at a reduced pressure of 150 millitorr and an oxygen flow rate of 50 cc/min was used to determine the durability of siloxane-imides in an atomic oxygen environment. Only the air-polymer surface of each film was exposed to the atomic oxygen. Exposure time was kept constant at 45 min and duplicate experiments were conducted for each sample. After the exposure, percent weight loss in the polymer films was determined.

RESULTS AND DISCUSSION

Synthesis and characterization of siloxane-imide copolymers have been investigated by various research groups in the last decade.[16-28]. The interest in these types of materials arises mainly due to the inherent unique properties possessed by each homopolymer. Polyimides are very well known for their thermal, electrical, oxidative and solvent stability, very high T_g, toughness and excellent adhesive properties. However, their processing using conventional solution or melt techniques is almost impossible due to their insolubility and infusibility. Polydimethylsiloxanes, on the other hand, have good thermal, oxidative, UV and atomic oxygen stability, very low T_g (-123°C) (as a result good flexibility even at extremely low temperatures), hydrophobic surface properties and good processability. Therefore, when combined in a block or segmented type of macromolecular architecture, siloxane-imide copolymers were expected to lead to the preparation of materials which would display many interesting properties on a single polymer chain and would possibly find applications in various fields requiring extreme service conditions and good flexibility.

In the earlier studies on the preparation of siloxane-imide copolymers, usually the low molecular weight aminoalkyl-terminated disiloxanes were incorporated into the imides at high levels.[16-18] This imparted good solubility and processability to the resulting copolymers. However, since thse disiloxanes did not lead to the formation of phase separated morphologies, overall properties of the copolymers, especially the thermal stability, were relatively poor. When high molecular weight siloxane oligomers were employed, they usually resulted in the formation of insoluble and infusible copolymers after complete imidization. This was mainly due to the formation of long imide segments and high overall molecular weight of the copolymers. It was possible to obtain soluble siloxane-imide segmented copolymers by limiting the overall molecular weight of the copolymers or by increasing the amount of siloxane incorporated (so decreasing the amount and length of the imide blocks). Most of these materials were evaluated for possible applications in electronic coatings.[16,17] Some siloxane modified addition-type polyimides were also evaluated as high temperature, toughened adhesives for aerospace applications.[26] In a recent Space Shuttle (STS-8) experiment siloxane-imide copolymers were found to show very good resistance against atomic oxygen degradation.[10,11] Therefore, these types of materials are potential candidates for applications as protective coatings in spacecraft and other space structures orbiting in LEO environments.

One of the primary requirements for a good coating is the ease of application, possibly from a solution. Therefore in this study our main objective was to investigate the synthesis of high molecular weight, soluble and thermally stable poly(siloxane-imides) for possible use as protective coatings against atomic oxygen in aerospace applications. Physical, chemical and engineering properties of the materials synthesized were also characterized.

<u>Synthesis of Siloxane Oligomers and Siloxane-Imide Copolymers</u> α,ω-Aminopropyl-terminated polydimethylsiloxane oligomers (PSX) were prepared by the base-catalyzed equilibration reactions of octamethylcyclotetrasiloxane in the presence of 1,3-bis(3-aminopropyl)tetramethyldisiloxane end blockers.[37,38] Table 1 provides a list of PSX oligomers synthesized and some of their physical characteristics. Formation of the predicted structures were confirmed by IR and ^{1}H-NMR spectroscopy. As expected, T_g values of PSX oligomers are all around -120°C as determined by DSC. These oligomers were later used in the preparation of siloxane-imide copolymers.

Siloxane-imide copolymers synthesized in this study were all based on BTDA, which is known to provide better solubility than PMDA (pyromellitic dianhydride). DDS was chosen as the diamine because of its 3,3'-substitution and due to the presence of bulky, somewhat flexible SO_2 bridging units, which are known to inhibit the formation of ordered structures and therefore improve solubility and processability of polyimides, while at the same time providing good thermal and oxidative stability.[31,32]

The generalized reaction scheme for the synthesis of siloxane-imide segmented copolymers is given in Fig. 1. One of the major problems in the synthesis of siloxane-imide copolymers starting from high molecular weight polydimethylsiloxane oligomers, is the proper choice of the reaction solvent. This is due to the great difference between the solubility parameters of polydimethylsiloxanes (δ=7.5 $(cal/cm^3)^{1/2}$] and aromatic polyimides [$\delta \approx 10$ $(cal/cm^3)^{1/2}$] which makes it very difficult to find a common solvent during the reactions. The use of a poor reaction solvent results in the premature precipitation of the polymeric products formed and therefore makes the synthesis of high molecular weight copolymers impossible. This type of behavior is usually observed in the copolymerization of siloxanes with many other organic monomers to synthesize segmented copolymers such as in siloxane-urethanes and siloxane-ureas.[13,14,21] As a result, in these types of systems control of the polymer structures (backbone composition) is also fairly difficult. Therefore, in our investigations special emphasis was given to the proper choice of the reaction solvent(s) and the curing conditions in order to obtain high molecular weight, soluble products with controlled structures.

During our studies a combination of THF and DMAC as the reaction solvent has been very successful. In many cases it is possible to substitute DMAC with NMP or other polar aprotic solvents. These solvent combinations have also been reported by McGrath and co-workers[26] for siloxane-imide synthesis. Highly polar solvents like DMAC and NMP are commonly used for imide reactions, however, PSX oligomers are not soluble in DMAC. Addition of THF, a good solvent for siloxanes, provides a clear homogeneous solution during the reactions, and the solution is essential in producing high molecular weight copolymers.

Table 1. Characteristics of α,ω-Aminopropyl Terminated Polydimethylsiloxane Oligomers

$$H_2N-(-CH_2)_3-[-\underset{CH_3}{\overset{CH_3}{|}}Si-O-]_n-\underset{CH_3}{\overset{CH_3}{|}}Si-(-CH_2)_3-NH_2$$

M_n (g/mole)	n	T_g (°C)
600	7	-115
910	11	-118
1500	19	-120
2500	32	-123
5000	66	-123
10000	133	-123

Special curing procedures were also followed for the conversion of poly(amic acid-siloxane) intermediates into fully imidized and soluble poly(imide-siloxanes). For our system it was necessary to remove the solvents and the major portion of the water released during imidization in a vacuum oven at temperatures around 80-90°C, in order to obtain transparent, high molecular weight polymers. This suggests that high levels of water evolved in these reactions may be participating in the competing hydrolysis reactions. At the end of this procedure usually there was 50-65% imidization in the copolymers. Complete conversion into imide was achieved by final curing of the films produced in a furnace at 200 and 250°C, 1 hour each. The control BTDA-DDS based polyimides were prepared in DMAC and the films were cured at 100, 200 and 300°C, 1 hour each, after removing the solvent under IR lamps.

The formation of poly(amic acid-siloxane) intermediates and their conversion to poly(imide-siloxanes) were followed by IR spectroscopy monitoring the disappearance of hydroxy (3300 cm^{-1}) and amide and acid carbonyl (1750-1600 cm^{-1}) bands and the formation of imide peaks (1776 and 700 cm^{-1}). The incorporation of siloxanes into these systems was

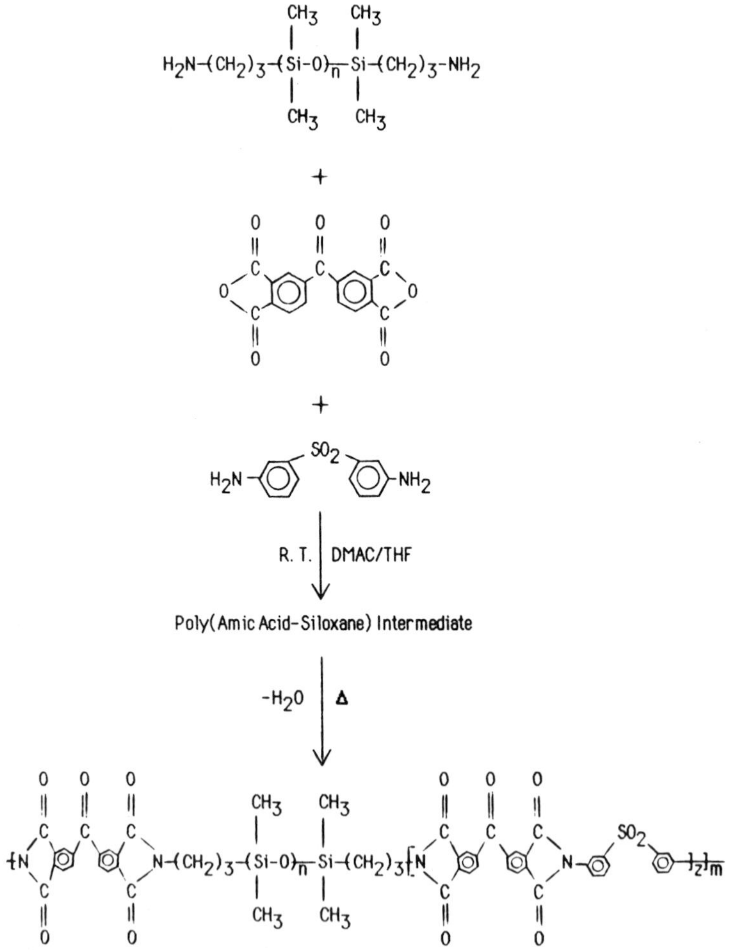

Fig. 1. Reaction scheme for the synthesis of siloxane-imide segmented copolymers.

evidenced by the formation of broad bands between 1100 and 1000 cm^{-1}, which is due to the (Si-O-Si) stretching.

Table 2 provides compositional data and various other characteristics of the siloxane-imide copolymers synthesized. All of the polymer films obtained were transparent and had light to dark brown colors. During the reactions three α,ω-aminopropyl terminated siloxane oligomers with molecular weights of 910, 2500 and 5000 g/mole were chemically incorporated into aromatic polyimides at levels varying between 10 and 50 percent by weight. The control polyimide based on BTDA and DDS was insoluble in the solvent systems investigated. However, all of the siloxane-imide copolymers were soluble in NMP and DMAC and at high level (50%) of siloxane incorporation they were also soluble in DMF. As expected, this clearly shows the dramatic effect of siloxane incorporation on the improvement in the solubility of the resulting siloxane-imide copolymers. None of the copolymers were soluble in methylene chloride. The intrinsic viscosity data given in Table 2 strongly indicates the formation of fairly high molecular weight products.

Thermal Characteristics of Siloxane-Imide Copolymers Results of the DSC analysis on siloxane-imide copolymers are also included in Table 2. The control polyimide based on BTDA-DDS displays a T_g value of 271°C. All but one of the siloxane-imide copolymers show two distinct glass temperatures, which indicate the formation of multiblock copolymers having two phase morphologies. The low temperature transitions around -120°C are due to the presence of siloxane groups in the copolymers.

Table 2. Characteristics of Poly(siloxane-imide) Segmented Copolymers Synthesized

Sample	PSX		$[\eta]^*$ (dl/g)	T_g(°C)	
	M_n(g/mole)	(Wt%)		PSX	Imide
1	BTDA-DDS Control	BTDA-DDS Control	---	---	271
2	910	10	0.6	---	252
3	910	20	0.8	-117	246
4	910	30	0.4	-119	226
5	910	50	0.5	-119	211
6	2500	10	0.7	-121	256
7	2500	30	0.3	-123	252
8	2500	50	---	-123	258
9	5000	10	0.7	-123	263
10	5000	20	0.5	-123	258

*NMP, 25°C

In Sample 2, which contains 10% by weight of PSX-910, no siloxane T_g was observed because it was beyond the sensitivity of the DSC instrument. T_g values observed in the high temperature region (211-263°C) indicate the formation of imide hard segments with varying block lengths resulting from the backbone composition. Since the incorporation of higher levels of siloxane into the copolymers will result in lowering the amount and the molecular weight of the imide segments, a decrease in its T_g value is expected. This trend is clearly seen in Table 2, for Samples 2-5 which contain 10, 20, 30 and 50% by weight of PSX-910 and high temperature T_g values of 252, 246, 226 and 211°C respectively. However, as the block length of the polysiloxane is increased to 2500 g/mole keeping the overall weight composition constant, the microphase separated polyimide segments are also lengthened and as a result the imide glass transitions become quite high again. This suggests that incorporation of low levels of a high molecular weight polysiloxane into DDS chain-extended polyimides may be useful to obtain optimum overall properties in these copolymers.

Thermogravimetric analyses under both air and nitrogen atmosphere have indicated the formation of materials with fairly good thermal stability, Representative TGA curves are reproduced in Fig. 2. BTDA-DDS control polyimide shows excellent thermal stability under air up to 500°C in TGA as in Fig. 2. Siloxane containing polyimides are stable up to about 350°C (in air) where they start degrading depending on the amount and molecular weight of the siloxane oligomers incorporated into their backbone structures. In a siloxane-imide copolymer synthesized using aminopropyl-terminated polydimethylsiloxane oligomers, the weakest bonds in terms of thermal stability are the methylene units in the aminopropyl groups. As a result, the degradation starts from these linkages. This can be observed in Fig. 2. When we compare PXS-910 and PSX-2500 based siloxane-imides containing the same weight percent of siloxane in their backbone composition (e.g., Samples 4 and 7), it is clear that PSX-910 containing polyimides start degrading earlier, due to higher levels of methylene units present in their structures. In order to prevent the premature degradation of siloxane-imide copolymers due to the presence of aliphatic methylene groups in siloxane oligomers, one can synthesize aromatic amine-terminated PSX oligomers, where the aromatic ring is directly bonded to the silicon atom or via a (Si-O-C) linkage, which is known to be thermally more stable.

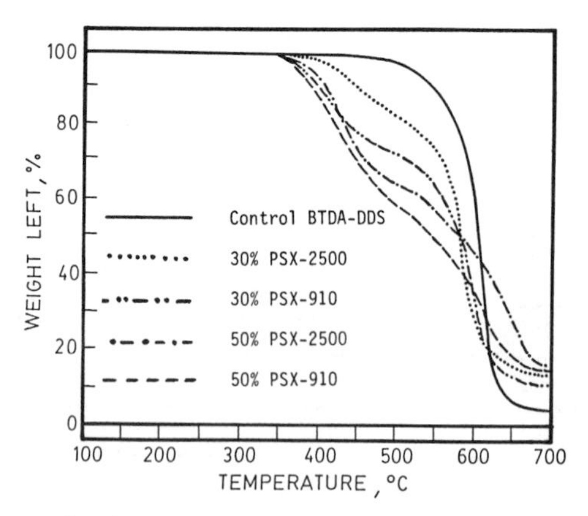

Fig. 2. Thermogravimetric analysis of siloxane-imide copolymers based on BTDA and DDS.

Mechanical Properties of Siloxane-Imide Segmented Copolymers Stress-strain behaviors of control polyimides and segmented siloxane-imide copolymers were also investigated. The results are given in Table 3. The control polyimides and commercial Kapton samples show extremely high modulus and tensile strength values and low elongation at break, typical of polyimides. Incorporation of siloxanes into these strong structures results in reductions in the modulus and ultimate tensile strengths and improvements in the ultimate elongations, as expected. The extent of changes in specific properties is a function of the level of siloxane incorporation and the molecular weight of the siloxane oligomers used. However, as segmented copolymers, all of these siloxane-imide systems, even at very high levels of siloxane incorporation, show very high modulus and strength when compared with other siloxane copolymers.[12-14] Since adhesion to a substrate and abrasion resistance are important for coating systems and are often related to modulus, it is suggested that the mechanical properties must now be evaluated in those terms to optimize the coating performance without affecting other desirable properties of these copolymers.

Surface Properties and Water Absorption In addition to the flexibility, solubility and processability, incorporation of siloxanes into polyimides also imparts hydrophobicity to the resulting copolymers. This is a very important contribution, especially when these siloxane-imide copolymers are used as protective surface coatings. It is known that imide bonds are susceptible to hydrolysis. Depending on their backbone structures, aromatic polyimides usually absorb a few percent of water. Under actual application conditions this may lead to a slow cleavage of the imide bonds, which, in turn, affects the overall properties and performance dramatically. Siloxanes are known for their hydrophobicity and find various applications as water repellants. It has also been shown by ESCA studies that in siloxane-containing block or segmented copolymers, due to their very low surface energies, siloxane segments thermodynamically tend to occupy or migrate to the polymer-air or polymer-vacuum interface, thereby providing hydrophobic surface properties to the resulting systems.[14,31] A simple but effective way of determining surface hydrophobicity or hydrophillicity is by contact angle measurements using distilled water. A low contact angle value (e.g., 70-75° or lower) usually indicates a

Table 3. Mechanical Properties of Poly(siloxane-imide) Segmented Copolymers

Sample	PSX		Initial Modulus (ksi)	Tensile Strength (ksi)	Elongation (%)
	M_n(g/mole)	(wt%)			
1	Kapton	Kapton	430	25	40
2	BTDA-DDS	BTDA-DDS	400	19	5
3	910	10	103	6.0	20
4	910	30	17	2.9	100
5	2500	10	170	5.4	23
6	2500	30	21	1.7	50

hydrophillic surface and higher values (e.g., between 90 and 110°) show hydrophobic surfaces, characteristics of polysiloxanes. The results of our investigations on water contact angle measurements of the control polyimides and siloxane-imide copolymers are given in Table 4. DDS-based control polyimide shows a water contact angle of 72°. However, when only 10 weight% of PSX-910 is incorporated, a sharp increase in the water contact angle from 72 to 98° is observed (Table 4), reflecting the formation of siloxane-rich surfaces. As the amount of siloxane is increased to 50% by weight, the contact angle goes to 104°, which generally shows a surface completely saturated with siloxane. The same type of behavior is observed when PSX-2500 is used.

Another major effect of siloxane incorporation into polyimides is the dramatic reduction in the amount of water absorbed by the resulting copolymers compared with the control polyimide. The results obtained are given in Fig. 3 and Table 4. The control BTDA-DDS polyimide absorbs about 2.70% water at equilibrium when immersed into a distilled water bath at 23°C. Incorporation of siloxane reduces the amount of water absorption to very low levels, 0.80 to 0.30 weight percent. Table 4 indicates that the reduction in water absorption by siloxane-imide copolymers is a function of bulk concentration and also the molecular weight of the siloxane incorporated into the system, which is an expected behavior.

Reduction in water-uptake by the imide copolymers through the incorporation of siloxanes is a very important improvement. Because it is well established that absorbed water in addition to hydrolyzing the imide linkages also plasticizes the polymers, thus lowering the T_g and as a result the upper use temperatures of these types of polymers.

Stability of Siloxane-Imide Copolymers Against Oxygen Plasma Atomic oxygen resistance of control polyimides (Kapton and BTDA-DDS), siloxane-imide copolymers and two Kapton samples coated with thin films of siloxane-imide copolymers were studied in an oxygen plasma generator, under a reduced pressure of 150 millitorr, at a fairly high oxygen flow rate of 50 cc/min. All the samples were exposed to the

Table 4. Water Contact Angle Measurements and Equilibrium Water Uptake Levels of Poly(siloxane-imide) Segmented Copolymers

Sample	PSX		Contact Angle (degree)	Eq. Water Uptake (wt%)
	M_n(g/mole)	(wt%)		
1	BTDA-DDS Control	BTDA-DDS Control	72	2.70
2	910	10	98	---
3	910	30	97	0.70
4	910	50	104	0.30
5	2500	30	100	0.80
6	2500	50	105	0.40

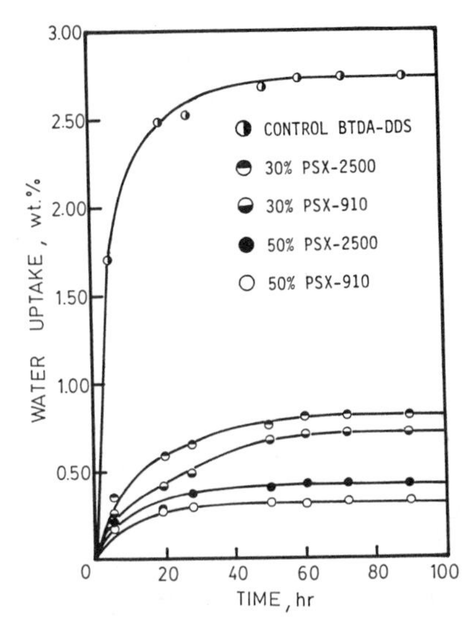

Fig. 3. Water uptake-time curves for siloxane-imide copolymers (T = 23°C).

plasma for 45 minutes at room temperature. However, at the end of each run a rise in sample temperature to 40-50°C was observed. Duplicate experiments were run for each polymer sample. Percent weight loss was determined gravimetrically.

Data obtained is summarized in Table 5. This data suggests that under the experimental conditions Kapton undergoes the highest amount of weight loss, higher than 8% of its original weight. BTDA-DDS seems to be more stable, losing only 3.0% of its original weight upon exposure to oxygen plasma.

Table 5. Effects of Oxygen Plasma on Poly(siloxane-imide) Copolymers and Coated Kapton Films*

Sample	PSX		Weight Loss %
	M_n(g/mole)	(wt%)	
1	BTDA-DDS Control	BTDA-DDS Control	3.0
2	910	30	0.9
3	910	50	<0.05
KAPTON			8.0
KAPTON	Coated with	Sample 2	1.3
KAPTON	Coated with	Sample 3	0.8

*Coating thickness 5-8 μ

Incorporation of siloxanes into polyimides increases their atomic oxygen stabilities very dramatically.[10,21] The siloxane-imide copolymer containing 30% of PSX-910 shows only 0.9% weight loss compared to 3.0% for the control polyimide. An increase in siloxane content of the copolymer to 50% reduces the weight loss to less than 0.05% (Table 5). In fact, in three different runs for this specific sample, we were able to observe weight loss in only one of the experiments. In the other two runs no weight change in the copolymer films was detected within 0.1 mg of the sample weight. This is a dramatic improvement and shows the effectiveness of siloxanes against atomic oxygen attack. It may be possible to further reduce the weight loss by incorporating higher molecular weight polydimethylsiloxane oligomers into these siloxane-imide copolymers. Increasing the PSX molecular weight will effectively reduce the amount of organic (propyl) end-group contribution, the weakest linkages in the copolymers.

In a related experiment, thin (5-8μ) coatings of fully cured, soluble siloxane-imide copolymers were applied onto Kapton films and oxygen plasma resistance of these coated samples was investigated. The copolymers used as protective coatings were BTDA-DDS based siloxane-imides containing 30 and 50% by weight of PSX-910 (Table 5). They were spray-coated on Kapton films from NMP solutions and the films were dried in a vacuum oven at 80°C overnight. These siloxane-imide copolymers showed excellent adhesion to Kapton. Thickness of the coatings obtained were about 5-8μ. The results of oxygen plasma exposure are provided in Table 5. As can be seen from this Table, uncoated Kapton film loses more than 8% of its weight upon exposure to oxygen plasma. Under the same conditions Kapton films coated with two different siloxane-imide copolymers lose only 0.8 and 1.3% of their original weights respectively. As expected, an increase in the amount of siloxane results in a decrease in weight loss. From these results it is clear that even an extremely thin coating of a siloxane imide copolymer improves the oxygen plasma resistance of a Kapton film dramatically. A more controlled coating technique and use of a siloxane-imide copolymer with higher PSX molecular weight may provide better atomic oxygen resistance.

CONCLUSIONS

Synthesis of high molecular weight, fully imidized and soluble poly(siloxane-imide) segmented copolymers has been demonstrated. These copolymers which show two-phase morphologies have a very broad service temperature range and display good thermal and mechanical properties. Due to their solubilities in various solvents, these copolymers can easily be coated onto polyimides or other polymeric composites. Their adhesion to Kapton, qualitatively, appears to be excellent without the aid of any adhesives or adhesion promoters. As evidenced by water contact angle measurements, the air surface of these copolymers is predominantly polydimethylsiloxane, whereas the mechanical properties are indicative of the strong polyimide segments. Incorporation of siloxanes into polyimides also results in substantial reductions in the amount of the water absorbed by these polymers, and water is usually detrimental to their performance.

Overall results of our preliminary studies on the interactions of various polyimides and poly(siloxane-imides) with oxygen plasma indicate that the introduction of siloxanes into polyimides either as a part of the polymer backbone (segment) or as a protective coating provided substantial improvements in the stability against atomic oxygen degradation. Incorporation of 50% by weight of a polydimethylsiloxane oligomer (Mn=910 g/mole) into a BTDA-DDS based polyimide reduced the weight loss about 60 times when compared with the control polyimides. The stability may further be improved by optimizing the backbone structure, end-group type and molecular weight of the polysiloxane oligomers which will be the focus of our future studies. Other important points which need further investigation for a better understanding of the atomic oxygen degradation and consequently the protection methods against its attack are, (i) the identification of reaction rates and mechanisms of siloxanes with atomic oxygen, (ii) determination of the structure (chemical bonds), morphology and thickness of the protective coating formed upon exposure, and (iii) the role and fate of the hard (organic) segments (imides) during atomic oxygen attack.

ACKNOWLEDGEMENT

We thank NASA-JPL for supporting this work under Contract NAS7-957.

REFERENCES

1. L.J. Leger, Proceedings, Third European Symposium, Spacecraft Material Space Environment, (ESA SP-232) p.75 (1985).
2. D.G. Zimcik, R.C. Tennyson, J.L. Kok and C.R. Maag, Proceedings, Third European Symposium, Spacecraft Material Space Environment, (ESA SP-232), p.81 (1985).
3. D.R. Tenney, G.F. Sykes, Jr., and D.E. Bowles, Third European Symposium, Spacecraft Material Space Environment, (ESA SP-232) (1985).
4. L.J. Leger, "Oxygen Atom Reaction With Shuttle Materials at Orbital Altitudes," NASA Techical Memorandum 58246, (1982).
5. L.J. Leger, AIAA Paper 83-0073, 21st Aerospace Science Meeting, (1983).
6. L.J. Leger, J.T. Visentine, J.F. Kuminecz and I.K. Spiker, AIAA Paper 84-0548, 22nd Aerospace Science Meeting, (1984).
7. A.E. Hadin, C.A. Reber, G.P. Newton, N.W. Spencer, H.C. Brinton, H.G. Mayer and W.E. Pottes, J. Geophysics Research, 88, 10170 (1983).
8. P.W. Knopf, R.J. Martin, R.E. Damman and M. McCargo, AIAA Paper 85-1066, A1AA 20th Thermophysics Conference, (1985).
9. G.N. Taylor, T.M. Wolf and J.M. Moran, J. Vacuum Science Technology, 19(4), 872 (1981).
10. W.S. Slemp, B. Santos-Mason, G.F. Sykes, Jr. and W.G. Witte, Jr., AIAA Paper 85-0421, 23rd Aerospace Science Meeting, (1985).
11. L.J. Leger, J.T. Visentine, J.F. Kuminecz and I.K. Spiker, AIAA Paper 85-0415, 23rd Aerospace Meeting, (1985).
12. A. Noshay and J.E. McGrath, Block Copolymers: An Overview and Critical Survey, Academic Press, New York, NY (1977).
13. I. Yilgör and J.E. McGrath, Advances in Polymer Science, 86, 1 (1987).
14. I. Yilgör, A.K. Sháaban, W.P. Steckle, Jr., D. Tyagi, G.L. Wilkes and J.E. McGrath, Polymer, 25(12), 1800 (1984).
15. J.S. Riffle, Ph.D. Thesis, VPI&SU, Blacksburg, VA (1981).
16. A.J. Yerman, U.S. Patent 4,017,340 General Electric Company (1977).
17. A. Berger, U.S. Patent 4,030,948, General Electric Company (1977).
18. J.T. Hoback and F.F. Holub, U.S. Patent 3,740,305, General Electric Company (1973).
19. A. Berger ad P.C. Juliano, U.S. Patent 4,011,279 General Electric Company (1977).
20. A.K. St. Clair, T.L. St. Clair and S.A. Ezzell, NASA Technical Memorandum, 83172, (1981).

21. B.C. Johnson, Ph.D. Thesis, VPI&SU, Blacksburg, VA (1984).
22. H.S. Ryang, U.S. Patent 4,404,350, General Electric Company (1983).
23. I. Yilgör, et al., Polymer Preprints, 24(1), 170 (1983).
24. I. Yilgör, J.S. Riffle, G.L. Wilkes and J.E. McGrath, Polymer Bulletin, 8, 535 (1982).
25. C.J. Lee, SAMPE Symposium, 30, 52 (1985).
26. J.D. Summers, C.A. Arnold, R.H. Bott, L.T. Taylor, T.C. Ward and J.E. McGrath, Polymer Preprints, 27(2), 403 (1986).
27. A. Berger, U.S. Patent 4,395,527 M&T Chemicals (1983).
28. A.K. St. Clair and T.L. St. Clair, U.S. Patent 4,497,935 NASA (1985).
29. J.S. Riffle, I. Yilgör, A.K. Banthia, C. Tran, G.L. Wilkes and J.E. McGrath, in Epoxy Resin Chemistry II, Ed., R.S. Bauer, ACS Symposium Series, No. 221, Washington, D.C., Chapter 2 (1983).
30. I. Yilgör, J.S. Riffle and J.E. McGrath, in Rective Oligomers, Eds. F.W. Harris and H.J. Spinelli, ACS Symposium Series, No. 282, Washington, D.C., Chapter 14 (1985).
31. V.L. Bell, U.S. Patent 4,094,862, NASA, (1978).
32. A.K. St. Clair and T.L. St. Clair, SAMPE Symposium, 26, 165 (1981).

PART FOUR:

SEALANTS FOR SPACE AND HARSH ENVIRONMENTS

Introductory Remarks

Robert M. Evans

Center for Adhesives, Sealants and Coatings.
Case Western Reserve University
Cleveland, Ohio 44106

At first glance, the topics which will be discussed today seem to lack a common thread. But there is, indeed, such a thread. The harsh effects of harsh environments. The twin papers by Paul on polysulfides are a paradigm, how harsh environments can injure or destroy polysulfide aircraft sealants. One paper deals with those effects of the environment which cause failure of polysulfides, particularly depolymerization. The other with producing an environment which will most effectively depolymerize the sealant.

The paper by Klosowski and Owen gives us a rationale for understanding the strong points and some of the weak points of silicone sealants when they are confronted by such environmental insults as ultraviolet, heat, water, ozone, electricity and soiling media. The paper by Chu describes a thermoplastic resin which can withstand not only ultraviolet and water, but also the rigors of automotive application.

The contribution from the China Academy of Railway Sciences deals with a very terrible environment--that which produces desertification. Their paper shows that a byproduct of little value--residual oil--when emulsified and sprayed on the desert floor--allowed the desert to bloom at a cost of only $.10/sq. meter.

My own paper deals with the little recognized hazards of plain old portland cement as a harsh environment. Since its problems were not recognized by most chemists, there was room for small companies to find a niche with coatings, membranes and sealants designed to resist this problem surface's attack.

In all, Dr. Lee has put together an interesting session.

EFFECTS OF ENVIRONMENT ON PERFORMANCE OF POLYSULFIDE SEALANTS

D. Brenton Paul, Peter J. Hanhela and Robert H.E. Huang

Materials Research Laboratories,
Defence Science and Technology Organization
PO Box 50, Ascot Vale
Victoria, Australia, 3032

ABSTRACT

The effects of exposure of polysulfide aircraft sealants to air, moisture and extremes cf temperature (together with various combinations of these conditions) are reviewed with reference to the cured sealants, the mixtures during cure and the stored sealant components. In order to properly understand the reasons for property changes resulting from exposure to these effects, the underlying chemical processes must be identified. This in turn requires detailed knowledge of the composition of the sealant formulations and of the reactive sites in the polymer backbone. Examples will be given in which the environmental resistance of polysulfides can be rationalized in these terms.

INTRODUCTION

One of the most extensive aircraft repair programs in recent years arose from the failure of a new integral fuel tank sealant. This material was introduced because it had been anticipated that the aerodynamic heating during flight would produce aircraft skin temperatures exceeding the thermal stability limits of the traditional polysulfide sealants. Operation of the aircraft in the tropics, however, caused the firm elastomer to degrade and form a mobile paste with virtually no sealing capability. Examination showed the sealant to be a polyester derived from neopentyl alcohol (I) and sebacic acid (II) and cured by an acyl aziridine (III) as shown in Eq. 1. It was rapidly established that the degradation of the polyester was simply due to hydrolysis in the hot, humid conditions.[1] This situation arose since the specification for the sealant was based primarily on thermal performance and insufficient consideration was given to the chemistry of such materials.

$$\begin{array}{c} CH_3 \\ | \\ HOCH_2\text{-}C\text{-}CH_2OH \\ | \\ CH_3 \end{array} \qquad\qquad HO_2C(CH_2)_8CO_2H$$

(I) (II)

$$NCO\text{-}C_7H_{14}\text{-}CON + 2\ HO\overset{O}{\overset{\|}{C}}{\sim} \rightarrow {\sim}\overset{O}{\overset{\|}{C}}O(CH_2)_2NHCO\text{-}C_7H_{14}\text{-}CONH(CH_2)_2O\overset{O}{\overset{\|}{C}}{\sim} \qquad (1)$$

(III)

This extreme example of the effect of imposed conditions on performance illustrates the nexus between structure of sealants, which governs their chemical reactivity, and their resistance to environmental influences. These interrelationships will be considered with reference to polysulfide aircraft sealants in the cured state, during cure and also the unmixed components of two-part systems.

REACTIVE CENTERS IN POLYSULFIDE FORMULATIONS

Physical methods allow behavioural trends to be identified and developed to provide guides to probable performance under various conditions. A proper predictive capability, however, requires interpretation of the underlying chemical processes leading to property changes. This, in turn, demands knowledge of the composition of sealant formulations. Two-part polysulfide sealants are essentially comprised of a liquid base component which is interacted with a separately packaged curing system. The reactive sites in liquid polysulfides (IV) are the formal groups in the backbone and the thiol terminals. The formal groups are susceptible to both autoxidation (by a free radical mechanism) and hydrolysis[2,3] and are often implicated in polysulfide degradation processes. The thiol groups are readily oxidized by inorganic oxidants and this process is commonly used to effect sealant cures. They are, however, also susceptible to air oxidation and despite packaging precautions, such reactions can occur during storage. Under normal conditions the disulfide linkages are stable to both oxidation and heat, but can participate in thiol-disulfide interchange reactions.[4] This has formed the basis of controlled degradation[5] but is also possible in the presence of fuels with a significant thiol content.

$$-\!\!\left[S\text{-}CH_2CH_2\text{-}OCH_2O\text{-}CH_2CH_2\text{-}S\text{-}S\text{-}CH_2CH_2OCH_2OCH_2CH_2S \right]\!\!-_n$$

(IV)

Any assessment of reactivity, however, must involve consideration of the total sealant formulation which can comprise additives such as plasticisers, fillers, cure retarders or accelerators, adhesion promoters, viscosity adjusters and thixotropic agents. While many of these components are chemically inert, some can participate in reactions which modify sealant performance. For example, hydroxymethyl groups of phenolic resins, or the epoxide rings in epoxy resins, both introduced to improve adhesion, can interact with thiols[6] (Eqs. 2 and 3). In the context of long-term storage, slow reactions of this type are possible. Many fillers are inert but one commonly used material is calcium carbonate which under acidic conditions can generate carbon dioxide.[3] An acidic or basic filler can also catalyse other reactions or act as a buffer.

$$\text{HOC}_6\text{H}_4\text{CH}_2\text{OH} + \text{HSRSS}\sim \xrightarrow{H^+} H_2O + \text{HOC}_6\text{H}_4\text{CH}_2\text{SRSS}\sim$$

$$\sim RS^- + \underset{\diagdown O \diagup}{CH_2 — CH}\sim \longrightarrow \sim RS{-}CH_2{-}\underset{\displaystyle |\atop O^-}{CH}\sim$$

ANALYSIS OF SEALANT COMPOSITIONS

Convenient procedures to identify the major components have now been developed.[7] The first step involves collection of volatile liquids under vacuum. The residues are then extracted with appropriate solvents to isolate the liquid polymer and activators, and the insolubles are collected for further assessment. Treatment with c. hydrochloric acid separates unreactive clay from calcium carbonate-based fillers and dissolves manganese dioxide leaving a mixture of inert filler and impurities from the MnO_2. The digestion in acid has the additional effect of removing any remaining oxidant from the mixture of fillers which is a prerequisite for determining carbon black in the residue. Thermogravimetric analysis first in nitrogen and then in air enables the proportion of carbon to be assayed through loss of carbon dioxide.[8] As residual oxidant will also interact with carbon at high temperatures, its removal is essential. Results obtained from a carbon black-clay mixture are shown in Fig. 1 and analysis of a typical manganese dioxide-cured sealant is given in Table 1.

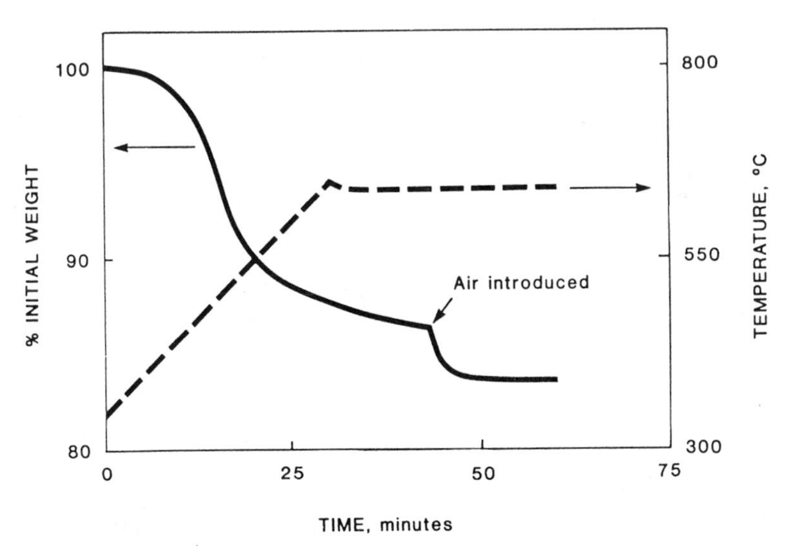

Fig. 1. Thermogravimetric analysis of acid washed fillers from a commercial dichromate-based cure paste. Initial weight loss in nitrogen is ascribed to dehydration of clay material.

Table 1. Composition of a MnO_2-Cured Sealant (Class B-2)

Base Polymer Component		Cure Paste	
Water	(0.3%)	Water	(1.9%)
$CaCO_3$ filler	(35%)	Carrier (hydrogenated terphenyls)	(27.4%)
Polysulfide	(64.7%)	N,N-diphenylguanidine	(4.8%)
		Sodium stearate	(4.6%)
		MnO_2	(47%)
		Carbon black	(11.5%)
		MnO_2 impurity	(2.8%)

EFFECT OF ENVIRONMENT ON UNCURED POLYSULFIDE SEALANT COMPONENTS

Aircraft sealants have a shelf life which is generally stipulated as 9 months at 27°C or below. For sealants procured from overseas this storage life can be further reduced by long delivery times. In a study designed to extend shelf life, the effect of lowering storage temperature was examined using standard performance criteria (viscosity of base, tack-free time, cure rate, hardness, peel strength and application life) supplemented by measurements of cure paste activity (concentrations of Mn(IV) or Cr(VI)), changes in thiol concentration, and preparation of materials using the aged components with either a reference liquid polymer or cure paste.[7] The key observations were:

(i) Even for sealants with significantly elapsed service life, storage at 2°C extended shelf life to approximately 3 years and even better performance could be attained at -16°C. Significant variations were observed at 13°C and 25°C with only 6 to 12 months satisfactory storage obtained.

(ii) The critical changes which occurred at the higher temperatures were in viscosity of the base (accompanied, eventually, by skinning and a drop in thiol concentration) and in the rate of cure. Decline in activity of the curing agent was small with MnO_2 pastes but for dichromate-based systems there was solvent loss from the polythene containers at higher storage temperatures. Sealants which resulted from slow cures had satisfactory properties, however, and the slower rates are attributed to changes in the base. The presence of surface skinning indicated an oxidation rather than interaction with an adhesion additive.

EFFECT OF ENVIRONMENT ON CURING OF POLYSULFIDE SEALANTS

Factors which influence the curing of polysulfide sealants are humidity and temperature. Although indications of the effects of these conditions are often reported, factual data is difficult to obtain. We have undertaken studies with two representative sealants with varying ratios of curing agent to base polymer, humidities ranging from 33% to 100% RH and temperatures from 25° to 60°C. Typical results are shown in Figs. 2 and 3 for which cure time is defined as the period required to reach a hardness of 35. Whereas the effect of temperature is predictable, the increase in cure rate with humidity is less obvious and has been attributed[9] to aqueous catalysis of the oxidation. If it is assumed that at a given temperature and fixed concentrations of polymer and curing agent the rate of cure is dependent on the humidity then

$$\text{rate} = K[H_2O]^n. \quad (4)$$

Since time, t, for a reaction to occur is inversely proportional to the reaction rate (i.e., t = c/rate), equation (4) can be rewritten as

$$t = \frac{c}{K}[H_2O]^{-n}, \quad (4')$$

or

$$\log t = \log \frac{c}{K} - n \log [H_2O]. \quad (5)$$

In fact a plot of log t_{35} (where t_{35} = time to cure to a hardness of 35 units) against log $[H_2O]$ is linear for both systems. The cure rates are therefore dependent on water concentration although the orders of reaction are less than one. In the absence of water the cure by either manganese dioxide or dichromate is very sluggish. The role of water is difficult to establish but on the basis of these results appears to be chemical rather than physical (as for example in stabilizing a transition state complex, or physical in assisting the diffusion of reactants). Further study is necessary to define the mechanisms of the processes which are involved.

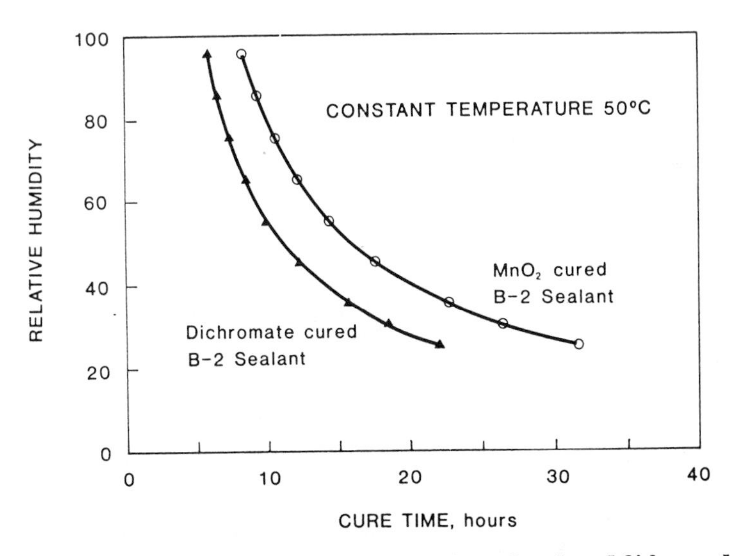

Fig. 2. Effect of humidity on cure rate of polysulfide sealants.

By application of the Arrhenius equation a relationship between temperature and the cure rate constant can also be derived as follows:

$$\log \frac{c}{K} = - \log A + \frac{E_a}{2.303\,R} \times \frac{1}{T} \qquad (6)$$

A linear relationship was also found between log c/K, 1/T and this permits calculation of the activation energy E_a.

One interesting feature is that sealants can be cured while in contact with hot water without significant detriment to their performance. This was examined in an endeavour to establish the necessity to wait for the prolonged cure period before exposing a sealant to the fluid it was designed to contain.[10]

CURED SEALANTS

A. Thermal Effects. Polysulfide sealants have excellent low temperature properties.[2] The commercial formulations based on poly(ethyl formal disulfide), which has a T_g of -59°C, show little tendency to crystallize and exhibit good flexibility down to -55°C. The thermal stability of polysulfide aircraft sealants, while considered acceptable for many years, is now recognized as the factor which may restrict their future use due to continuing increases in aerodynamic heating. The use of dichromate and manganese dioxide curing systems has produced sealants with stability up to 120°C (continuous exposure) and in some cases up to 180°C (short-term exposure). Thermal failure is often indicated by embrittlement, cracks and weight loss.

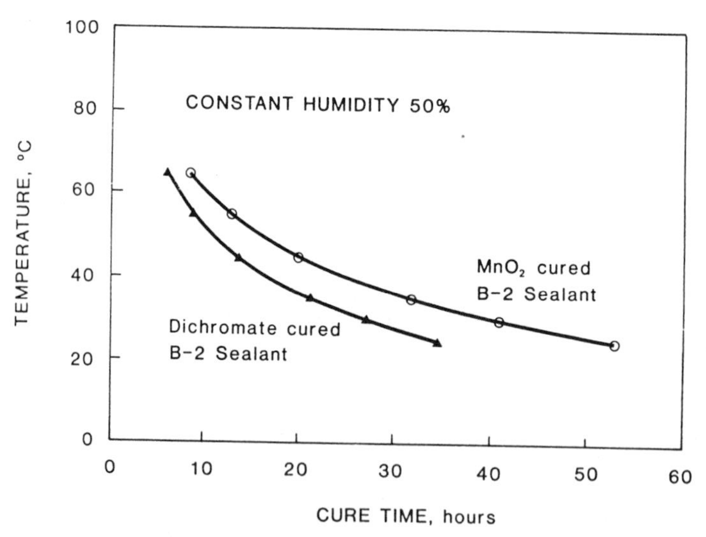

Fig. 3. Effect of temperature on cure rate polysulfide sealants.

The standard rationalization of thermal degradation is due to Rosenthal and Berenbaum[11] who examined the effect of exposing liquid polymers to temperatures of 100°C and 150°C. They proposed the sequence of reactions (7)-(9) to explain their observations and suggested that an initial autoxidation was necessary to generate the acid catalyst required in the first step. Although these reactions account for the weight loss on heating, the suggested explanation for the embrittlement was restricted rotation of the monosulfide products (Eq. 8) relative to disulfides.[2] This is not only unlikely on theoretical grounds but the development of the thioether polymers,

$$-SCH_2CH_2OCH_2OCH_2CH_2S- + H_2O \xrightarrow{H^+} -SCH_2CH_2OH + CH_2O + HOCH_2CH_2S- \qquad (7)$$

$$-CH_2CH_2SSCH_2CH_2- + CH_2O + H_2O \xrightarrow{H^+} 2-CH_2CH_2SH + HCOOH \qquad (8)$$

$$-CH_2CH_2SH + HOCH_2CH_2S- \xrightarrow{H^+} -CH_2CH_2SCH_2CH_2S- + H_2O \qquad (9)$$

which are elastomeric sealants with superior thermal properties to the polysulfides, has demonstrated that this view is incorrect. The probability of Eq. (8) must also be questioned: in our hands this process could not be realized using model compounds.[7]

A more recent study involving product analysis using pyrolysis-GC-MS indicated that at 358°C decomposition of liquid polysulfides was radical in nature and was explicable in terms of cleavage of the formal C-O bond.[12] With some manganese dioxide-cured polysulfides some indications were found to support the operation of a concurrent ionic decomposition pathway, due to the presence of metal mercaptides. Various backbiting mechanisms which produce cyclic structures can be envisaged for this process and a typical example is given in Eq. (9). Although of general interest, this approach is

$$\sim S-S-CH_2-O-CH_2-O-CH_2-CH_2-S^{-}\ (\text{cyclic}) \longrightarrow \sim S^{-} + \text{cyclo}(S-CH_2-CH_2-O-CH_2-O-CH_2-CH_2-S) \qquad (9)$$

of limited relevance to the practical situation since it was carried out in the absence of oxygen at high temperatures. Under the conditions experienced in service (120-180°C) direct thermal cleavage of the formal C-O or even the disulfide bond is improbable.

A preliminary examination of thermal degradation[7] has revealed that the process is complex. In Table 2 the effects on hardness of two commercial sealants (both containing calcium carbonate filler) are shown after exposure at 150°C for 160 h.

The effect of fillers, air and cure system were examined further using laboratory prepared sealants. Hardness changes of several sealants based on LP-32, after being maintained at 130°C for 60 h, are shown in Table 3.

It would appear, on the basis of these preliminary results, that more than one decomposition process is involved. Moreover, variations in the mechanisms of degradation are evident between sealants cured by manganese dioxide and dichromate salts and these are considered to result from differences in the pH of the curing systems. Ammonium dichromate is an acidic curing agent[3] and the thermal degradation of unfilled sealants cured with this reagent is ascribed to acid catalyzed hydrolysis (Eq. 6). In the presence of the basic filler calcium carbonate this degradation is suppressed. No obvious differences in response were noted when such sealants were heated under air or nitrogen which suggests that autoxidation is minimal with dichromate-cured polysulfides. A similar conclusion was reached after consideration of the effect of hot water on these sealants[3] as described below.

Table 2. Variation in Hardness of Commercial Sealants After Heating at 150°C for 160 Hours[a]

Conditions	Dichromate-Cured Sealant	Manganese Dioxide-Cured Sealant
Initial	60	55
Air	70 (3)	76 (6)
Nitrogen	68 (1.5)	70 (6)

[a]% weight loss given in parenthesis, hardness is Rex A

Table 3. Effect of Thermal Degradation at 130°C for 60 h on Hardness of Some Laboratory Prepared Sealants[a]

Cure System	Manganese Dioxide			Ammonium Dichromate		
Filler	$CaCO_3$	$CaSO_4$	None	$CaCO_3$	$CaSO_4$	None
Initial	51	50	37	53	53	38
Air	16	5	13	25	14	5
Nitrogen	30	26	18	26	20	6

[a] hardness is Rex A

Manganese dioxide is an alkaline curing agent[3] and for sealants prepared with this reagent the presence of a basic filler has little influence on the mode of degradation. In this case there was a trend to greater decomposition in air and the possibility of a contribution from autoxidation can be inferred. This is considered to proceed by an oxidation of either the formal methylene group or a methylene group adjacent to an ether link. Alkoxy radical formation would then occur followed by β-scission to generate a carboxylic acid (Eq. 10). The radical (V) could react further to either form volatile cyclization products[12] or an alcohol capable of participating in the processes shown in Eqs. (8) and (11). For manganese dioxide cured sealants a further degradation mechanism has been suggested[13] which involves the abstraction of a sulfur atom from a trisulfide centre in the polymer to produce hydrogen sulfide. Under the basic conditions hydrosulfide anions would be formed and these would cleave the polymer backbone by the interchange process.

$$RCH_2OCH_2R' \xrightarrow{(O)} RCH(OOH)OCH_2R' \rightarrow \underset{(V)}{RCH(O^{\bullet})OCH_2R'} \rightarrow RCO_2H + R'CH_2^{\bullet} \tag{10}$$

$$2RR'CHOH \xrightarrow{H^+} RR'CHOCHRR' + H_2O \tag{11}$$

Production of a carboxylic acid (Eq. 10) is of interest since it has been shown in another context[1] that exposure of cured polysulfides to dilute ethanolic solutions of sebacic acid results in marked embrittlement (Fig. 4). It is conceivable that the presence of carboxylic acids may contribute to degradation through hydrolysis of formal linkages.

B. Heat and Humidity. Polysulfide sealants are stable to contact with cold water and have performed valuable service in sealing reservoirs and pools. Some applications involving hot water contact, however, can arise (e.g. water tank in aircraft heat-exchanger systems) and here the stability of the sealant has been shown to depend on the nature of the curing system.[3,10]

Dichromate-cured polysulfide sealants have superior resistance to swelling in hot water than those cured with manganese dioxide (Fig. 5). Formation of large voids in MnO_2-cured sealants after prolonged immersion at 70-90°C suggested the formation of a water-soluble acid which then reacted with the $CaCO_3$ filler and this was supported by pH measurements. The acid was shown to arise through a free radical oxidation of formal groups (Eq. 10; compare with thermal degradation) and this process was suppressed by radical scavengers, including ammonium dichromate. When MnO_2-cured sealants were examined under nitrogen or with carbon black as filler a similar time-swell curve to that of dichromate-cured materials was produced (Fig. 6A). The performance differences were therefore concluded to result from the respective abilities of these sealants to counter radical-based oxidation. Since the manganese dioxide-cured sealants cannot exhibit protection, acidic degradation products are formed. In itself the autoxidation does not greatly affect swelling (Fig. 6B) but if a component is present which is capable of releasing carbon dioxide through reaction with the acidic oxidation products the normal swelling of the sealant will be augmented either by osmosis or carbonation.[10]

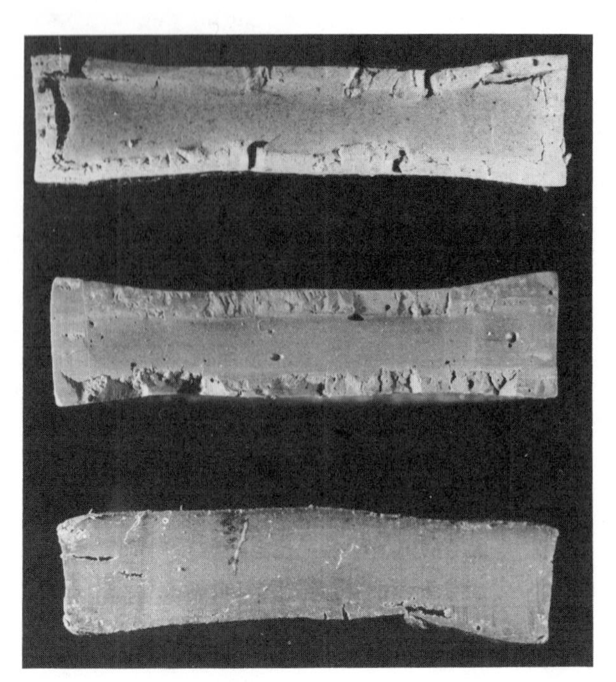

Fig. 4. Cross sections of polysulfide sealants following immersion at 60°C in a 1M ethanolic solution of sebacic acid for 180 days. Top and middle specimens were MnO_2-cured and bottom sealant was dichromate-cured.

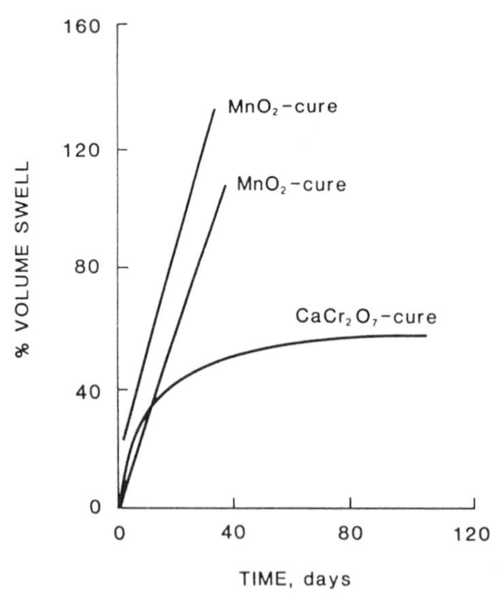

Fig. 5. Time-swell relationships for manganese dioxide and dichromate-cured sealants at 90°C.

Table 4. Effect of Thiol in Simulated Fuel Mixtures Acting on a MnO_2-Cured Sealant for 40 Days at 60°C.

% [Thiol] in JRF	Hardness	Weight Loss	Tensile Strength (MP)	Ultimate Elong-ation (%)	Peel Strength (N)	Final [Thiol]%
0	62	5.0	1.25	180	68	0
0.004	61	5.0	1.21	199	71	0.001
0.02	58	4.9	0.83	226	82	0.006

C. Aircraft Fuels. Polysulfides are excellent fuel tank sealants as they are resistant to hydrocarbon-based solvents. Unless suitable precautions are taken, however, aircraft fuels will contain thiols and in concentrations exceeding 0.001% have been shown to have a detrimental effect on physical properties such as elongation and tensile strength of sealants cured by lead dioxide. This would be explicable in terms of the thiol-disulfide interchange process since such fuels may be considered to be very dilute desealing mixtures.[5,14] We have now conducted an examination[7] which shows that sealants cured by manganese dioxide or dichromate salts are very little affected after interaction for 40 days at 60°C with fuel containing up to 0.02% thiol (Table 4).

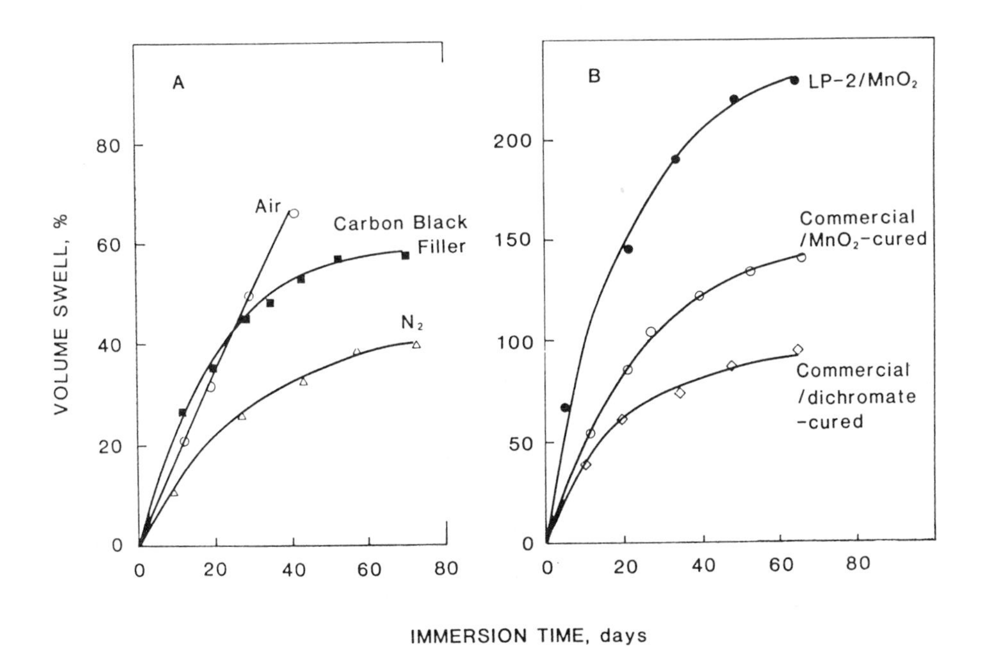

Fig. 6. (A) Influence of air and filler type on volume-swell curves for unfilled sealants in water at 90°C and

(B) Time-swell curves for unfilled sealants in water at 90°C.

CONCLUSIONS

While much useful data can be obtained by carefully monitoring the effects of different environmental conditions on the properties of polysulfide sealants, better understanding the chemical processes will assist attempts to improve performance. Progress through this approach has been achieved despite the complex nature of polysulfide formulations.

REFERENCES

1. P.J. Hanhela and D.B. Paul, MRL Report 657 (1983); MRL Report 658 (1984).
2. R.H. Gobran and M.B. Berenbaum, in High Polymers, Vol. XXIII, Part II, Ch. 8C, Interscience, New York (1969).
3. P.J. Hanhela, R.H.E. Huang, D.B. Paul and T.E.F. Symes, J. Appl. Polym. Sci., 32, 5415 (1986).
4. M.B. Berenbaum, in High Polymers, Vol. XXIII, Part III, Ch. XIII, Interscience, New York (1962).
5. W. Mazurek and D.B. Paul, ACS Poly. Mater. Sci. Eng. Preprints 56 (1987).
6. E.R. Bertozzi, Rubber Chem. Technol., 41, 114 (1968).
7. P.J. Hanhela, R.H.E. Huang and D.B. Paul, unpublished work.
8. B.C. Ennis, P.J. Hanhela and D.B. Paul, unpublished work.
9. J.R. Panek, in reference 4, Ch. XIV, p.167.
10. P.J. Hanhela, R.H.E. Huang and D.B. Paul, I & EC Prod. Res. & Dev., 25, 321 (1986).
11. N.A. Rosenthal and M.B. Berenbaum, paper presented at 131st meeting of Am. Chem. Soc., Miami, Fla., April 1957.
12. M. Rama Rao and T.S. Radhakrishnan, J. Appl. Polym. Sci., 30, 855 (1985); T.S. Radhakrishnan and M. Rama Rao, J. Appl. Pyrolysis, 9, 309 (1986).
13. L. Morris, Products Research Corporation, private communication (1987).
14. W. Mazurek, Aust. Pat. Appl. PH 02485 (1985); MRL Report 954 (1985); MRL Report 1002 (1986).

DURABILITY OF SILICONE SEALANTS

Michael J. Owen and Jerome M. Klosowski

Dow Corning Corporation
Midland, Michigan 48686

ABSTRACT

The durability of silicone sealants is outstanding. This valuable behavior is a direct result of the unusual organic/inorganic hybrid nature of silicones and the consequent chemistry and physics of their interaction with the environment. Responses to a variety of environmental stresses are reviewed including those associated with weathering such as ozone and temperature extremes as well as electrical stresses such as corona discharge. The emphasis is on relating this behavior to the surface and bulk properties of silicones and particularly to the low intermolecular forces and high chain flexibility that are the fundamental consequence of the unique molecular architecture of silicones. Attention is paid to aspects such as the low surface energy and poor adhesion of silicone surfaces and to dynamic effects such as surface reorientation to changing environnments.

INTRODUCTION

The long service life of silicone sealants and coatings in a variety of environments is widely recognized and has led to numerous applications. Exceptional attributes include high temperature resistance, low temperature handleability, excellent UV and ozone resistance and long durability over 20 years.[1] This outstanding environmental durability is a general consequence of the hybrid nature of silicones occupying the regime between silicate materials and organic polymers and exhibiting attributes of each. Specific material performance naturally depends on the silicone product composition including type of organic substitution and crosslinking chemistry and other components such as fillers and catalysts as well as on the nature of the environment to which the silicone is exposed. Because of these complications it is not the intent of this review to provide performance data of specific formulations. However, references to such data have been included for those requiring more quantitative information to support the generalizations presented here. Many of the selected references also contain comparison with other materials for those interested in this aspect. To reduce the topic to manageable proportions, this discussion will be mostly limited to polydimethylsiloxane (PDMS) materials which are the most widely used and readily available silicones, with occasional reference to polyphenylmethylsiloxane (PPMS) and the fluorosilicone, polytrifluoropropylmethylsiloxane (PTFPMS). The structural formulas of these polymers are:

$$\text{PDMS} \quad \left[-\underset{CH_3}{\overset{CH_3}{\underset{|}{\overset{|}{Si}}}}-O- \right]_n$$

$$\text{PPMS} \quad \left[-\underset{CH_3}{\overset{C_6H_5}{\underset{|}{\overset{|}{Si}}}}-O- \right]_n$$

$$\text{PTFPMS} \quad \left[-\underset{CH_3}{\overset{CF_3CH_2CH_2}{\underset{|}{\overset{|}{Si}}}}-O- \right]_n \quad \text{or} \quad \left[-\underset{CH_3}{\overset{CF_3(CH_2)_2}{\underset{|}{\overset{|}{Si}}}}- \right]_n$$

Environmental factors that will be considered are oxygen, ozone, ultraviolet (UV) radiation and other radiation, temperature extremes, water, electrical discharge and corrosive solids.

A qualitative list of the features of PDMS that contribute to its durability is given in Table 1. Some relevant numerical data are given in Table 2. Almost all of these attributes are a consequence of two fundamental aspects of PDMS, the low intermolecular forces between the methyl groups and the unique flexibility of the siloxane backbone. These properties are a direct consequence of the particular molecular architecture of PDMS and have been discussed in detail elsewhere.[2] Because of the configurations adopted by the siloxane chain, the pendant methyl groups dominate the molecular interaction between PDMS chains and with other molecules. These are London dispersion forces amongst the lowest known interaction energies, only those of aliphatic fluorocarbons are lower. Evidence for these low intermolecular forces comes not only from the low surface tensions but also from the lower boiling points when compared to organic materials of similar molecular weight. The maintenance of liquid nature to unusually high molecular weights of linear PDMS polymers is a further manifestation of this weak molecular interaction. The flexibility of the siloxane backbone is unique. This is reflected in several characteristics of PDMS, notably the low glass temperature, the large free volume, the low apparent energy of activation for viscous flow, and the free rotation about the siloxane bond.[3]

Table I. Silicone Attributes Contributing to Durability

Low surface tension
High water repellence
Partially ionic backbone
Large free volume
Low apparent energy of activation for viscous flow
Low glass temperature
Freedom of rotation about bonds
Small temperature variations of physical constants
High gas permeability
High thermal and oxidative resistance
Low reactivity
Low environmental hazard
Insolubility in water
High silicon-oxygen bond energy

Table 2. Selected PDMS Properties

Critical surface tension of wetting	24 mN/m
Dispersion force component of solid surface Tension	21.7 mN/m
Polar component of solid surface tension	1.1 mN/m
Water contact angle	110°
Glass temperature	150°K
Energy of rotation	0 kJ/mol
Activation energy of viscous flow	14.7 kJ/mol
Oxygen permeability	4.54×10^{-11} cm^3(STP)cm $cm^{-2}s^{-1}$Pa
Si-O bond energy	445 kJ/mol
% Polar bond contribution	41%

These two fundamental properties of PDMS provide physical explanations of much of the behavior of PDMS in various environments. There are also chemical consequences of the hybrid organic/inorganic nature of PDMS, notably the high bond energy of the siloxane backbone and its partially ionic nature, that result in the higher thermal and oxidative stability of a methyl group on silicon compared to an all hydrocarbon chain. The interplay of these chemical and physical factors on the durability of PDMS will be apparent in the following discussion of PDMS behavior in a variety of environments.

OXYGEN AND OZONE

The resistance of PDMS to oxygen and ozone attack is primarily chemical not physical. Residual carbon-carbon double bonds in polymers are the major locus of attack.[4] The absence of such bonds in silicones such as PDMS has much to do with their resistance to these gases. Saturated polymers such as polyethylene do oxidize but at a much slower rate than unsaturated materials such as natural rubber. Methyl groups on silicon are even more resistant to oxidation but will eventually be oxidized to silanol groups. This increased oxidation resistance is a consequence of the partially ionic nature (40-50% polar according to Pauling electronegativity calculations) of the siloxane backbone, the positively polarized silicon atom acting as an electron drain polarizing the methyl group and rendering it less susceptible to attack. Such oxidation becomes serious over 200°C although the upper working temperature can be raised by using appropriate oxidation inhibitors. Suitable silicone products can be heated in air at 200°C for a year without any appreciable change in properties and even up to 300°C for short periods.[5] The effect of oxygen and ozone is inextricably mixed with the effect of elevated temperatures. A useful paper which supports the preceding generalization for a specific type of silicone sealant is given by Beers.[6]

Some commercial sealants however, have stability up to only 130°C; others up to 150°C. The largest contributing factor is impurities which catalyze the oxidative decomposition of silicones. These often arise from the various fillers used in silicone sealants. The cure catalysts themselves can sometimes be the cause of the destabilization. The precise chemistry of degradation will depend on the particular impurity or catalyst involved. Not only can there be oxidation of the pendant hydrocarbon groups, but sometimes nucleophilic or electrophilic attack occurs at the siloxane backbone resulting in chain scission and production of lower molecular weight fragments. Thus, silicone sealants and coatings exist with excellent thermal stability in air but these do not represent most of the silicone products available. Careful matching of product with performance characteristics is always required. Conventional acidic and basic catalysts particularly sodium and potassium hydroxide, if left unneutralized, have very adverse effects on thermal stability and even salts such as potassium chloride can have a detrimental influence.

The high gas permeability of silicones is also of interest because both oxygen and ozone attack appear to be primarily a surface phenomenon. For example, wax surface films formed from waxes added to the rubber mixture are used as anti-ozonants because they reduce the diffusion of gas into the material. In the case of PDMS, diffusion will occur very readily causing more uniform damage. Rapid oxidation of the surface region, as will be experienced in a fire, causes oxidation to silanol and on to silica which can form a very gas impermeable crust providing a type of self-protecting mechanism in extreme exposures. The high gas permeability can be a significant benefit in areas where a protective yet breathable coating is desired. It is also fundamental to the cure of most silicone sealants which require water vapor for the cure reaction. The excellent permeability ensures uniform curing throughout the sealant.

UV AND OTHER RADIATION

In order for the light energy to initiate chemical change it must be absorbed by the molecules of the material in question. Silicones do not absorb near-UV radiation in the 300-400 nm region which is the wavelength range that causes problems with polymers at or near ground level. This is consistent with the fact that even after 20 years of outdoor weathering in sunny climates, silicone elastomers show no significant changes in physical properties. A typical example of this kind of study is given by Karpati.[7]

UV radiation in the far-UV and vacuum UV regions which are encountered in aerospace applications, is a different matter. UV radiation at wavelengths lower than 290 nm in the presence of air induces the cleavage of methyl radicals from the silicon, producing methane, carbon monoxide and carbon dioxide.[8] However, it is only the C-H and Si-C bonds that are ruptured for both PDMS[9] and PPMS[10] and cleavage of the Si-O bonds is not observed. Thus, the polymer backbone is unaffected. For sealants this implies that products crosslinked via siloxane bonds will be more useful in such environments than those crosslinked by silicon-carbon bonds.

The transparency to UV in the normal atmosphere must be considered carefully when using a silicone coating or sealant. The clear silicone products usually provide no protection from UV to the surfaces beneath them. Consequent decay at this interface can lead to loss of adhesion and detachment of the coating. Some very useful and durable silicone sealants can be obtained by extensive pigmenting to give them hiding power from UV radiation. Common UV absorbers can also be incorporated. Such opaque products are extensively used to protect surfaces from UV degradation. Examples of UV absorbers used in silicone sealants are 2-hydroxybenzophenone, benzotriazole, hindered amines, cinnamates and 4-aminobenzoic acid.

The effect of nuclear radiation on elastomeric materials is much more severe than UV radiation. The two main results are crosslinking and chain scission. Examples of polymers which crosslink when exposed to gamma radiation are styrene-butadiene, natural rubber and PDMS. Those which primarily undergo chain scission include polysulfide and butyl rubber. Being in the former class it might be expected that PDMS might survive gamma radiation rather well. However, its properties are affected by lower levels of radiation than many other elastomers. Some quantitative data on this aspect are given by Sieron and Spain.[11] PPMS is considerably less affected by ionizing radiation than PDMS. Presumably the aromatic system is able to dissipate more harmlessly the absorbed energy in nonchemical ways such as heat. A useful table putting the gamma radiation stability of dimethyl and phenylmethyl silicones in perspective is given by Robinson and coworkers.[12]

There have been a number of specific studies to simulate the space radiation environment. An example is the study of Jones[13] who examined the effect of 147 nm UV radiation on the volatile condensable material from Dow Corning Silastic 140 RTV. This was a simulation of synchronous earth orbit to evaluate the radiation-induced darkening of contaminants on satellite optical components (Titan IIIC payload application). UV transmittance has also been used to measure the effect on silicone elastomers of another environment, high energy particles particularly electrons, such as are trapped in the radiation belts of Jupiter.[14] Both these studies highlight the outgassing concern which is discussed in more detail next.

OUTGASSING IN SPACE

Even the most casual perusal of the "silicones in space" literature reveals that outgassing causing subsequent contamination of other space vehicle components is perhaps the major concern in using silicone sealants. NASA has invested considerable resources in tackling the problem. A comprehensive example is the work of Allen, Hughes and Price[15] who used a variety of techniques including ellipsometry, quartz crystal microbalance and an electrobalance to study contamination kinetics. They also used internal reflectance IR spectroscopy to identify silicone contaminants.

Thorne and Whipple[16] provide a useful summary of the origin and identification of outgassing components. They suggest three pre-conditioning techniques for reducing or even totally eliminating vacuum outgassing:

i. Curing at temperatures beyond conventional levels normally used to obtain desired physical and electrical properties.

ii. Vacuum purification

iii. Solvent extraction

(ii and iii: Both of components and of formulated sealant)

A good study of vacuum purification (distillation of unvulcanized components) is given by Guillaumon and Guillin.[17]

TEMPERATURE EXTREMES

The high siloxane bond energy and its considerable ionic character are clearly responsible for the substantial thermal stability of silicones. A temperature of 250°C is considered the safe limit for extended use of PDMS elastomers. This upper temperature limit is significantly better than most organic elastomers. The remarks on impurities in the above section on oxygen and ozone apply equally well to this discussion of thermal stability. Both PPMS and PTFPMS are somewhat more thermally stable than PDMS. A useful article that considers silicone elastomer thermal stability in both air and nitrogen is that of Critchley, Knight and Wright.[18] At the other end of the temperature range Hauck has provided useful low temperature data.[19] The low temperature performance of phenylmethyl-dimethyl silicones is superior to both fluorosilicones and dimethyl silicones. This is due to the stiffening of PDMS caused by crystallization that is disrupted by the phenyl groups.

One property of PDMS that sets it apart from all other polymers is the shallow slope of the viscosity-temperature curve.[20] This is because viscosity is just another manifestation of the physical interaction between molecules, as is surface tension. Increasing temperature increases the separation distance between molecules reducing interaction and thus reducing viscosity. However, the ultra-flexible PDMS can readily re-configure to compensate for the temperature separation effect. Moreover, its very low intermolecular forces inevitably imply a correspondingly low temperature coefficient.

Viscosity was chosen as an illustrative example but there is in general for PDMS a smaller temperature variation of physical constants than for other materials. This has benefits when the material is required to operate over a wide range of temperatures. In particular, with the uniquely low glass temperature, elastomeric properties are retained to lower temperatures than other sealants. A similar lack of variation occurs with pressure, or lack thereof, which may have importance in space applications. There are further implications. Since the silicone sealants do not stiffen to the degree exhibited by most other materials, the forces at the bondline are less. The reliability of the adhesive bond is thus improved for products used in applications which experience a considerable amount of movement due to extension and compression because of thermal cycling.

WATER

In some ways, this is the weak spot of silicone's armory of defences against environmental attack. The considerable ionic nature of the siloxane bond renders it susceptible to hydrolysis. The problem is normally evident at extremes of acidity or alkalinity--the very conditions under which the bonds were formed in the first place. The reaction between water and siloxane shown in the following equation is an equilibrium that results in a redistribution of siloxane linkages during silicone formation that is known as the equilibration of siloxanes.[21] This is the most significant difference in chemical reactivity between silicones and organic polymers

$$2(\text{-}\overset{|}{\underset{|}{\text{Si}}}\text{-OH}) \rightleftharpoons \text{-}\overset{|}{\underset{|}{\text{Si}}}\text{-O-}\overset{|}{\underset{|}{\text{Si}}}\text{-} + H_2O$$

Mass action effects keep this equilibrium well over to the right for small amounts of water, a condition much aided by the water-insoluble nature of PDMS. However, if the amount of water is increased as when a monolayer of PDMS is spread on water[22] or when the silicone is solubilized in water in surfactant form as in a short chain silicone-glycol copolymer, aqueous instability is more evident at highly acidic or alkaline conditions. In the extreme aqueous exposure case of the silicone surfactants, a pH between 4 and 9 is considered the safe working range.

Another aspect of the problem is that some of the re-equilibrated siloxane is the low molecular weight hexamethyldisiloxane which can be lost from the system. A rise in temperature increases both the rate of redistribution and evaporation so hot, damp environments, for example exposure to super-heated steam, are very detrimental to PDMS integrity. This is sometimes referred to as the silicone "reversion" problem. This has ramifications in adhesion to materials with a natural alkalinity or acidity. The reversion at the interface is prevented by priming with suitable coupling agents as well as including coupling agents as an integral part of the sealant composition. The effect is to form other types of bonds along with the siloxane bonds at the interface. Thus, while reversion is important, it can be overcome in many situations.

This susceptibility to water is shared by virtually all inorganic polymers.[23] Based on bond energy considerations most of the feasible linkages should have higher thermal stability than carbon chain polymers. However, these inorganic bonds are usually ionic in nature and this, together with the availability of bonding orbitals in many of the elements, provides low-energy routes to degradation reactions such as hydrolysis. The majority of inorganic polymers are susceptible to nucleophilic or electrophilic attack and one of the most serious problems encountered has been the ease with which many of them hydrolyze in the presence of water.

ELECTRICAL DISCHARGE

In practical applications of dielectrics in electronic and electrical systems, surface effects are vitally important. When an electrically stressed dielectric is placed in a wet environment, an ionized film may cause leakage currents and surface discharges that may seriously affect the material. The ability to maintain or recover its hydrophobicity under stresses like partial discharge is a key to the successful use of dielectric polymers. In the case of silicones exposure to corona causes the surface to become hydrophilic but hydrophobicity is recovered when the exposure ceases.[24]

The two most likely general mechanisms for the hydrophobic recovery of insulating polymers are:

1. Reorientation of surface hydrophilic groups away from the surface (also described as the "overturn" of polar groups in the polymer surface).
2. Migration of untreated polymer chains from the bulk.

Both these mechanisms require molecular mobility and the silicones with their lowest glass temperature, freedom of rotation and high free volume, should display the most rapid recovery as seems to be the case in practical applications. Note that this hydrophobic recovery has been reported when other environmental stresses are removed, for example in UV exposure[8] and appears to be a significant factor in the durability and weatherability of silicones.

CORROSIVE SOLIDS

The effect of solid particles is related to the time of their contact with the materials. PDMS has a number of properties that cause it to have low adhesion to many materials. These release properties are widely exploited in areas such as pressure sensitive adhesive release coatings and integral plastics additives. Release agents usually exhibit several of the following features:[25]

1. Barrier to mechanical interlocking.
2. Prevention of interdiffusion.
3. Poor adsorption.
4. Low surface tension and work of adhesion.
5. Low electrostatic attraction.
6. Non-setting.
7. Provision of weak boundary layer.

Liquid and solid silicone surfaces have most of these characteristics. They both have low surface tensions in the 20-24 mNm^{-1} range and exhibit poor adsorption by virtue of the low dispersion forces of the methyl groups that dominate their molecular interactions.

PDMS, although having a partially ionic backbone, has no fully charged ionic entities to cause appreciable electrostatic attraction. Acid-base interactions have been proposed as the predominant force at interfaces other than dispersion forces.[26] The all-methyl surface of PDMS should be neutral in this respect. The solid silicone surface will provide a barrier to mechanical interlocking and prevent interdiffusion of most other polymers. The liquid PDMS is non-setting and provides an excellent low cohesion weak boundary layer. Some silicone sealant compositions contain silicone fluid that enhances this synergy. They have a certain affinity for oily dirt, which is usually non-corrosive, but often the free silicone fluid, by virtue of its lower surface tension, will spread over and effectively encapsulate the adhered particle.

Cuddihy[27] has provided the following evolving requirements for low-soiling characteristics--hard, smooth, hydrophobic, low surface energy, chemically clean of sticky materials (surface and bulk), weather-stable (resistance against oxidation, hydrolysis, UV reactions). Highly crosslinked silicones, with no free fluid, fit this set of criteria well. Fluorosilicones such as PTFPMS with lower solid surface tension should be even more satisfactory. Such materials are much more oil resistant than PDMS which should also diminish oily dirt pick-up.

CONCLUSIONS

An attempt has been made to relate fundamental chemical and physical properties of silicones to their well recognized durability in a variety of environments. Emphasis has been placed on the importance of the partially ionic nature of the siloxane backbone, its unique flexibility, and the low intermolecular forces between silicone molecules dominated by the London dispersion forces associated with the pendant methyl groups.

Compared to organic polymers, silicones are more thermally stable and perform over a wider range of temperature both high and low. They are also more resistant to oxygen and ozone attack and the effect of corona discharge. Weak aspects of silicone environmental stability are aqueous stability particularly at extremes of pH and ability to resist gamma radiation. The low surface energy of silicones and its rapid ability to recover hydrophobicity are important factors in its good behavior under partial electrical discharge, UV radiation, weathering and contamination resistance.

The discussion has of necessity been very general. Focus has been on the silicone polymer and has ignored the important factors of crosslinking mechanisms, filler type and catalysts. With regard to crosslinking both siloxane and organic crosslinks are used in silicone sealants. The former will obviously maintain silicone behavior whereas the latter could introduce some organic-like weaknesses in the environment. Filler particles are usually not found in the surface and will have no direct effect on many of the environmental factors considered which are primarily surface effects. However, should the filler/polymer bond be very sensitive to oxygen or water attack, this could be very detrimental. Similarly small amounts of residual catalysts could have both positive and negative effects on thermal and oxidative stability. Precise attention to all components of the sealant must be considered when being considered for more extreme environments. Another warning that must be applied to this general review is that often envirornnents are mixed, for example, far-UV, oxygen, ozone and water in the upper atmosphere, and this too must be taken into consideration in specific applications.

REFERENCES

1. J.R. Panek and J.P. Cook, Construction Sealants and Adhesives, 2nd Edition, John Wiley and Sons, New York, 1984, p.127.
2. M.J. Owen, CHEMTECH, 11, 288 (1981), Ind. Eng. Chem. Prod. Res. Dev., 19, 97 (1980).
3. A.V. Tobolsky, Properties and Structure of Polymers, John Wiley and Sons, New York, 1960, p.67.
4. P.G. Burstrom, Aging and Deformation Properties of Building Joint Sealants, Report TVBM-1002, Lund, Sweden, 1979, p.10.
5. W. Noll, Chemistry and Technology of Silicones, Academic Press, New York, 1968, pgs. 458 and 494.
6. M.D. Beers, in Handbook of Adhesives, 2nd Ed., Ed. I. Skeist, Van Nostrand Reinhold Co., New York, 1977, p.631.
7. K.K. Karpati, Adhes. Age, 23, 41 (1980).
8. C.W. Lentz, Ind. Res. and Dev., April, 139 (1980).
9. A.D. Delman, M. Landy and B.B. Simms, J. Polym. Sci., A1, 7, 3375 (1969).
10. S. Siegel, R.J. Champetier and A.R. Calloway, J. Polym. Sci., A1, 4, 2107 (1966).
11. J.K. Sieron and R.G. Spain, in Environmental Effects on Polymeric Materials, Vol. 2, Eds. D.V. Rosato and R.T. Schwartz, John Wiley and Sons, New York, 1968, p.1662.
12. G.L. Robinson, R.I. Akawic, M.N. Gardos and K.C. Krening, "Development and Characterization of Lubricants for Use Near Nuclear Reactors in Space Vehicles," NASA-CR-123966, Sept., (1972).
13. P.F. Jones, "Radiation Effects on Contaminants From the Outgassing of Silastic 140 RTV", NASA Spec. Publ. NASA SP-298, p.173 (1972).
14. S.F. Pellicori, Appl. Opt., 9, 2581 (1970).
15. T.H. Allen, T.A. Hughes and B.C. Price, "A Study of Molecular Contamination," NASA Spec. Publ. NASA SP-379, p.609 (1975).
16. J.A. Thorne and C.L. Whipple, "Silicones in Outer Space," 11th Natl. Symp. of the Soc. of Aerospace Mater. and Process Engr., St. Louis, April (1967).
17. J.C. Guillaumon and J. Guillin, "Reduction of the Outgassing of Silicones by Purification," Misc. Symp., Conf. Meet. (France) 1974 (C.A. 83: 98806D).
18. J.P. Critchley, G.J. Knight and W.W. Wright, Heat-Resistant Polymers, Plenum Press, New York, 1983, p.337.
19. J.E. Hauck, Mater. Eng., 61, 118 (1965); 65, 108 (1967).
20. B.B. Hardman and A. Torkelson, in Kirk-Othmer Encyclopedia of Chemical Technology, 3rd ed., Vol. 20, John Wiley and Sons, New York, 1982, p.938.
21. C. Eaborn, Organosilicon Compounds, Butterworths, London, 1960, p.255.
22. H.W. Fox, E.M. Solomon and W.A. Zisman, J. Phys., Chem., 54, 723 (1950).
23. Reference 18, p.441.

24. M.J. Owen, T.M. Gentle, T. Orbeck and D.E. Williams, to be published in proceedings of Symposium on Dynamic Aspects of Polymer Surfaces, ACS Rocky Mountin Regional Meeting, Denver, CO, June, 1986.
25. M.J. Owen, accepted for publication in the 2nd edition of the Encyclopedia of Polymer Science and Engineering, John Wiley and Sons, Inc.
26. F.M. Fowkes, in Physicochemical Aspects of Polymer Surfaces, Vol. 2, Ed. K.L. Mittal, Plenum Press, New York, 1983, p.583.
27. E.F. Cuddihy, "Surface Soiling: Theoretical Mechanisms and Evaluation of Low Soiling Coatings,; JPL Publication 84-72, DOE/JPL 1012-102, p.379, Nov. 15, 1984.

DEPOLYMERIZATION OF POLYSULFIDES: THE DEVELOPMENT OF IMPROVED CHEMICAL DESEALERS

Waldemar Mazurek and D. Brenton Paul

Materials Research Laboratories
Defence Science and Technology Organization
PO Box 50, Ascot Vale
Victoria, Australia, 3032

ABSTRACT

Chemical desealing of polysulfide aircraft sealants has posed a serious maintenance problem. Such polymerizations are usually based on the thiol-disulfide interchange process for which aryl thiols are particularly reactive reagents. The only effective commercial desealant contains thiophenol but this presents potential environmental and health problems and recently its use has been either avoided or abandoned by some aviation authorities. It has now been demonstrated that by conversion of thiophenol to the potassium or tetraalkyl ammonium salt, the noxious odour is removed while desealing activity remains undiminished. Relative desealing rates have been determined with cations of both alkyl and aryl thiolates and compared with that obtained with thiophenol. Suitable solvent mixtures were defined and the effect of added monomeric disulfide on reaction rates was also evaluated. An optimized formulation has been suggested following examination of the relationships between desealing performance and concentrations of the various components.

INTRODUCTION

The periodic need for the removal and replacement of polysulfide fuel tank sealants has proved a long-term aircraft maintenance problem since by design these materials are required to be durable and exhibit good adhesion. As sealants possess a network structure, solvents alone cannot accomplish their removal. Chemical degradation is, however, possible and a desealant in common use contains thiophenol as the reactive constituent, together with a swelling agent and anticorrosion additive in a hydrocarbon solvent. The basis of this approach is the thiol-disulfide interchange reaction (Eq. 1) in which the disulfide bonds in the polysulfide backbone (I) undergo sequential attack by the thiol. This leads to depolymerization with production of soluble oligomers and release of fillers. Although thiophenol is effective as a desealant it is malodorous and its prolonged use has been claimed to pose environmental and health problems. In Australia its use has been abandoned and in the UK, the tedious method of mechanical desealing is preferred. The many advantages offered by a safe and effective desealant have prompted the present study.

$$-[-SCH_2CH_2OCH_2OCH_2CH_2S-SCH_2CH_2OCH_2OCH_2CH_2S-]_n$$

(I)

Typical polysulfide polymer

Many reagents are known to cleave the disulfide bond and these include hydroxide ion both alone or in the presence of metal ions[1-3], tertiary phosphines[4-5] and metal carbonyls.[6] These processes, however, are either very slow or require a homogeneous phase. As the polysulfides are insoluble a heterogeneous reaction is mandatory. Other requirements for a practical formulation are that the reagents and reaction products be odourless and soluble in a readily available solvent. Attention was therefore directed to the thiol-disulfide interchange process and a variant, the disulfide-disulfide interchange reaction (Eq. 2), which is promoted by the presence of thiolate anion.[7] This latter reaction has been used synthetically[8] but not for depolymerization of polysulfides and the suitability of this process for desealing purposes has now been evaluated.

$$R'SH + RSSR \overset{R'S^-}{\rightleftharpoons} RSSR' + RSH \quad (1)$$

$$R'SSR' + RSSR \overset{R'S^-}{\rightleftharpoons} R'SSR + RSSR' \quad (2)$$

EXPERIMENTAL

Sealants were cast on a Teflon sheet to a thickness of 3-4 mm, allowed to cure at 24°C for 48 h and then maintained at 40°C for 7 days. Test specimens were cut from the sealant sheets using a 20.6-mm diameter die.

The tetraalkylammonium thiolates were prepared from an excess of the thiol and a solution of the tetraalkylammonium hydroxide. The potassium thiophenolate was prepared from an ethanolic solution of potassium hydroxide. All products were dried _in vacuo_ and stored under dry nitrogen at -4°C.

The desealants were evaluated by suspending the sealant by a wire from a rod resting on a top-loading balance, such that the specimen was totally immersed in the desealing mixture. The change in weight of the specimens with time was expressed as a percentage loss of weight with time and plotted using a computer program.

RESULTS AND DISCUSSION

Survey of Relative Reactivities of Thiolate-Disulfide Systems

Preliminary investigations were conducted using a B-1/2 grade manganese dioxide-cured polysulfide sealant (meeting MIL-S-83430) and dimethylformamide (DMF) as the solvent. Weight losses were recorded for the following reagents: thiophenol (reference desealant), potassium thiophenolate alone and with added diphenyl disulfide, tetramethylammonium thiophenolate alone and with diphenyl disulfide, and tetramethylammonium _t_-butylthiolate both alone and in the presence of di-_t_-butyl disulfide. The relative efficiencies of these reagent

Table 1. Comparison of Various Depolymerization Mixtures

Reagent Combination	Optimum Concentrations (M)	Time to 75% weight loss (min)
PhSH	0.13	20-25
PhSK	0.016	45
PhSK/PhSSPh	0.01/0.13	20
$PhSNMe_4$	-----	-----
$PhSNMe_4$/PhSSPh	0.008/0.2	30
t-$BuSNMe_4$	0.018	75
t-$BuSNMe_4$/*t*-BuSSBu-*t*	0.018/0.007-0.03	60
$PhSNEt_4$	0.01-0.02	50
$PhSNBu_4$	0.005-0.01	55

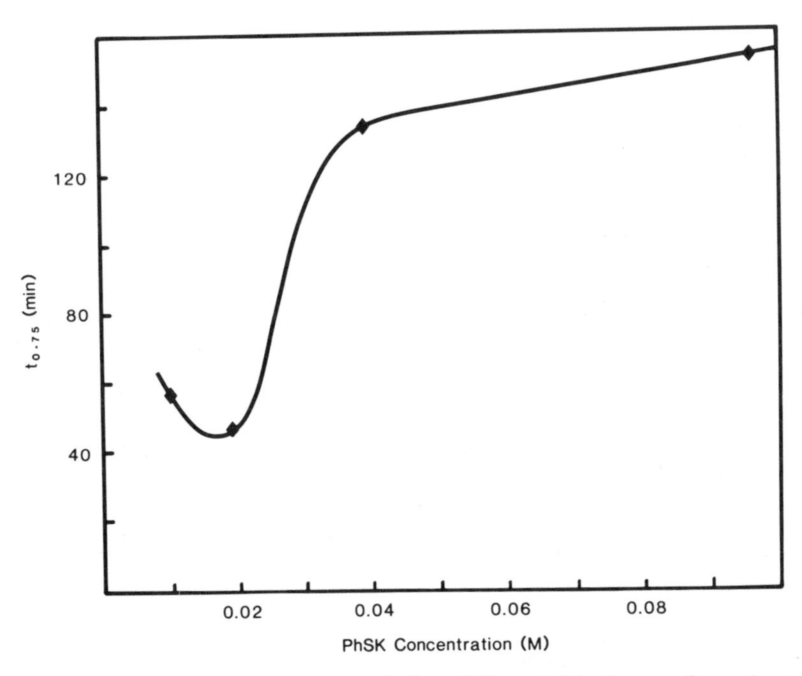

Fig. 1: Variation in time required for 75% weight loss of sealant as a function of potassium thiophenolate concentration.

combinations, based on time to achieve a 75% weight loss, are summarized in Table 1.

The following observations were made:

(a) Effective desealing occurred with thiophenol, potassium thiophenolate, tetraethyl- and tetrabutyl-ammonium thiophenolate and tetrabutylammonium t-butylthiolate. Consistent with relative nucleophilicities, the aryl thiolates were more reactive than t-butylthiolate. Moreover, in solution the thiophenolates oxidize slower and generate a nonvolatile product (diphenyl disulfide). The effect of the cation on depolymerization rates was small.

(b) The rate of desealing with thiophenol and the thiolates was concentration dependent. Above an optimum concentration range the rate decreased and eventually became insensitive to further changes in reagent concentration (Fig. 1).

(c) All combinations of thiolates with disulfides produced effective desealing mixtures and addition of disulfide generally led to a significant rate increase.

(d) By using an optimized thiolate concentration (0.01 M to 0.03 M) and varying the proportion of disulfide, the most suitable concentration of disulfide (0.1 M to 0.2 M) was established. For the case of t-butylthiolate only marginal rate improvements resulted with added disulfide.

(e) The approximate size of the oligomers formed by depolymerization was estimated from the quantities of polymer and reagents used. In some cases these oligomers were calculated to contain more than 25 repeat units corresponding to a molecular weight in excess of 4000. The products therefore approximate to the original liquid polymer.

It was consequently concluded that thiolates and thiolate/disulfide mixtures are effective desealants as predicted by interchange reaction considerations and represent viable alternatives to thiophenol.

Development of a Practical Desealant

Further experiments were designed to refine the method for practical application. Choice of solvent was dictated by cost, toxicity and capacity to dissolve the reagents. Chlorinated solvents are undesirable in aircraft structures and therefore hydrocarbons (toluene, xylene) were examined. Although economical relative to DMF and readily available, these solvents do not dissolve the thiolates as smoothly. The tetrabutylammonium derivative, however, was substantially more soluble in toluene than the other thiolates and clearly offered the best compromise between reactivity and solubility.

The optimum ratio of DMF to hydrocarbon in the solvent mixtures was determined using 0.01 M $PhSNBu_4$. Increasing the hydrocarbon content beyond a given level caused a marked decrease in reactivity. This occurred with mixtures containing >90% of toluene or >85% of xylene (Fig. 2). Dimethylacetamide could be substituted for DMF with no change in desealing performance.

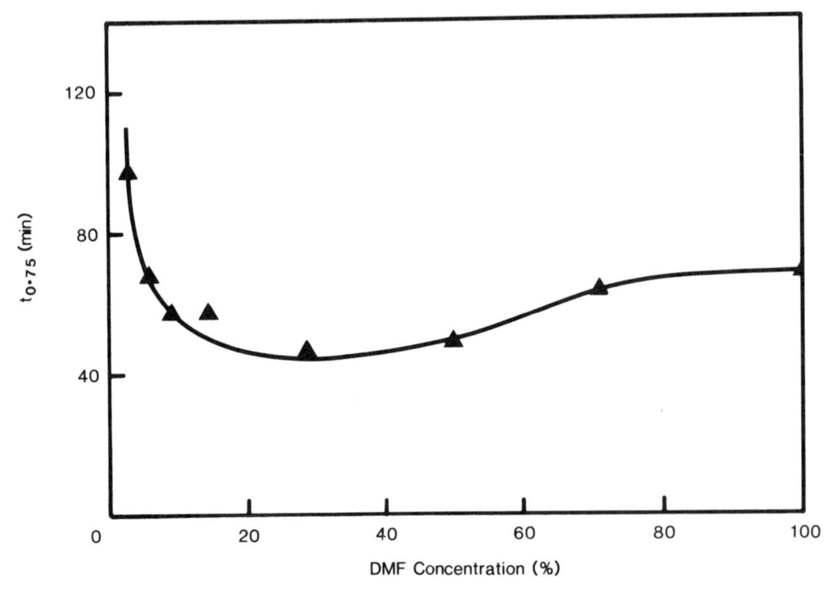

Fig. 2: Variation in time required for 75% weight loss of sealant in DMF/toluene solvent mixtures ($PhSNBu_4$ concentration of 0.01 M).

Effect of Disulfide

A two-fold rate improvement could be achieved by addition of 0.1 M diphenyl disulfide to $PhSNBu_4$ (0.01 M) in DMF-xylene (20:80). As with all the thiolate-disulfide interchanges, however, the penalty for the increase in reaction rate is the expense of extra reagent.

The increase in reactivity in the presence of disulfide may be attributed to additional reactions between the thiolate oligomers and PhSSPh (Eq. 6) in the bulk solution.

$$PhS^- + PhSSPh \rightleftharpoons PhSSPh + PhS^- \quad (4)$$

$$PhS^- + RpSSRp \rightleftharpoons PhSSRp + RpS^- \quad (5)$$

$$RpS^- + PhSSPh \rightleftharpoons PhS^- + RpSSPh \quad (6)$$

(where Rp = polysulfide polymer/oligomer)

This mechanism would effectively maintain a constant level of thiophenolate ions which would otherwise be depleted by the thiolate-disulfide interchange reaction with the polysulfide polymer. Hence, at relatively high concentrations, PhSSPh effects an increase in depolymerization rate.

Depolymerization of Dichromate-cured Sealants

The depolymerization rate dependency of $PhSNBu_4$ on the DMF (DMA/toluene (xylene)) ratio is attributable both to the dilution of the aprotic solvent and to the ability of the solvent to penetrate the sealant. Both DMF and DMA are able to swell the sealant whereas the hydrocarbon solvents are not as effective. This was exemplified by the slower depolymerization rate for a B-1/2 grade dichromate-cured sealant meeting MIL-S-8802, which had a hardness of 76 Rex. In this case the DMF content must be as high as 30 to 40% to obtain desealing rates comparable with those from the softer manganese dioxide-cured material (hardness 56) used in the preceding studies.

Reagent Concentration

The reagent rate dependency on concentration implies that there are major factors dominating the desealing process apart from the simple ionic thiol-disulfide interchange reaction for which the desealing rate would be expected to increase with reagent concentration. A common relationship is observed in all cases involving the reaction rate dependency on the thiol/thiolate concentration. At very low concentrations (<0.005 M) the kinetics of the interchange reaction appear to control the desealing rate. Furthermore, depletion of the reagent at the sealant surface would increase the dependency on the diffusion of the reagent from the bulk solution. As the reagent concentration is increased the desealing rate increases and after the optimum concentration of approximately 0.01 M is exceeded the desealing rate then decreases again. This is thought to arise from a decrease in sealant permeability (with respect to the reagent solution) which reaches a minimum at high concentrations and thereafter the desealing rate becomes concentration independent.

Desealing studies conducted on cured polysulfides without fillers or other additives do not show a desealing rate maximum at low reagent concentrations but rather an exponential decrease in desealing rates as the reagent concentration is decreased. This supports the view that the desealing processes at concentrations above 0.01 M are not attributable to the kinetics of the interchange reaction but are probably dependent on the physical characteristics of the sealant.

In the case of the DMF (DMA)/hydrocarbon solvent mixtures, the desealing processes are solely dependent on the permeability of the sealant and the rate of diffusion of the reaction products from the sealant surface. Here again the desealing rate dependence on the solvent mixture follows a similar pattern to that of the concentration dependence. This further supports the case for attributing the decline of desealing rates, at high reagent concentrations, to transport processes rather than to chemical processes.

CONCLUSIONS

DMF-toluene solutions of tetrabutylammonium thiophenolate, either alone or in combination with diphenyl disulfide, have been shown to depolymerize polysulfide sealants with comparable efficiency to thiophenol-based mixtures. The advantages of using thiolates over thiols lie mainly in the lack of volatility of the salts. This minimizes noxious odours and avoids environmental and health problems resulting from thiophenol vapour.

REFERENCES

1. J.P. Danehy and K.N. Parameswaran, J. Org. Chem., 33, 568 (1968).
2. J.M. Downes, J. Whelan and B. Bosnich, Inorg. Chem., 20, 1081 (1981).
3. W.G. Jackson and A.M. Sargeson, Inorg. Chem., 17, 2165 (1978).
4. B.J. Sweetman and J.A. Maclaren, Aust. J. Chem., 19, 2347 (1966).
5. R.A. Amos and S.M. Fawcett, J. Org. Chem., 49, 2637 (1984).
6. U. Berger and J. Strähle, Z.Anorg. Allg. Chem., 516, 19 (1984).
7. G.M. Whitesides, J.E. Lilburn and R.P. Szajewski, J. Org. Chem., 42, 332 (1977).
8. M. Kleiman, U.S. Pat 2, 510, 893 (1950).

HOT MELT SEALANTS BASED ON THERMOPLASTIC ELASTOMERS

Sung Gun Chu

Hercules Incorporated
Research Center
Wilmington, Delaware 19894

ABSTRACT

New hot melt sealants based on saturated mid-block styrene-ethylene/butylene-styrene (SEBS) were developed for various applications. Their physical properties are compared with the commercial hot melt butyl and silicone sealants. In this paper, we explain how we developed hot melt sealant formulations based on their viscoelastic properties.

INTRODUCTION

The introduction of UV stable styrenic thermoplastic elastomers, such as styrene-ethylene/butylene-styrene, (SEBS) and thermally stable tackifying resins accelerated the development of new sealant products. The thermoplastic sealants have advantages over the conventional thermoset sealants, such as polyurethane, polysulfide and silicones. Most of these thermoset sealants are cured by moisture and their curing reaction is slow (if it is too fast, it has a short shelf life). Since the thermoplastic sealant can be applied as a hot melt, the set time is much shorter than that of the conventional thermoset system. Consequently, we expect the applications for these hot melt sealants to expand rapidly into the automobile assembly and mobile home manufacturing industries.

The butyl-based insulated glass window sealant (IGS) is the only commercial sealant which can be applied by hot melt processing. It has a good market share in window assembly. However, its long-term stability is not adequate for high quality product assembly. In Table 1, the advantageous properties of sealants based on block copolymers are described.

The new hot melt sealants to be discussed in this paper are formulated primarily from the UV stable SEBS elastomer, KRATON G, and two low molecular weight resins (a tackifying resin and an end-block reinforcing resin). Other ingredients, such as UV stabilizer, filler, antioxidant, and adhesion promoter are also included.

In this paper, we will explain how we develop sealant formulations from these ingredients, specifically, how we modify the viscoelastic properties of block copolymer (SEBS) using low molecular weight resins. We also evaluated a hot melt sealant formulation for window assembly applications. Their properties are compared with those of the commercial hot melt butyl and silicone sealants.

Table I

Advantageous Properties of Block Copolymer Based Hot Melt Sealant
• Short set time and no waste (remeltable)
• Excellent UV, ozone and thermal stability
• Excellent mechanical properties (good recovery)
• Low melt viscosity compared to hot melt butyl sealant
• Low MVTR Compared to Polysulfide and silicone
• Stable in alkali solution (hydrophobic)
• Forming clear sealant

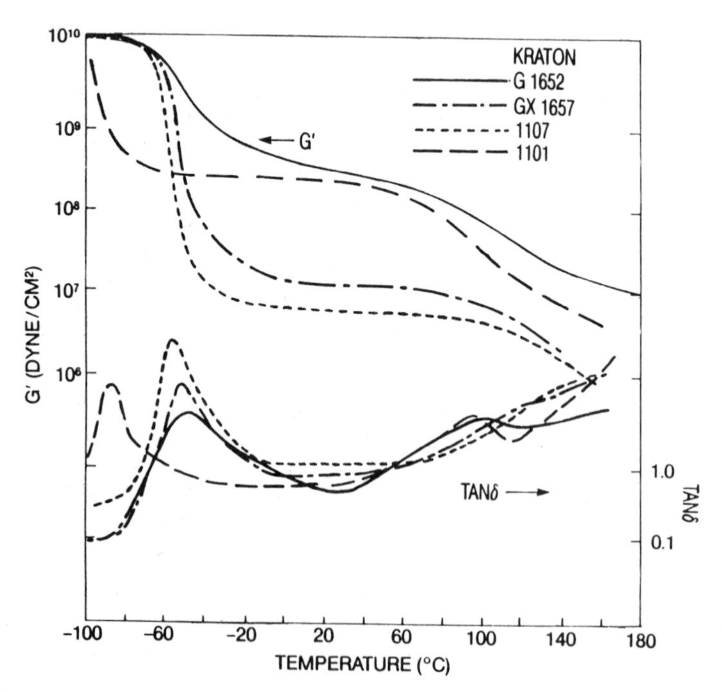

Fig. 1. Dynamic mechanical properties of KRATON samples.

Table 3. Physical Properties of Resin Samples

Resin	Saturation, %	M^*_w,	M^*_n	N_w/M_n	T_g(°C)**	Structure
Sample A	0	664	437	1.5	36	Aromatic
B	0	1020	604	1.7	48	
C	0	2420	590	4.1	53	
D	0	4000	1500	2.6	85	
E	0	6200	2380	2.6	100	
F	0	8200	2930	2.8	104	
G	30	1200	750	1.6	44	Aromatic Cycloaliphatic
H	60	1500	1000	1.5	49	Cycloaliphatic
I	100	410	370	1.1	-22	
J	100	490	390	1.3	-9	
K	100	720	600	1.2	16	
L	100	820	650	1.2	26	
M	100	850	650	1.3	38	
N	100	1100	730	1.5	63	

* by SEC
** by DSC

Resins are based on coplymers of styrene, alpha-methylstyrene and other styrenic monomers. Some have been hydrogenated as indicated.

EXPERIMENTAL

Materials

KRATON samples used in this study were obtained from Shell Chemical Company. KRATON D types are styrene-isoprene-styrene (SIS) and styrene-butadiene-styrene (SBS) block copolymers. The diene mid-blocks contain olefinic unsaturation which leads to poor UV stability. The "G" type KRATONs are styrene-ethylene/butylene-styrene (SEBS) block copolymers obtained by hydrogenation of SBS precursors. The G Polymers offer superior UV stability because of the saturated mid-block. Physical properties of the rubbers are described in Table 2 and the dynamic mechanical properties are plotted in Fig. 1.

Resins described in this paper are based on copolymers of styrene, alpha-methylstyrene and other styrenic monomers. Some have been hydrogenated as indicated in Table 3. Products of this type are available from Hercules Inc. under the following trade names; Piccolastic®, Piccotex®, Kristalex®, Endex™ and Regalrez®. The molecular weight and glass transition temperature data of resins shown in Table 3 were determined by SEC and DSC, respectively. Dynamic mechanical properties of low molecular weight resins are plotted in Fig. 2. Here, the polystyrene sample is the same polystyrene which is used as an end-block of KRATON G 1650. Dynamic mechanical properties were measured on a dynamic spectrometer (Rheometrics, Inc., NJ, USA) in the parallel plate mode using 8 mm plates. The detailed experimental procedures are described in published articles by the author.[1] Evaluation of the hot melt sealant is followed by ASTM E 773-83 and E 774-84a.

RESULTS AND DISCUSSIONS

A. Polymer

The success of thermoplastic elastomer hot melt sealant compounding depends mainly on how much one can lower the melt viscosity (enough to generate perfect wetting) without destroying high temperature performance. The other properties, such as thermal and long-term UV stability of the hot melt sealants, can be obtained easily because of the chemical nature of the saturated thermoplastic elastomer KRATON G and low molecular weight resins. In the early stages of the research program on hot melt sealants based on KRATON G, we found it difficult to develop good sealant formulations which had high elongational properties and low modulus. Most formulations based on KRATON G1650, G1652 and GX1657 exhibit adhesive failure rather than the cohesive failure which is one of the basic requirements for sealants. In order to solve this problem, we used a new polymer, KRATON G1726X, which contains 30% triblock polymer, KRATON G1652, and 70% diblock copolymer (SEB). The room temperature modulus and melt viscosity of the blends of diblock and triblock copolymer are shown in Fig. 3.

This new polymer, KRATON G1726X, has much less cohesive strength than KRATON G1652. Therefore, we can change the failure mode of the sealant from adhesive to cohesive failure with a combination of various KRATON G's. The property of this polymer, KRATON G1726X, is well explained in the recent paper written by G. Holden and S. Chin of Shell.[2]

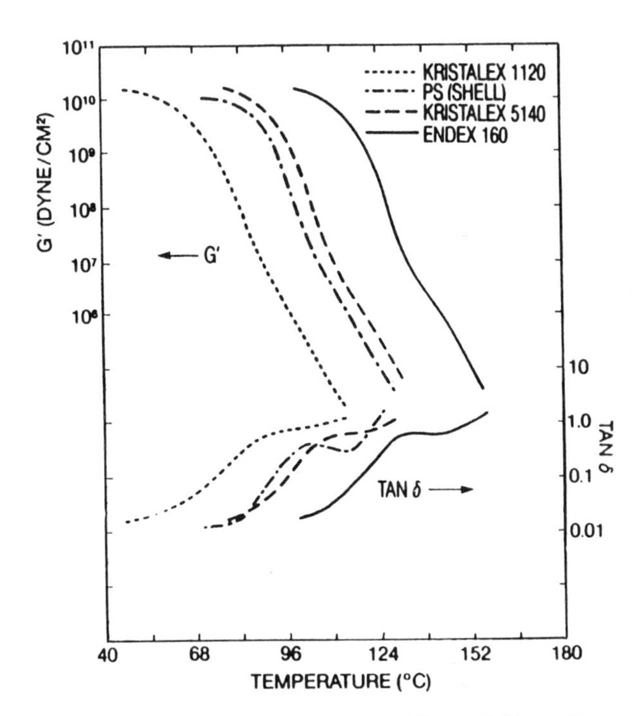

Fig. 2. Dynamic mechanical properties of Hercules resins.

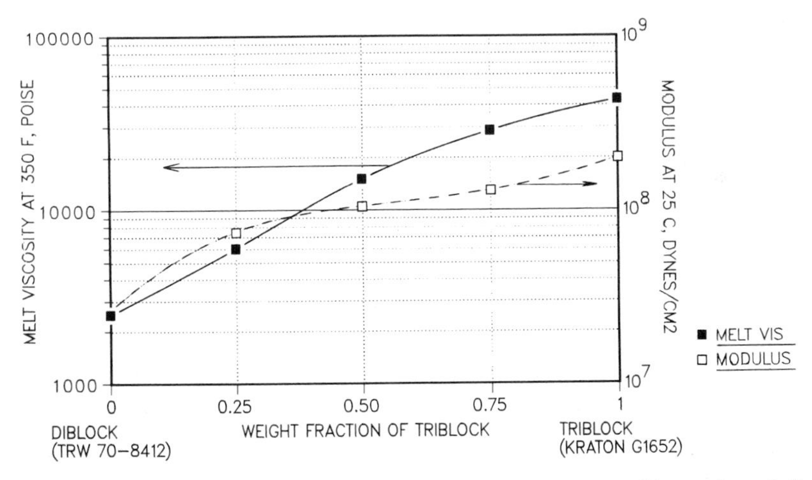

Fig. 3. Effect of triblock polymer on melt viscosity and modulus at 25°C.

B. Resin

Low molecular weight thermoplastic resins are the most important ingredients next to the rubber. The resin generates tack and/or good flow properties which are necessary for making a good sealant. There are two different kinds of resins which are necessary for compounding good sealants from KRATON rubber. These are a tackifying resin and an end-block reinforcing resin. The resins used for compounding hot melt sealants should have excellent UV and thermal stability.

C. Rubber-Resin Blends

The rubber-resin compatibility data can be obtained from the dynamic mechanical properties of the blend.

Figure 4 shows the viscoelastic properties of blends of KRATON block copolymers with low molecular weight resins. The figure presents the $\tan\delta$ and G' values of KRATON block copolymer as a function of temperature. In general, three different phenomena can be observed from blends of KRATON with low molecular weight resins.

Resin Group I: (mid-block compatible resin). These resins increase the mid-block T_g of KRATON samples without changing the T_g of the styrene domains, as they are compatible only with the mid-block. These resins also simultaneously decrease the plateau modulus values. These are so-called tackifying resins or mid-block compatible resins because they reduce the room temperature modulus and, consequently, generate tack. Most aliphatic and cycloaliphatic resins, such as resins I-N of Table 3 belong to this group. The experimental data with this resin is shown in Fig. 5 KRATON G 1650 was blended with Regalrez® 1078 at various loadings. The resin is only compatible with mid-blocks of KRATON G 1650 and raises the glass temperature of mid-block as shown in Fig. 5.

Resin Group II: (end-block compatible resin). These resins increase or decrease the end-block T_g of KRATON samples without changing the T_g of the mid-block domain. These resins also simultaneously increase the modulus values of the KRATON rubber. Aromatic resins, A-F in Table 3 belong to this group. The changes of end-block Tg depends on the T_g of resin. The resins with higher T_g than polystyrene will raise the end-block T_g of KRATON and increase high temperature performance of KRATON. These resins are so-called end-block reinforcing resin. Resins, D, E and F, belong to this group. Example is shown in Fig. 6. Endex™ 160 raised glass temperature of styrene domains of KRATON 1101 from 100°C to 113°C at various loadings.

Resin Group III: These resins are compatible with both blocks of KRATON. Therefore, they increase the T_g of mid-block and decrease the T_g of the end-block. Resins such as G and H, belong to this group. Example is shown in Fig. 7. Resin I is compatible with both domains of KRATON G 1650 and changes their glass temperatures as shown in Fig. 7.

Table 4 summarizes the experimental data on the blends of KRATON G1650 (SEBS) and various resins. These data were obtained from a temperature sweep using the Rheometrics Dynamic Spectrometer (ω = 10 radians/sec.). The data can be used for the other KRATON SEBS rubbers, such as GX1657, G1652 and G1726X. The data indicate that the rubber-resin compatibility depends on the resin chemical structure and molecular weight. Cycloaliphatic and aliphatic resins are compatible with ethylene-butylene mid-blocks while aromatic resins are compatible with end-blocks of KRATON rubber. However if the molecular weight of resin is low, the above rule is not valid. The low molecular weight hydrogenated resins (Samples I, J and K) are soluble in both blocks of KRATON rubber.

Figures 8 and 9 show the effect of resin molecular weight on the T_g and room temperature modulus, G′, of KRATON G1650 at various resin concentrations. These resins are all hydrogenated and are compatible with the mid-block domains of KRATON (tackifying resin). A tackifying resin, which has a higher T_g than KRATON G, will increase the glass temperature of the rubbery domain. It decreases the room temperature modulus value at low loading and then increases the modulus at high loading. The modulus vs. resin loading curve depends on the structure and molecular weight of tackifying resin used.

Table 4 Effect of Resin Type on Midblock and Endblock T_g of KRATON G-1650/Resin Blends

	Resin Sample	Resin* T_g(°C)	Mid-block** T_g(°C)	End-block** T_g(°C)
KRATON G1650			-50	100
Endblock Resin (Group III)	A	36	-50	72
KRATON/Resin(70/30)	B	48	-50	83
	C	53	-50	102
	D	85	-50	108
	E	100	-50	110
	F	104	-50	113
Resins partially compatible with both blocks (Group II) KRATON/Resin (50/50)	G	44	-45	85
	H	49	-10	87
Midblock Resin Group I KRATON/Resin (50/50)	I	-22	-23	84
	J	-9	-20	85
	K	16	-8	95
	L	26	2	100
	M	38	8	100
	N	63	28	100

* T_g obtained from DSC data of resins.
** T_g obtained from RDS data of the KRATON/Resin blends.

D. Formulation Work

In this section, we will explain how we can develop a hot melt sealant formulation using the above mentioned KRATON G and resin blends.

We used the viscoelastic properties of sealant formulations as guidance for developing hot melt sealant formulations from KRATON G.

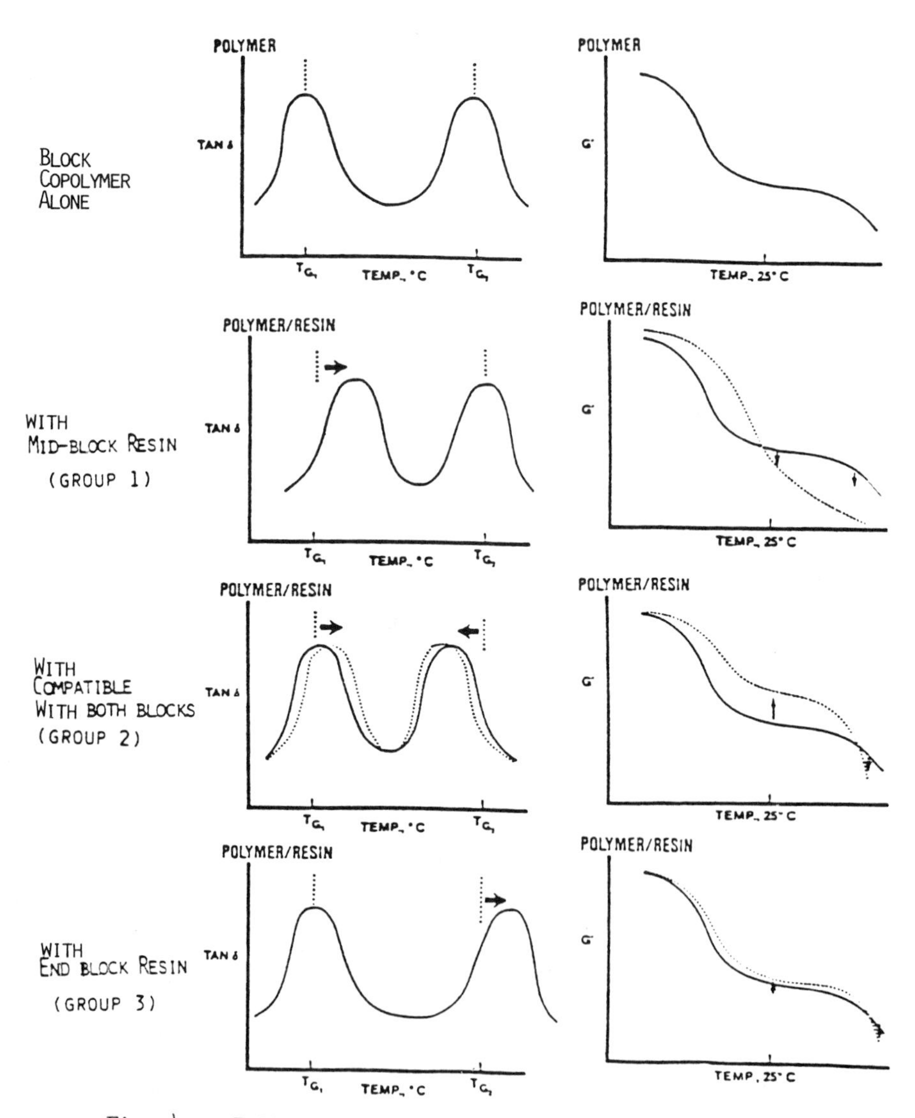

Fig. 4. Rubber - resin compatibility (Block copolymer)

The dynamic mechanical properties of commercial silicone, and hot melt butyl sealants are shown in Fig. 10. Silicone sealant has long plateau modulus value (excellent high temperature performance) with very low glass transition temperature (T_g) compared to hot melt butyl sealant. Our objective is to develop a hot melt sealant formulation with lower temperature flexibility, higher temperature creep resistance and lower application temperature than those of the currently available hot melt butyl sealant.

The low temperature flexibility is related to the glass transition temperature of the sealant and can be adjusted by tackifying resins. High temperature creep resistance and melt viscosity of the sealant can be adjusted by end-block reinforcing resins. The ratio of elastomer and low molecular weight resin in the sealant formulation is the main determining factor which controls the cohesive strength of sealant.

In order to generate a sealant mass with hot tack and low T_g properties we used Regalrez® 1018 resin. Regalrez 1018 is a commercial version of the resin sample I in Table 2 from Hercules. Two different grades of KRATON G elastomers were blended in order to balance the cohesive strength and melt processability of the sealant (see Fig. 3). The KRATON and Regalrez 1018 blend (100 parts KRATON G/250 parts Regalrez 1018) is a nontacky mass at room temperature with acceptable hardness and elongational properties for sealant application. It has a hot tack property with glass temperature about -15°C. It can be easily extruded at 350°F using a commercial hot melt gun. However, it still has poor creep resistance at 100°C,

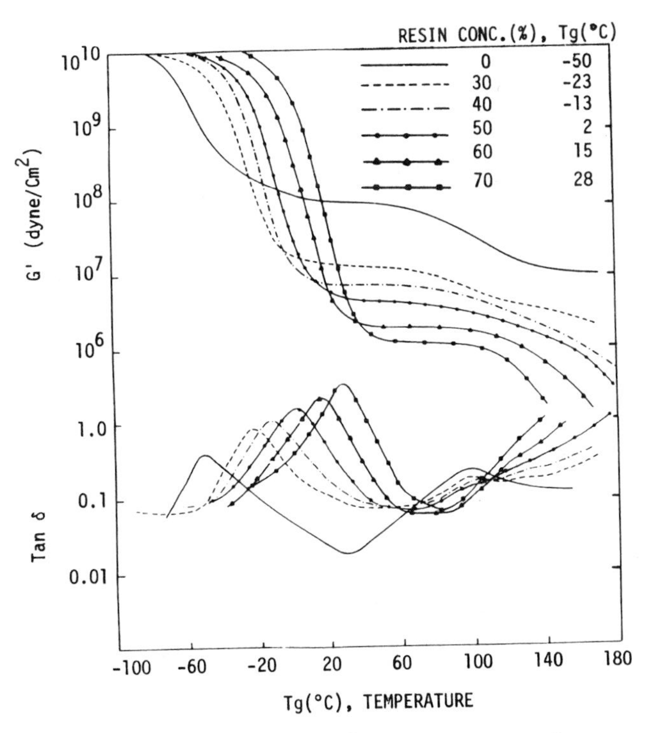

Fig. 5. KRATON G-1650/regalrez 1078.

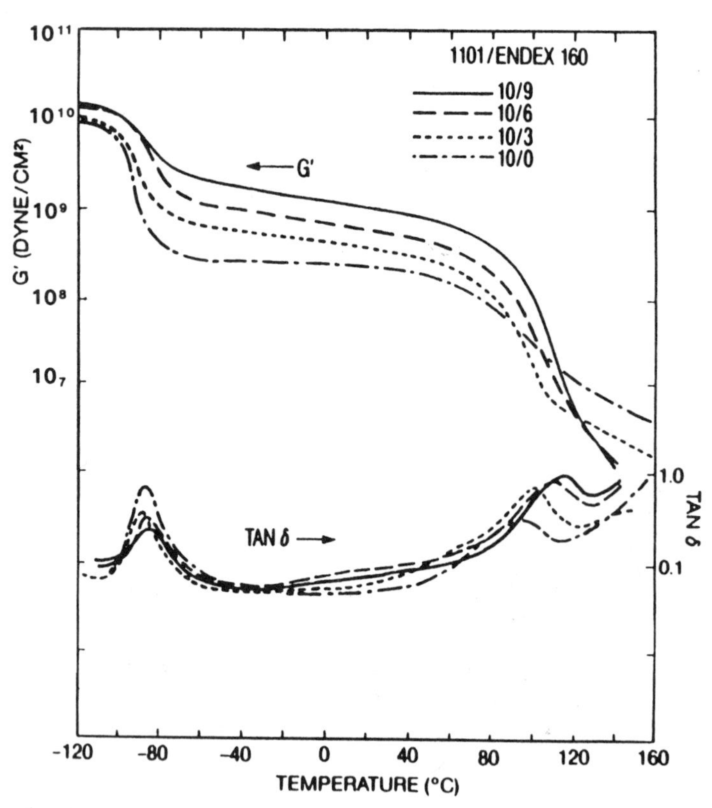

Fig. 6. Dynamic mechanical properties of KRATON 1101 and its blends with Endex 160.

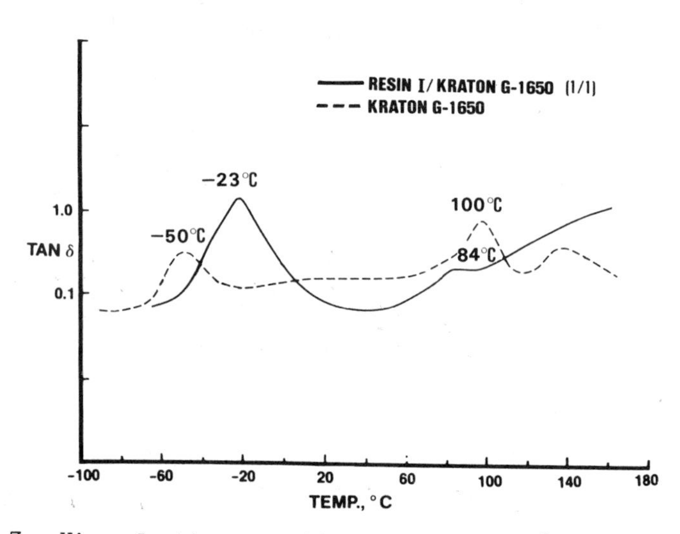

Fig. 7. Viscoelastic properties of KRATON G-1650 and its blends with Hercules resin.

therefore, we used Endex™ 160 which is a commercial version of the resin samples F in Table 3.

The improvement of high temperature creep resistance of the sealant was confirmed by the slump test shown in Fig. 11.[3] The sealants without Endex™ 160 or with other commercial polyalphamethylstyrene resin has poor creep resistance at 100°C. The sealant including Endex 160 is a clear mass due to their excellent compatibility with KRATON G. However, the sealant including polyalphamethylstyrene is opaque due to their partial incompatibility with KRATON G. This incompatibility is due to its higher molecular weight (Mw=30,000) and broader molecular weight distribution (Mz/Mw=30) than those of the styrene domains of KRATON G.

The reason for the good creep property of the sealant with Endex 160 is also confirmed by a morphological analysis. Figure 12 is a picture showing the morphology which was obtained from KRATON 1107 and its blends with Endex 160. The transmission electron micrographs were taken after staining the samples with OsO_4 for 36 hours. The samples were microtomed at cryogenic temperature. The magnification factor of these micrographs is 15,000X. It has a spaghetti-and-meatball morphology with a continuous rubber phase and a dispersed polystyrene phase. The discontinuous phase is polystyrene (end-block) and the electron-dense continuous phase is an isoprene rubber (mid-block). These dispersed domains are quite small, less than 200Å

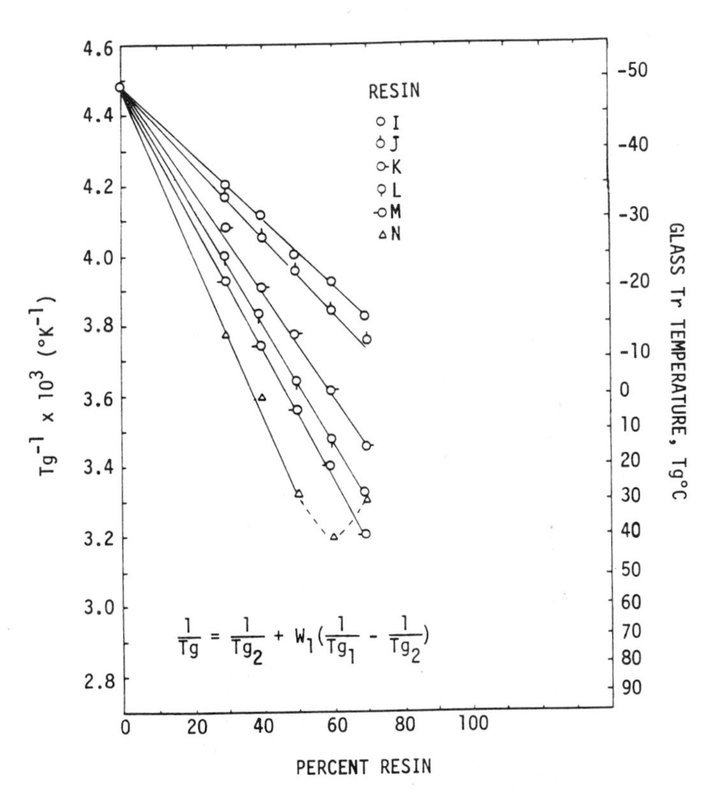

Fig. 8. T_g of KRATON G1650/resin blends as a function of resin concentrations.

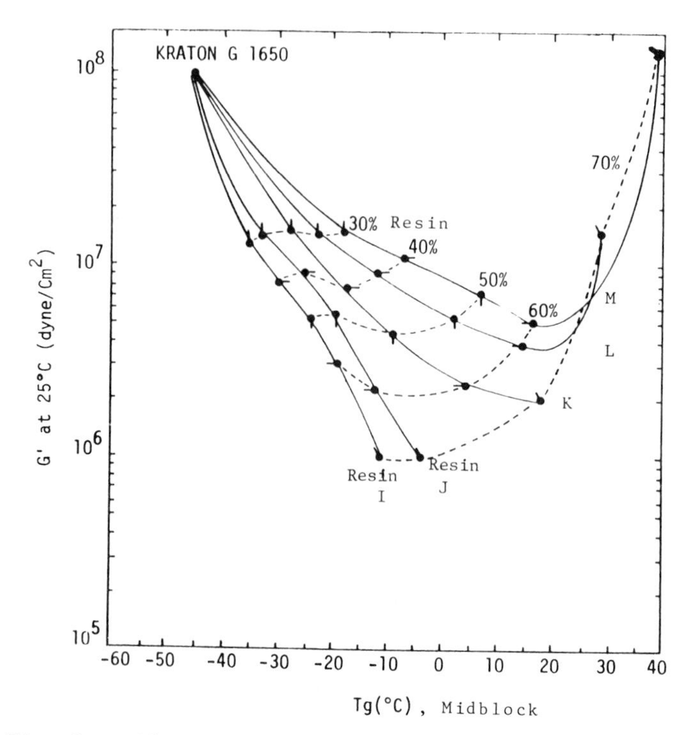

Fig. 9. G' and T_g of the KRATON G1650/resin blends.

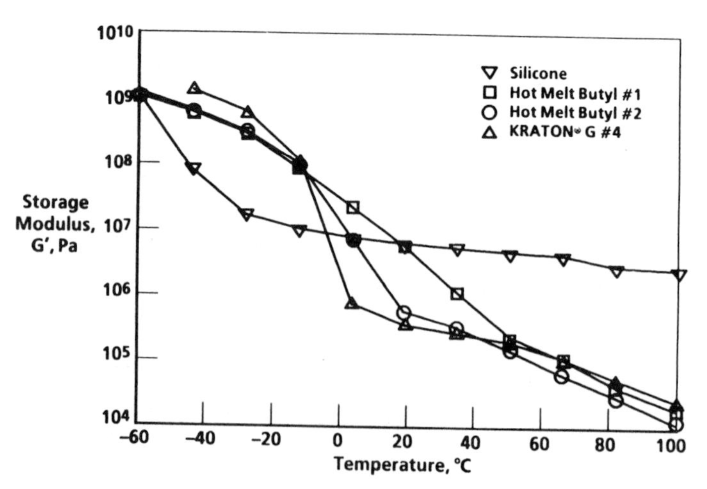

Fig. 10. Storage modulus as a function of temperature KRATON®G rubber based sealant compared.

in size. In Fig. 12-B, the morphological change of KRATON D1107 blends at various loadings of Endex 160 is shown. By blending 4.5 g of ENDEX 160 with 10 g of KRATON D1107, the size of the dispersed styrene phase increased and the shape changed from spherical types to jellybean types while the size of the dispersed phased homogenous. From this, we know that all of the end-block reinforcing resin, Endex 160, entered into the polystyrene domains. So by blending with Endex 160, we can get a bigger dispersed domain size. Consequently, it had a better creep resistance at room temperature. Also it had better high temperature performance due to the higher glass transition temperature of the dispersed domains shown in Figs. 6 and 13. The T_g of styrene domains in Fig. 12 was measured using the RDS from KRATON D1107 and Endex 160 blends. If one adds more Endex 160 (7 g to 10 g of KRATON D1107), one will see a much bigger domain size. However, one will also see a small domain size due to overloading of Endex 160 in KRATON D1107. This sample is very cloudy. Endex 160 starts to form its own domains. From this level (7 g to 10 g) Endex 160 is overloaded. If one adds more Endex 160, it will undergo a phase inversion where the dispersed phase will be rubber and the continuous phase will be polystyrene. From this micrograph, we know that 4.5 g of Endex 160 and 10 g of KRATON 1107 is the ideal concentration, (about 3 times of the styrene content of KRATON D1107).

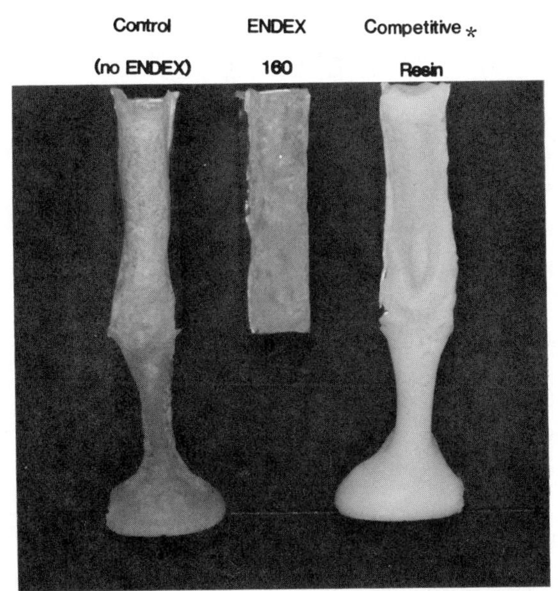

* Polyalphamethylstyrene (Mw=30,000 Mz/Mw=30)

Fig. 11. Slump test of KRATON G based hot melt sealant with various endblock resins (one week at 100°C (212° F).

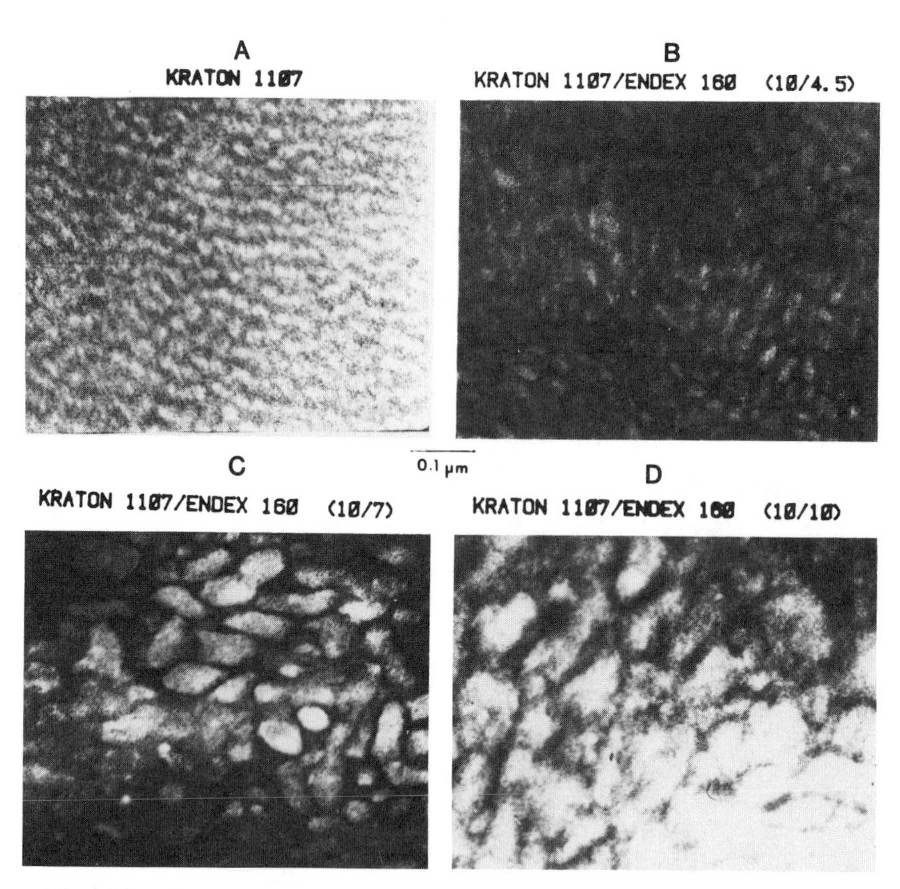

Fig. 12. Morphology picture of KRATON 1107/ENDEX 160 blends.

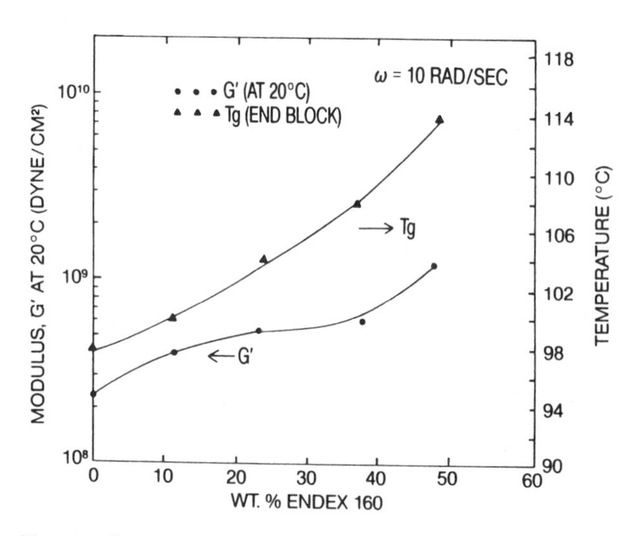

Fig. 13. Physical properties of KRATON 1101 blends with Endex 160.

The addition of Endex also changed the melt viscosity of KRATON. The melt viscosity of KRATON and Endex 160 blend at 350°F is shown in Fig. 14. The addition of Endex 160 reduced melt viscosity of the sealant at application temperature, and consequently improved mixing and extrudability of the sealant.

E. Hot Melt Sealant for Insulated Glass Window Assembly

In Table 5, a hot melt sealant formulation based on KRATON G was evaluated for insulated glass window sealant applications. The adhesion properties of the sealant onto glass or anodized aluminum was excellent (cohesive failure) compared to the commercial hot melt butyl and silicone sealants (Table 6). It had a higher tensile strength and a lower modulus value with good high temperature performance than the butyl sealant. In Figs. 15 and 16, dynamic mechanical properties of this sealant are compared with those of the commercial butyl sealants A and B. KRATON G based sealant had lower temperature flexibility (low T_g) along with better higher temperature performance (for 100°C) than the butyl sealants. It also had a lower melt viscosity with thixotropy at 350°F than the butyl sealant. The most interesting properties of this hot melt sealant is its low moisture vapor transmission rate (MVTR) which is comparable to the butyl sealant (Table 6). It is 20 times better than the silicone sealant. It also has excellent moisture resistance (stable) because it contains hydrophobic elastomer and resin. Since it has an excellent adhesion and low moisture vapor transmission rate, it will not have a fogging problem when it uses as a double layer glass window sealant. We also evaluated its UV resistance using QUV for 200 hours and did not detect any deterioration of adhesion and mechanical properties of the sealant.

The formulation in Table 5 should be optimized and their long-term durability evaluated before commercialization. Since its raw material cost is much lower than that of silicone sealant and it has better performance than that of the butyl sealant, it is expected that the market potential will be growing in the future.

Table 5 Hot Melt Sealant Formulation for Window Assembly

Ingredients	PHR
Kraton G1707X Rubber	95
Kraton G1726X Rubber	5
Regalrez™ 1018	250
Endex® 160	50
Piccofyn® T125	25
Silane	4
Irganox 1010	1.0
Tinuvin 770	1.0
Tinuvin P	1.5

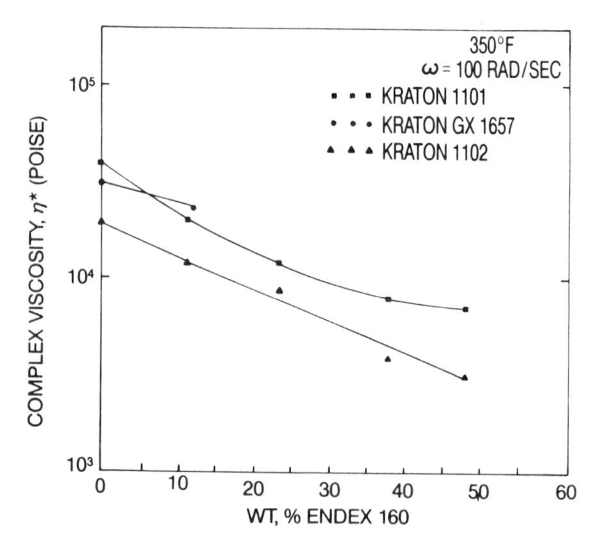

Fig. 14. Melt viscosity of KRATON samples blended with Endex 160

Table 6. Evaluation Data of Various Sealants

Test Results	Kraton G Sealant (Table 5)	Hot Melt Butyl	Silicone
Tensile Strength, MPa	1.0	0.1	2.8
100% Modulus, MPa	0.2	0.2	0.6
Elongation, %	1850	1500	425
Hardness, Shore A	20	45	33
Melt Viscosity[a)], Pas	110	540	NA
180° Peel to Glass, KN/m	11C	4C	NA
180° Peel to Glass/Soak[b)], KN/m	12C	4C	NA
180° Peel to Aluminum, KN/m	12C	4C	NA
180° Peel to Aluminum/Soak[b)], KN/m	12C	4C	NA
SAFT[c)], °C	73	68	NA
Slump, °C	110	115	
MVTR[d)] at 23°C, 50% R.H.	0.25	0.34	NA
MVTR[d)] at 38°C, 90% R.H.	2.46	1.11	>40

a) Measured at 177°C.
b) Immersed in 23°C water for one week.
c) Shear adhesion failure temperature.
d) Moisture vapor transmission rate, g/m2/day.

C: Cohesive Failure
NA: Not Available

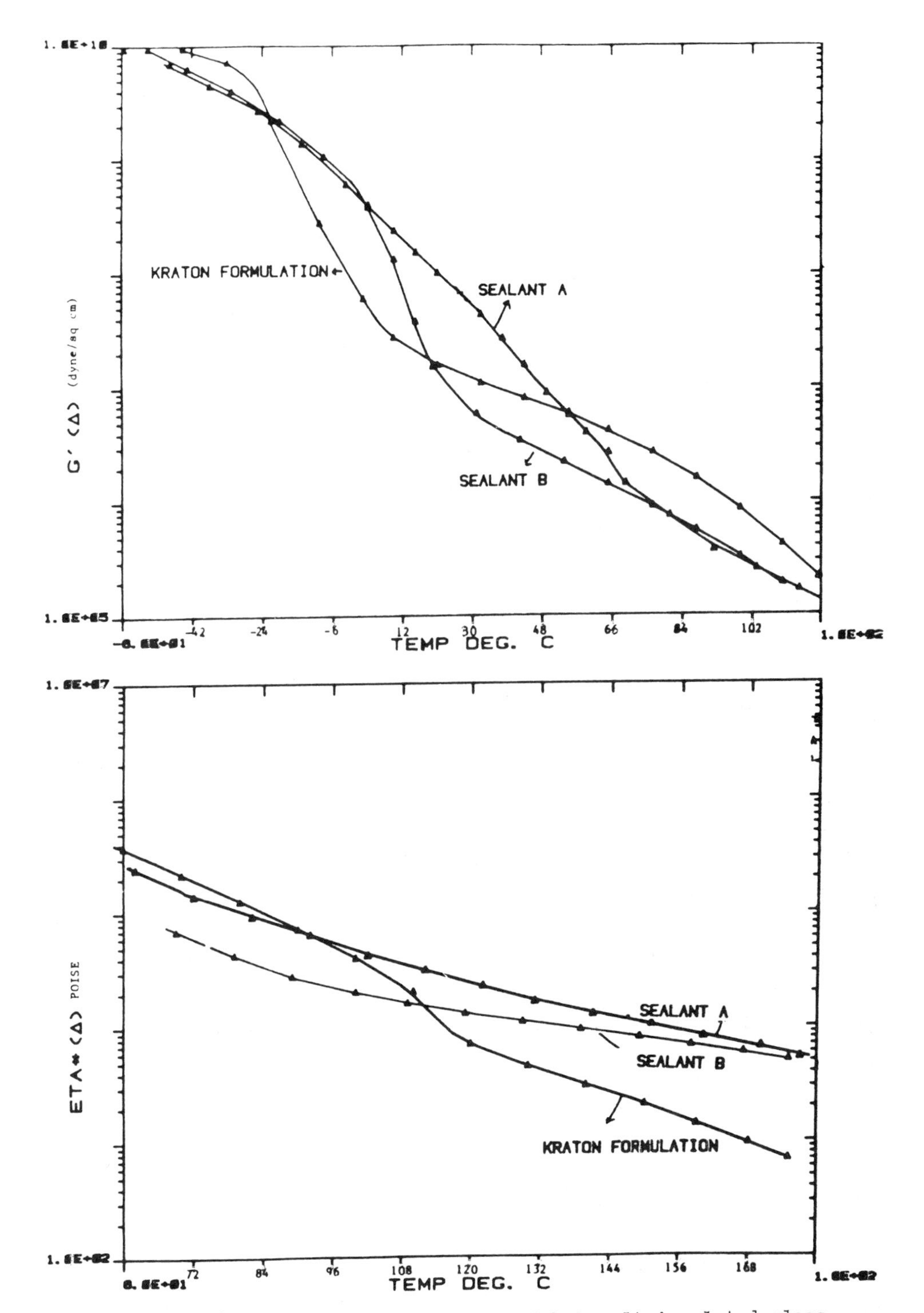

Fig. 15 (a) Viscoelastic properties of hot melt insulated glass sealants.
(b) Melt viscosity of hot melt insulated glass sealants.

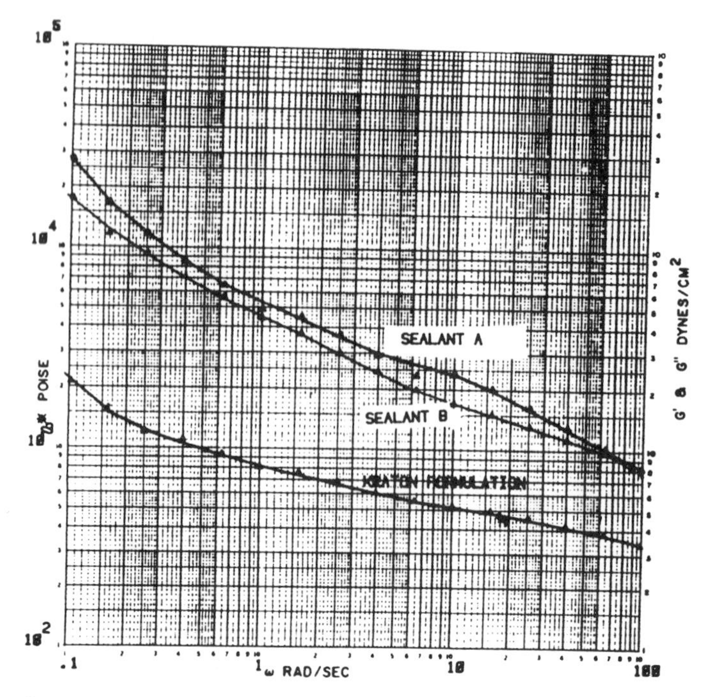

Fig. 16. Complex viscosity at 177°C as a function of frequency. (Strain: 99, 50, 10%; test mode: RS/12.5PP).

CONCLUSIONS

Hot melt sealants can be made from saturated mid-block KRATON G (SEBS) and hydrogenated low molecular weight hydrocarbon resins. The dynamic mechanical properties of thermoplastic elastomer Kraton G resin and their blends were studied. The principles of tackification and formulation work for KRATON G were explained. A good starting sealant formulation was developed and its properties are evaluated. The hot melt sealant based on KRATON G had good adhesion to glass or aluminum along with excellent mechanical properties. It also had a low moisture vapor transmission rate along with excellent UV and thermal stability. It's potential for commercialization is expected to be great.

ACKNOWLEDGEMENTS

The author would like to thank Mr. R. Sample and Mr. B. Hodgson for the experimental work, and Ms. Vicki Lancello for the preparation of this manuscript. The author wishes to thank Shell Chemical Company for the permission to use their valuable information.

REFERENCES

1. J. Class, S.G. Chu, "The Viscoelastic Properties of Rubber-Resin Blends," J.Appl. Polym. Sci., 30, 825, 1982.
2. G. Holden and S. Chin, "Styrenic Block Copolymers in Sealant." The proceedings of the Adhesive and Sealant Council Meeting, March 24-26, 1986.
3. S.G. Chu, "A New Reinforcing Resin for KRATON Block Copolymer." The proceedings of the Adhesive and Sealant Council Meeting, March 28, 1984.
4. Technical Bulletin SC: 198-83, Shell Chemical Co., Houston, TX, 1983.

SEALANTS, WATERPROOFINGS AND COATINGS FOR CONCRETE

Robert M. Evans

Center for Adhesives, Sealants and Coatings
Case Western Reserve University
Cleveland, Ohio, 44106

ABSTRACT

Concrete, particularly when exposed to moisture, represents a harsh environment for sealants, waterproofings and coatings. This is a result of its high alkalinity, the fact that it can form an osmotic cell with semipermeable membranes, its weak powdery surface and its tendency to crack when curing. This paper discusses how the chemical and physical nature of concrete is the source of these problems. It also discussed some of the materials that industry has developed to meet these requirements and the test methods that have are employed to predict behavior under these conditions.

INTRODUCTION

The hazards of concrete as a substrate for adhesives, sealants and coatings have been vastly underrated by industry. There are four factors which make it so: (1) Its extreme alkalinity destroys hydrolysis sensitive materials at the interface. (2) It forms, when coated, osmotic cells (3) It has a weak, powdery surface layer which must be penetrated or removed. (4) It tends to crack for a considerable period of time after it has been formed. This paper will deal with (a) the tests for (1) coatings (2) waterproofing membranes and (3) sealants which are designed to withstand these harsh conditions and (b) the materials which have evolved to meet the requirements of the conditions and of the tests.

THE CHEMICAL PROBLEM OF THE CONCRETE SURFACE

Wet Concrete Produces Alkali My interest in this substrate came about "naturally." Our family company specialized in paints for the food industry. They did so because small companies, then as now, had to find a particular niche which was somewhat protected from the big guys. We manufactured coatings made with China wood oil (tung oil). This oil was employed because it polymerized easily to a high enough viscosity to accept respectable amounts of solvent and because it gave a hard, quick drying coating. It also turned out that coatings made with tung oil had a unique ability to resist the attacks of the environment of food processing plants.

Rumination showed that the happy marriage of our paints with food processing plants came about because tung oil produced paints which have high alkali resistance and the walls of food plants were mostly concrete which was perpetually moist. Doing what is now called operations research but was then merely extinguishing of fires had shown that there was an almost catastrophic correlation between lack of resistance to alkali and coatings failures.

Why must coatings for concrete which is wet for long periods withstand alkali? It comes from the nature of portland cement--which forms cement paste, the binder for the aggregate in concrete. One of the silicates contained in portland cement hardens by the hydration reaction shown below:

$$2Ca_2SiO_5 + 4H_2O \rightarrow Ca_{3.3}Si_2O_{7.3}\cdot 3H_2O + 0.7Ca(OH)_2 \qquad (1)$$

With alkali salts present, the following reaction produces NaOH:

$$Na_2SO_4 + Ca(OH)_2 \rightarrow CaSO_4 + 2NaOH$$

The NaOH produced will attack any material--particularly esters--which is alkali sensitive.

Coated Concrete Forms an Osmotic Cell Mere production of NaOH is bad enough, but because the walls are subject to condensation and because a paint is a semipermeable membrane, osmotic cells are formed as shown in Fig. 1. These set up extremely high pressures which lift the coating from the surface in, ordinarily, the form of aqueous blisters. Because the liquid is strongly alkaline, (a pH of 13 is not uncommon) the interfacial surface tends to be destroyed.

The Physical Problem of the Concrete Surface

Ordinarily, the surface of concrete is covered with laitance, a chalky dust. This forms a weak and powdery surface layer. It is defined by the American Concrete Institute as a layer of weak and nondurable material containing cement and fines from aggregates, brought by bleeding water to the top of overwet concrete. The amount of laitance can be increased by overworking or overmanipulating concrete at the surface by improper finishing.[2] The surface weakness could extend to some depth beneath the surface because the "bleed water" that comes to the surface with overwet concrete will be worked into the concrete to a considerable depth. Since the strength of concrete is inversely proportional to the ratio of water to cement, this will be a zone of weakness. Hence either the laitance must be removed or the material that is applied to it must penetrate and harden the layer. Tung oil enamels, for instance, were endowed with this capability. To penetrate this layer, many sealants require a primer.

The Problem of Cracking of Concrete Concrete is formed by mixing together cement paste and aggregate. The idealized reaction for cure of the cement paste was shown above. One of the reaction products was $Ca(OH)_2$. This forms nodular crystals which constitute a zone of weakness. As Fig. 2 from Grudemo[3] shows, these crystals are the locus of cleavage surfaces.

Many concrete surfaces, e.g., floors, pavements, are restrained. They tend to crack because of the tensile stress which develops when the shrinkage which occurs upon cure of cement paste is restrained. Other causes are temperature stresses, frost action or excessive loading. However, shrinkage is particularly important because it occurs while the tensile strength of the concrete is far lower than the design strength.[4] Fig. 3 shows this graphically.

COATINGS FOR ALKALINE SURFACES

<u>Requirements of Coatings for Alkaline Surfaces</u> The two most important requirements are resistance to alkali and ability to bond to the powdery surface (laitance) characteristic of concrete.

<u>Coatings for Plaster Surfaces</u> My own theories as to the cause of failure of paints on alkaline surfaces arose from work that was done by the "Plaster Committee" of the Cleveland Production Club back in 1951-4.[5] At that time, walls were still plastered with gypsum-lime plasters. There was much peeling of paint from these surfaces. Consequently, the

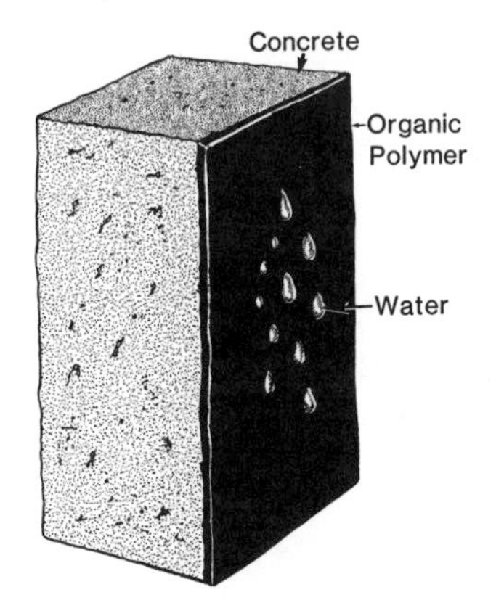

Fig. 1. Coated concrete as an osmotic cell. The pores of the concrete are filled with a salt solution which is highly alkaline. The organic coating is a semipermeable membrane. Condensed water, on the surface, has virtually no ionic solute.

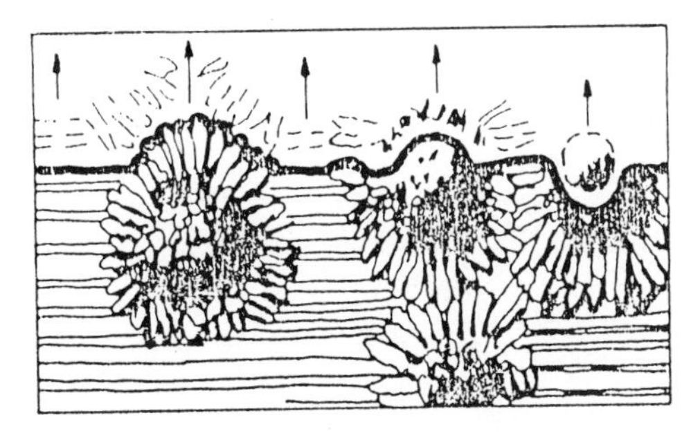

Fig. 2. Crack growth in cement paste travels through weak $Ca(OH)_2$ nodules.

Cleveland Production Club (as it was known in those unenlightened days) had organized a technical committee whose charge was to develop a test which could predict failure on plaster surfaces. The test we developed required coating plaster discs made with one third gypsum (gaging) plaster and two thirds hydrated lime. These were placed on water saturated cotton in a covered petri dish (Fig. 4). Surprisingly, even those paints which were known bad actors did not fail in a few weeks. (They did fail in a few months) This verified the work of Llewellen,[6] a British investigator. He showed that plasters made with only lime--no gaging (gypsum) plaster or portland cement--caused no paint failures. Plasters made with gypsum did cause paint failures. Llewellen reasoned that the source of the failure was Na^+ impurity ions in the gypsum which reacted with the OH^- ions of the hydrated lime to form NaOH.

As a consequence, the Cleveland Club's Plaster Research Committee concluded that it required a dilute solution of NaOH to bring about failure in the test structure as described above. With this modification their results were rapid and in the proper order e.g., high-styrene-butadiene solution polymer best; calcicoater made with limed rosin worst and various alkyds and varnishes in between.

Test Methods for Coatings for Concrete

An Osmotic Cell on Concrete Since the same reactions that Llewellen postulated for plaster were operative in concrete, I tested applicability of the Cleveland Club method to coatings for concrete surfaces (Fig. 4). However, I substituted neat white cement for plaster discs and KOH for NaOH (to reduce the excessive efflorescence volume characteristic of Na_2SO_4). The petri dish provided an osmotic cell: the concrete substrate and the KOH provided the salt filled pores and alkali; the paint film, a semipermeable membrane; the covered petri dish generated sufficient condensation on the paint surface to form an osmotic cell.

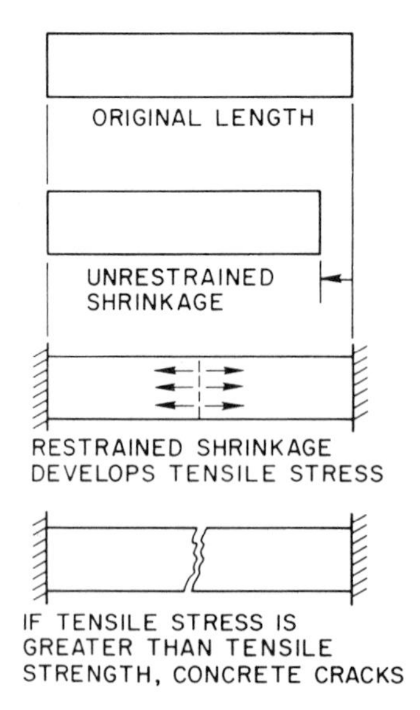

Fig. 3. Shrinkage of weak, restrained concrete slab causes cracking.

Failures were by saponification, by peeling and by formation of large blisters. As in the Cleveland Club's work, failure on this test correlated well with field exposure.

<u>ASTM Method Testing High Performance Architectural Coatings</u> The Standard Practice for Testing High Performance Architectural Wall Coatings, ASTM D 3730,[7] includes a test for adhesion. Since most architectural coatings are applied to concrete, most of the tests listed are conducted on a substrate of either lightweight concrete block or smooth asbestos cement board. The cement asbestos board test panel for the adhesion test is coated with the system being tested and allowed to cure. 25 mm diameter dowels are then cemented to the surface. After further cure, they are immersed in water to a depth that will wet the panel but not the coated surface. After 7 days, they are removed and the tensile adhesion of the dowels to the surface is determined. It must exceed 90 N. (20.2 lbs.)

Types of Coatings

<u>Solvent-Born</u> Two types of coating have achieved remarkable success--chlorinated rubber and high styrene-butadiene solution polymers. Both can wet out laitance and both have excellent alkali resistance. A factor which may have added to their success is their very low moisture vapor transmission rate. I believe that this discourages adhesion loss due to the development of osmotic pressure. Excellent results have also been achieved with amine-cured epoxies, polyether-based polyurethanes (polyether polyols are very hydrolysis resistant) and with those polyesters which have excellent alkali resistance. The marketplace soon banished those polyesters which did not.

<u>Water-Based</u> Certain latex based materials have also been successful. In some experiments which I conducted, I found that when these were used as primers for solvent-born coatings in the test described above, the top coat was often saponified, while the latex primer was unharmed. I reasoned that the primer was so porous that Na^+OH^- ions passed through the film to do their destructive work.

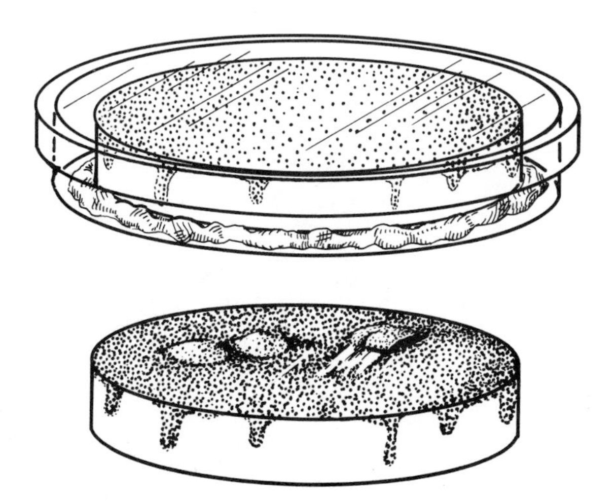

Fig. 4. Test cell for coatings to be applied to concrete surfaces subject to condensed moisture.

Requirements of Sealants for Alkaline Surfaces Preformed concrete building components and highway pavements have this in common; both require sealants to prevent passage of water through joints which are moved by seasonal and daily temperature variations. As Fig. 5 shows movement[8] can be greater than ±35% from summer to winter. In the case of highway pavements the situation is confounded by ponded water and by extremes of cyclical movement due to loading by traffic. In all cases, it is essential that the sealant retain its adhesion to the substrate. Preferably, this will be without a primer. The most common configuration of these sealants is as a butt joint (Fig. 5).

Tests for Elastomeric Joint Sealants

ASTM Test The key test for these materials in the U.S. is ASTM C-719 for Adhesion and Cohesion of Elastomeric Joint Sealants Under Cyclic Movement (the Hockman test).[9] This test uses the equipment and the concrete specimens shown in Fig. (6). The cycle employed is as follows:

1. Cure samples for two weeks under ambient conditions.
2. Immerse for one week in distilled water.
3. Following immersion, hand flex each specimen 60° to check the bond. (this test is very important for concrete specimens).

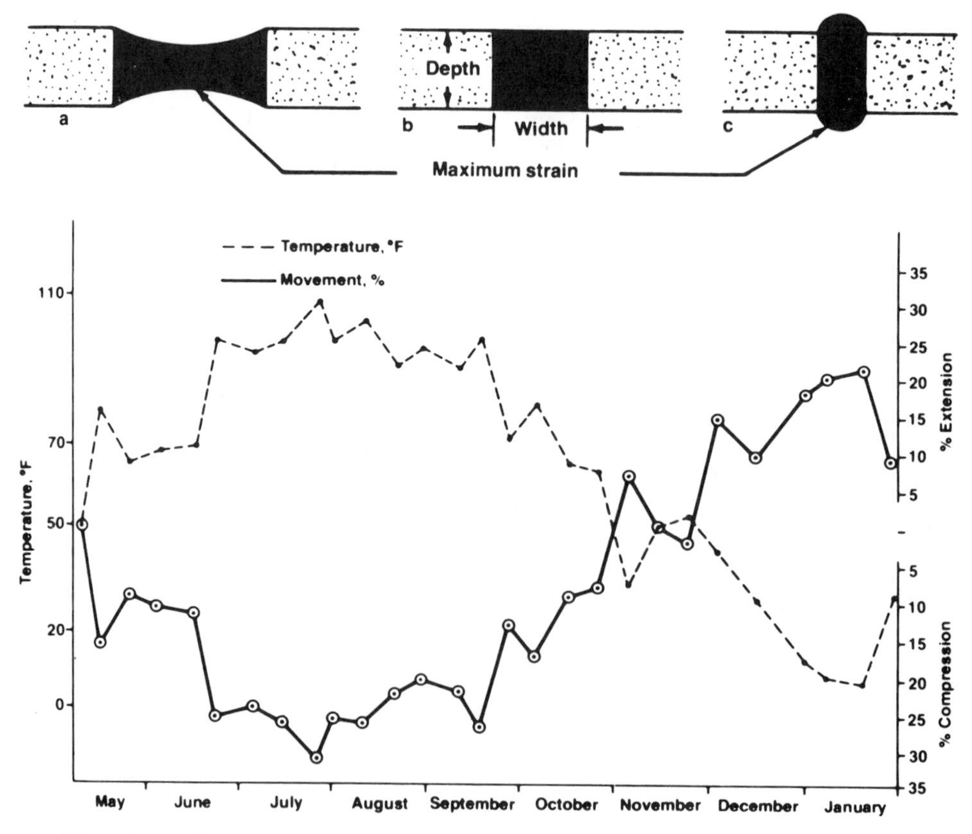

Fig. 5. Movement of a butt joint due to temperature variations.

4. If the bond is still firm, compress the joint 25%.
5. Place the compressed specimens in an oven at 70°C for one week.
6. Remove from oven, allow to recover from compression.
7. Cycle 10 times ±25% at 3.2mm/hr.

If the sample passes these tests, it must go through 10 cycles in which the sample is held extended after cooling to -25°C and compressed at 70°C.

Insofar as concrete substrates are concerned, there are two crucial elements to this test; the 60° bend test after immersion and compression for one week at 70°C. The bend test on the samples, which is executed after the concrete has been water soaked, stresses surfaces which have been exposed to alkali and osmotic pressures. The importance of the compression test will become evident in the discussion of the evolution of these tests and these sealants in the sections which follow.

The Issue of Compression While Heated The requirement that compressed test samples be held at 70°C clearly approximates the stresses to which construction sealants are exposed in the summertime. In fact, Hockman's original test method which did not have such a requirement,[10] qualified materials which failed in the field. In my discussions with him,[11] he told me that the compression requirement had improved the quality of sealants purchased by the Government. Nonetheless it has become an important issue in discussions of a qualification test for elastomeric sealants by the building sealants subcommittee of the International Standards Organization (ISO).

To determine the effect of compression, the German and French delegations developed data running a cyclical test followed by determination of tensile properties of the bonded cement specimens. Some of the samples were compressed 7 days at 70°C, some were not. Table 1 shows the results of this test.[12] In this test, compression did make a substantial difference. This difference, however, would apply whether the test specimen was concrete, glass, or aluminum. But because a substantial percentage of butt joints (which would be most affected by compression set) occur in concrete structures and pavements, this result is significant for concrete joints.

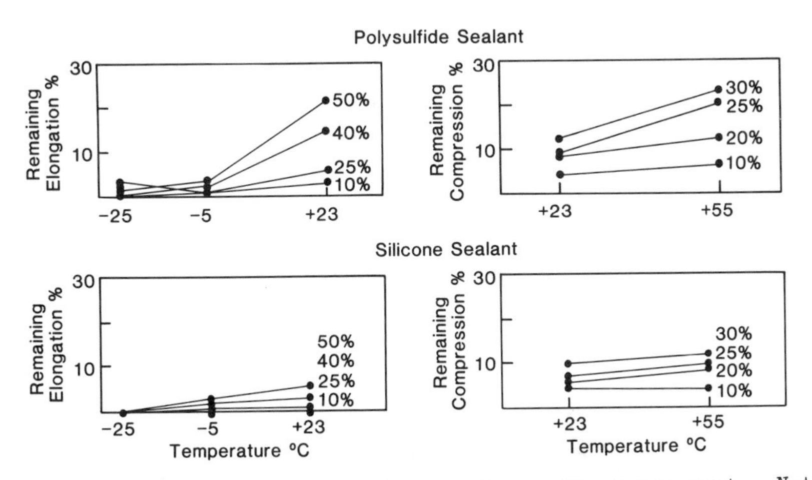

Fig. 6. Sealant tester for resistance to cyclical movement. Note the configuration of the test specimen. All samples aged 7 days at 20°C, 56 days at 70°C.

Table 1. Effect of Elevated Temperature Compression on Tensile Elongation

	Control,%	Without Compression,%	Conditioning + Compression
Polysulfide I	300	275	fail
Polysulfide II	250	200	200
Silicone I	300	225	200
Silicone II	275	175	150
Acrylate I	fail	fail	fail
Acrylate II	fail	fail	n.a.
Urethane I	600	225	fail
Urethane II	200	150	150

Types of Elastomeric Joint Sealants This discussion is limited to elastomeric joint sealants because they are designed to withstand extremes of temperature and movement. Three types are in common use, namely: polysulfides, silicones and polyurethanes. The polysulfides were the first in the field. They are characterized by a polysulfide bond in their chain. When applied to concrete, they require a primer. The lability of the disulfide bond causes poor recovery of these sealants after elongation or compression.[13] Figure 7 shows the results of tests in which sample joints which had been filled with polysulfide or silicone sealants were either elongated or compressed the amount shown at the temperature and then allowed to recover for one hour. It is apparent that polysulfide elastic recovery from either tension or compression was far poorer than that of silicone. Some newer polysulfides, which combine a polyoxypropylene-urethane backbone with mercaptan chain extension to achieve a less labile bond.[14]

Silicones, which Fig. 7 shows have excellent recovery properties, cure when their end-group hydrolyzes. The $Si(OH)_3$ groups remaining then condense with each other or with the substrate to which they are bonding. In either event, the result is a bonded cross-linked elastomer. At first, the only sealants available were terminated with an acetoxy as shown below. These released acetic acid when the sealant cured. This was obviously a "no-no" when the substrate was concrete--

```
                                                  |
CH3  OH              CH3 OH          CH3  O
 |    |               |   |           |   |
-Si-O-SiOX + H2O → -Si-O-SiOH  →  -Si-O-Si-O---- + HX
 |    |               |   |           |   |
 CH3 OH              CH3 OH          CH3  O
                                          |
```

WHERE X=CH_3COO-or Et(Me)C=N-

Cure of a Silicone Sealant

the acetic acid released ate away its substrate. Silicone manufacturers solved this problem with, among others, an oxime cure. This improvement eliminated the need for a primer, making it possible for the silicones to achieve considerable inroads into the sealing of concrete surfaces--a substantial percentage of building joint caulking.

Silicones, however, have a weak spot. The considerably ionic nature of the silicone bond renders it susceptible to hydrolysis.[15] This could be a source of failure, particularly in the case of pavement joints which are notoriously subjected to ponded water.

Because polyurethane sealants had no such problems, their sales growth rate is the greatest of the elastomeric sealants.[16] Their elastic recovery properties are good (though not so good as the silicones), their adhesion to unprimed concrete, according to Evans et al. is excellent.[17] This was probably due to the fact that their sealant consisted of a mixture of polyurethane prepolymer and a modified coal tar (CP 524) or analog (Table 2). The modification was neutralization of its active hydrogens with polyisocyanates. This improved hydrolysis resistance and, by reducing the moisture vapor transmission rate (MVTR), reduced osmotic pressure at the joint. The tests (results shown in Table 1) were run on concrete samples which were joined together in a butt joint, cured, water immersed for one week, and then tested in tension while wet.

WATERPROOFING MEMBRANES

Split Slab Construction Modern construction often requires what is called "split slab construction" for underground parking garages, malls, etc. Prior to the development of this technique it was found that, when the load-bearing slab of concrete cracked during its cure, the cracks provided channels for water to penetrate through the slab. To obviate this problem and to prevent leaks at the joints, split slab construction was developed. This consists of a structural slab, a waterproofing membrane, and a wearing slab (Fig. 8).

Crackbridging membranes are also used as a wearing surfaces for concrete. These are defined in ASTM C957 as "elastomeric waterproofing membrane(s) with (an) integral wearing surface."[18] This requires many different characteristics which are discussed below.

Table 2. Effect of Tar Adduct Modification on Wet Adhesive Joints[17]

Prepol. Wt.	Adduct Wt.	Adduct Type	Strength (Dry)	Elong (Dry)	Strength (Wet)	Elong (Wet)
100	---	None	133	72	---	48
100	100	CP524/MDI	119	144	101	140
100	300	" "	41	431	49	619
100	250	Dehyd 524 MDI Adduct	166	754	33	180
100	125	(Cumarone-Tar acid-MDI)	66	399	35	496

Testing Waterproofing Membranes ASTM has two specifications for waterproofing membranes. These are for (a) Split Slab Construction[19] and (b) Construction With an "Integral Wearing Course"[18] in which the membrane also constitutes a wearing course. The latter is most often used with structures such as parking garages which require a high degree of abrasion resistance. While certain tests are the same in both specifications, there are some differences. Table 3 lists the tests. The rationale for differing test methods is given below:

ASTM C 836 for split slab membranes requires only 1 pli of peel adhesion, as contrasted to the 5 pli required by C 957. This is because a membrane in split slab construction is kept in place by the wearing slab on top of it.

ASTM C 957 for Construction with "An Integral Wearing Course" was designed around the requirements of parking garages. Consequently it required abrasion resistance of the wearing surface and resistance to liquids which would leak from automobiles or which they would bring into the garage. This meant:

Fig. 7. Elastic recovery from tension and compression of polysulfide and silicone sealants.

Table 3. Requirements of ASTM C 836 and C 957

	C 836 Split Slab	C 957 Parking Garages
*solids content, %	80	60
*Abrasion resistance	----	50 mg, Taber 1000 cycles, CS-17
*Resistance to liquids	----	water, ethylene glycol mineral spirits
*adhesion in peel after water immersion, N/m	175 (1#/in)	875 (5#/in)
*crack bridging	see text	see text
*extensibility after heating aging	open coated crack 1/4" (see text)	retain 90% of elongation after 500h weatherometer exposure.

(1) Lower solids content is required for the membrane used as a wearing surface. This reduction was permitted because a higher molecular weight polymer is necessary, in the case of many useful membranes, to have the abrasion resistance required of the surface.

(2) Resistance to deterioration by liquids leaking from automobiles. To determine this resistance, samples which were immersed in test liquids for 14 days are required to have a tensile strength retention of 70% after immersion in ethylene glycol or water; of 40% after mineral spirits immersion.

In the C 836 extensibility test at the bottom of Table 3 the materials are coated onto the notched block shown in Figure 9 and allowed to cure for one week. They are then heat-aged two weeks at 70°C to expel volatile plasticizers or retained solvents which also plasticize the coating. The notched block is broken in such a way that the coating remains in compression as the crack is generated. It is then extended 0.25 inches. This test is not included in the parking garage specification (C 957); instead, a weatherometer exposure test on the free unadhered membrane is run. This is probably less significant than the extended heat aging in the C 836 extensibility test.

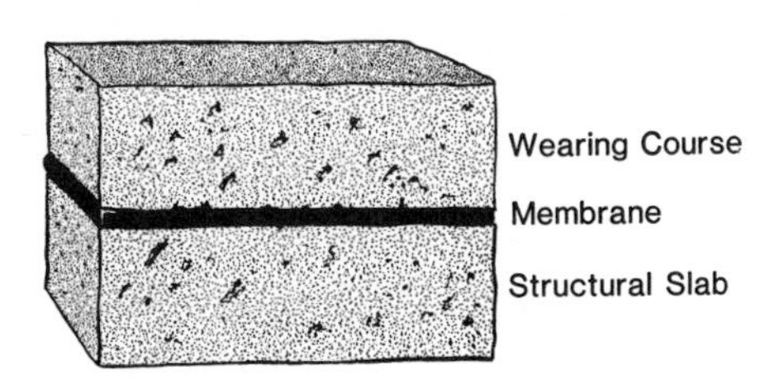

Fig. 8. Split slab construction. The structural slab is separated from the wearing course by a waterproofing membrane.

Table 4. Comparison of Cyclic and Transient Failure Elongation in Crack-bridging Test on Cement Asbestos Board

Elastomer	Transient Elongation, %	After 2000 Cycles, %
Acrylic Resin "A"	4.0	1.5
Acrylic Resin "B"	4.6	1.5
Neoprene	2.2	0.4
Silicone	10.5	0.75

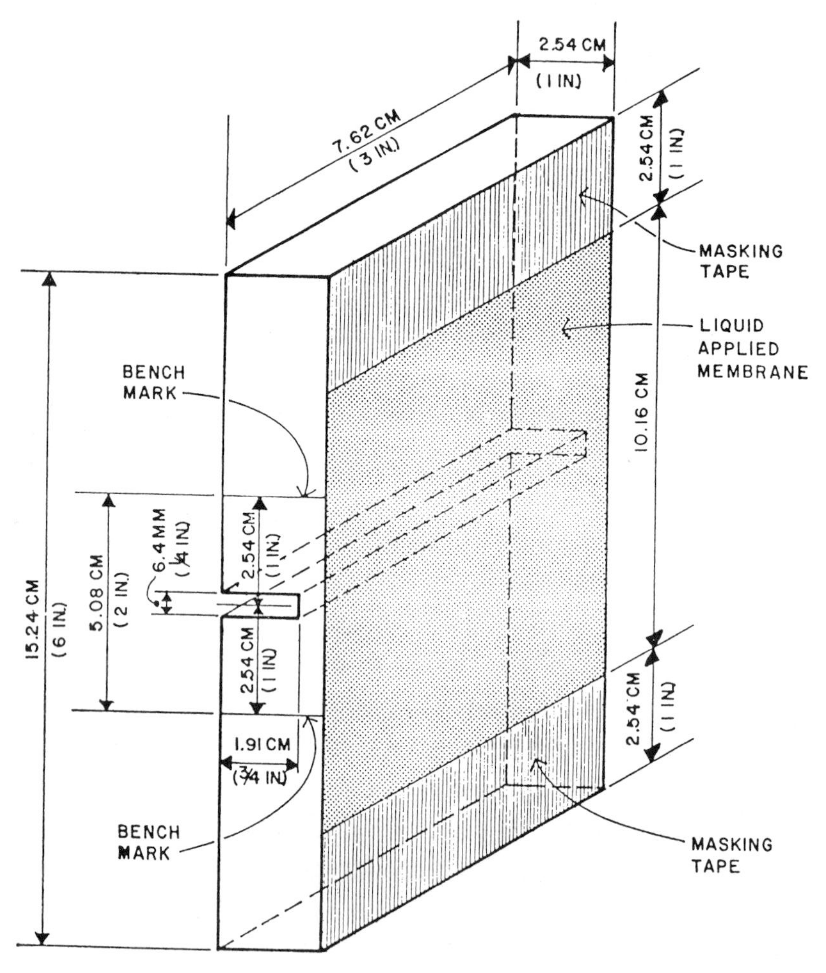

Fig. 9. Notched mortar block to test resistance to cracking after heat aging.

Both C 836 and C 957 use a crack bridging test. The membrane is coated on preformed blocks, heat aged one week at 70°C, and then cycled 10 times at -26°C.

The difficulty with this test and most such tests is the lack of the extensive cycling that a joint suffers in real life. In an interesting study, Iwai reported[20] on the effect of cycling a cracked, liquid membrane coated cement asbestos board. As Table 4 shows, 2000 cycles of crack opening and closing drastically altered both the elongation capacity and the rank of various waterproofing membranes. Another significant result of his work was to show that, in a membrane, the thicker the application, the greater the resistance to failure. However, space requirements forbid quoting him at greater length.

Types of Waterproofing Membranes Originally the waterproofing membrane was a cured elastomer such as neoprene. This was rolled over the top of the concrete bearing slab. The concrete wearing slab was poured over this membrane. There were two problems with this approach: (1) if stones were left on the surface of the bearing slab, in time the membrane would be penetrated by the weight of the bearing slab on top of it and (2) once the membrane permitted water to pass it would travel under the membrane until it found a crack to pass it through the bearing slab. This made it extremely difficult to determine the source of the leak.

The answer to this problem was a liquid applied coating which bonded to the surface upon cure. There are three types of material in use: (1) the tar modified one-component urethane described in the Evans patent,[17] (2) asphalt modified one-component coatings;[21,22] and (3) two-component materials made with hydroxyl-terminated polybutadiene (poly BD) in the asphalt component and a free isocyanate containing prepolymer in the other component. All of these are characterized by hydrolysis resistance as well as a low MVTR. The latter would reduce buildup of osmotic pressure. In the case of the one-component asphalt modified material, the asphalt must have enough active hydrogens present to insure compatibility with the urethane.[21]

The "integral wearing surface" was, as mentioned above, required to withstand more than opening cracks. Liquids leaking from the automobiles parked on it and abrasion--particularly at turns and stopping places--were its lot. The formulations would tend to be more viscous--either using solution polymers, or abrasion resistant polyurethanes. An alternative would be to have the wearing surface a separate coat of, say, an epoxy, which is applied over the urethane membrane. This is permitted by C 957.

SUMMARY

The manufacturers of coatings, sealants and waterproofing membranes which are designed to be applied to concrete surfaces have evolved materials for these surfaces and tests which predict the performance of their products. This has been a pragmatic process. However it would be useful, for future development of both products and test methods, if the harsh nature of the concrete substrate were more generally understood.

REFERENCES

1. I. Jawed and J. Skalney, in Structure and Performance of Cements; J.P. Barnes, Applied Science Publishers (1983).

2. Cement and Concrete Terminology 116R, Vol. 1, Manual of Concrete Practice, American Concrete Institute (1984).
3. A. Grudemo, Cement and Concrete Research, Vol. 9, pp. 19-34, Pergammon Press (1979).
4. Control of Cracking, ACI 224R-9, Part 3, Manual of Concrete. Practice, American Concrete Institute, Detroit, MI (1984).
5. Cleveland Club; Fed of Paint and Var. Prod. Clubs, Official Digest, p. 1025 (1952).
6. Llewellen, cited in reference (5).
7. ASTM D 3730, Testing High Performance Intereior Architectural Wall Coatings; ASTM, 1981 Book of Standards, Vol. 27.
8. E.A. Peterson, et al., Building Seals and Sealants, ASTM STP 606, pp. 19-23, J. Panek, Editor (1976).
9. Annual Book of ASTM Standards, Vol. 4.07 (1985), ASTM, Philadelphia.
10. U.S. Government Federal Specification TT-S-00230.
11. Arthur Hockman, Private Communication.
12. International Standards Organization Committee 59, Subcommittee 8, Documents N136, 138 and 139 (1985).
13. Per Gunnar Burstrom, Report TVBM-1002, University of Lund, Lund Sweden (1979).
14. L. Morris, Caulks and Sealants Short Course, pp.85-113, Adhesives and Sealants Council (1985).
15. J.M. Klosowski and M.J. Owen, Polym. Mater. Sci. and Eng., 56, p.472 (1987).
16. Chemical Week, Sealant Advertising Section, p.15, (1986).
17. U.S.Patent 3,196,122, R. Evans and B. Brizgys, to Master Mechanics Company (1968).
18. ASTM C 957 Elastomeric Waterproofing Membrane With Integral Wearing Surface. ASTM, Book of Standards (1985), Vol. 4.02 ASTM, Philadelphia, PA.
19. ASTM C 836 High Solids Cold Applied Liquid-Applied Elastomeric Waterproofing Membrane for Use With a Separate Wearing Course.
20. T. Iwai, Crackbridging Properties of Elastomeric Waterproofing Wall Coatings, Proceedings of the 1985 International Symposium on Roofing Technology, National Roofing Contractors Assn., Chicago (1985).
21. J. Kozakiewicz and A. Lendzion, Chapter 9 in K. Allen, Adhesion 9, Elsevier New York (1984).
22. J. Regan, Adhesives and Sealants Council Short Course, p.137 (1985).

THE STUDY AND USE OF SEALANT ON QUICK SAND SURFACE FOR STABILIZATION AND AFFORESTATION OF DESERT LAND

Lu Jiqing, Yan Tingju, Pei Zhangqin and Lu Chunyang

Northwest Institute
China Academy of Railway Sciences
Lanzhou, Gansu, People's Republic of China

ABSTRACT

This paper presents the study and use of the residual-oil emulsion (O/W type) as a sealant to aid afforestation and prevent the sand damage along railway lines in desert areas in the Inner Mongolia Autonomous Region of China. The compositions and properties of the residual-oil, emulsifiers, the formulations and properties of the residual-oil emulsion are discussed separately. So are the physical and mechanical characteristics of the sand layer stabilized by the residual-oil emulsion sealant. The benefits and economic value of the sand-control and the sand-stabilization by the sealant are summarized. Our laboratory test and field application have shown that it is an effective way to combine the residual-oil emulsion sealant with afforestation for sand stabilization. The technology is simple and the cost is low.

INTRODUCTION

Loose and fine sand grains can be cemented to become a large porous and solidified sand layer by a chemical sealant so that they will not be blown by the wind. Thus, the sand hazard can be prevented. This method can be called "sand stabilization with a chemical."

Asphalt emulsion was the original sand stabilization chemical. Its compositions, properties and the applications in various engineering fields, such as soil stabilization, road stabilization, pavement technology, waterproof and antiseepage and erosion-control have been well studied.[1-7] Their applications were the most widespread in the Soviet Union, Australia, etc.[8-11] In the early 1930's, the study of sand stabilization with asphalt emulsion began in the Soviet Union and it was then carried out in China in the 1960's.[12-13] The test results indicated that this method could play a good role in controlling drift sand.

Scientists have investigated many chemical agents for sand stabilization, for example, S.A. Zallar[11] examined over 20 species of them. These agents were mainly inorganic salts, organic compounds and their salts, polymers, rubber emulsions, etc. R.S. Zahirov[14] introduced sand stabilization methods with asphaltous compound and emulsion, crude oil, Nerosine, rubber emulsions, synthetic polymers, etc. The benefits of the above-mentioned materials are different, and their costs could be high. For example, some polymers can be very expensive. The application aspect of asphalt is very wide-ranging. However, it is in short supply in China. Thus, we have selected a residual-oil,[15] which is a byproduct of petroleum industry. It has been emulsified as a stabilization agent. Our study has shown that sand drift stabilization with the emulsion of residual-oil is economical and favorable to plant growth. For application, the residual-oil emulsion has good dispersability, good stability upon dilution, heat resistance, cold resistance and static stability. The compositions of emulsion and the conditions of preparation are considered to be very important in achieving these results.

The aim of our study is to combine organically emulsified residual-oil sealant with afforestation, i.e., the seedling of plants in the desert which can grow before the sealant loses its effectiveness, thereby preventing the sand damage. We have already had successful experience in afforestation.[16-17] However, we have not obtained sufficient data of combining afforestation with sand stabilization by a chemical sealant. Thus, we have made a further study of significance, i.e., to sprinkle the residual-oil emulsion as a sealant over 366 thousand square meters along the Baotou-Lanzhou railway line between 1982 and 1985. Two hundred thousand desert plants were planted before the sprinkling.

This paper describes the compositions of residual-oil, emulsifying machines and tools, the process of producing the emulsion, physical and mechanical properties of the residual-oil/sand layer, methods of planting trees in the desert, and the economic benefits and the future prospect of this study.

NATURAL SURVEY IN THE TESTING SITE

The testing site is located at a place 1050 meters above sea level, 39°30′ N and 107° E, i.e., it is also on both sides of K381-K386, the Baotou-Lanzhou Line. The site is on the southeastern edge of the Wulanbuhe Desert, which is a semi-desert region. It is situated west of the Zhouzi Mountain and east of the Huanghe River. The annual rainfall there is from 20 mm and 250 mm and the annual evaporation capacity is about 3000 mm. The mean annual temperature is 9.2°C (the maximum from 38.4 to 39.4°C and the minimum from -25.4 to -28.6°C). The mean annual wind speed is from 3.6 to 5.1 m/s (the maximum 28 m/s). The windy seasons have eight months, sometimes 6 to 9 months. Along the Line, there are primarily crescent and semi-fluid dunes. The natural plant coverage is small and the species are also monotonous. They are distributed mainly on the lowland among the sand dunes. The common plants in the desert are Salix matsudana, Calligonum arborescens Litv., Artemisia sphaerocephala and Agriophyllum arenarium (M.B.A.) Bge., etc. Their coverage is only 0.16-13.7 percent. In brief, the testing site has been critically exposed to damage by the wind sand.

MATERIALS AND EMULSIFYING EQUIPMENTS

Residual-oil The residual-oil belongs to the asphaltous compound family. Their compositions and properties are similar to those of the asphalt. The major components affecting emulsifiability and physical and chemical properties include asphaltous acid, acid anhydride, resin and asphaltine. In comparison with the asphalt, the density of the residual-oil is lower and it flows easily. The oiliness in the residual-oil is caused by the presence of heavy oil. There are few volatiles, and the flash point is high (not less than 280°C). Thus, it is safe to apply. Since the speed of aging of the residual-oil is slower than that of the asphalt, the service life of the sand stabilization can be long. The brittle point of the residual-oil is about -30°C, thereby its fracture resistance is quite good making it suitable for the sand stabilization in a severely cold region.

Surfactant To prepare an ideal emulsion, the key is the selection of a suitable surfactant. The H.L.B. value of an emulsifier and that of the material to be emulsified should not be too different; otherwise, the affinity between them is small, and the emulsifying effect will be poor. To select a surfactant of a H.L.B. value between 8-22, a good residual-oil emulsion (O/W) must be prepared. The cationic surfactant used in this study was Aerosol OT, a quaternary ammonium salt with a C-18 chain; the anionic one, SZF (Chinese abb. of paper pulp acidic waste liquor) and the nonionic one, Aerosol OP which is aromatic. In addition, a secondary emulsifier, an inorganic acid or salt, and other stabilizers, e.g., polyvinyl alcohol (PVA) and $CaCl_2$ were also used in the formulations.

Equipments used for emulsification The residual-oil emulsion was prepared by an emulsor designed and made by us and a JTM 50 AB type of colloid mill made in China for the tests in our laboratory. While the equipments used for emulsion on the site were mainly a KZG 0.5-8 type steam boiler (the minimum evaporation capacity was 0.5 T/h and the maximum work pressure was 700 KPa), a KCB 300-1 type pump of raw oil, a residual-oil pool filled with about 100 tons, a reservoir for heated residual-oil, a heated kettle (tank) for compounded aqueous solution of emulsifying agent, and a colloid mill, etc. The colloid mill used at the site could produce 1.5T-2.5T/hr. of emulsion. Eighty-five percent of the particle size of the emulsion were between 3 to 5 microns which were in accordance with the requirements for this study.

THE RESIDUAL-OIL EMULSIONS

The quality of the emulsion for sand stabilization is a decisive factor in combining emulsified residual-oil sealant with planting trees in a large area of the desert. Of course, the selection of saplings in the paper is also important. This will be discussed below. Various surfactants have been tested in our laboratory, such as, the cationic surfactants, Aerosols OT and SN; the anionic ones, Aerosol SZF, sodium oleate, sodium benzenesulphonate, and waste liquor of paper pulp with various basicity and acidity; the nonionic ones, Aerosols OP and PVA. PVA and $CaCl_2$ have also been used as stabilizers. Sulphuric and hydrochloric acids were used as the secondary emulsifiers. A polyether was used as a defoaming agent. The testing of various compositions was done with a blender (mixer) colloid mill in our laboratory. The residual-oil was heated to 90°-110°C, and the aqueous solution of the emulsifying agent 80°-90°C. The results show that the emulsions prepared by these emulsifiers, i.e., sodium oleate, sodium benzenesulphonate, and paper pulp basic waste liquor etc., are unstable so that they were not suitable for the sand stabilization sealant. Several hundred tests have been made for other surfactants. Seventeen kinds of formulations have been selected (see Table 1). The emulsion

Table 1. The Compositions of Residual-Oil Emulsions

No.	R-Oil	SN[1]	OT[2]	SZF[3]	OP[4]	PVA[5]	$CaCL_2$	HCL	H_2SO_4	Water	Costs Yuan/T
1	50	----	0.4	----	----	----	----	----	----	49.6	66.16
2	50	0.5	----	----	----	----	----	----	----	49.5	106.00
3	50	----	----	----	0.4	----	----	----	----	49.6	51.16
4	50	----	0.2	----	0.2	----	----	----	----	49.6	54.77
5	50	0.3	----	----	0.12	----	----	----	----	49.6	75.66
6	50	----	0.2	----	----	0.5	----	----	----	49.3	62.45
7	50	----	----	----	0.2	0.5	----	----	----	49.3	58.86
8	50	----	----	----	0.2	----	----	0.12	----	49.7	41.66
9	50	----	----	2.5	-----	0.5	----	----	0.15	46.8	52.50
10	50	----	0.2	----	0.2	0.5	----	----	----	49.1	74.45
11	50	----	----	2	0.2	0.5	----	----	0.1	47.2	81.86
12	50	----	----	2	0.04	0.1	----	----	0.1	47.4	47.54
13	50	----	----	----	0.2	0.2	0.1	----	----	49.5	47.10
14	50	0.5	----	----	0.2	----	----	----	----	49.3	132.00
15	50	0.5	----	----	-----	----	----	----	----	49.0	120.00
16	50	----	----	2.5	-----	----	----	----	0.15	47.3	32.80
17	----	----	----	2.5	0.4	----	----		0.15	47.3	35.20

Notes: 1 and 2 are C18-alkyl quaternaryammonium salt; 3, Chinese abb. of paper pulp acidic waste liquor; 4, C_{18} alkyl derivatives and 5, polyvinyl alcohol.

with high quality was pale brown, and its stability was also satisfactory.

The characteristics of the residual-oil emulsion sealant can be identified by various tests, e.g., the resistance to high and low temperatures, the stability upon dilution under different conditions and the residue on the sieve. The detailed results are listed in Table 2. The 17 kinds of prepared emulsions were all basically stable under the conditions of various tests. In addition, although there was a tendency to separate during the storage of emulsion, its dispersability was still very good after it was restirred. The demands for the sand stabilization were satisfied. On site, No. 11 residual-oil emulsion has been used and a process which produced the emulsion on site is shown in Fig. 1. The quality of the residual-oil emulsion sealant was judged by its color, dispersability, and stability in water on the site, i.e., it was of high quality if the appearance of the emulsion was pale brown, and the emulsion did not break after being diluted with 5 times water. The diluted emulsion was applied by a sprinkling truck. The formulation required strictly to be sprinkled evenly.

AFFORESTATION*

According to our investigation, the priority plants were mainly Artemisia sphaerocephala, Agriophyllum arenarium (M.B.A.) Bge., and Psammochloa villosa Bor., etc. On quick sand in the Lihuazhongtan region, their capacities of sand-fixation were not strong. To enhance the degree of the plant coverage and stability of sand, the shrub and semi-shrub were taken as the sand-fixation forest belts in the area and the mixed forest type were used. Five kinds of saplings have been selected. They were Caragana opulens Kom, Calligonum arborescens Litv., Hedysarum scoprium Flisch. et Mey., Hedysarum mongolicum Turcz., and Haloxylon ammodendron Bge. They were disposed by 3 kinds of row spacing, i.e., 2x2 m, 2x1 m, 2x3 m, etc. (25, 50 and 17 trunks of trees in 100 m^2). The planting sapling was mainly the afforestation of direct seeding. Some seedlings grown in vessels were tested. Annual seedlings of selected shrubs were suitable for afforestation of direct seeding, and this work had to be carried out during the spring and autumn while the afforestation of seedlings growing in vessels had to be carried out in the summer. Table 3 shows a statistical table of the survival and conserving rates for the various plants.

The adaptability of the planted seedling is measured by the adapting coefficient and survival rates:

$$K = 1 - \frac{Survival\,rate - Conserving\,rate}{Survival\,rate},$$

where, K is the adapting coefficient. The greater the K value is, the better the adaptability of plant will be; the longer the time to calculate the conserving rate, the greater the reliability of K value. The adapting coefficients of 5 kinds of plants are calculated by the above-mentioned formula in Table 4.

* See a Research Report by Zhou Shouyi, et al., who cooperated with us.

Table 2. The Properties of Residual-oil Emulsions

No.	Residue on sieve	Heat resistance		Dilution stabilities PH						Dilution stabilities $CaCL_2$ (%)			Dilution stabilities $MgCl_2$ (%)			Dilution stabilities Clay (%)		
		65°-70°C	-10° zero° C	2	4	6	8	10	12	0.1	0.5	1.0	0.1	0.5	1.0	0.1	0.2	0.5
1	+	+	90a	+	+	+	+	+	57	+	+	+	+	+	+	+	+	+
2	+	+	90a	+	+	+	+	+	32	+	+	+	+	+	+	+	+	+
3	+	+	90a	+	+	+	+	+	/	+	+	+	+	+	+	+	+	+
4	+	+	90a	+	+	+	+	3	20	+	+	+	+	+	+	+	+	/
5	+	+	20a	+	+	+	+	/	/	+	+	+	+	+	+	+	+	3
6	+	+	3b	+	+	+	+	+	58	+	+	+	+	+	+	+	+	13
7	+	+	90a	3	6	8	+	+	+	+	+	+	+	+	+	+	+	+
8	+	+	/	+	+	+	+	+	+	+	+	+	+	+	+	+	+	+
9	+	+	/	+	+	+	+	+	+	+	+	+	+	+	+	+	+	+
10	+	+	/	+	+	+	+	+	+	+	+	+	+	+	+	+	+	+
11	+	+	90a	+	+	+	+	+	+	+	+	+	+	+	+	+	+	+
12	+	+	+	14.4	+	+	+	+	+	+	+	+	+	+	+	+	+	+
13	+	+	–	+	+	+	+	+	–	+	+	+	+	+	+	+	+	+
14	+	+	10b	+	+	+	+	+	/	+	+	+	+	+	+	+	+	7
15	+	+	90a	+	+	+	+	/	/	+	+	+	+	+	+	+	+	4
16	+	+	10b	+	+	+	+	6	6	+	+	+	+	+	+	+	+	+
17	+	+	10b	3	3	3	3	3	4	+	+	+	+	+	+	+	+	+

Notes: The mark "+" shows that the amount of broken emulsion is less than 1 percent; "-", broken entirely "/", unmeasured. "90" in the amount of broken emulsion (percent in weight) under the condition of low temperature; a. emulsion was frozen; b. emulsion was unfrozen.

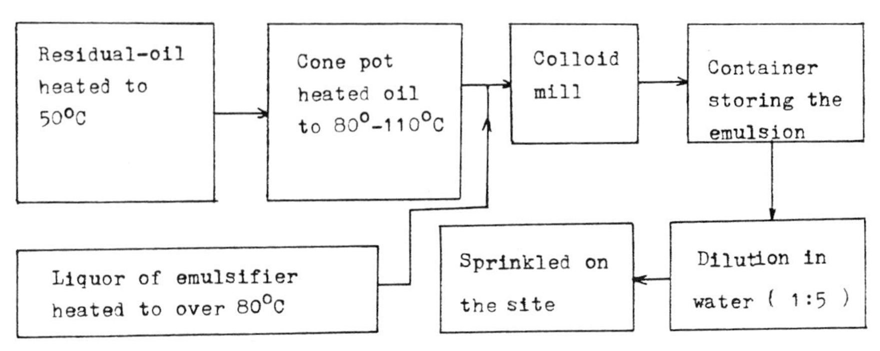

Fig. 1. Scheme of the Process of On-site Produced Emulsion

Table 3. The Survival and Conserving Rates for Various Plants

Species of plant	Number of transplantings	Survival rates (%)[a]	Conserving rates (%)[b]
Calligonum arborescens Litv.	1650	51.3±4	47.5±3
Hedysarum scoprium F.et M.	1550	54.7±4	51.9±5
Haloxylon ammodendron Bge.	1650	35.5±4	10.2±3
Hedysarum mongolicum Turcz.	1550	58.6±4	55.7±5
Caragana opulens Kom.	850	37.1±11	28.1±4

Notes: The various plants were transplanted in October 1983; a and b are the percentages of transplanting number; a, investigated in June 1984; b, investigated in September 1985; and the regional reliability was 95 percent according to our estimation.

Table 4. The Adapting Coefficients of Plants

Species of plants	Coefficient, K
Hedysarum mongolicum Turcz.	0.951
Hedysarum scoparium Flisch. et Mey	0.949
Calligonum arborescens Litv.	0.926
Caragana opulens Kom.	0.757
Haloxylon ammodendron Bge.	0.287

Table 5. The Even Crown, Coverage Area and Degree of Planting.

Species of plants	Conserv-ing rate(%)	Even crown(m^2)	Coverage area(m^2)	Density of transplanting (in 100 m^2) Degree of plant coverage(%)			
				17	25	50	100
Calligonum arbor-escens Litv.	48	2.12	101.76	16.96	25.44	50.08	
Hedysarum scop-arium F. et.M.	52	2.03	105.56	17.59	26.36	52.78	
Haloxylon ammo-dendron Bge.	10	0.26	2.60	0.43	0.65	1.30	
Hedysarum mon-golicum Turcz.	56	1.58	88.48	14.75	22.12	44.24	
Caragana opulens Kom.	28	0.02	0.56	0.09	0.14	0.25	0.56

Notes: Planted in October 1983, and investigated in September 1985.

The ability of a plant sand-fixation is measured by the degree of coverage of the plant crown which is the total area of the crown divided by that of sand land. The detailed results are listed in Table 5. According to Table 5, the ability of sand-fixation of several plants reduces in the proper order, i.e., Hedysarum scoparium E. et M., Calligonum arborescens Litv., Hedysarum mongolicum Turcz., Haloxylon ammodendron Bge., and caragana opulens Kom.

The following numerical models have been obtained by the test for the relationships between the capacity and the cost of the sand-fixation forest:

$$C = 0.163\ X_1 + 0.160\ X_2 + 0.127\ X_3;$$

$$D = 0.676\ X_1^{-1} + 0.691\ X_2^{-1} + 0.888\ X_3^{-1}\ ,$$

Where: C, the degree of coverage of the plant crown; D, the cost of the sand-fixation (Yuan/m^2); X_1, the survival rate of Calligonum arborescens Litv.; X_2, one of Hedysarum scoparium F. et. M. and X, one of Hedysarum mongolicum Turcz. The above-mentioned models should provide a controlled way to improve the success of sand-fixation by plant.

PHYSICAL AND MECHANICAL PROPERTIES OF THE SAND STABILIZATION LAYER

Our study of sand stabilization also refers to the effects of the sealant on the desert plants. Here, first, we discuss several results on crack resistance, permeability, evaporation resistance and wind erosion resistance of the sand stabilization layer containing the residual-oil sealant. The crack resistance was carried out with a test unit of crack resistance. The unit was made up of a tinplate box with sand, pulley, chest-developer and weight tray. The permeability test was made by "static water state." The equipment included a bank with black sheet iron, container of water supply and support. The wind erosion resistance was carried out with a device of wind erosion connected by an artificial source of wind, air-blower, wind tunnel, funnel with sand etc. The tests mentioned above were made after the samples were air-dried. Figure 2 shows the relationship between tensile strength (Kpa) of the sand-stabilization layer and the emulsion used (kg/m^2). The results of the penetration test are tabulated in Table 4, while those of the wind erosion of the sand-stabilization layer in Table 5. In Table 6, the evaporation resistance of the sand-stabilization layer is listed. In addition, the relationship between the holding water ratio of the sand-stabilization layer and the residual-oil drawn by analyzing data of evaporation resistance test is shown in Fig. 3, while that between the mean penetration speed (ml/s) of the sand treated and the amount of the residual-oil (kg/m^2) is shown in Fig. 4. Figure 5 shows the relationship between the benefit of the wind erosion resistance for the sand-stabilization layer (%) and the amount of residual-oil (kg/m^2) at a wind speed of 30 m/s, the sand discharge of 1432 g/m^2/min and the wind erosion time of 24 hr.

Figures 6 and 7 are the untreated and treated sand land, respectively. The growth of 5 plants, after the residual-oil emulsion sealant was sprinkled, is shown in Fig. 8-11. Figure 12, a clear line, shows that this way is able to prevent the railway line from the harm of drift sand.

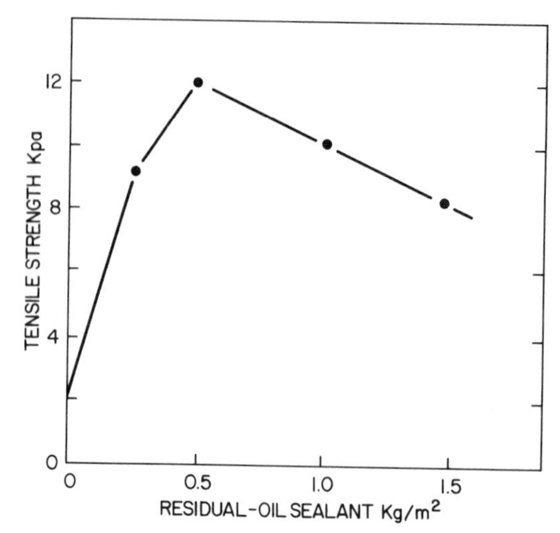

Fig. 2. The relationship between the tensile strength of sand-stabilization layer and the residual-oil amount.

Table 6. The Results of the Evaporation Test

Dates (in May, 1985)	The amount of sealant used (kg/ m^2) Day-evaporation* (g)					
	0.00	0.50	0.75	1.00	1.25	1.50
4	149	7.0	8.0	5.3	2.7	2.7
5	73	4.3	2.3	2.7	2.3	2.7
6	113	8.0	3.3	2.7	1.7	2.0
7	135	9.7	7.0	2.7	2.7	0.7
8	140	11.7	7.7	7.0	5.7	5.7
9	225	15.0	9.7	6.0	5.7	2.3
10	212	20.3	10.3	7.7	5.3	7.7
11	113	11.7	10.3	6.7	4.0	3.0
12	170	8.0	2.0	0.0	0.0	0.0
Daily Mean(g)	147.8	10.6	6.7	4.5	3.3	2.9

* Mean value of three samples on a day-evaporation.

DISCUSSIONS AND CONCLUSIONS

We have found that the residual-oil could be emulsified. The quality of the residual-oil emulsion sealant depended mainly on its composition. A selected standard should be economical and keeping with the necessity of production. According to the results shown in Table 2, the emulsions prepared as in Table 1 were relatively stable, in terms of the heat resistance, the low temperature resistance and the stability upon dilution. They were suitable for the stabilization of a large area.

The residual-oil emulsion sealant prepared according to Formulation No. 11 was sprinkled over 366 thousand square meters along the Baotou-Lanzhou line between 1982 and 1985. Meanwhile, 200,000 desert plants were planted before sprinkling. Our tests and observations indicated that the residual-oil emulsion did no harm to the growth of the plants.

The residual-oil emulsion sealant did not depend on the quality of water.

Table 7. The Results of the Penetration Test

Water (ml)	The amount of sealant used (kg/ m^2) Time for penetration* (m)					
	0.00	0.50	0.75	1.00	1.25	1.50
100	0.4	33	56	71	92	107
200	0.8	54	91	93	114	131
300	1.2	68	95	99	123	142
400	1.6	70	98	104	135	152
600	2.2	74	103	114	147	169
800	2.5	78	108	122	156	185
1000	2.8	82	112	129	166	199
1200	3.3	85	115	136	176	212
1400	4.0	88	118	142	184	223
1600	4.4	91	121	146	191	232
1800	4.9	95	124	152	198	242
2000	5.4	98	127	157	203	253
2200	6.0	101	129	163	209	261
2400	6.7	104	132	167	219	268
2600	7.3	107	135	171	229	176

* The mean value of three tests.

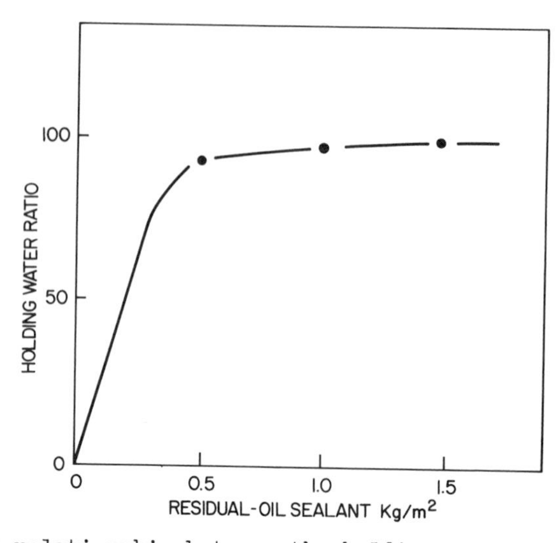

Fig. 3. The relationship between the holding water ratio% of sand-stabilization layer and the residual-oil.

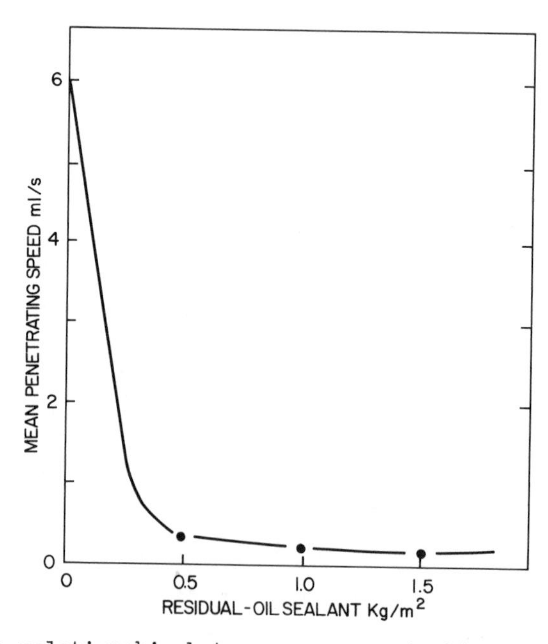

Fig. 4. The relationship between mean penetrating speed of sand-stabilization layer and the residual-oil.

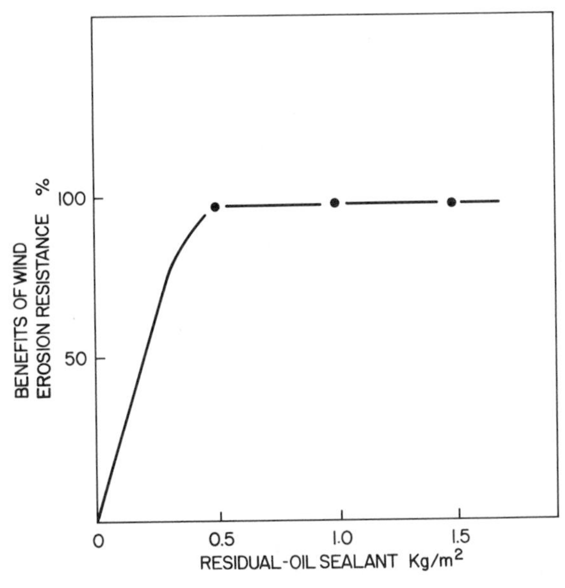

Fig. 5. The relationship between the benefits of wind erosion resistance for sand stabilization layer and the residual-oil (wind speed, 30 m/s; sand discharge, 1432 g/m^2m; and wind erosion time of 24 hrs.).

Fig. 6. Desert land not coated with sealant.

Fig. 7. Desert land coated with sealant (Photographed 3 years later).

Fig. 8. Calligonum arborescens litv. on treated desert land.

Fig. 9. Hedysarum scoparium Flisch.et.May. on treated desert land.

Fig. 10. Hedysarum mongolicum turcz. on treated desert land.

Fig. 11. Haloxylon ammodendron bge. On treated desert sand.

Fig. 12. Caragana opulens Kom. On treated desert land.

Table 8. The results of the wind erosion resistance.

R-oil sealant (kg/m^2)	Sand discharge ($g/m^2/min$)	Mean wind speed (m/s)	Time (hr.)	Amount wind erosion (g)	Apparent descriptions on the samples after wind erosion
0.50	1432	10	24	0	unchanged
0.50	1432	15	24	Trace	A few pits
0.50	1432	20	24	25	Spots of the fish scales
0.50	1432	30	24	25	1/6 surface eroded
0.75	1432	20	24	Trace	1/3 surface eroded
0.75	1432	30	24	/	2/5 surface eroded
1.00	1432	30	24	10	A few pits
1.25	1432	30	24	10	A few pits
1.50	1432	30	24	Trace	1/4 surface eroded
0.00	1432	30	0.5	2100	Eroded seriously

Generally speaking, both surface water and groundwater could be used to dilute the emulsion. This has been shown by the dilution stability test.

After the emulsion was sprinkled over the sand surface, the sand-stabilization layer had higher mechanical strengths. Our results have shown that the strength of the sand-stabilization layer reached a maximum when the amount of the residual-oil was 0.5 kg/m^2.

The residual-oil/sand layer in contrast to the untreated layer in the desert showed a reduction of the evaporation (about 93%) for that sprinkled with 0.5 kg/m^2. This reduction alone was beneficial to the plants growing in the arid desert area.

The residual-oil sand layer had a specific permeability to water, but the permeating speed varied with the amount of the residual-oil.

The stabilized layer could greatly reduce the wind erosion. For example, under the conditions of a wind speed of 30 m/s and the time of blowing for 24 hours, only 25 g of the sand on the surface treated with 0.5 kg/m^2 of the residual-oil was eroded, while the untreated surface, 100 kg/m^2, i.e., the ability to resist the wind erosion was increased about four thousand times. The sand layer consolidated with the residual-oil emulsion could raise the temperature by about 1°C and humidity by about 50 to 104% of the sand ground, and it could also reduce the inroad of plant diseases and insect pests on the seedlings. Thus, the survival and conservation rates of plants greatly increased. Consequently, the artificial irrigation could substantially be reduced. The sealant was retained for 3 to 5 years. During that time the seedlings have gradually grown up. As a result, the damage by drifting sand to the railway was effectively prevented.

The conservation rate of the seedlings planted on the desert land sprinkled with the sealant was about 40 percent. Some of the plants have already grown taller than 2 meters. In general, the degree of plant coverage has reached over 40 percent. This indicates that this method can be employed for the afforestation of arid or desert land.

Fig. 13. This line was free from damage by drifting sand (Photographed 3 years later).

Since the unit engineering material cost by this method was only 0.3 Yuan per square meter (or about 0.1 dollar (U.S.)/m^2), this appears to be a rather low-cost method in the stabilization and afforestation of desert land. In conclusion, the method of sand stabilization with the residual-oil as a sealant in conjunction with the tree plantings has been proven to be effective and economical in China.

ACKNOWLEDGEMENTS

We are very thankful to Prof. Hu Zhide, President of Lanzhou University in Gansu, and Mr. Wang Zhugue, Deputy Director and Senior Engineer of the Northwest Institute, China Academy of Railway Sciences in Lanzhou, Gansu, for reviewing the paper.

REFERENCE

1. C.N. Forrest, Proc. Assoc. Asphalt Paving Tech., (1929).
2. E.S. Ross, Proc. Assoc. Asphalt Paving Tech., (1929).
3. U.S. Pat., 3,389,090 (1968).
4. U.S. Pat., 3,466,247 (1968).
5. K.J. Lissant, Ed., Emulsions and Emulsion Technology, Pt. 1, New York, Dekker, p.440 (1974).
6. O.G. Ingles and J.B. Metcalf, "Bituminous Stabilizations," Soil Stabilization, Australia, 145-168 (1972).
7. N.T. Ralph, Asphalt, Its Composition, Properties and Uses, New York (1961).
8. S.R. Mehra, C.G. Swaminathan and B.C. Mazumdar, "Bituminous Stabilization of Sandy Soils," Jour. Indian Roads Congress, 29, No. 3, 413-440 (1965).
9. V.G. Zacharov, "Strengthening of Mobile Sand With Bituminous Emulsion," Forest and Plains (in Russian), No. 4, (1954).
10. Wu Zhen and Pen Xigu, et al., Road Engineering in Desert Region, (In Chinese) Publishing House of Chinese Communication (1981).
11. S.A. Zallar, Soil Stabilization and Revegetation Manual, Australia (1980).
12. Pei Zhangqin, et al., "The Test of Sand Stabilization With Asphalt Emulsions," J. Desert Res., 3, No. 3, (1983).
13. He Yueqiang, "A Preliminary Report of Sand Stabilization With a Chemical," (In Chinese) Geography Institute of Academia Sinica (1960).
14. R.S. Zahieov, "Sand Drift Stabilization With Physical and Chemical Methods", Reports of Int. Sanddrift Stabilization, Part 1, Moscow (1979).
15. Hulan University, Road Surface of Residual-oil, (In Chinese) Publishing House of Chinese Communication, Peking, (1974).
16. Bureau of Construction, Ministry of China Railway, Experience of the Building Line in Desert Region, (In Chinese) Publishing House of Chinese Railway (1961).
17. Team of Controlling Sand, Chinese Academy, A Report of Investigation in Desert Region, (In Chinese) Part 1, Scientific Publishing House, China (1958).

Discussion

On the Paper by D.B. Paul

Dr. Brossaso (University of Louis Pasteur, Strasbourg, France):

(a) Do you observe differences when you cured the polymer polysulphide by different metal dichromates?

(b) What is the best?

D.B. Paul (Materials Research Laboratories): Choice of the dichromate is important since it can influence cure rate and also the general applicability of the cured sealant. The cure reaction leads to formation of hydroxide ions and therefore metal cations which form highly alkaline hydroxides (e.g., sodium, potassium) are avoided since they could eventually cause corrosion of metal substrates. Calcium and magnesium dichromates are commonly used for aircraft sealants. In our laboratory we have examined both calcium and ammonium dichromates and, in the context of the work discussed here, we have noted no significant differences in performance of sealants produced from these curing agents.

On the Paper by M.J. Owen and J.M. Klosowski

Tinh Nguyen (National Bureau of Standards): How about adhesion properties? Does it need an adhesion promoter?

M.J. Owen (Dow Corning Corp., Midland, MI): Silicone sealants have natural adhesion with covalent bonding to many hydroxylated substrates such as those having silanol, carbinol and various metal-OH groups. However, the maintenance of adhesion with water exposure to these substrates, as well as unprimed adhesion problems with other substrates, requires the use of adhesion promoters. These adhesion promoters are most often silane coupling agents or derivatives of them.

J.C. Saukaitis (American Hoechst, Coventry, R.I.):

(a) Will increasing fluorine-containing groups in a silicone sealant decrease the water absorptivity of the sealant?

(b) Do fluorinated silicones have as good adhesive properties as the nonfluorinated silicones?

M.J. Owen (Dow Corning Corp.): I do not know of any data pertaining to the water absorption of fluorosilicone sealants but I would speculate that it would be reduced compared to polydimethylsiloxane sealants.

Similarly I know of no direct comparison of their adhesive properties. Our experience would indicate that in general the adhesion of polytrifluoropropylmethylsiloxane sealants and polydimethylsiloxane sealants is quite similar.

Sung Gun Chu (Hercules Inc., Wilmington, DE): How can we improve (increase?, or decrease?) moisture transmission rate of silicone sealant?

M.J. Owen: This is a seldom-asked question. The most common question along this line is how to decrease the very high moisture vapour transmission rate. The patent literature indicates that certain fillers, such as talc, at high loadings decrease the moisture vapour transmission rate. Providing bulkier hydrophobic pendant groups such as butyl and higher branched alkyl groups in place of the methyl group would also be expected to have a similar effect. It has been our experience that the degree of change achievable in these ways is not very great.

PART FIVE:

COATINGS FOR CORROSIVE ENVIRONMENTS

Introductory Remarks

Joseph W. Holubka

Ford Motor Company
Research Staff
Dearborn, MI 48121

The session on coatings for corrosive environments included papers dealing with theoretical studies on the mechanics of coatings, surface chemistry of coated substrates and the effects of environments on the polymer chemistry of the coatings.

Farris, Bauer and Vratsanos described the relationship between stresses in coatings and coating properties. Two techniques, impulse viscoelasticity and a vibrating wave method, were described in the study of residual stresses in thermosetting and coating materials.

Jain, Rosato, Kolania and Agarwala described a new approach to corrosion prevention. A built-in electronic barrier coating at a metal-semiconductor or metal-thin insulator-semiconductor interface is used to limit corrosion instead of conventional corrosion preventing techniques involving passive barriers, inhibitors and paints. The paper describes the application of both inorganic and organic semiconductor coatings for corrosion protection.

Sugama reported on the adhesion properties of polyelectrolyte-chemisorbed zinc phosphate conversion coatings. Sugama described how the molecular structure and functionality of polyelectrolytes act to suppress the crystal growth of zinc phosphate and thereby improve the adhesion between zinc phosphate and polymers. Treatment of the metal surface with the polyelectrolyte was reported to improve the bond durability of adhesive joints to zinc phosphate conversion coatings.

Nguyen and Byrd reported on corrosion-induced degradation of amine-cured epoxy coatings on cold rolled steel. They differentiated between the polymer degradation that results from the corrosion reactions from that due to thermal degradation reactions. They concluded that amine-cured epoxy coatings on steel undergo dehydration and chain scission in corrosion environments, and that this is the mechanism that is primarily responsible for the degradation of the coating in the corrosion environment.

Rogers and Nguyen reported on the effects of mechanical deformation on the photodegradation of acrylic-melamine coatings. A non-destructive technique to follow the photodegradation and hydrolysis reaction of these coatings was described.

Editor's Note: Dr. Tinh Nguyen and E. Byrd's paper on "Corrosion-Induced Degradation of Amine-cured Epoxy Coatings on Steel" will be published elsewhere.

DETERMINATION OF THE STRESSES AND PROPERTIES OF POLYMER COATINGS

Charles L. Bauer, Richard J. Farris and Menas S. Vratsanos

Polymer Science and Engineering Department
University of Massachusetts
Amherst, MA 01003

ABSTRACT

New techniques such as Impulse Viscoelasticity and a "vibrational wave" method are used to characterize the solidification process in polymer coatings. In general, solidification may be the result of polymerization, crystallization or solvent removal. Many of the rheological and mechanical properties associated with these processes can be measured or calculated. Most notable among these is the determination of the stresses developed during solidification. In particular, residual stresses due to solvent evaporation in poly(amic acid) films (precursors to polyimide films) and the curing of thermosetting systems were investigated.

INTRODUCTION

Polymer coatings are widely used in industry, serving as protective or insulative layers. In many applications, the development of residual stresses in the coating causes failure either by cracking or delamination. Classifying coatings merely by their strength or adhesion to substrates does not provide a reliable usage criterion since strong, high modulus coatings develop greater residual stresses than soft coatings. To determine the reliability of coatings, stress levels in the coatings must be compared to the strength of the material.

Stresses in a coating may develop from solidification of the coating and differences in thermal expansion and swelling between the coating and substrate. Solidification may be the result of polymerization, crystallization or solvent removal. Many of these processes occur at elevated temperatures and on cooling to room temperature, thermal stresses develop. Additional exposure to temperature and solvents can also change the stress state in the coating.

In solvent-based coatings, stresses do not develop until a critical amount of the solvent has evaporated, at which point, the material gels or solidifies. Further solvent removal results in formation of stresses in the plane of the film. At a given temperature, it may be impossible to remove all of the solvent because of the strong interplay between stress level and swelling in these materials.[1-3] Many materials which undergo further processing at elevated temperatures usually have higher shrinkage stresses due to additional solidification.

Clearly the residual stresses are dependent upon the solidification history. In order to characterize the solidification process, the stress and volumetric change due to solidification, and properties such as the thermal expansion coefficient and glass temperature must be determined. While these properties are changing as a result of solidification they can be assumed to have no viscoelastic character. It is then possible to describe the stress state mathematically using the equations of incremental linear elasticity. Such an approach assumes dimensional changes due to pressure, temperature, crystallization, polymerization or solvent removal, material isotropy, no viscoelastic behavior and short duration deformations.

Several methods have been developed to measure residual stresses in organic coatings. Many of these methods involve plate or beam deflection.[2,4-8] Two new techniques, Impulse Viscoelasticity and the vibrating wave method, have been utilized to characterize the solidification of coatings. These methods along with force-temperature experiments were used to investigate solvent removal and curing of thermosetting systems.

STRESSES IN CONSTRAINED COATINGS

For a sample held at constant strain (one-dimensional constraint) it is simple mathematically and experimentally to determine the thermal stresses (in the absence of polymerization or solvent removal) or solvent removal stresses (in the absence of polymerization or temperature changes). Similarly, the cure stresses (in the absence of temperature changes or solvent removal) may be measured for a polymerizing sample held at fixed strain. In particular for polymer coatings, a biaxial stress state is developed during solidification. The effect of biaxial constraint is to increase the residual stresses by a factor of $1/(1-\nu)$, where ν is Poisson's ratio, relative to uniaxial constrained samples. These stresses are in the plane of the coating and for a thinly coated planar substrate, there are no interfacial shear or normal stresses in the middle of the plate. At the edges, shear and normal stresses exist as dictated by equilibrium, but these have been shown to decay to zero at a distance from the edge into the film equal to approximately one coating thickness.[9]

IMPULSE VISCOELASTIC TECHNIQUE

Many techniques have been developed for studying the changes which occur during the solidification of polymers. The most prevalent mechanical techniques have been dynamic mechanical methods. They are, however, limited with regard to the information that can be obtained from such tests. In order to more completely study the liquid to solid transition, Farris[10] has developed the technique of Impulse Viscoelasticity. With this technique it is possible to measure many of the mechanical and rheological changes associated with solidification. The technique examines the response of an aging linear viscoelastic material subjected to deformation. Among the properties which can be measured or calculated using Impulse Viscoelasticity for a polymerizing system are: gel time, gel temperature, equilibrium modulus, cure stress, cure shrinkage, steady state elongational viscosity, mean relaxation time, thermal expansion coefficient, glass temperature and dynamic mechanical properties. In addition, efforts are currently underway to develop a method for calculating the relaxation spectrum.

Experimentally, the mechanical characterization of the solidification process involves short duration deformations applied periodically to a sample during its solidification history. During the deformation the properties are considered constant. Some time later in the solidification another short duration pulse deformation is applied to the sample. The properties have now changed but they are again considered constant. The requirement of short duration deformation ensures that the assumption of constant mechanical properties (during the deformation) is valid. By sequentially combining the information obtained from each pulse it is possible to reconstruct and monitor the solidification process. While any deformation path is viable, rapidly applied and removed deformations which induce time effects are more suitable than slowly applied deformations. Typically, uniaxial box-strain deformations have been used. It is important to note that this technique is independent of a particular piece of equipment or mode of deformation. Figure 1 is a typical viscoelastic stress response to a uniaxial deformation during cure for an epoxy copolymerized with a ring opening monomer. The strain disturbance and the resulting stress response contain all of the information needed to calculate the relevant mechanical properties. Specifically, time weighted moment, Laplace and Fourier transform analyses of the stress and strain responses, are used to determine many of the above-mentioned properties. Some of the mathematical derivations have been published by Vratsanos and Farris.[11] Typically, about 50-150 such deformation pulses comprise a single polymerization experiment.

In order to obtain uniaxial mechanical properties, a method has been developed for constraining initially liquid materials. The choice of a uniaxial geometry is based on the fact that a shear geometry is insensitive to the dimensional changes which can occur during solidification. Basically, a soft, thin-walled rubber tube is stretched between

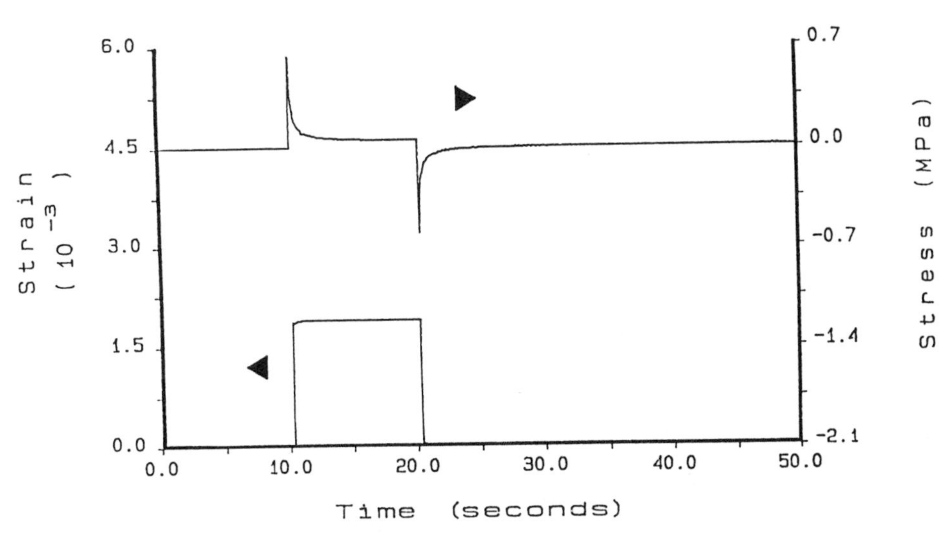

Fig. 1. Viscoelastic stress response to a uniaxial deformation for a curing epoxy.

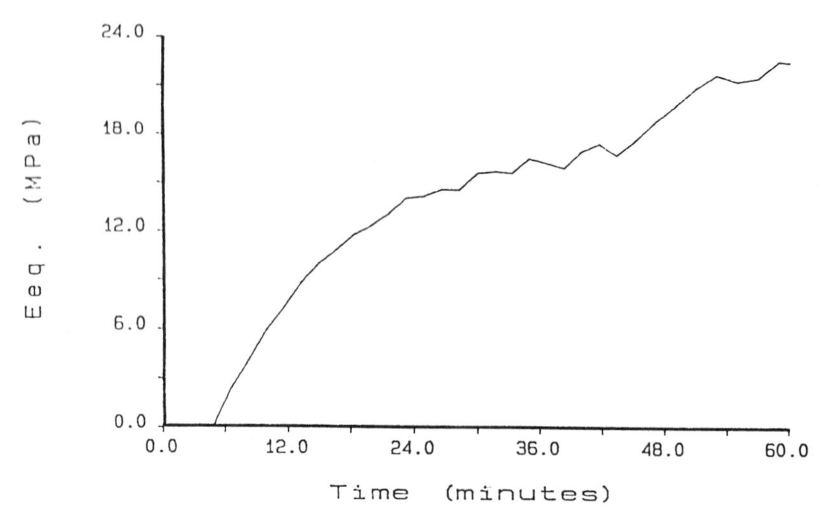

Fig. 2. Equilibrium tensile modulus (Eeq.) as a function of polymerization time for a curing epoxy copolymer.

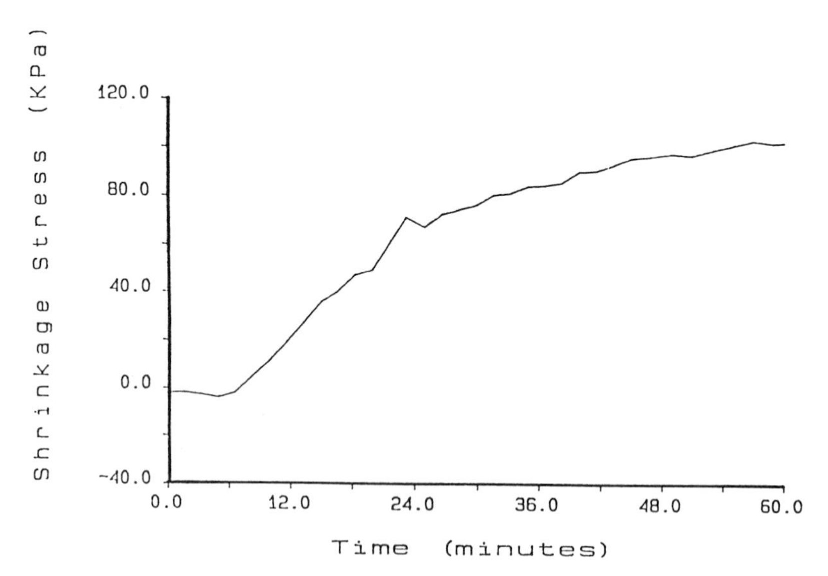

Fig. 3. Shrinkage stress behavior as a function of polymerization time for a curing epoxy copolymer.

aluminum tabs and physically constrains liquid samples in a uniaxial geometry. Details regarding this sample geometry have been published.[12]

As examples of the types of data which can be obtained, Figs. 2 and 3, respectively, plot the equilibrium tensile modulus (Eeq) and shrinkage stress behavior during the copolymerization of an epoxy with a ring opening monomer. Thus far, the Impulse Viscoelastic technique has been applied to the curing of epoxy, polyester, polyurethane and polycarbonate resins.

FORCE-TEMPERATURE AND VIBRATING WAVE TECHNIQUES

Residual stresses due to solidification of coatings are usually determined by beam deflection techniques. These methods are an indirect means of determining the stresses in a coating. Using techniques such as force-temperature experiments and the vibrating wave method, the development of stresses in solvent-based coatings has been directly measured. These techniques monitor the stresses in a coating and the mass of the coating as a function of temperature and time.

The advantage of these techniques arises from using solvent-based coatings applied to rubber substrates (sheets or threads). These samples are attached to aluminum tabs and then mounted in the appropriate apparatus. The shrinkage forces due to solvent evaporation are then directly measured. For stiff coatings the contribution of the rubber is negligible for the force changes measured. In particular, poly(amic acid) (PAA) solution coatings were used (15 wt% PAA in NMP). This material is formed by the solution reaction of pyromellitic dianhydride (PMDA) and oxydianiline (ODA) (Fig. 4). Upon solvent removal, the material may be thermally cyclized to form the polyimide, poly[N,N'-bis(phenoxyphenyl)pyromellitimide]. Both the PAA and polyimide films have tensile modulus around 1 GPa compared to the rubber substrates where the modulus is about 1 MPa.

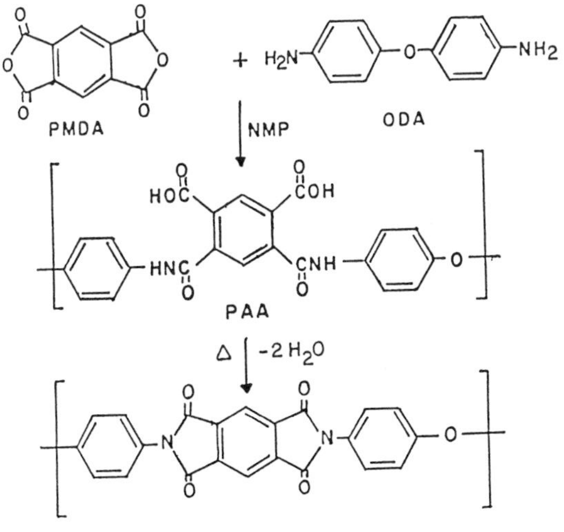

Fig. 4. Reaction scheme for formation of polyimide (PI).

Force-Temperature experiments are simply the evaluation of the force as a function of temperature for a sample held at constant strain. A coated rubber sheet (6 cm x 1.5 cm x 0.25 mm) is placed in the apparatus and the force monitored via a load cell as the temperature is changed. The vibrating wave technique is similar to this but samples consist of long, coated rubber threads (diameter = 0.2 mm) instead of sheets. In addition to measuring the tension on the sample, the time of flight of a traveling wave in the sample is determined. These quantities are then related to the mass per unit length using the wave equations for a flexible string giving:

$$\text{mass/length} = \text{tension/(velocity)}^2$$

Experimental details of these two techniques have been described previously.[13,14]

Results of the force-temperature experiments are shown in Fig. 5 (forces are corrected to a stress based on final coating dimensions). In the first cycle, A, the sample was heated to 80° C and kept at this temperature until no further increase in stress was detected. After cooling to room temperature if the sample was heated a second time, the curve would trace the cooling portion of path A (the line between 3.5 MPa at 30° C to 3.2 MPa at 80° C) and reversibly follows the same path on cooling.

In this case, after the first cycle the load was manually increased above any previous level. On heating to 80° C (cycle B), little shrinkage occurred after 20 minutes and upon cooling, the path was reversible between 30° and 80° C. For the next cycle (C), the load was manually reduced to 1.5 MPa, the initial starting value. On heating the sample, additional solvent evaporation occurs creating shrinkage stresses which approach the level of the first cycle. Thus the heating/cooling cycles are irreversible.

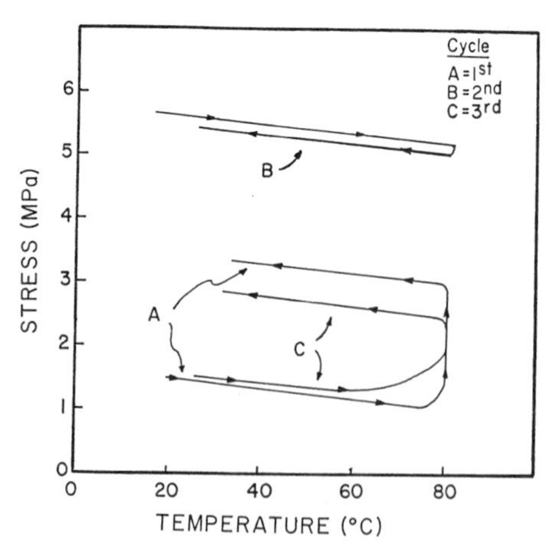

Fig. 5. Development of shrinkage stress for PAA coated on rubber sheet.

These results indicate that solvent removal is dependent on the stress level. To confirm that the observed behavior is due to solvent loss, the vibrating wave technique was used to follow mass changes. Due to experimental difficulty the coated fiber was partially dried in air and then placed in the apparatus. Results are shown in Fig. 6. The first cycle of the coated rubber represents a heating to 87° C and immediate cooling to room temperature. In the second cycle additional shrinkage occurs due to solvent loss as indicated by the mass/length curve in Fig. 6B. This is expected since the sample did not reach an equilibrium solvent state in cycle 1. During the second cycle, the sample was held at the elevated temperature until the tension appeared constant and then it was cooled to room temperature. For the third cycle, the load was manually reduced to 8.3 MPa and the sample was heated as in the previous cycle. The corresponding mass/length traces for the stress-temperature curves show much scatter between points; however, the trends do indicate solvent loss for each successive step.

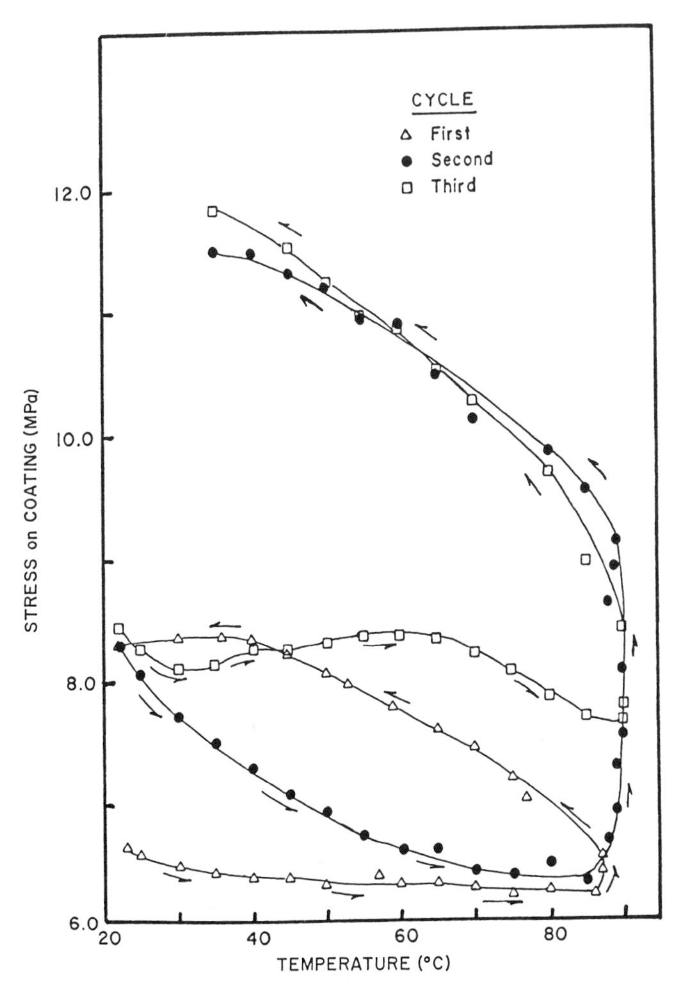

Fig. 6. Drying of PAA coated rubber fibers. Load manually reduced to 8.3 MPa at the start of the third cycle.
6A: Stress changes with temperature.

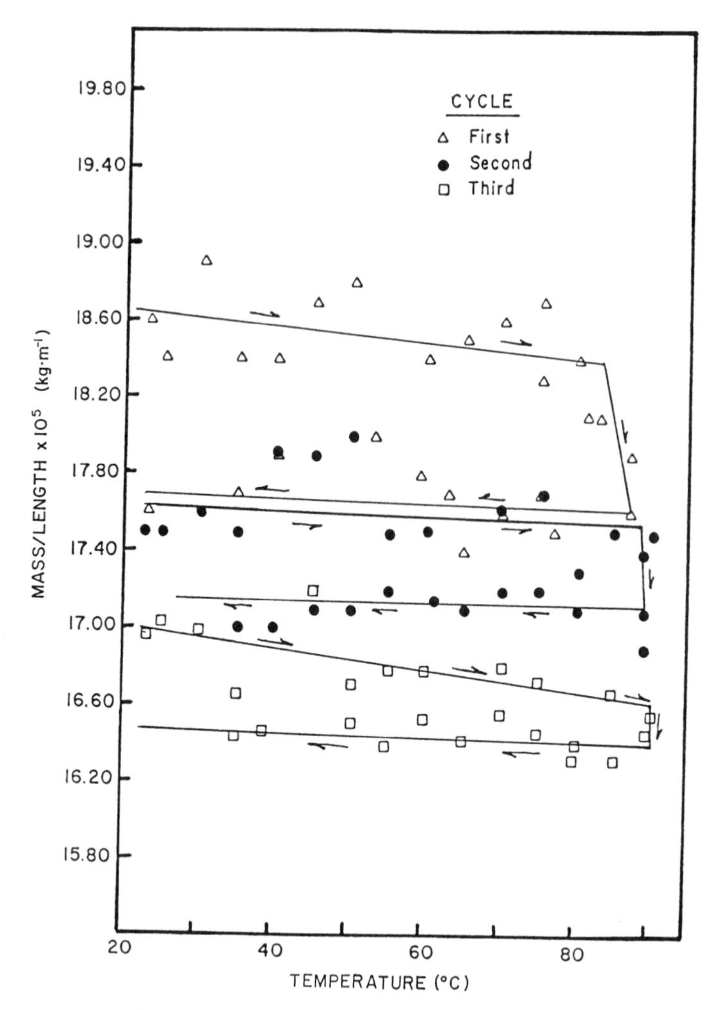

Fig. 6B: Corresponding mass/length changes.

In converting the PAA to polyimide, water is formed as a by-product and evaporated. This, along with structural changes, leads to shrinkage stresses. Using free standing PAA films and those constrained on rubber, force as a function of temperature was obtained for the sample up to 160° C. At this high temperature a noticeable increase in force was observed, attributed to cyclization and additional solvent loss.

CONCLUSIONS

For flat coatings, stresses only develop in the plane of the coating. There are no interfacial shear or normal stresses except at the edges. New techniques such as Impulse Viscoelasticity and the vibrating wave method were used to study these residual stresses in thermosetting materials. Using Impulse Viscoelasticity, it is possible to measure many of the relevant engineering properties during solidification. Using the vibrating wave technique, corresponding stress and mass changes due to solvent removal in PAA coatings were measured. Both methods involved the use of rubber as a constraining support.

ACKNOWLEDGEMENT

This work was supported in part by the Center for University of Massachusetts-Industry Research on Polymers.

REFERENCES

1. K. Sato, Prog. Org. Coat., 8, 143 (1980).
2. D.Y. Perera, J. Coat. Technol., 56 (716), 111 (1984).
3. S.G. Croll, J. Appl. Polym. Sci., 23, 847 (1979).
4. E.M. Corcoran, J. Paint Technol., 41 (538), 635 (1969).
5. S.G. Croll, J. Coat. Technol., 50 (638), 33 (1978).
6. T.S. Chow, C.A. Liu and R.C. Penwall, J. Polym. Sci., Polym. Phys. Ed., 14, 1311 (1976).
7. R.N. O'Brien and W. Michalik, J. Coat. Technol., 58 (735), 25 (1986).
8. H.M. Tong, C.-K. Hu, C. Feger and P.S. Ho, Polym. Eng. Sci., 26, 1213 (1986).
9. B.J. Aleck, J. Appl. Mech., 16, 118 (1949).
10. R.J. Farris, J. Rheol., 28 (4), 347 (1984).
11. M.S. Vratsanos and R.J. Farris, in Composite Interfaces, H. Ishida and J.L. Koenig, eds., (Elsevier, New York) 1986 p.71.
12. M.S. Vratsanos and R.J. Farris, J. Appl. Polym. Sci., 33, 915 (1987).
13. C.L. Bauer and R.J. Farris, ACS Symposium Series 346 Polymers for High Technology, Ed. M.J. Bowden nd S.R. Turner, American Chemical Society, New York (1987), Chapter 23.
14. M. Chipalkatti, J. Hutchinson and R.J. Farris, accepted for publication, Rev. Sci. Instr., 58, 112 (1987).

INFLUENCE OF COMPOSITION AND FILM THICKNESS ON ANTIFOULING PAINTS BIOACTIVITY CONTAINING CASTOR OIL AS THE THIXOTROPIC AGENT

Carlos A. Giúdice, Beatriz del Amo and Vicente J.D. Rascio

CIDEPINT, Research and Development Center for Paint Technology (CIC-CONICET) - 52 entre 121 y 122, 1900 La Plata, Argentina

ABSTRACT

The object of this paper is to establish the behaviour of high-build antifouling paints. The influence in the bioactivity of the WW rosin/chlorinated rubber ratio, binder percentage, cuprous oxide content, castor oil amount incorporated as gel, dispersion time of the toxicant and also dry film thickness were the variables studied.

Samples preparation was achieved on a laboratory scale using a high-speed disperser and a ball mill with a 3.3-liters jars. The bioactivity was performed by means of trials in a marine zone with aggressive fouling. Fouling attachment on the panels was determined after 25-month immersion.

The values of antifouling efficiency were statistically treated according to the factorial design 3 x 2 x 2 x 2 x 2 x 2. The variance of the mentioned effects and of the experimental error (residual variance) permitted the calculation of F values. Then the antifouling coat with best performance was selected.

INTRODUCTION

Antifouling paint represents the least reliable variable in any scheme for the protection of ship hulls. Due to its complex formulation it is difficult to carry out a satisfactory quality control on antifouling paint at the laboratory. The most convenient procedure to establish the practical performance is by the exposure of painted panels on experimental rafts. This method has the disadvantage of the long period of time required to obtain a conclusion related to the bioactivity of the product. Furthermore, the permanent change in the conditions of the marine environment (pH variation, different fouling organisms, modification of the settlement periods) may influence the anticipated results based on formulation characteristics.

In previous papers,[1,2] the authors have proposed some antifouling paints effective for 24-26 month immersion. However, considering only a 12-month period, more reliable antifouling paints, applied with 60-80 µm dry film thickness and based on films of high dissolution rate, were attained. After the 12-month immersion, the film lost its ability to prevent fouling settlement.

An increase in the useful life of the paints could be attained with higher dry film thicknesses, i.e. 120-150 µm. It is possible to reach these thicknesses with several coats of a conventional paint or with one or two coats of a thixotropic product, with lesser application cost and reducing the dry-docking period.[3,4]

The main objective of this paper is to establish the raft-test behaviour of high-build antifouling products prepared with binders based on WW rosin (gum rosin) and grade 10 chlorinated rubber, using castor oil as the thixotropic agent.

VARIABLES UNDER ANALYSIS

Experimental formulations are shown in Table 1. The most important aspects considered in this paper are the following:

WW rosin/chlorinated rubber ratio Binder dissolution rates were varied employing two WW rosin/grade 10 chlorinated rubber ratios: 2/1 and 1/1 w/w; the first ratio corresponds to the matrix with the higher dissolution rate.

Binder content To evaluate the influence of binder content for each one of the above-mentioned rosin/chlorinated rubber ratios, three percentages of binder were selected: 50.3, 57.7 and 64.0% by volume on the total solids of the paints.

Toxicant content Red cuprous oxide is a well-known and efficient toxicant to cirripedia, serpulids, bryozoa, etc. These organisms are considered to be highly resistant species. In the tested samples, cuprous oxide was used as the fundamental toxicant in two concentrations: 9.8 and 15.2% by volume.

Thixotropic agent content The castor oil was incorporated in the paint as a gel, after pigment dispersion occurred. This gel was previously manufactured by activating this substance in xylene (15% w/w), with the application of shear stress at 40-45°C until a stable colloidal structure was reached. Experimental paints were formulated employing 1.3 and 2.6% w/w of thixotropic agent based on solids of the paint.

Dry film thickness Depending on the dry films to be obtained, the sample viscosity was adjusted before application[5] and panels with two different dry film thicknesses were prepared, 60-80 and 120-140 µm.

Dispersion time of the toxicant

The main component of WW rosin is abietic acid. Its carboxylic group is responsible for the reaction with the divalent ions, forming resinates of a lesser sea water dissolution rate than the original resin.[6-8] The cupric ions, generated during cuprous oxide dispersion, produce the neutralization reaction before mentioned.[9] In order to study this influence, toxicant dispersion times of 3 and 4 hours were selected.

EXPERIMENTAL PART

The following aspects were considered:

Paints manufacture Sample preparation was achieved on a laboratory scale, using a ball mill (3.3-liters jars). Operating conditions of the ball mill were such that an efficient dispersion was obtained.[10]

Afterwards the rheological agent was added to the paints as a gel, employing a high-speed disperser and selecting a work temperature of 40-45°C. Each experimental paint was manufactured in duplicate.

Raft trials In order to establish the toxic behaviour of the paints in sea water, a 25-month immersion test was carried out on a raft anchored at Puerto Belgrano (38°54' S; 62°06' W), an area whose hydrological and biological conditions were previously studied.[11,12]

SAE 1010 steel plates were used for the test. These plates were previously sandblasted to A Sa 2½ (SIS Specification 05 59 00/67), with 40 µm maximum roughness (Rm).

Before applying the antifouling paints, the panels were protected with an anticorrosive coating based on basic zinc chromate as inhibiting pigment and a chlorinated rubber-phenolic varnish (1:1 ratio w/w) as binder (150-180 µm of dry film thickness). After 24 hours drying, the panels were coated with the antifouling paints employing airless spray equipment, to give 60-80 and 120-140 µm dry film thicknesses. Drying time prior to immersion was 24 hours (panels were placed vertically on the raft frames).

RESULTS AND CONCLUSIONS

In judging the results of the immersion test, fouling fixation was evaluated employing a scale which ranges from 0 (surface without fouling settlement, 100% effectiveness) to 5 (surface completely fouled, 0% effectiveness). Fixation value 1 (little or rare, 80% efficiency) was considered as the maximum acceptable limit of fixation for a good antifouling product.

Values obtained were statistically treated using a factorial design 3 x 2 x 2 x 2 x 2 x 2 (48 compositions and two thicknesses). Each composition was prepared and tested in duplicate.[13,14]

The main effects can be related to the variables under analysis, that is WW rosin/chlorinated rubber ratio (R), binder content (L), toxicant content (C), thixotropic agent content (A), dry film thickness (E) and toxicant dispersion time (T).

The results of fouling attachment shown in Table 2 were transformed into efficiency values and then the value 50 was subtracted from each of them (this is close of the efficiency average). Later, the sum of squares and the degrees of freedom of each effect were calculated. By dividing each sum of squares by the corresponding degrees of freedom, the estimated variance was calculated.

The variance of the mentioned sources and of the experimental error (residual variance) permitted the calculation of F values (Table 3).

For the zero hypothesis all the effects shown in Table 3 are considered equal to zero. Thus, all the variance estimations would be independent and would refer to the same amount estimated by means of the residual variance, that is the magnitude of the experimental error.

If the variance of the mentioned sources is bigger than that based on the residual error (experimental), the F test indicates that there is little probability that the observed variances ratio occurred randomly.

If the F test provides a positive result, hypothesis zero fails. In such cases, it will be evident that the variance does not simply arise from the experimental error, but also from an additional variance introduced by the fact that the designs modification was significant.

It was deduced that the third-order interaction ERLT provides a 5.7 variance ratio (F=1483.0/259.8) for 2 and 108 degrees of freedom. Consultation of the relevant table[14] proves that the interaction is significant at the 0.5% level.

The third-order interaction ERLT is significant as are ELT, ERL, ERT and RLT (second-order); EL, ET, RL, RT, LT and ER (first-order), as well as E, R, L and T (main effects). This is also demonstrated by the F test results.

Based on the statistical study, it is concluded that in order to reach the best bioactivity in 25-month immersion in raft test, in the area of Puerto Belgrano selected for this research, it is necessary that the dry film thickness E is 120-140 µm, the WW rosin/chlorinated ratio R = 2/1 w/w, the toxicant dispersion time T = 3 hours and the binder content L = 64.0%. This combination led to the most efficient toxic coat (highest mean value corresponding to the ERLT third-order interaction, Table 4). Further, it is important to note that similar bioactivity was reached with the level L = 57.7% when the above levels of the other effects constituting the ERLT interaction were used.

The same conclusion is reached when the significant interactions of the second and of the first-order, as well as the main effects, were analyzed.

Regarding cuprous oxide content (main effect C), this was shown to be important since the F test provided a positive result (F = 3588.0/259.8 for 1 and 108 degrees of freedom, significant at the 0.1% level). Considering the mean values (Table 5) it is observed that the most efficient coatings were those prepared with the highest toxicant content.

Nevertheless, a separate analysis taking into account only the more efficient samples (that is paints 3, 4, 5 and 6), applied with 120-140 µm film thickness, after 25-month immersion in raft trials, shows a similar biocidal power of the above materials (mean value of 60 considering an efficiency average of 90, for the two cuprous oxide content). Then, the lesser cuprous oxide content should be selected for those paints formulated with binder content L = 57.7 or 64.0% and ratio R = 2/1, and prepared dispersing cuprous oxide during only T = 3

Table 1. Antifouling paint composition (% by volume).

Paint*. . . .	1	2	3	4	5	6
Cuprous oxide	9.8	15.2	9.8	15.2	9.8	15.2
Zinc oxide	1.0	1.5	1.0	1.5	1.0	1.5
Calcium carbonate	38.7	32.8	31.3	25.4	25.0	19.1
WW rosin (gum rosin)	32.0	32.0	36.7	36.7	40.7	40.7
Chlorinated rubber (grade 10)	10.8	10.8	12.4	12.4	13.8	13.8
Chlorinated paraffin (42%)	7.5	7.5	8.6	8.6	9.5	9.5
Additives	0.2	0.2	0.2	0.2	0.2	0.2
Paint	7	8	9	10	11	12
Cuprous oxide	9.8	15.2	9.8	15.2	9.8	15.2
Zinc oxide	1.0	1.5	1.0	1.5	1.0	1.5
Calcium carbonate	38.7	32.8	31.3	25.4	25.0	19.1
WW rosin (gum rosin)	23.5	23.5	26.9	26.9	29.9	29.9
Chlorinated rubber (grade 10)	15.9	15.9	18.2	18.2	20.2	20.2
Chlorinated paraffin (42%)	10.9	10.9	12.6	12.6	13.9	13.9
Additives	0.2	0.2	0.2	0.2	0.2	0.2

* Base composition. The castor oil was incorporated to 100 g of the base composition in the following quantities: 1.3 and 2.6 g.

Table 2. Fouling fixation after 25-month immersion in raft trials*

Paint	1	2	3	4	5	6	7	8	9	10	11	12
Castor oil 1.3%												
Dry film thickness, 60-80µm: 3 hours of dispersion	0 1-2	0 0	2-3 1-2	2-3 1	3-4 4	4-5 3	3 4-5	4 2-3	2-3 2	2-3 1	0-1 2	0 1-2
4 hours of dispersion	4 2-3	2-3 3	1 2-3	0-1 1-2	0 1-2	0 0	5 5	3-4 5	3 4-5	4 2-3	2-3 3	2 2-3
Dry film thickness, 120-140µm: 3 hours of dispersion	0 1-2	0 0-1	0-1 0	0 0	0 0	0 0	3 4	4 2-3	3 2-3	2-3 4	0 2	0 1
4 hours of dispersion	3 4-5	2-3 2-3	2-3 1	2 0-1	0 1	0 0-1	5 3-4	3-4 5	2-3 4	3-4 2-3	2-3 1-2	1 2-3
castor oil 2.6%												
Dry film thickness, 60-80µm: 3 hours of dispersion	0 2	0 0-1	1 2-3	1 2-3	4-5 3	3-4 4-5	4-5 3	2-3 4	1-2 2-3	2 1-2	0-1 2-3	0 2
4 hours of dispersion	4 2-3	3-4 2-3	1 2	1-2 0-1	0 1	0 0-1	4-5 5	3-4 4-5	3-4 4-5	3-4 2-3	2-3 2-3	2-3 2
Dry film thickness, 120-140µm: 3 hours of dispersion	1-2 0	0 1	0-1 0	0 0	0 0	0 1	3 4-5	2-3 4	2-3 2-3	2-3 4	1-2 0	1 0
4 hours of dispersion	4-5 2-3	2 2-3	1 2-3	0-1 2	1 0	0 0	5 4	3-4 5	2-3 4-5	4 2-3	1-2 2-3	1 2

*Key of the Table:
0 Without fixation (100% effectiveness)
1 Little (80%)
2 Rare (60%)
3 Regular (40%)
4 Abundant (20%)
5 Completely fouled (0%)

Intermediate values were also considered

Table 3. Analysis of variance

Nature of the effect	Source	Sum of squares	Degree of freedom	Estimated variance	F test
Main effects	A	1	1	1.0	
	T	12192	1	12192.0	46.9
	E	5742	1	5742.0	22.1
	R	34938	1	34938.0	134.5
	L	30064	2	15032.0	57.8
	C	3588	1	3588.0	13.8
Interaction between pair of effects	EL	4447	2	2223.5	8.6
	EC	13	1	13.0	
	ET	2776	1	2776.0	10.7
	RA	4	1	4.0	
	LA	13	2	6.5	
	CA	25	1	25.0	
	TA	42	1	42.0	
	EA	0	1	0.0	
	RL	10651	2	5325.5	20.5
	RC	26	1	26.0	
	LC	294	2	147.0	
	RT	2480	1	2480.0	9.5
	LT	11878	2	5939.0	22.8
	CT	151	1	151.0	
	ER	2067	1	2067.0	8.0
Interaction among three effects	ERA	14	1	14.0	
	ELA	39	2	19.5	
	ECA	14	1	14.0	
	ETA	5	1	5.0	
	LCA	21	2	10.5	
	RTA	26	1	26.0	
	LTA	116	2	58.0	
	CTA	26	1	26.0	
	ECT	4	1	4.0	
	ELT	2732	2	1366.0	5.2
	RLA	22	2	11.0	
	RCA	42	1	42.0	
	ERL	1747	2	873.5	3.4
	ERC	42	1	42.0	
	ELC	8	2	4.0	
	ERT	5104	1	5104.0	19.6
	RLC	257	2	128.5	
	RCT	274	1	274.0	
	RLT	15590	2	7795.0	30.0
	LCT	26	2	13.0	

Table 3 (cont). Analysis of variance

Nature of the effect	Source	Sum of squares	Degree of freedom	Estimated variance	F test
Interaction among four effects	ELTA	3	2	1.5	
	ECTA	25	1	25.0	
	ERTA	63	1	63.0	
	LCTA	12	2	6.0	
	ERCA	4	1	4.0	
	ERLA	101	2	50.5	
	ELCA	5	2	2.5	
	ELCT	209	2	104.5	
	RLTA	7	2	3.5	
	RLCA	4	2	2.0	
	RCTA	5	1	5.0	
	RLCT	41	2	20.5	
	ERCT	190	1	190.0	
	ERLC	116	2	58.0	
	ERLT	2966	2	1483.0	5.7
Interaction among five effects	ERCTA	12	1	12.0	
	ELCTA	66	2	33.0	
	ERLCA	28	2	14.0	
	ERLTA	7	2	3.5	
	ERLCT	174	2	87.0	
	RLCTA	21	2	10.5	
Interaction among six effects	ERLCTA	2	2	1.0	
Replica	Residual	27750	95	292.1	
Revised residual		12	1		
		66	2		
		28	2		
		7	2		
		174	2		
		21	2		
		2	2		
		27750	95		
	Total	28060	108	259.8	

hours, since they avoid the fouling fixation and in addition they lead to a more economic paints and a lower sea water pollution.

Concerning the thixotropic additive content, the two levels considered (that is 1.3 and 2.6% of castor oil) led to paints of similar behaviour. This effect was not statistically significant.

ACKNOWLEDGEMENTS

The authors are grateful to CIC (Comisión de Investigaciones Científicas) and CONICET (Consejo Nacional de Investigaciones Científicas y Técnicas) and to SENID (Servicio Naval de Investigación y Desarrollo) for their sponsorship for this research, and also to Puerto Belgrano Naval Shipyard and Control Laboratory for the assistance given during painting and observations.

Table 4. Main values of third-order interaction ERLT*

	E_1						E_2					
	R_1			R_2			R_1			R_2		
	L_1	L_2	L_3	L_1	L_2	L_3	L_1	L_2	L_3	L_1	L_2	L_3
T_1	320	110	-210	-160	90	220	310	380	380	-150	50	290
T_2	-90	190	340	-320	-160	10	-80	160	350	-290	-120	110

* Key of the Table: E_1 = 60-80μm, E_2 = 120-140μm; R_1 = 2/1 ratio, R_2 = 1/1 ratio; L_1 = 50.3%, L_2 = 57.7%, L_3 = 64.0%.

Table 5. Main values of effect C*

C_1	C_2
450	1280

* Key of the Table: C_1 = 9.8%, C_2 = 15.2%

REFERENCES

1. C.A. Giúdice, J.C. Benítez and B. del Amo.--Bioactivity of chlorinated rubber antifouling paints tested in sea water. In: Proc. 6th International Congress of Marine Corrosion and Fouling, II, Marine Biology, 283-292, Athens, Greece (1984).
2. C.A. Giúdice, B. del Amo, V.J. Rascio and O. Sindoni.--Composition and dissolution rate of antifouling paint binders (soluble type) during their immersion in artificial sea water. J. Coat. Technol., 58, 733 (1986).
3. T.C. Patton.--Fundamentals of paint rheology. J. of Paint Technol., 40, 522 (1968).
4. P. Grandou et P. Pastour.--Peintures Marines. In: Peintures et Vernis, Germann, Paris (1966).
5. H.L. Beegerman and D.A. Bergren.--Practical applications of rheology in the paint industry. J. Paint Technol., 38, 492 (1966).
6. F.H. de la Court and J.J. De Vries.--The leaching mechanism of cuprous oxide from antifouling paints. J. Oil Col. Chem. Assoc., 56, 388 (1973).
7. A. Partington and P.F. Dunn.-The use of paint extenders in antifouling compositions. Paint Technol., 26, 6 (1962).
8. A. Partington.--Antifouling compositions. Paint Technol., 28, 3 (1964).
9. C.A. Giúdice, B. del Amo and J.C. Benítez. Determination of metallic copper, cuprous oxide and cupric oxide during the manufacture and storage of antifouling paints. J. Oil Col. Chem. Assoc., 64, 1 (1981).
10. B. del Amo, C.A. Giúdice and V.J. Rascio.--Influence of binder dissolution rate on the bioactivity of antifouling paints. J. Coat. Technol., 56, 720 (1984).
11. R. Bastida, E. Spivak, S.G. L'Hoste and H. Adabbo.--Las incrustaciones biológicas de Puerto Belgrano. I. Estudio de la fijación sobre paneles mensuales, período 1971/72. Corrosión y Protección, 8, 11 (1977).
12. R. Bastida and V. Lichtschein.--Las incrustaciones biológicas de Puerto Belgrano. III. Estudio de los procesos de epibiosis registrados sobre paneles acumulativos. Corrosión y Protección, 10, 7 1979).
13. W.E. Duckworth.--Factorial Designs. In: Statistical techniques in technological research, Methuen & Co. Ltd., London (1968).
14. J.C.R. Li.--Analysis of variance; one way classification. In: Statistical Inference, Eduards Brothers Inc., Michigan (1964).

CORROSION PREVENTION IN METALS USING LAYERED SEMI-CONDUCTOR/INSULATOR STRUCTURES FORMING AN INTERFACIAL ELECTRONIC BARRIER

F.C. Jain, J.J. Rosato and K.S. Kalonia

Electrical and Systems Enginering Dept. and
Institute of Materials Science
The University of Connecticut
Storrs, CT 06268

V.S. Agarwala

U.S. Naval Air Development Center
Warminster, PA 18974

ABSTRACT

A new approach to corrosion prevention involving the use of layered semiconductor/insulator films on metal surfaces is described. It is shown that the improved corrosion protection is due to the existence of a built-in electronic barrier at the metal-semiconductor (MS) or metal-(thin) insulator-semiconductor (MIS) interfaces. This is in contrast to the conventional techniques which rely on physical barriers (e.g., paints) or high resistivity oxide/nitride films at the exposed metal surfaces. The electronic barrier, which arises due to charge redistribution at the MS or MIS interface, serves to impede the transfer of electrons from the metal surface to foreign oxidizing species, thereby preventing oxidation. Specific structures fabricated and tested include: Al-Indium Tin Oxide (ITO) for MS and Al-SiO_2-ITO for MIS configurations. A comparison with Al-Si_3N_4 (passive barrier) is also made. High purity (single and polycrystalline) and commercial purity aluminum and aluminum alloy (7075-T6) samples were used in this study. Cathodic and anodic polarization data, weight-loss measurements, and the results of physical, optical and electronic characterizations are presented for numerous aluminum samples. It is shown that the magnitude of the electronic barrier height, and the resultant corrosion protection, is enhanced by: (1) the presence of a thin (20-100Å) SiO_2 (insulator) layer, and (2) an increased indium/tin ratio in ITO films, which results in a larger energy gap. The application of the active electronic barrier concept to semiconducting polymers such as doped polyacetylene, phthalocyanine and chlorophyll is also discussed.

INTRODUCTION

This paper presents a new approach to corrosion prevention which has recently been described by the authors.[1,2] Whereas conventional corrosion coatings generally rely on the use of passive barriers (e.g., oxides, nitrides), inhibitors, and paints for protection, the films described in this work utilize a built-in (active) electronic barrier at a metal-semiconductor (MS) or metal-thin insulator-semiconductor (MIS) interface. This active electronic barrier exists due to the presence of an interfacial space charge distribution. The magnitude of the barrier depends upon the work function of the metal, the conductivity type, the carrier concentration and the energy gap of the semiconductor coating.[3] Both inorganic and organic semiconductor coatings can be used in the MS and MIS configurations to provide protection against corrosion.

The oxidation of a metallic surface involves a net transfer of electrons from the metal to a foreign species such as oxygen, sulfur, etc. Oxidation is energetically favored as it results in a decrease in the free energy of the system. The reaction is commonly expressed as[4]

Oxidation:

$$Me \rightarrow Me^{z+} + z \text{ electrons}$$

Reduction:

$$O_2 + z \text{ electrons} \rightarrow \frac{z}{2} O^{2-}$$

Net:

$$Me + O_2 \rightarrow MeO_{z/2}$$

A schematic illustration of the oxidation process is shown in Fig. 1.

Reactions involving chloride, nitride and sulfide formations, which generally occur in the environment, can be expressed similarly. In some cases (e.g., Ni/NiO and Cr/Cr_2O_3), the native oxide does not grow beyond a thickness value of 10 to 15Å at room temperature, regardless of exposure time to the atmospheric conditions. In other cases, the oxidation occurs continually. For some metals, the oxide itself can be used as a protective coating to inhibit corrosion. SiO_2, NiO and Cr_2O_3 are well-known examples of naturally occurring protective thin oxides

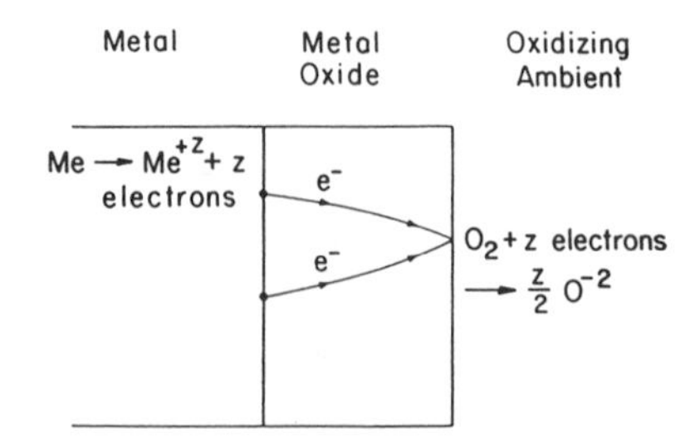

Fig. 1. Electron transfer during oxidation. (Only the electron transfer is specifically shown.)

on Si, Ni and Cr surfaces, respectively. In addition, growing oxide on Al, Ti and Ta surfaces using anodic oxidation is a common practice for protecting these metals from corrosion.[4-7] Coatings of silicides,[8] and aluminides and nitrides have also been used as protective layers.

The ability of an oxide or nitride layer to prevent corrosion is dependent upon its electrical resistivity. The higher the electrical resistance of the oxide layer, the more effective the coating will be in preventing electron transfer, thereby reducing the current flow due to corrosion at the metal surface. The electrical resistance of an oxide coating, grown naturally or intentionally, depends on the chemical and electronic structure (defect states in the energy gap, etc.), thickness, and porosity of the oxide layer. Figure 2 shows the energy band diagram of an oxide with energy states in the forbidden gap arising due to defects. The energy states within the band gap result in a reduced electrical resistance, causing an increased corrosion current flow. This can be viewed as a lowering of the magnitude of effective electron barrier from an ideal value of $q\Phi_b$ to $q\Phi_{be}$. Fromhold[9,10] has summarized the mechanisms responsible for electronic and ionic current flow during oxidation in oxides ranging in thickness form 10 to 100 microns. This type of protection against corrosion can be viewed as a passive electronic barrier at the metal surface.

THEORY OF THE ACTIVE ELECTRONIC BARRIER AT METALLIC SURFACES

Metallic surfaces host positive dipole layers when they are interfaced with appropriately doped semiconductors to form metal-semiconductor (MS) and metal-oxide (insulator)-semiconductor (MIS) structures.[3] These interfacial space charge layers result in a built-in electric field and cause bending of the electronic energy bands. The net band bending is defined as the active electronic barrier. This barrier, in turn, impedes the transfer of electrons from the metal surface to oxidizing species present in the ambient. This corresponds to a lowering of the corrosion current. It should be emphasized that this current reduction is due to the existence of the active electronic barrier at the interface, and not due to the electrical resistance of the semiconductor film.

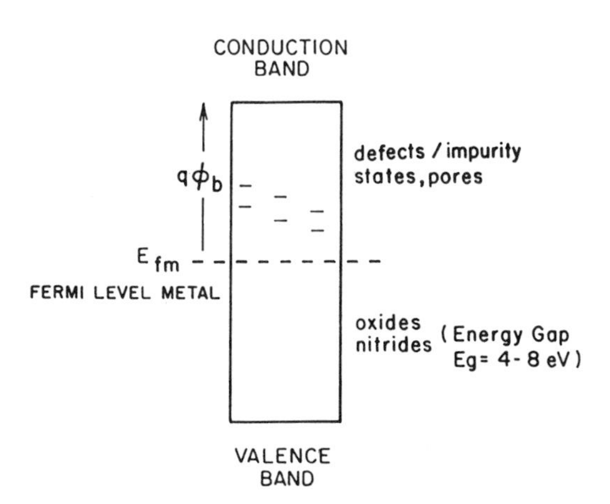

Fig. 2. Electronic energy band diagram of metal-oxide system (passive electronic barrier).

Metal-Semiconductor Interfaces The electronic barrier of a metal-semiconductor interface, commonly called a Schottky barrier, is known to depend on the work function of the metal, conductivity type, carrier concentration and energy gap of the semiconductor.[3,11-13] For a given metal, the barrier height of a metal-semiconductor interface depends on the doping levels and energy gap of the semiconductor. In general, wider energy gap semiconductors yield higher barriers.

The space charge distribution and the electric field at the metal-semiconductor interface is shown in Fig. 3. The presence of the field E results in an electrostatic force F which impedes electron transfer from the metal surface. As shown in Fig. 3(a), oxidation occurs when an electron is transferred from the metal surface, through the protective coating, to the oxidizing ambient. In the space charge region of width x_d, where the interfacial electric field exists, the electrons will see an electronic barrier which prevents their transfer.

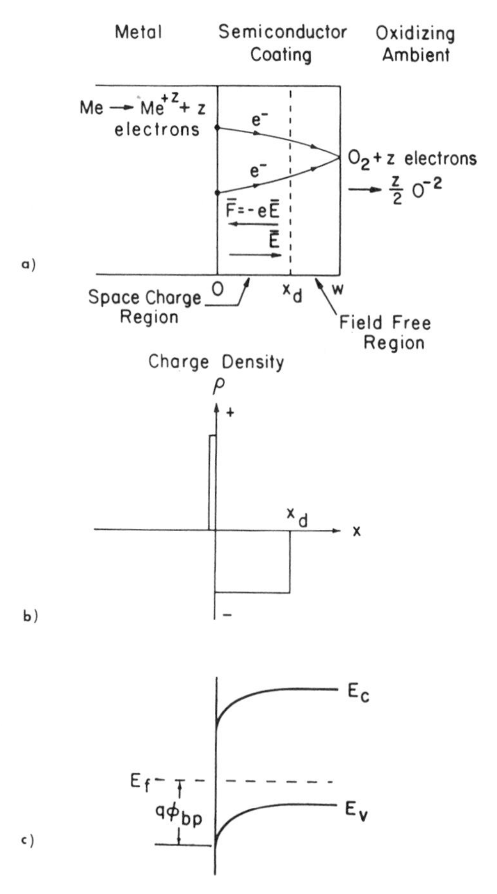

Fig. 3. Metal/p-type semiconductor interface showing a) space charge region and built-in electric field which impedes electron transfer, b) charge density at the interface, c) energy band and the active electronic barrier.

This results in a lower probability of oxidation/corrosion. Fig. 3(b) shows the charge density at the metal/semiconductor interface, and Fig. 3(c) illustrates the band bending which results in the active electronic barrier.

Under ideal conditions, with no surface states present, the electronic barrier height $q\Phi_{bp}$ can be expressed as

$$q\Phi_{bp} = E_g - q(\Phi_m - \chi) ,$$

where E_g is the band gap of semiconductor, Φ_m the metal work function, and χ is the electron affinity of the semiconductor. This simple treatment does not include the correction due to image forces. When surface states are present on the semiconductor surface, the effective barrier height is determined by the properties of surface states. It should be noted that n-type semiconductors will also form barriers with a given metal. In this case the limiting value of barrier height $q\Phi bn$ is given by, ($q\Phi_{bn} = q(\Phi m - \chi)$.

The magnitude of the electronic barrier is dependent upon the choice of metal (i.e., its work function) and the properties of the semiconductor. In general, wider energy-gap semiconductors yield higher barriers. Brillson[14] and others[15,16] have investigated the barrier formation in its initial phases using electron loss and other spectroscopic techniques. Recently, Monch[17], and Doniach et al.[18] have proposed improved models for Schottky Barrier interfaces.

It should be emphasized that the interfacial electric field does not result from any externally applied voltage; rather, it is built-in at the metal-semiconductor interface. However, the magnitude of this field can be changed by applying an external voltage such as in polarization experiments. In contrast to a passive barrier, an active electronic barrier may help in regions having micropores and pinholes in the semiconductor layer. In these regions, a finite electric field is expected as a result of field fringing effects which will act to retard the transfer of electrons. By properly selecting the semiconductor coating material for a metal surface, both the traditional passive as well as the novel active barriers can be realized.

Metal-Oxide (Insulator)-Semiconductor Interfaces A metallic surface can also develop a positively charged dipole layer and the associated active electronic barrier in a Metal-Insulator-Semiconductor (MIS) configuration. The physics of the barrier formation depends on the thickness of the oxide. The manifestation of the electronic barrier in thick oxide (≥150Å)systems is similar to those of conventional MOS devices used in Si integrated circuit technology.[3,19] Thick oxide MOS structures are not treated in this paper. Structures using thin oxide (insulator) films with thicknesses ranging from 20 to 100Å are referred as MIS interfaces and they behave similarly to the MS systems described above. However, the barrier has been found to be larger in MIS than in MS configurations.[20-22] For instance, it has been found that the barrier height is increased at Al-nSi and Al-pSi and Au-nGaAsP interfaces with the introduction of a thin 20 to 30Å SiO_2 layer. Reference is made to Nicollian and Brews,[20] and Jain and Marciniec[21] for detailed conditions of barrier enhancement in Schottky-type MIS structures.

Figure 4 shows a typical energy band diagram for a MIS device with thin insulator. Here, the barrier height Φ_{bp} is higher than that shown in Fig. 3 for the same metal and semiconductor combination. MIS structures are particularly advantageous for metal-semiconductor systems which yield low values of barrier heights.

The active electronic barrier inhibits the net transfer of electrons from the metal surface to the oxidizing species, resulting in a lower probability of oxidation/corrosion. Additionally, the electronic barrier can help in regions with micropores and pinholes in the semiconductor layer. In these regions, a finite electric field (due to field fringing effects) acts to inhibit the transfer of electrons.

EXPERIMENTAL

Selection of Semiconductor/Insulator Coatings The fabrication of MS and MIS structures for corrosion prevention requires the deposition of semiconductor and insulator coatings on metal surfaces. The selection of the semiconductor material is crucial in determining the magnitude of the active electronic barrier at the metal-semiconductor interface. A large energy bandgap semiconductor is desired, and the doping properties must be controllable during film deposition. In addition, the semiconductor film coatings must be stable, durable and utilize relatively simple fabrication techniques.

Semiconductor films of indium-tin-oxide (ITO) were chosen for use in this study. ITO films exhibit large energy bandgaps (3.5 to 4.0 eV depending on the In:Sn ratio) and controllable n- and p-type doping can be achieved in a chemical vapor deposition (CVD) process. In addition to forming the active electronic barrier at the metal interface, an ITO film (having a lightly doped neutral region) also acts as a passive barrier due to its high resistivity (the resistivity/passivity of metals and semiconductors has been the topic of a special conference.)[23] The ITO coatings are also chemically stable and physically rugged.

In a previous report, Jain[2] has proposed the use of polymeric semiconductor coatings for realizing the active interfacial electronic barrier at metal surfaces. Polymeric semiconductor films, including polyacetylene and phthalocyanine, have been developed for use in solar cell devices.[24-26] Further work is needed in utilizing these films for corrosion prevention since the doping in these materials is not controllable, and it has been reported that phthalocyanine[24] is

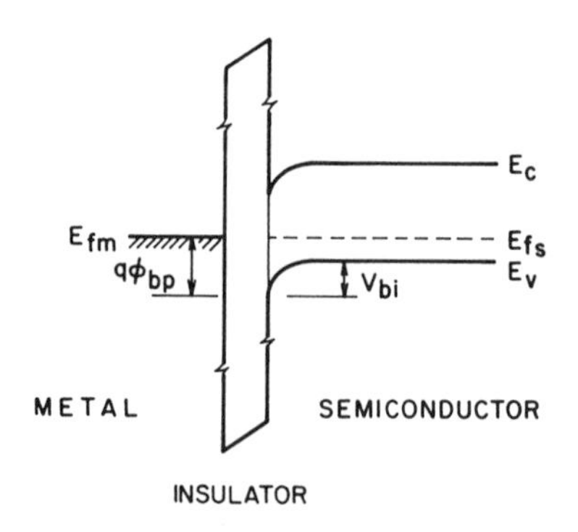

Fig. 4 Energy band diagram for a MIS interface.

unstable under illumination while polyacetylene[16] is unstable in oxidizing environments. The use of chemisorbed porphyrins has also been reported by Agarwala and Longo.[27] (Here, the charge transfer mechanism is somewhat different in nature.)

The use of a thin insulator layer (20-100Å) in the MIS configuration enhances the electronic barrier height at Al-ITO interfaces and also serves to protect Al alloys [e.g. Aluminum Association (AA) 7075-T6] from being attacked by reaction products during the deposition of ITO. Silicon dioxide (SiO_2) was chosen for the insulator since it is easily deposited in varying thicknesses and exhibits few defects and pinholes. In addition, it features a large energy bandgap which has been shown to enhance the magnitude of effective electronic barriers at metal surfaces.

<u>Structures Fabricated</u> The semiconductor (ITO) and insulator (SiO_2) films were deposited in a variety of configurations. Figure 5 shows the MS and MIS structures fabricated using the ITO, SiO_2 and Si_3N_4 films. The final Si_3N_4 coating was used to physically protect the ITO surface against wear and abrasion. It will be shown in the next section that the active electronic barrier is the primary mechanism preventing the corrosion of Al substrates, and that the Si_3N_4 coating without the ITO film is a relatively poor corrosion inhibitor. The ITO semiconductor coatings were deposited in the thickness range of 0.1 to 0.3 μm. The SiO_2 coating thickness varied between 20 and 200Å. The outermost Si_3N_4 coatings were about 0.75 μm thick. The aluminum substrates were prepared from several different grades of aluminum and aluminum alloys. For example, commercial-purity as well as high-purity alluminum were used for polycrystalline substrates. High-purity, single-crystal aluminum substrates were also used. In the case of aluminum alloy, AA 7075-T6 was used.

<u>Fabrication</u> Aluminum and aluminum alloy sample surfaces were prepared using mechanical, chemical-mechanical, and chemical polishing techniques. A thin layer of native aluminum oxide was grown (in many instances prior to ITO or SiO_2 deposition) in a dry oxygen ambient at the relatively low temperature range of 300-375°C. Aluminum substrates with a thin native oxide were then deposited with either thin SiO_2 or ITO films depending on the structural configurations desired.

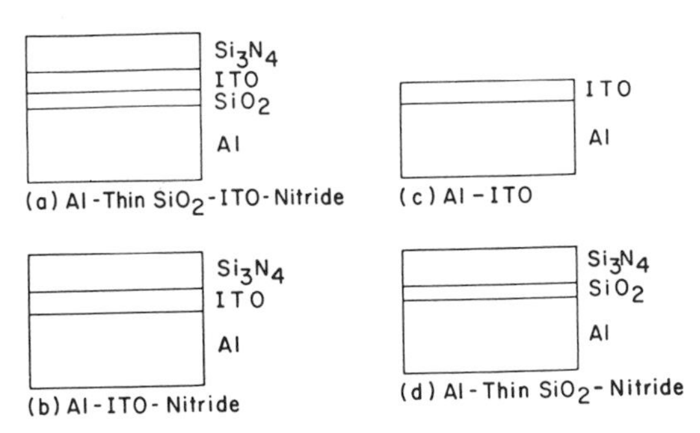

Fig. 5. MS and MIS structures fabricated.

The semiconductor films of ITO and tin oxide (TO) were deposited by three techniques:[28] (1) CVD, (2) spray injection CVD and (3) air spray pyrolysis. The CVD reactor was developed in the initial phases of the project to allow for precise control of the ITO film composition and to insure high quality films. The spray injection CVD reactor was utilized to increase the ITO film thickness and indium content. The air spray pyrolysis reactor, yielding the thickest films, was developed to simulate a commercial deposition process.

Chemical Vapor Deposition (CVD) Reactor Figure 6 shows the chemical vapor deposition (CVD) reactor which was utilized in the growth of ITO. The system features separate feed lines for the reactant chemicals to allow for precise control of the film deposition. Indium chloride, water, and ethanol were mixed to form one reactant solution, while anhydrous tin chloride was used for the second solution. By controlling the flow rates of the two reactant solutions, either SnO_2 or ITO films could be fabricated. It should be noted that the anhydrous $SnCl_4$ was contained in a methylene chloride base, which initiated a competing reaction in the CVD reactor. The chemical reactions involved in the growth are listed below. The A1 samples were located in a temperature zone of 380-430°C. The ITO films grown in this reactor were of very high quality. However, the In/Sn ratios were lower than desired.

Chemical Reactions

ITO Deposition Reaction

Oxidation

$$CH_3CH_2OH \rightarrow CH_3\underset{\underset{:O:}{\|}}{C}\text{-}H + 2H^+ + 2e^-$$

Reduction

$$Sn^{+4} + 2e^- \rightarrow Sn^{+2}, \; [E° = +\,0.15V \text{ at } 25°C \text{ for } Sn^{+4}]$$

Total Reaction (Aldehyde)

$$2In^{+3}Cl_3 + Sn^{+4}Cl_4.5H_2O + CH_3CH_2\text{-}OH \xrightarrow{400°C,\ N_2} 2In^{+3} + Sn^{+4} + 10Cl^- + 10H^+$$

$$+2H^+ + 2e^- + 5O^- + CH_2\underset{\underset{:O:}{\|}}{C}\text{-}H$$

$$\rightarrow In_2^{+3}\; Sn^{+2}\; O_4^{-2} + CH_3\underset{\underset{O}{\|}}{C}H + 10\ HCl + H_2O$$

Methylene Chloride competing reaction (when anhydrous $SnCl_4$ is used)

$$2CH_2Cl_2 + H_2O \rightarrow CH_2CH_3OH + 2Cl_2 + \text{heat}$$
$$\text{and } 2Cl_2 + 2H_2O \rightarrow 4HCl + O_2$$

SnO_2 Formation in absence of significant $InCl_3$

$$SnCl_4 + 2H_2O \rightarrow SnO_2 + 4HCl$$

Spray Injection Chemical Vapor Deposition Reactor Figure 7 shows a schematic diagram of the reactor using the spray deposition technique developed by Ashok et al.[29] This system utilizes a controlled nitrogen ambient and a precisely regulated temperature. The starting solution, containing varying amounts of $InCl_3$, $SnCl_4.5H_2O$, was CH_3CH_2OH, and H_2O aspirated through the injector nozzel using a high-pressure nitrogen gas stream. The resultant mist was sprayed into the reactor where it is preheated at the nozzel tip and then vaporized in zone I. The aluminum substrates are placed in a quartz holder which is located at the zone I/zone II boundary. The temperature of zone I was typically maintained at 575°C while that of zone II was kept at 500°C. This difference created a temperature gradient which enhances the chemical vapor deposition on the aluminum substrate. The actual substrate temperature was typically in the range of 390-460°C depending on the linear gas stream velocity. The composition of the ITO films could be varied by changing the spray solution makeup, altering the concentration of $InCl_3$ in the starting solution, and by varying the zone temperature. This allowed for an effective means of tailoring the bandgap of the semiconductor film. The thickness of the film was controlled by the quantity of solution injected and the precise location of the substrate.

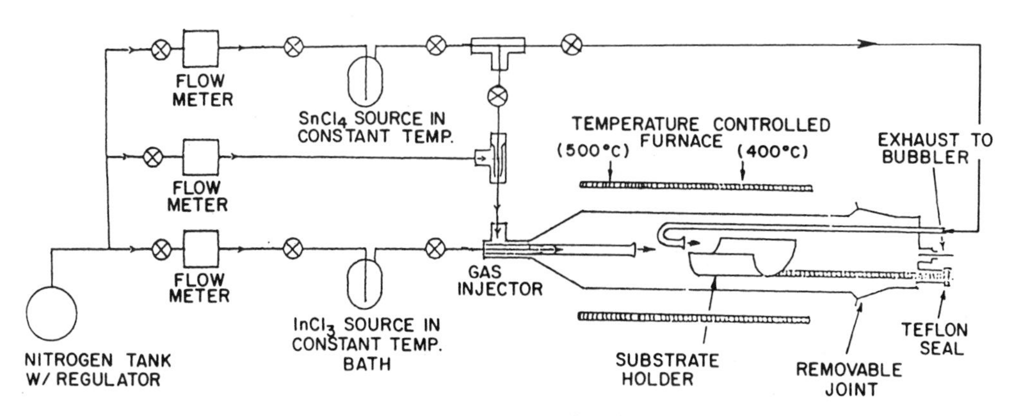

Fig. 6. Schematic of indium-tin-oxide (ITO) chemical vapor deposition (CVD) reactor.

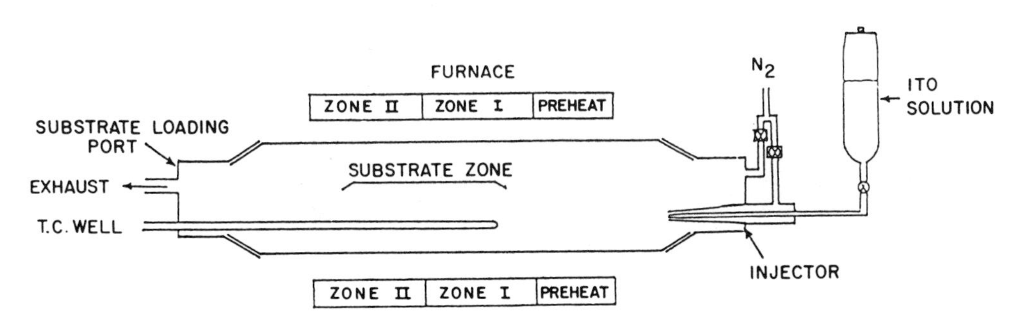

Fig. 7. Controlled ambient spray injection CVDE ITO reactor.

Air Spray Pyrolysis This reactor system was developed to simulate a commercial application process. The system utilized an open-air environment with an aspirated spray gun. The aluminum substrates were placed on a heated plate which was maintained at 400-420°C. The films grown in this system were very thick, although of somewhat lower quality.

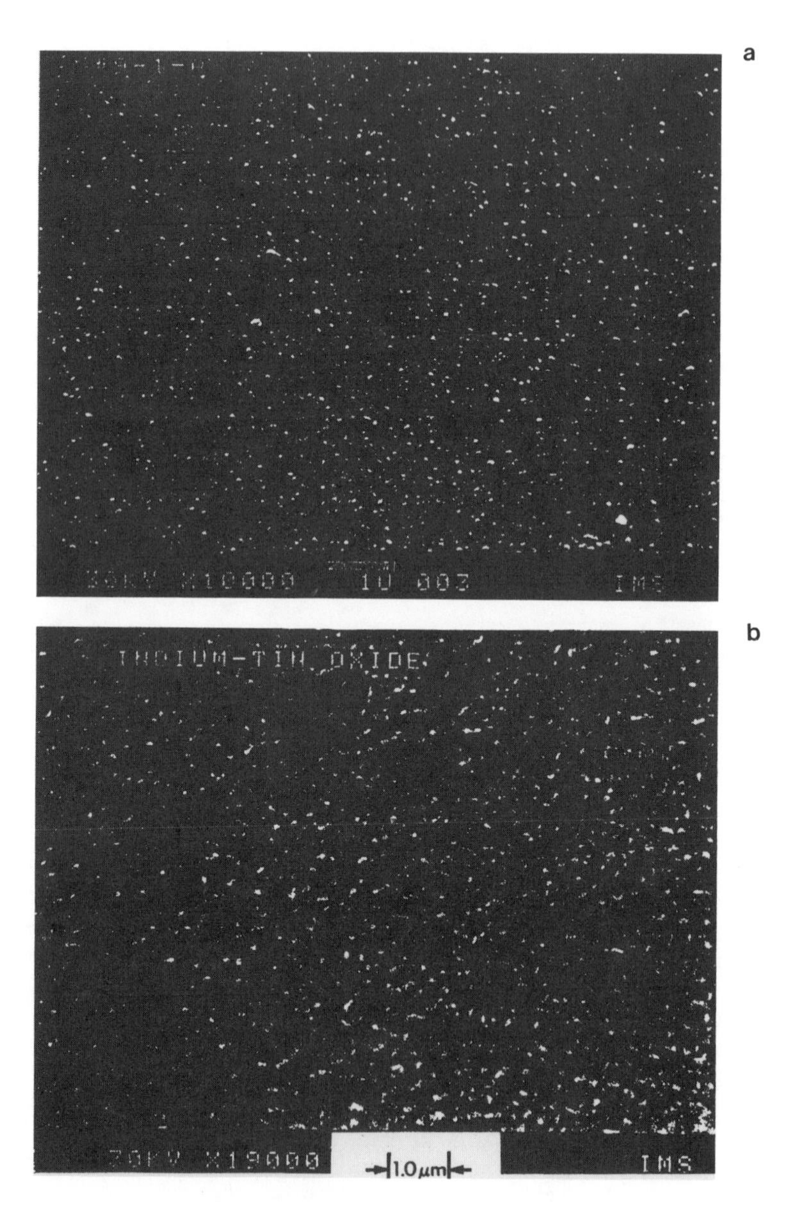

Fig. 8(a). SEM micrograph of CVD deposited ITO film.
(b). SEM micrograph of spray CVD deposited ITO film.

Deposition of Insulator Films High quality insulating layers of SiO_2 were grown in our laboratory using the pyrolytic decomposition of tetraethylorthosilicate (TEOS)[30] at about 700°C. The aluminum samples were, however, kept in a cooler temperature zone of 390-430°C. The chemical reaction resulting in the growth of SiO_2 is described below.

```
         R
         |
         O        Δ                       SiO and C
         |
     R-O-Si-R     →     SiO2 + R-R +      depending on
         |                                the conditions
         O
         |
         R
```

where R represents the hydrocarbon radical.

Finally, the outermost Si_3N_4 films were grown in a plasma CVD system. This system utilized ammonia and silane gases for the feed.

RESULTS

Characterization of ITO and SIO_2 Coatings

Physical Characterizations Physical characterizations of the ITO and SiO_2 films were performed to determine their chemical composition, grain size and film thickness. Some of these results have been previously reported.[28] The film thickness for the ITO coatings were determined using interferometric techniques, and were found to be typically in the range of 0.1-0.3 µm for the CVD and spray injection CVD films. The air spray pyrolysis reactor yielded much thicker films with faster deposition rates.

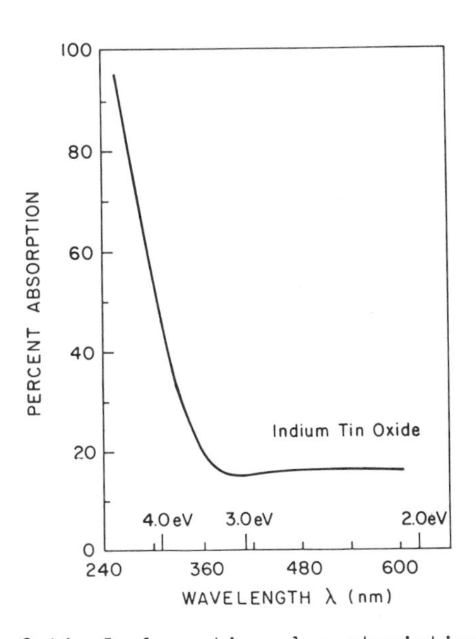

Fig. 9. Optical absorption characteristics of ITO.

SEM micrographs of ITO films grown in different reactor systems are shown in Fig. 8. Figure 8(a) shows an ITO film deposited by the CVD technique. The film surface appears extremely fine grained with little evidence of pinholes. Figure 8(b) shows a spray injection deposited ITO film. The film structure in this case is coarser and shows some pinholes.

The chemical composition of the ITO films was determined using energy and wavelength dispersive techniques such as EDX and WDX. These analytical techniques are available as an attachment to a scanning electron microscope. In addition, an Auger depth profiling setup was used for additional information on the deposited films. ITO films deposited by CVD techniques yielded very low In/Sn ratio ($\simeq$0.002) whereas the spray CVD method resulted in In/Sn ratios as high as 0.6.

Optical Characterization of SnO_2 and ITO Coatings The energy bandgap of the ITO films was determined using spectrophotometric techniques. The optical absorption studies on ITO films were done using a Varian spectrophotometer (Model 17D). SnO_2 and ITO films were grown under identical growth conditions (CVD/CVD Spray) on quartz substrates. The quartz samples were located in the same zone as the aluminum samples which were used for corrosion testing. The absorption data for a typical ITO sample is shown in Fig. 9.

The bandgap of ITO (In/Sn ratio of 0.62) was found to be approximately 3.5 eV. In contrast, pure SnO_2 films showed a lower energy gap of about 1.5 eV. The optical data is a clear indication that the films are semiconducting. In addition, the results show that the energy bandgap is significantly larger in ITO having a high In/Sn ratio.

Electrical Characterization of Tin Oxide (TO) and ITO Films Electronic measurements were performed on ITO films grown on Si and quartz substrates used as control samples to eliminate the adverse effects of a conducting aluminum substrate. These included resistivity, conductivity-type, and voltage-current (dark and illuminated) characteristics. Both n- and p-type films were tested, with the conductivity type being determined by the reactor growth conditions.

Table 1 Four Point Probe Resistivity Measurements

Control Sample	Method of Deposition	ITO Conductivity Type	Resistivity Ω-cm	Growth Temp. °C
ITO/Si	Open-Air Spray	n	3.7×10^{-3}	485
ITO/Si	Open-Air Spray	p	5.9×10^{-3}	485
ITO/Si	Open-Air Spray	n and p	8.5×10^{-3}	485
ITO/Si	CVD Spray	p	5.2×10^{-4}	460
ITO/Quartz	CVD Spray	n	3.6×10^{-4}	465
ITO/Quartz	Open-Air Spray	n	5.9×10^{-3}	480

A four-point setup was used to determine the electrical resistivity of the ITO films. The tested ITO films included films grown on Si as well as quartz substrates. The resistivity of CVD deposited ITO films was to be less than that of air spray deposited samples. Table 1 lists the results of some resistivity measurements.

The V-I characteristics, performed under both dark and illuminated conditions, gave direct evidence of the electronic barrier formation at Al/ITO interfaces. ITO/quartz and ITO/Si samples were used. In the case of ITO/quartz samples, Al and Ag were deposited for metal contacts. Here, the aluminum contact simulates the Al substrate (as in the Al/ITO samples used for corrosion testing) and Ag serves as the ohmic contact to the ITO film.

The V-I characteristics of two samples are shown in Fig. 10. Figure 10(a) shows the results for a spray injection CVD/ Al/ITO/Quartz sample having an In/Sn ratio of 0.6. The reverse characteristics (negative voltages) show the rectification behavior typical of a Schottky diode with an electronic barrier. In the absence of a barrier, the characteristics (both forward and reverse) would exhibit straight line behavior (diagonal lines). The forward characteristic does indicate the presence of a series resistance component, probably originating

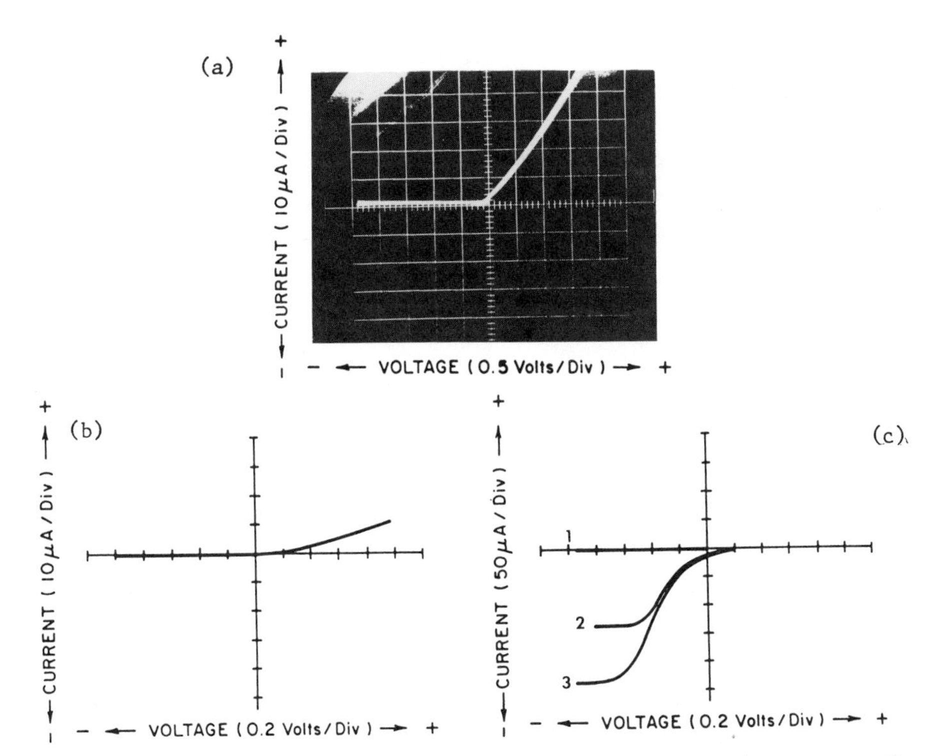

Fig. 10(a). Current-voltage characteristic of an Al-ITO/quartz sample.
10(b). Current-voltage characteristic of a CVD al-ITO/Si sample. under dark conditions.
10(c). Current-voltage characteristic of a CVD Al-ITO/Si sample under dark ('1') and illuminated ('2' and '3') conditions.

from the ohmic contact. Figures 10(b) and 10(c) show the V-I characteristics for a CVD deposited Al/ITO/Si sample under dark and illuminated conditions, respectively. Figure 10(b) shows the dark V-I characteristics, again indicating strong rectification with virtually no current flow in the reverse direction. Figure 10(c) shows the results for the same sample with the dark and illuminated reverse characteristic curves superimposed. Computations based on the dark V-I and photovoltaic measurements yielded effective barrier heights in the range of 0.15-0.4 eV.

Corrosion Testing

Fabricated structures were tested using weight-loss, and cathodic and anodic polarization techniques.

Weight-loss Measurements Weight-loss measurements (Table 2) were performed in our laboratory using 1-3% NaCl (with pH=2) solutions. The weight-loss measurements were conducted on Al-ITO samples coated with SiO_2 or Si_3N_4. High purity aluminum substrates with a surface area of 5 cm^2 were coated with CVD deposited ITO films. A typical weight loss for a Al-ITO-Si_3N_4 sample was only 0.27 $mg/dm^2/day$. This is to be compared to a value of 7.705 $mg/dm^2/day$ for the uncoated Al.

The ITO films deposited by the spray deposition technique, having higher indium content, showed different weight loss results. Films deposited on commercial grade aluminum substrates exhibited high weight losses. The rate of weight loss was small initially but accelerated with time. In contrast, films which were spray-deposited on quartz and mirror smooth silicon substrates showed negligible weight loss. For example, a typical Si sample (surface area $\simeq$7.84 cm^2) coated with ITO having a total weight of 0.83714 gms showed no measurable weight-loss over a time period of 48 hours in a 1% NaCl solution with pH=2. Similar results were found for ITO coated quartz samples. These results show that the spray deposited ITO films are chemically impervious to the test environment, and chemical dissolution of the films is not responsible for the high weight loss on aluminum substrates. The higher weight loss of spray injection deposited ITO films (supposedly having higher electronic barriers) is attributed to the relatively poor surface qualities of the films grown by this method on Al substrates. Surface morphological analysis of weight-loss samples revealed that the weight-loss was concentrated at selective sites. It is believed that these areas hosted large pinholes in the ITO films (possibly at etch-pit sites on Al substrate) where selective undercutting occurred. Figure 11 shows photographs of these areas

Table 2. Weight Loss Measurement on CVD Deposited Samples of Al in 3% NaCl Solution at pH 2

Sample #	Initial Wt gm	Wt after 70 Hrs. gm	Weight loss gm
Al-ITO-Nitride	2.08867	2.08863	0.00004
Al-ITO-SiO_2	0.75436	0.75412	0.00024
Al	0.81532	0.81307	0.00225

(Surface area=5cm^2)

magnified by 420X. The substrates were commercial grade aluminum alloy (#7075-T6) with controlled ambient CVD spray deposited ITO films. The sequence of photographs dramatizes the growth of a pinhole due to undercutting at the pinhole site. It is further believed that the undercutting was reduced, if not eliminated, at small pinhole sites due to electric field fringing. Additional confirmation of this theory is provided by the weight-loss results of Si-ITO and quartz-ITO samples.

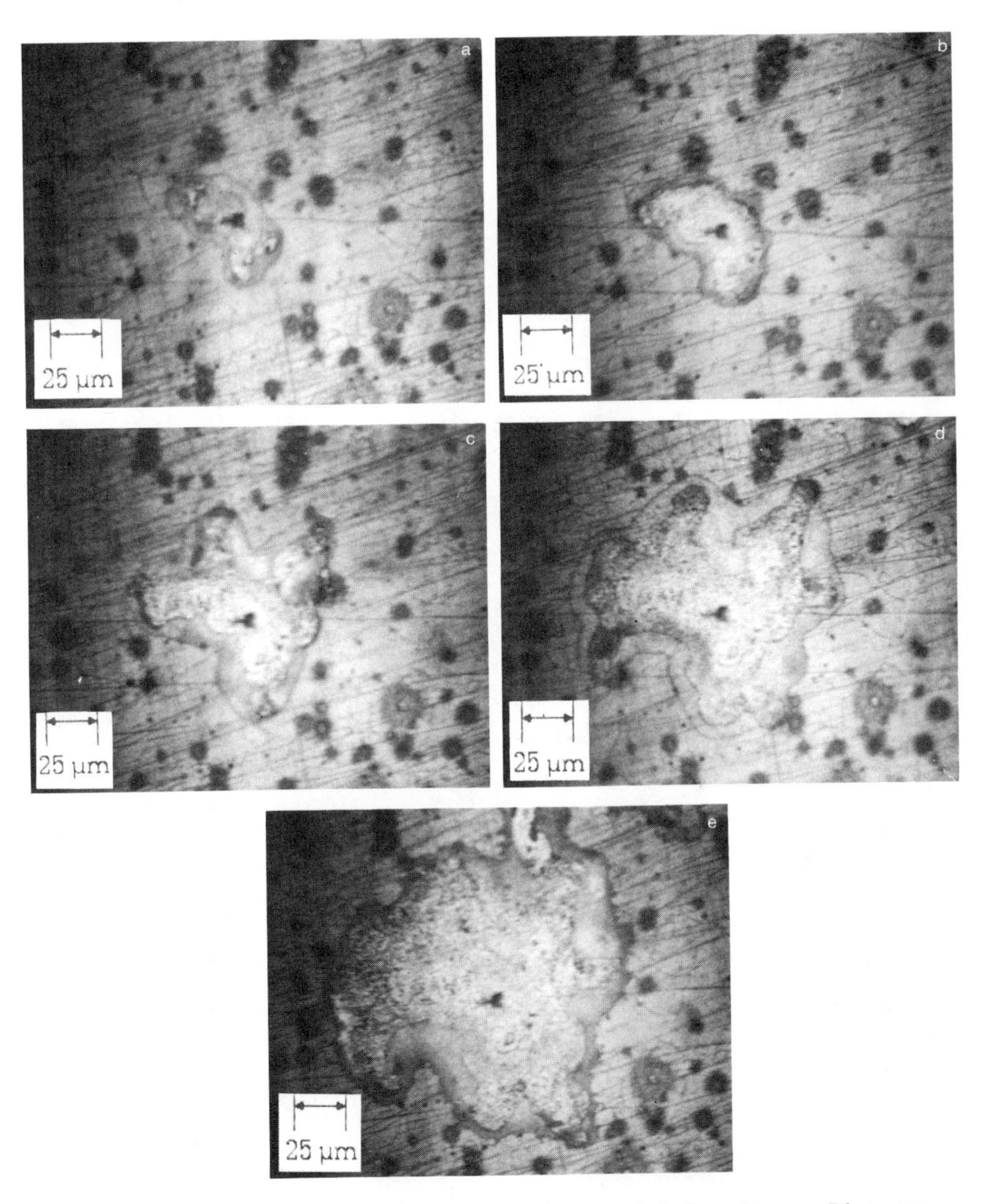

Fig. 11. Undercutting of ITO films at large pinhole sites. Pictures taken after exposure to 1% NaCl (pH=2) solution for following durations: a) 15 minutes, b) 45 minutes, c) 75 minutes, d) 105 minutes adn e) 135 minutes. Magnification=420X; Sca.

Polarization Measurements Layered structures were also tested for anodic and cathodic polarization characteristics using standard potentiostatic and potentiodynamic techniques.[31] Table 3 summarizes the rest potentials (with respect to SCE) of some samples employed in our study. Values range from -0.7 V for uncoated Al to -0.98 V for Al-SiO_2-ITO structures.

Figures 12 through 15 present the anodic polarization results for these samples. A ramp rate of 10-20 mV/min was used. Figure 12 shows the anodic polarization results of spray deposited ITO films on commercial grade aluminum. The polarization data of high purity aluminum samples with CVD deposited ITO films is shown in Figs. 13 and 14. The ITO samples of Figs. 13 and 14 have low In/Sn ratio ($\simeq$0.002) whereas the ITO films in Fig. 12 have higher In/Sn ratios ($\simeq$0.6). Figure 15 shows the results for SiO_2-ITO films on aluminum alloy (AA 7075-T6) substrates.

Cathodic Polarization The polarization data (both anodic and cathodic) for samples with CVD grown ITO layers and low In/Sn ratios are shown in Figs. 16 and 17. The substrates used in these samples were high purity aluminum. The experimental plots were taken using EG&G's Model 351 Corrosion Measurement System (parts of the anodic curve are smoothened for clarity). The anodic characteristic of the MIS sample in Fig. 16 is similar to that of two samples (spray injection CVD) of Fig. 12 which exhibit a kink (that is, a lowering of current as the voltage is increased). The cathodic polarization behavior shows significantly lower current densities. Figure 17 shows the cathodic characteristic for an Al/ITO MS sample) to give slightly higher currents than the MIS sample of Fig. 16.

Table 3. Electrochemical Rest Potentials in 1% NaCl at pH 2

Aluminum Type	Structure	Rest Potential vs. SCE (-) V	Method of Growth
CP(1)	Al	0.72	Control Sample
CP	Al-SiO_2-ITO	0.980	CVD Spray
CP	Al-SiO_2-ITO-Si_3N_4	0.845	Air Spray
CP	Al-SiO_2-ITO/TO-Si_3N_4	0.965	CVD Spray
HP(2)	Al-Si_3N_4	0.7	Plasma
Hp	Al-SiO_2-ITO	0.66	CVD Spray
Hp	Al-SiO_2-ITO-Si_3N_4	0.85	CVD
Hp	Al-SiO_2-ITO	0.707	Air Spray
Alloy(3)	Al-SiO_2-ITO	0.825	CVD Spray
Alloy	Al-SiO_2-ITO	0.9	CVD Spray

(1) CP = Commercial Purity
(2) HP = High Purity
(3) Alloy = AA 7075-T6

Figures 18 and 19 illustrate primarily the cathodic polarization data on Al-SiO_2-ITO, and Al-SiO_2-ITO-TO samples grown on commercial grade aluminum substrates. These samples were prepared using the spray injection CVD technique with high In/Sn ratios ($\simeq$0.6). Since the NaCl solution used was not de-aerated, the initial part of the V-I plot signifies the exhaustion of oxygen in the solution.

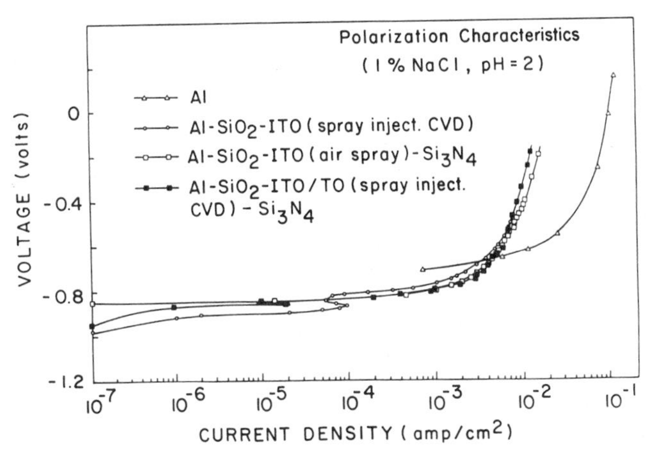

Fig. 12. Polarization characteristics of ITO coatings deposited on commercial purity Al substrates using spray deposition.

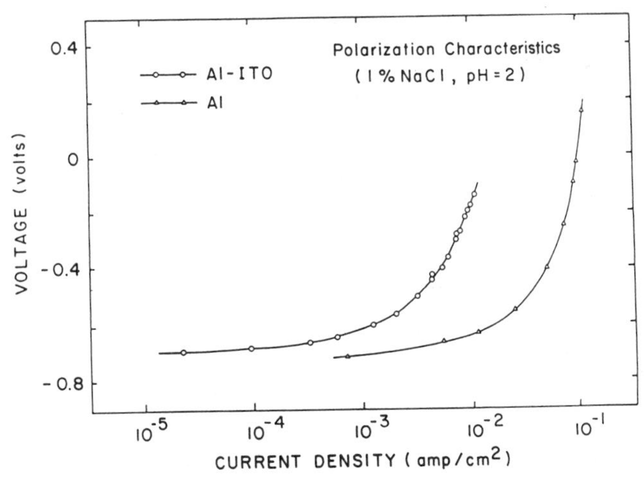

Fig. 13. Polarization characteristics of uncoated and ITO coated (CVD deposited with low In/Sn ratio) high purity aluminum substrates.

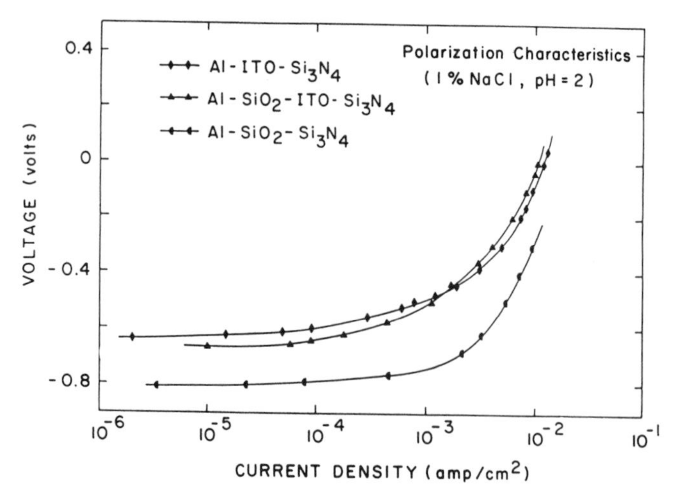

Fig. 14. Polarization characteristics comparing MS and MIS active barrier structures to passive barrier insulator coatings. The ITO films were CVD deposited with a low InSn ratio.

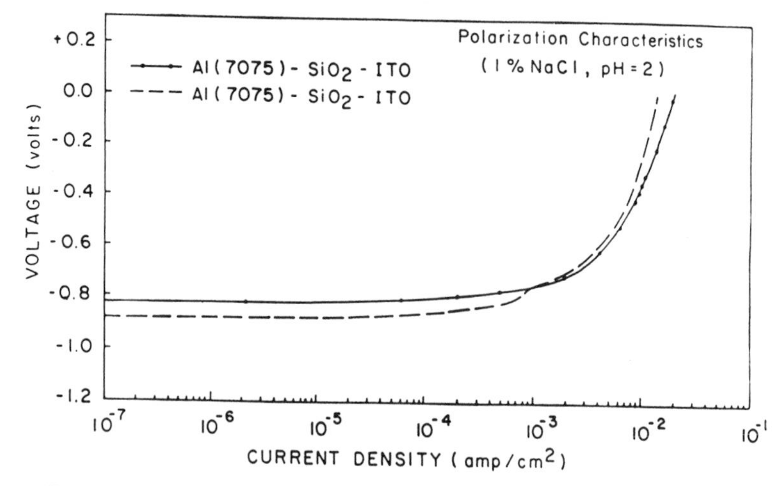

Fig. 15. Polarization characteristics of spray CVD deposited ITO films on aluminum alloy (#7075-T6) substrates.

DISCUSSION

The weight-loss and polarization data presented in the preceding section provide a useful means of evaluating the protective nature of the ITO films. The significant increase in the magnitude of the rest (corrosion) potential in the presence of ITO semiconducting films is evident from Table 3. This increase, in conjunction with the results of the electronic V-I characterization confirms the existence of the electronic barrier at metal-semiconductor interfaces.

Al-ITO samples which exhibit large shifts in rest potential also show reduced corrosion currents, as is evident from Figs. 12-19. This is particularly true at lower voltages (i.e., within the range of 100 to 200 mV of the rest potential).

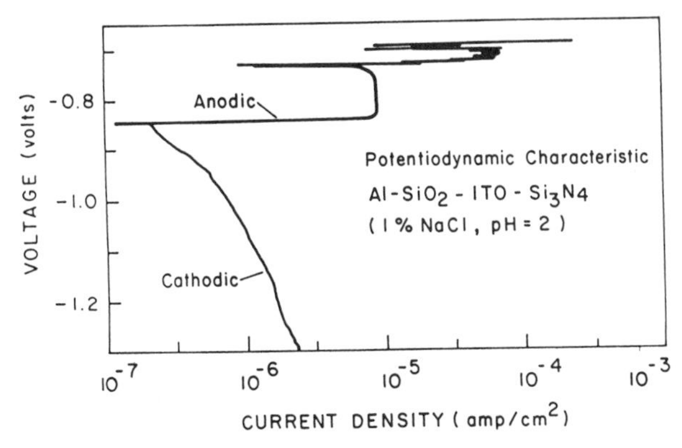

Fig. 16. Anodic and cathodic potentiodynamic characteristics of a CVD grown low In/Sn ratio $Al-SiO_2-ITO-Si_3N_4$ sample (high purity aluminum).

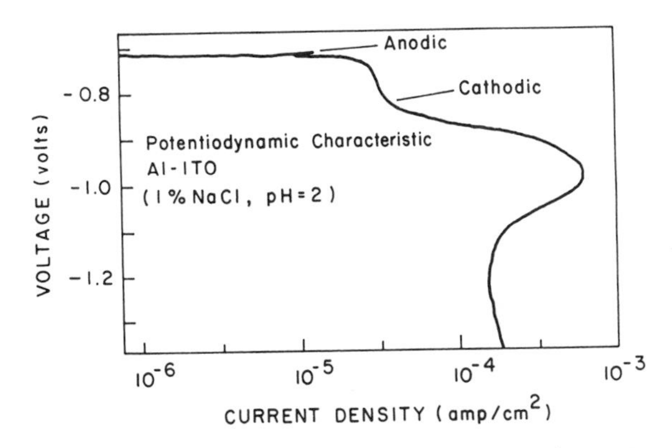

Fig. 17. Cathodic polarization characteristic of a CVD grown Al-ITO sample (high purity aluminum).

The importance of the electronic barrier and its usefulness in corrosion prevention is further demonstrated by Fig. 20. This figure plots the change in anodic current Δi as a function of voltage differential ΔV ($\Delta V = V_{Applied} - V_{Rest\ Potential}$). In this type of plot, a highly protective coating structure will show a small change in the current Δi for a given ΔV. Figure 20 compares ITO coatings prepared using different fabrication techniques and In/Sn ratios, as well as plasma deposited Si_3N_4 control samples. In addition, data on ITO/TO composites is also presented. The Al-SiO_2-ITO sample, prepared using the spray injection CVD, and having a high In/Sn ratio ($\simeq$0.6), is shown to yield the lowest increments in corrosion current for a given ΔV. However, the ITO/TO double electronic barrier (composite) sample shows

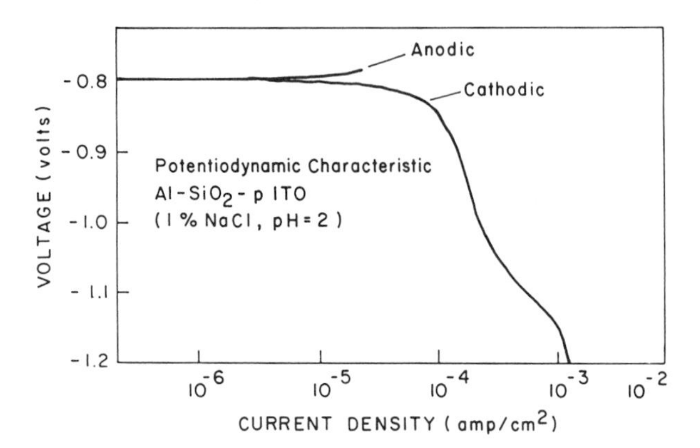

Fig. 18. Cathodic polarization characteristic of a spray injection CVD grown Al-SiO_2-p-ITO sample (commercial purity aluminum).

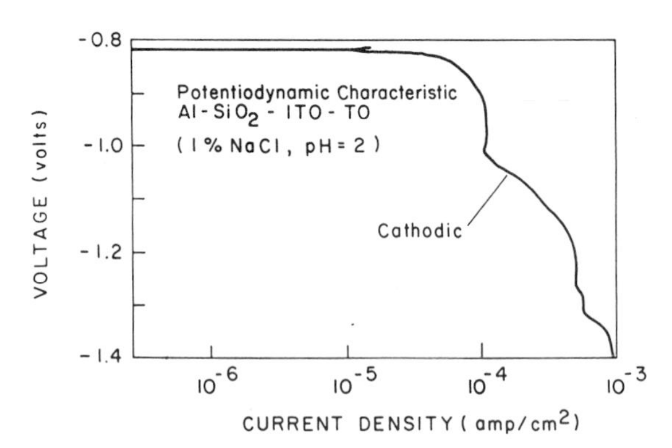

Fig. 19. Cathodic polarization characteristic of a spray injection CVD grown Al-SiO_2-ITO-TO sample (commercial purity aluminum).

better characteristics in the low current ranges. The $Al-SiO_2-ITO-Si_3N_4$ sample, having a low In/Sn ratio (≈.002), prepared by CVD deposition, shows lower Δi values than the $Al-SiO_2-Si_3N_4$ insulator coating structure. This confirms the effectiveness of the active electronic barrier in comparison to passive barrier films. It should be noted that the ITO coating is only 0.2 microns thick, while the Si_3N_4 film 0.75 microns.

Another important feature apparent from the comparative data is that the presence of a thin SiO_2 layer (≈50-100Å) does indeed further increase the effective barrier height. This effect is particularly evident in Fig. 14. In addition, the SiO_2 film enabled the growth of ITO layers on Al alloys which are susceptible to adverse chemical reactions during the CVD growth. Figure 15 demonstrates the protective nature of the SiO_2-ITO films on alloy substrates.

The cathodic polarization data, particularly for Figs. 15 and 16, exhibit very low current densities for a given change in the applied voltage (above the rest potential). In addition, the extrapolation of the V-I data yields significantly lower corrosion currents.

The preliminary results on ITO/TO double barrier coatings show promising results. The motivation for using a composite coating of ITO and TO is two-fold: (1) to increase the film thickness and minimize the occurrence of pinholes/micropores; and (2) to achieve a double active electronic barrier (one at the $Al-SiO_2$-ITO interface and the other at

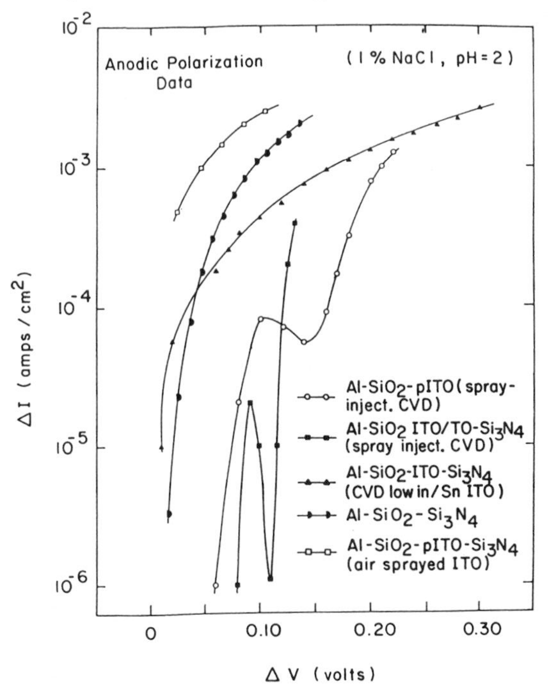

Fig. 20. Change in anodic polarization current density as a function of the applied voltage increment with respect to the rest potential.

the ITO/TO hetero-interface). It is expected that double barrier films will perform better than single barrier films in all ΔV ranges. Further work is needed to verify the effectiveness of double electronic barrier structures.

A theoretical model of the current transport in $Al-SiO_2-ITO$ structures is needed to characterize the effective height of the barrier formed at the interface. A relationship to predict effective barrier height in terms of the ITO energy gap E_g and the location of defect trap levels is desirable. This model could then be used to obtain relationships for the corrosion current. In addition, the nature of the ITO-electrolyte interface (as in polarization testing) must be studied. This information would be useful in understanding the effects of the ITO-ambient interface.

CONCLUSIONS

The results presented in this study utilizing ITO semiconducting films in Al-ITO and $Al-SiO_2-ITO$ structural configurations demonstrate the effectiveness of active electronic barriers in corrosion prevention. The protective nature of the active electronic barrier in $Al-SiO_2-ITO$ films is clearly evident when the polarization results of these samples are compared with $Al-Si_3N_4$ (plasma-deposited) passive barrier structures. The ITO coated samples with a film thickness of only 0.2 microns yielded lower currents than the Si_3N_4 coated (0.75 microns) samples. Weight-loss measurements further verify the concept of an active electronic barrier for use in corrosion prevention.

It is expected that the concept of active electronic barriers can be applied to other types of semiconducting films, including polymer semiconductor coatings, such as doped polyacetylene, phthalocyanine, chlorophyll and porphyrins. The recent advances[26,32] in the plasma CVD deposition of polymers ensure their increased future usage in corrosion protection applications.

ACKNOWLEDGEMENTS

This work was supported by the US Naval Air Development Center, Contract Nos. N62269-81-G-0288 and N62269-83-M-3295. The University of Connecticut group acknowledges the assistance of J.G. Bouchard and J. MacDougall (Sprague Electric Co., Worcester, Massachusetts) for the deposition of nitride films, C. Souchuns for technical assistance, and R. Blanchette and J. Porter for manuscript preparation.

REFERENCES

1. F.C. Jain, J.J. Rosato, K.S. Kalonia and V.S. Agarwala, "Formation of an Active Electronic Barrier at Al/Semiconductor Interfaces: A Novel Approach in Corrosion Prevention," Corrosion J., 42, 700 (1986).
2. F.C. Jain, "Semiconductor/Insulator Films in Corrosion Prevention," private communications, U.S. Naval Air Development Center, Warminster, Pennsylvania, September (1980).
3. S.M. Sze, Physics of Semiconductor Devices, 2nd ed., John Wiley and Sons, Inc., New York, New York, (1981).
4. J.C. Scully, The Fundamentals of Corrosion, Pergamon Press, New York, New York, (1975).
5. N.D. Tomashov, Theory of Corrosion and Protection of Metals, MacMillan Publishing Co., New York, New York, (1966).

6. K. Barton, Protection Against Atmospheric Corrosion: Theories and Methods, Wiley Interscience, New York, New York, (1976).
7. V.E. Carter, Metallic Coatings for Corrosion Control, Newnes-Butterworths, Boston, Massachusetts, (1977).
8. K. Schneider, R. Bauer and H.W. Grunling, "Corrosion and Failure mechanisms of coatings for gas turbine applications," Thin Solid Films, Vol. 54, 349 (1978).
9. A.T. Fromhold, Theory of Metal Oxidation, Vol. 1--Fundamentals, Series on Defects in Crystalline Solids, Vol. 9, North-Holland Publishing Co., Amsterdam, The Netherlands, (1976).
10. A.T Fromhold, Theory of Metal Oxidation. Vol. 1--Space Charge, Series on Defects on Solids, Vol. 12, North-Holland Publishing Co., Amsterdam, The Netherlands, (1980).
11. A.G. Milnes and D.L. Feucht, Heterojunctions and Metal-Semiconductor Junctions, Academic Press, New York, New York (1972).
12. R.S. Muller, T.I. Kamins, Device Electronics for Integrated Circuits, John Wiley and Sons, Inc., New York, New York (1977).
13. J.M. Andrews and J.C. Phillips, "Chemical bonding and structure of metal-semiconductor interfaces," Phys. Rev. Lett., 35, 56 (1975).
14. L.J. Brillson, "Chemically induced charge redistribution at Al-GaAs interfaces," Phys. Rev. Lett., 42, 397 (1979).
15. G. Ottviani, K.N. Tu and J.W. Mayer, Interfacial reaction and Schottky barrier in metal-silicon system," Phys. Rev. Lett., 44, 284 (1980). (Reference is made to the Proceedings of the Conference on Physics and Chemistry of Semiconductor Interfaces (PCSI) (1983, 1984).
16. A. Goetzberger, E. Kalusmann and M.J. Schulz, "Interface states on semiconductor/insulator surfaces," CRC Critical Rev. in Solid State Sciences, 6:1 (1976).
17. W. Monch, "Role of Virtual Gap States and Defects in Metal-Semiconductor Contacts," Phys. Rev. Lett, 58, 1260 (1987).
18. S. Doniach, K.K. Chin, I. Lindau and W.E. Spicer, Microscopic Metal Clusters and Schottky-Barrier Formation, Phys. Rev. Lett., 58, 591 (1987).
19. R.S. Muller and T.I. Kamins, Device Electronics for Integrated Circuits, John Wiley and Sons, Inc., New York, New York (1977).
20. E.H. Nicollian, B. Schwartz, D.J. Coleman, Jr., R.M. Ryder and J. R. Brews, "Influence of thin oxide layer between metal and semiconductor on Schottky diode behavior," J. Vac. Sci. Technol. 13:1047 (1976).
21. F.C. Jain and J.W. Marciniec, "A single-heterostructure MOS injection laser," IEEE J. Quantum Electron. QE-14; 398 (1978).
22. F.C. Jain and C.S. Nichols, Electronic transport and emission of light in Au-SiO_2-GaAsP interfaces. Bull. Am. Phys. Soc. 25, 586 (1979).
23. M. Froment, Passivity of Metals and Semiconductors," Elsevier, 1983 (Proc. 5th Int. Symp. Passivity, Bombanness, France, May 30-June 3, 1983).
24. J.P. Dodelet, "Characteristics and behavior of electrodeposited surfactant phthalocyanine photovoltaic cells," J. Appl. Phys. 53, 4270 (1982).
25. N. Koshida and Y. Wachi, "Application of ion implanation for doping of polyacetylene films," Appl. Phys. Lett. 45, 436 (1984).
26. Y. Osada and A. Mizumoto, "Preparation and electrical properties of polymeric copper phthalocyanine thin films by plasma polymerication," J. Appl. Phys. 59, 1776 (1986).
27. V.S. Agarwala and F.R. Longo, Proc. Int. Symp. Honoring Dr. Norman Hackerman (October 19-24, 1986, San Diego); Surfaces, Inhibition and Passivation, Eds. E. McCafferty and R.J. Budd, J. Electrochem. Soc., 86-7, 658 (1986).

28. F.C. Jain, "Semiconductor/Insulator Films for Corrosion," Technical Report NADC #8602860, Naval Air Development Center, Warminster, Pennsylvania, October 1985.
29. S. Ashok, P. Sharma and S. Fonash, "Spray-deposited ITO-Si SIS heterojunction solar cells," IEEE Trans. Electron Devices ED-27, 725 (1980).
30. E.L. Jordan, "A diffusion mask of SiO_2," J. Electrochem. Soc., 108, 478 (1961).
31. M.G. Fontana, N.D. Greene, Corrosion Engineering, McGraw-Hill Book Co., New York, New York (1978).
32. R.K. Sadhir and H.E. Saunders, Plasma processes for Electrical and Electronics Applications, IEEE Electrical Insulation Mag., 2, 8 (1986).

ADHESION PROPERTIES OF POLYELECTROLYTE-CHEMISORBED ZINC PHOSPHATE CONVERSION COATINGS

Toshifumi Sugama

Process Sciences Division
Department of Applied Science
Brookhaven National Laboratory
Upton, NY 11973

ABSTRACT

The outermost surface sites of polyelectrolyte complexed crystalline zinc phosphate (Zn·Ph) layers precipitated on cold-rolled steel surfaces act significantly to promote interfacial adhesive bonding to polymeric topcoats. This is the result of electrostatic internal diffusion and segmental chemisorption mechanisms of polyelectrolyte macromolecules either on newly precipitated crystal nuclei or on crystal growth sites during the primary Zn·Ph conversion process. The nature of the polymer-to-polymer chemical bond produced at the polymer-to-polymer complex Zn·Ph adherend interfacial joints plays the key role in the achievement of long-term bond durability upon exposure to chemically corrosive environments.

INTRODUCTION

Large quantities of organic-coated high strength cold-rolled steel plates were presently used by major consuming industries such as automotive and appliance manufacturers. Suppliers and users are becoming more aware that the modification and treatment of steel surfaces prior to application of the organic coatings are most important factors in improving the adhesion and corrosion resistance of the substrate surfaces. At present, the zinc phosphate (Zn·Ph) pretreatment conversion coatings are typically employed in automotive paint systems.

It has been reported[1] that when the cold-rolled steels are immersed in zinc orthophosphate dihydrate-based phosphating solutions, the resultant crystalline Zn·Ph conversion layer deposited on the steel substrate surface consists of a highly dense configuration of large rectangular-like crystals. The surface topographical features of the Zn·Ph conversion precoat were characterized by a dendritic microstructure array comprised of interlocking rectangular crystals of insoluble zinc phosphate hydrate. The open surface structure of the interlocked crystal layers contributed significantly to the formation of a strong mechanical interlocking reaction with the polymeric topcoat systems, thereby enhancing the magnitude of the adhesive force at the precoat/topcoat interfaces. Although the thick Zn•Ph precoat appeared to be suitable as a corrosion barrier for the substrate, the fragile characteristics of this bulky crystal structure lead to failure during flexure or other deformation of the substrate. Deformation failures of layers having low stiffness characteristics appear to be directly related to the development of micropores and fissures which reduce the effectiveness of the corrosion-resistant coatings. These characteristics lead to a high oxygen and moisture availability at the substrate surface and progressively promote a cathodic delamination reaction. The hydroxyl ions generated by the cathodic reactions induce an alkaline condition which causes delamination at the precoat/substrate interface.[2,3] Thus, an increase in the flexural modulus of the crystalline conversion layer itself is of considerable importance when the physical deformation characteristics of the metal substrate are considered.

The brittle characteristics of the Zn•Ph can be modified by the introduction of polyacrylic acid (PAA) macromolecules into the phosphating solutions.[4] The incorporation of PAA, which is generally expressed as a polyelectrolyte macromolecule, was shown to form a highly dense fine crystal topography, suggesting that the crystal formation results in an improvement in the stiffness and ductility characteistics of the normally brittle Zn•Ph layers. The stiff PAA-Zn•Ph complex crystal layer and surface not only provide a corrosion barrier on the substrates, but also possess the ability to promote adhesive bonds at polymeric topcoat-to-complex Zn•Ph adherend joints. The latter was due to the presence of organic functional species at the outermost surface sites of the Zn•Ph films.

Accordingly, the present paper describes the investigation of how the molecular structures and functional species of polyelectrolytes act to suppress crystal growh of Zn•Ph and improve interfacial bond durability of Zn•Ph-to-polymeric adhesive joints.

EXPERIMENTAL

Materials

The metal substrate used was a high strength cold-rolled sheet of steel supplied by the Bethlehem Steel Corporation. The steel contained 0.06 wt% C, 0.6 wt% Mn. 0.6 wt% Si, and 0.07 wt% P. The zinc phosphating liquid consisted of five parts zinc orthophosphate dihydrate and ninety-five parts of 10% H_3PO_4, and was modified by incorporating a water-soluble polymer at concentrations ranging from 0 to 4.0% by weight of the total solution. Four polyelectrolytes obtained from Scientific Polymer Products, Inc. were used; polyacrylic acid [PAA;-[$-CH_2-CH(COOH)-$]-n], polyitaconic acid [PIA;-[$-CH_2C[(COOH)]CH_2OCOH$]-]-n], polystyrenesulfonic acid [PSSA;-[$-CH_2CH(C_6H_4SO_3H)-$]-n], and poly-2-acrylamid-2-methylpropane sulfonic acid [PAMSA;-[$-CH_2C[CONH_2C(CH_3)_2 CH_2SO_3H-$]-n]. For the purpose of comparison with the polyacid macromolecules, two water-soluble amide type polymers were employed, polyacrylamide [PAM;-[$-CH_2-CH(CONH_2)-$]-n] and polyvinylpyrrolidone [PVP;-[$-CH_2CH(C_3H_6CON)-$]-n] also from Scientific Polymer Products, Inc. All of these macromolecules with an average molecular weight in the range of 40,000 to 120,000 were dissolved in water to prepare a 25% polymer solution.

An array of macromolecule-Zn•Ph composite conversion layers having crystal dimensions in the order of microns on the substrate surface was prepared in the following way: The metal substrate first was rinsed with an organic solvent to remove any surface contamination with mill oil. A typical Auger spectrum of the solvent-cleaned steel surface is shown in Fig. 1. The consequent quantitative data indicated that the predominant element at the outermost surface sites of the steel was carbon. Several investigators have reported that the presence of the surface carbon impedes the formation of high-quality Zn·Ph coatings.[5-7] Low zinc and phosphorous levels in Zn·Ph deposited on high carbon areas lead to a porous structure and poor bonding to the steel. These characteristics result in a higher availability of oxygen and moisture at the interface between the Zn·Ph and steel and promote a cathodic delamination reaction. Based upon these features, the quality of the steel surface used in our work upon which the Zn·Ph was deposited, can be categorized as inferior.

After rinsing with the organic solvent, the steel was immersed for up to 20 min in the coating solution described above at a temperature of 80°C. Then, it was placed in an oven at 150°C for 30 min to remove any moisture from the deposited conversion film surface and to solidify the polyelectrolyte macromolecules.

Commercial-grade polyurethane (PU) M313 resin, supplied by the Lord Corporation, was applied as an elastomeric topcoating. The polymerization of PU was carried out by incorporating a 50% aromatic amino curing agent M201. The topcoat system was then cured in an oven at a temperature of 80°C.

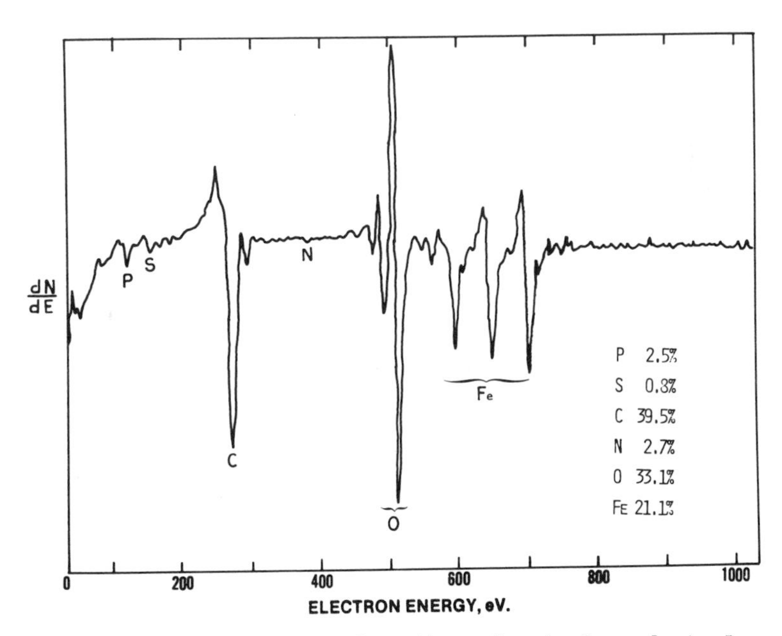

Fig. 1. Auger survey spectrum from the solvent-cleaned steel surface.

Measurements

An image analysis was conducted of the surface microtopography and the surface chemical components of the polyelectrolyte-modified Zn•Ph coatings using AMR 100Å scanning electron microscopy coupled with TN-2000 energy-dispersion X-ray spectrometry.

Auger electron spectroscopy (AES) combined with Ar ion sputter-etching was used in depth-composition profiling studies to detect interdiffusion of elemental composition at internal interfaces in the polyelectrolyte-adsorbed Zn•Ph layers. A Perkin-Elmer PHI Model 610 Scanning Auger Microprobe was used.

Electron spectroscopy for chemical analysis (ESCA) was employed for identifying the chemical states and elemental compositions at the surface sites of PAA-Zn•Ph composite layers. The spectrometer used was a V.G. scientific ESCA 3 MK II. The exciting radiation was provided by a magnesium Kα X-ray source operated at a constant power of 200 W (10 kV, 20 mA). The vacuum in the analyzer chamber of the instrument was maintained at 10^{-9} Torr throughout the experiments.

The Zn•Ph conversion products deposited on the treated metal surface were identified by X-ray powder diffraction analyses. To prepare the fine powder samples, the deposited conversion layers were removed by scraping the surfaces and were then ground to a size of ~325 mesh (0.044 mm).

The electrochemical testing for data on corrosion was performed with an EG and G Princeton Applied Research Model 362-1 Corrosion Measurement System. The electrolyte was a 0.5 M sodium chloride solution made from distilled water and reagent grade salt. The specimen ws mounted in a holder and then inserted into a EG and G Model K47 electrochemical cell. The exposed surface area of the samples was 1.0 cm^2. Prior to immersion of the test specimens the solution was deaerated by bubbling nitrogen through for 30 min. Nitrogen gas then flowed over the top of the NaCl solution containing the specimens for 60 min to stabilize the corrosion potential, before polarization was initiated. The cathodic and anodic polarization curves were determined at a scan rate of 0.5 mV/sec in the corrosion potential range of -1.2 to -0.2 Volts.

A Perkin-Elmer Model 257 spectrometer was employed for specular reflectance infrared (IR) spectroscopic analysis. To explore the interfacial interaction mechanisms at the polyelectrolyte-to-polyurethane joints, we recorded IR spectra from thin film samples overlaid on reflecting aluminum mirror surfaces.

Peel strength tests of adhesive bonds at the polyurethane topcoat-modified metal substrate interfaces were conducted at a separation angle of ~180° and a crosshead speed of 5 cm/min. The test specimens consisted of one piece of flexible polyurethane topcoat, 2.5 cm by 30.5 cm, bonded for 15.2 cm at one end to one piece of flexible or rigid substrate material, 2.5 cm by 20.3 cm, with the unbonded portions of each member being face-to-face. The thickness of the polyurethane topcoat overlaid on the complex crystal surfces was ~0.95 mm.

RESULTS AND DISCUSSION

Surface and Subsurface Characteristics

The surface microtopographies of unmodified and 2.0% polyelectrolyte-modified Zn•Ph conversion coatings deposited on the steel surface were studied by low resolution scanning electron microscopy (SEM). Typical micrographs are shown in Figs. 2 and 3. The images for both the unmodified and modified Zn•Ph crystalline films made after immersion of the steel in the zinc phosphating solution for 20 min, reveal interlocking and dense agglomerates of rectangular-like crystals completely covering the substrate surface. The only microscopically

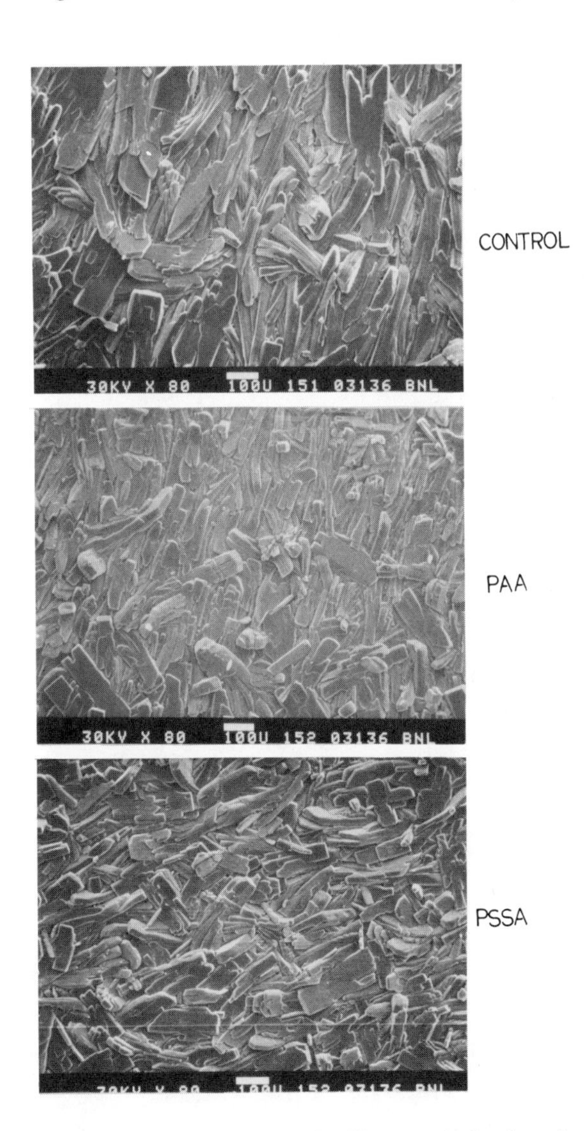

Fig. 2. Alteration in the control Zn.Ph crystal size by PAA and PSSA polyelectrolyte macromolecules.

discernible difference in the morphology of ummodified and modified crystals is a variation in size of the randomly growing crystals. Figure 2 shows that the crystal size of Zn•Ph converted by the PAA and PSSA phosphating solutions is notably smaller than that of the unmodified Zn•Ph. A smaller sized crystal was also precipitated when PIA macromolecules were included in the phosphating solution (not shown). There were no such alterations in the crystal dimensions of PAMSA, PAM, and PVP-modified Zn•Ph surfces (Fig. 3). The decreasing crystal size is believed to be due primarily to segmental chemisorption to the precipitated crystal surfaces of functional electrolyte groups such as carboxylic acid (-COOH) and sulfonic acid ($-SO_3H$).

Fig. 3. Photomicrographs of PAMSA, PAM, PVP-modified Zn·Ph crystal surfaces which suggest that no significant changes in the crystal size occur.

The magnitude of the chemisorption of organic macromolecules on Zn•Ph crystal faces was identified by studying the depth-composition profile in the conversion layers, using Auger electron spectroscopy (AES) in conjunction with Ar ion sputter-etching. The sputter rate for the depth profiling was ~1.0 nm/min. The profile obtained in ~50 min using simultaneous sputtering and analysis permits a rapid identification of layer constituents and structure. The thickness of all of the crystals deposited on the samples tested ranged from ~ 40 to ~10 μm.

Figures 4 through 6 depict the changes in atomic concentration versus sputter time for the unmodified, PAA-, and PAM-modified Zn•Ph layers, respectively. In the unmodified Zn•Ph samples (Fig. 4), the approximate concentrations of the elements occupying the outermost surface sites were 48% oxygen, 23% zinc, 16% carbon, 9% phosphorous, and 4% iron. The signal intensity for oxygen, the major component at the surface sites, was gradually reduced as the sputtering time was increased until the concentration stabilized at a value of 40% after 10 min. Similar trends were observed for the carbon and phosphorous concentrations, but the zinc and iron atom concentrations initially increased with sputtering time and then leveled off at a depth of ~10 nm. This seems to demonstrate the presence of a high quality and less contaminated zinc phosphate hydrate as a major conversion phase in conjunction with phosphophyllite [$Zn_2Fe(PO_4)_2 \cdot 4H_2O$] and iron phosphate [$Fe(H_2PO_4)_2$] as minor phases at depths >10 nm. The presence of surface carbon is due to the carbonates resulting from the adsorption of atmospheric carbon dioxide by residual hydroxide in the crystal. The constant value for carbon of ~4.2% at depths >10 nm may be associated with the contamination of carbon dissociated from the substrate during the conversion reaction processes.

The AES depth profile for the PAA-modified Zn•Ph film (Fig. 5) shows that carbon and oxygen were the predominant elements at depths up to ~5 nm from the surface. These reveal the presence of the PAA polymer and carbonaceous species. The carbon content slowly decreased with elapsed sputter time until it leveled off at 18% at a depth of ~35 nm. Of particular interest was the constant value of carbon reached after a sputtering time of 35 min, that was much higher than the value of contaminated carbon in the unmodified Zn•Ph layers. Therefore, even though the PAA-Zn•Ph composite layer contains a certain amount of carbon contaminant, it appears that the carbon element existing at depths >35nm is due mainly to the PAA macromolecules chemisorbed on the crystal faces throughout the conversion layers. The profiles for other elements such as O, Zn, P and Fe indicate that their concentrations increase monotonically within the first ~35 nm depth and then stabilize. This depth corresponds to the depth at which the carbon stabilized.

Although the actual data are not presented in any of the figures, the variation in atomic concentration obtained from the PIA- and PSSA-modified Zn•Ph conversion surfaces wre quite similar to those obtained from the PAA-Zn•Ph composites. Accordingly, the profiling structure of Zn•Ph layers modified with functional polyelectrolytes such as PIA, PAA and PSSA can be interpreted as follows: (1) highly concentrated polyelectrolyte macromolecules exist throughout the first ~5 nm of the layer, (2) poorly crystallized zinc and iron phosphate formations caused by chemisorption of the abundant polyelectrolyte are present at depths ranging from 5 to 35 nm, and (3) there is a well-crystllized phosphate phase in the presence of a small amount of polyelectrolyte at depths >35 nm below the surface.

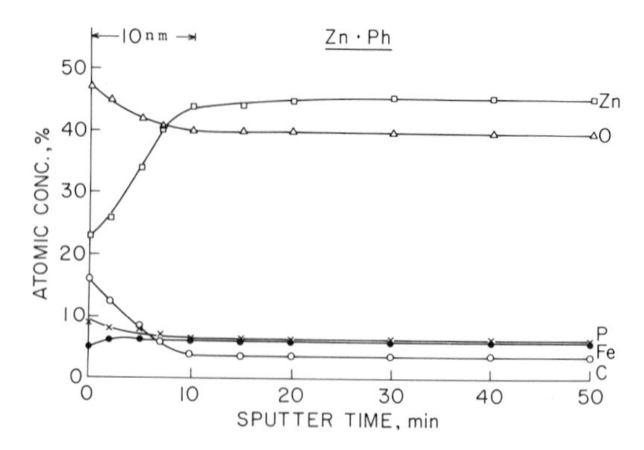

Fig. 4. Auger depth profiles for an unmodified Zn·Ph conversion layer.

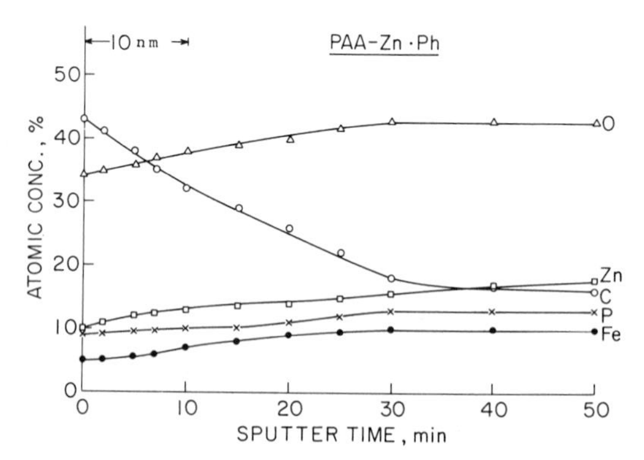

Fig. 5. Auger depth profiles for a PAA-chemisorbed Zn·Ph layer system.

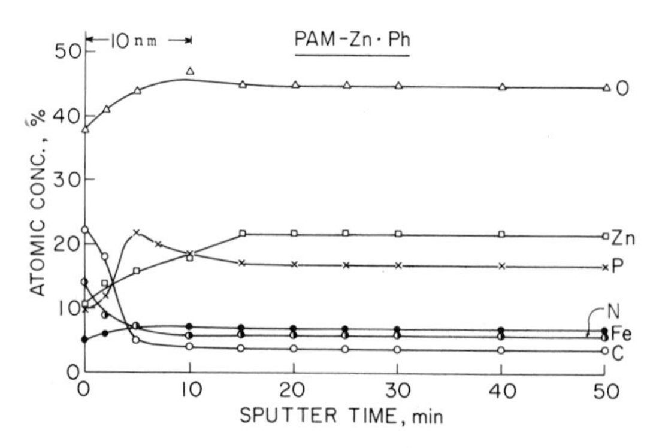

Fig. 6. Auger depth profiles for a Zn·Ph layer system modified with a water-soluble PAM macromolecule.

In contrast, the depth-composition profile of the PAM-Zn·Ph composite illustrates quite different features (Fig. 6). The major difference is the carbon atom profile, in which the carbon concentration at the outermost surface sites is considerably lower than that of the oxygen atom. Furthermore, the carbon concentration is conspicuously reduced with increased sputter time within the first five minutes, but subsequently levels off at a depth of ~10 nm. This concentration, of ~4.5%, is 25% less than that measured in the PAA-Zn·Ph systems, and is almost equal to that found in unmodified Zn·Ph layers. Therefore, it can be deduced that PAM, which is a water-soluble macromolecule, does not strongly chemisorb on the superficial and internal crystal faces. The ~4.5% carbon concentration at depths >10 nm is more likely to be associated with the carbon-contaminated Zn·Ph rather than resulting from chemisorbed organic macromolecules. Since nitrogen can be associated with the presence of pendant-NH_2 groups in PAM molecules, this conclusion is supported by the existence of an exiguous amount of N atoms at depths >10nm. Results similar to those for the PAM-Zn·Ph systems were obtained from an AES study of the Zn·Ph modified with PVP, another water soluble macromolecule. These profiles are not included in this paper. On the other hand, PAMSA macromolecules having two pendant groups, electrolyte COOH and non-electrolyte NH_2 groups, combined to the same backbone carbon atom, can be categorized as semi-polyelectrolyte macromolecules. The resulting profile for the semi-polyelectrolyte-modified Zn·Ph system is depicted in Fig. 7. When compared to those for the PAA- and PAM-Zn·Ph systems, the profiles for O, Zn, P and Fe atoms which represent the conversion crystal formations, exhibit unusual variations in concentration as a function of sputter time. For example, the signal intensities for O and P atoms slowly increase during the first ~5 min sputter time, and subsequently decrease with time. On the other hand, the intensities of the Zn and Fe signals progressively grow over the depth range from 0 to 50 nm. The reason for the uncommon features of the profile of these elements is not clear. The carbon depth profile, which relates directly to the presence of PAMSA macromolecules, is very similar to that for the PAM-Zn·Ph system. Therefore, a semi-electrolyte macromolecule appears to be less susceptible to the chemisorbing activity of the crystal faces.

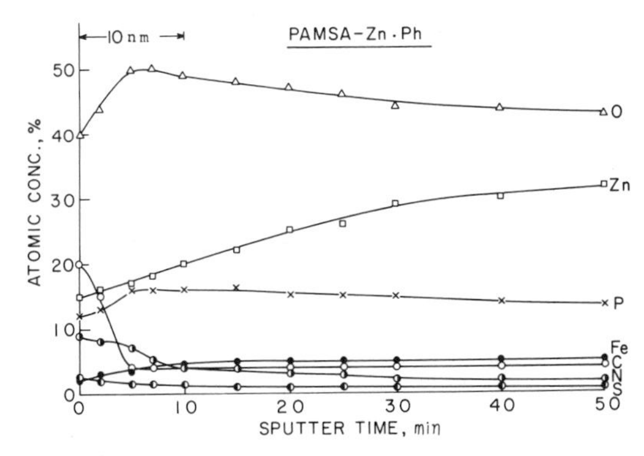

Fig. 7. Auger depth profiles for a semipolyelectrolyte PAMSA-Zn·Ph composite layer.

In conjunction with the results from the SEM surface survey, it was found that the presence of functional carboxylic acid (-COOH) and sulfonic acid ($-SO_3H$) pendant groups in the polyelectrolytes act significantly to suppress crystal growth. This is because of the strong chemisorption behavior of these functional groups on the precipitated crystal faces. In contrast, the magnitude in chemisorbing ability of the amide groups such as $-CONH_2$ and -CO-N is apparently very small.

The ESCA survey scan to identify elements at the outermost surface sites of PAA-modified Zn•Ph is shown in Fig. 8. Carbon oxygen atoms appear as major lines, and zinc and phosphorous as minor ones. The chemical states of Zn and P are assigned to the presence of the zinc and phosphorous compounds, and the C is attributed to the organic PAA macromolecules. The O atom is related to both the organic compounds and the polyelectrolyte. This suggested that certain areas of the Zn•Ph film surfaces were covered with a thin PAA layer no thicker than 5 nm, which corresponds to the escape depth of the magnesium photoelectron. Such areas probably represent the interface regions which are of primary interest in this study.

The C_{1s} spectrum of the PAA at the surface and interface is given in Fig. 9. The individual carbon components are indicated by the dashed lines. By comparison with the spectrum of the bulk PAA (not shown), the C_{1s} spectrum was characterized by shifting the C=O and COOH group signals to low BE sites. Other peaks on the spectrum can be ascribed to $-CH_2-$ at 284.6 eV, and C–O at 286.6 eV, both quite similar BE values to those obtained from the bulk PAA surface.

Of particular interest was focused on the decrease in C=O and COOH BEs of 0.4 eV. The shifts seem to verify that the possible two formations, a salt complex containing COO^- Zn^{2+} groups and a chemisorption of ionized PAA segments on the positively charged crystal surfaces, are present at the surfaces and interfaces. The former could be yielded by ionic interaction between carboxylate anions formed by

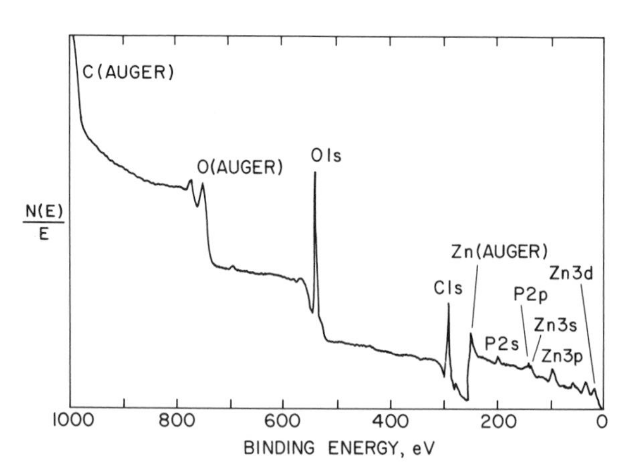

Fig. 8. ESCA survey spectrum taken from a PAA-complexed Zn.Ph conversion precoating surface.

the proton donor characteristics of the COOH groups and the active nucleophilic Zn ions in the zinc phosphating solutions. The complex formation would be produced during the conversion process of Zn·Ph overlaid on the steel substrates.

From the above results, the outermost surface appears to be composed of the Zn·Ph hydrate and the organic functional species such as ionic carboxylate and carboxylic acid groups.

On the basis of the assumption that the PAA-chemisorbed crystal conversion layer is composed of the hybrid phases of zinc phosphate, iron phosphate, and mixed iron and zinc phosphate compounds, both the

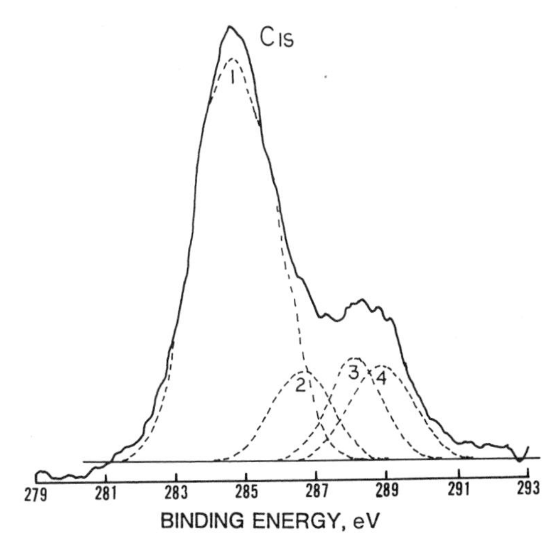

Fig. 9. Cls spectrum of PAA at surface and interface: (1) $-CH_2-$ at 284.6 eV, (2) C-O at 286.6 eV, (3) C=O at 288.1 eV, and (4) COOH at 288.9 eV.

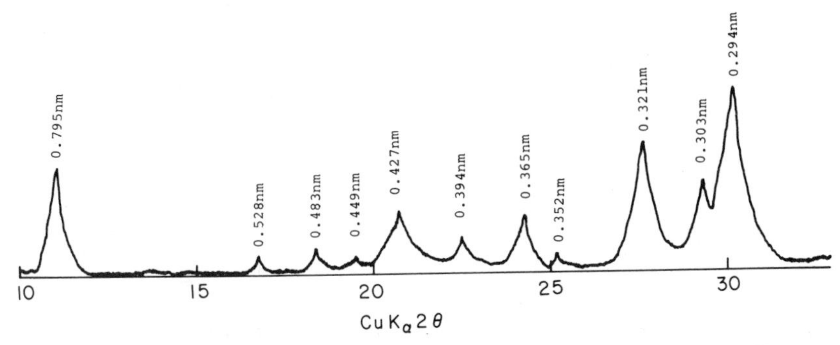

Fig. 10. Powder X-ray diffraction pattern of PAA-treated Zn·Ph crystal layer.

relative quantities and the identification of these conversion products were studied with X-ray powder diffraction (XRD). The resultant XRD pattern in the diffraction range of 0.883 to 0.167 nm for the 150°C-dried samples of a layer removed by scraping the surface is given in Fig. 10. These spacings suggest that the major conversion product is zinc phosphate dihydrate, $Zn_3(PO_4)_2 \cdot 2H_2O$.[8] The presence of hopeite, $Zn_3(PO_4)_2 \cdot 4H_2O$, phosphophyllite, and iron phosphate, which are reported by Ghali and Potvin,[2] could not be clearly identified in the layers derived from the zinc orthophosphate dihydrate-based phosphating solution. However, using AES, the presence of very small amounts of iron phosphate and phosphophyllite were found in the crystalline precoat layers.

On the basis of these results obtained using various analytical techniques, possible conversion mechanisms for polyelectrolyte-chemisorbed Zn·Ph crystal layers are presented schematically in Fig. 11. As depicted in the figure, the proposed mechanism consists of four steps. In step A, when the PAA electrolyte is introduced into the zinc phosphating liquid system, the divalent zinc cations are readily complexed with the proton-donating-type carboxylic acid groups in the PAA molecules. The extent of the complex is dependent upon the charge density of the polymer and the concentration of the counterion.[9] It must, therefore, be assumed in the present work that the PAA species are partly dissociated PAA molecules to which a number of Zn ions are bound. This amount of ion binding should permit extensive random coiling of polyanions. Upon immersion of the metal substrate in the acid medium, the substrate surfaces undergo an electrochemical dissolution reaction. The ferrous ions that migrated electrochemically from the substrate surface at the steel-solution interfaces which have a strong chemical attraction with the orthophosphoric acid. This reaction is followed by the precipitation of a thin layer of the first conversion product in terms of iron phosphate, $Fe(H_2PO_4)_2$.[2] The surface of the precipitated iron phosphate is charged positively because of the large quantity amount of Fe^{2+} ions adjacent to the iron phosphate cores. The surface cationic strength might be strong enough to chemisorb the polyelectrolyte in preference to that of Zn^{2+} ions (see stage B). It is, therefore, postulated that the most logical surface sites to electrostatically attack the ionized COOH groups would be Fe ions on the precipitated iron compound surfaces. As a result of chemisorption of ionized PAA segments, some Zn cations masked with PAA macromolecules are released from the PAA-Zn complex formations, and then the dissociated Zn^{2+} ions react with the $2H_3PO_4$ to form zinc phosphate, $Zn(H_2PO_4)_2$. Crystalline zinc iron phosphate hydrate, in terms of phosphophyllite, $Zn_2Fe(PO_4)_2 . 4H_2O$, is probably derived from a reaction between the $Zn(H_2PO_4)_2$ and $Fe(H_2PO_4)_2$ in the aqueous media.[10] Even though the newly precipitated phosphophyllite is superimposed on the PAA-chemisorbed iron phosphate, the segmental chemisorption to the precipitated crystal surfaces by PAA will occur consecutively, as illustrated in step C. As is evident from the AES and XRD studies, the amount of these iron-rich phosphates, which are precipitated at the beginning of the treatment, is much smaller than the zinc phosphate dihydrate, $Zn_3(PO_4)_2 \cdot 2H_2O$, which is the major conversion product of the precoat layers. During the final period depicted in step D, the residual zinc phosphate is converted into zinc phosphate dihydrate.

From the viewpoint of the polyelectrolyte chemisorption process, the strongly ionized segments of PAA are electrostatically diffused to the positive surface sites of zinc phosphate dihydrate embryos at the beginning of the precipitation. There is no doubt that the high

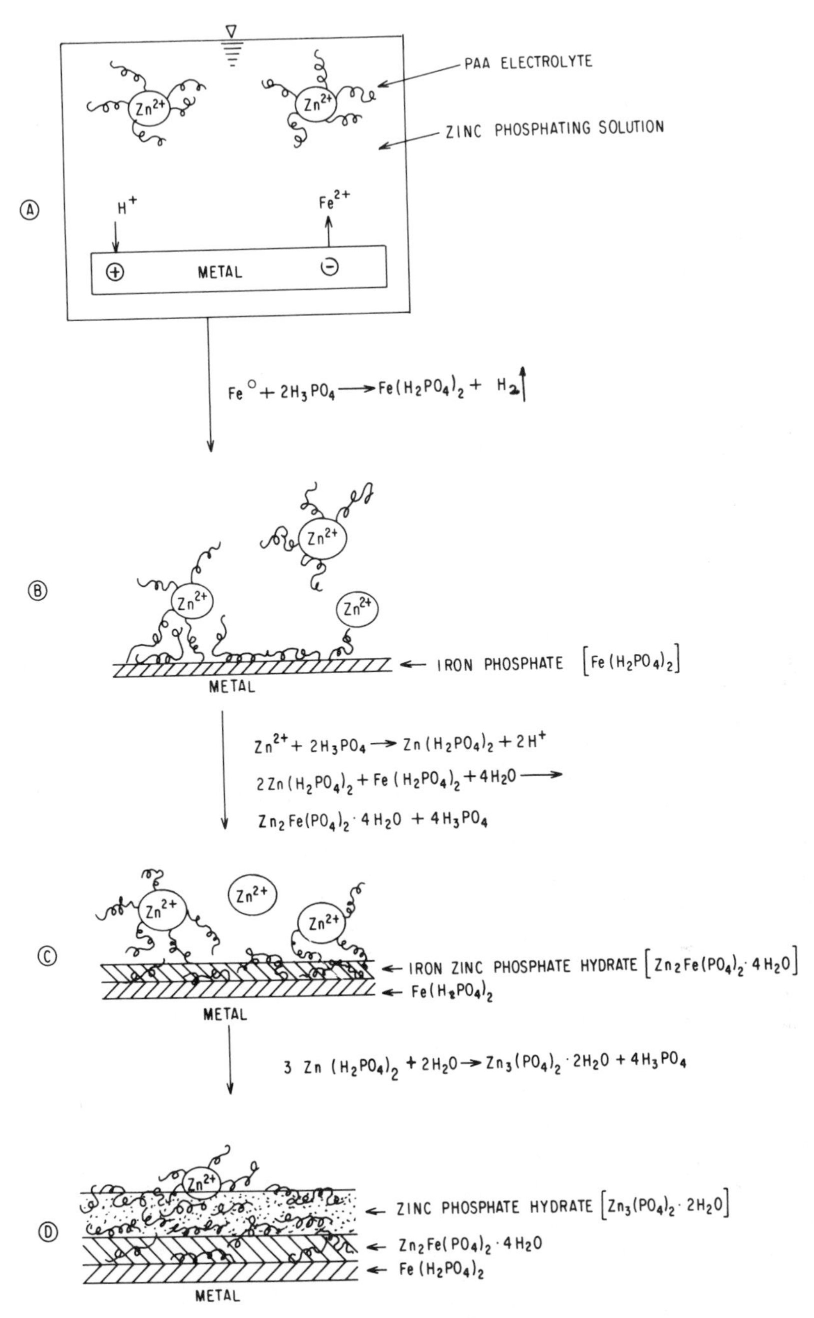

Fig. 11. Schematic conversion mechanism of polyelectrolyte-zinc phosphate systems deposited on steel surfaces.

efficiency of segmental chemisorption of organic macromolecules either on newly precipitated nuclei or on growth sites occurring during the primary crystallization processes acts significantly to inhibit the rate of crystal growth. The extent of chemisorption is increased with increasing molecular weight of the PAA. This is related directly to the decrease in size of the crystal or embryonic crystal. The precipitated crystal morphology and habit are also markedly transformed by a strong chemisorption of PAA with an appropriate molecular weight. However, as reported previously,[1] when a molecular weight >250,000 was used, the formed crystals were much smaller than those produced using a lower-molecular-weight PAA. The incorporation of extremely high-molecular-weight PAA completely suppressed the crystal growth by strongly chemisorbing the PAA on the embryonic crystal faces. An alternative explanation for the effect of molecular weight on the crystllization rate is that the high magnitude of entanglement in the PAA-Zn complex systems results in suppression of the electrostatic diffusion of segments to the cathodic crystal nuclei surfaces. In other words, it is very difficult for the Zn^{2+} ions existing in a tightly coiled configuration of PAA chains to migrate as free cations from the complex formations. This phenomenon can be expressed in terms of shielding effects which should occur at the strong ionic attractions of PAA with Zn^{2+}. Thus, it is rationalized that the presence of PAA having a particular molecular weight which is representative of less chain entanglement, contributes to the great accessibility of segmental diffusion to the crystal surface sites.

Corrosion Control

To gain information on the effectiveness of the Zn•Ph and PAA macromolecule-chemisorbed Zn•Ph conversion coatings as protective layers against corrosion for cold-rolled steel, we examined the electrochemical polarization behavior of the steel coated with Zn•Ph and PAA-Zn•Ph conversion films. Figure 12 shows typical polarization curves of log current density vs potential for a plain steel (blank), $Zn{\cdot}Ph^-$, and modified Zn•Ph-coated steels. The shape of the curves presents the transition from cathodic polarization at an onset of the most negative potential to the anodic polarization curves at an end of lower negative potential. The potential axis at the transition point from cathodic to anodic curves is normalized as the corrosion potential. The important features in a comparison of the cathodic polarization curves from Zn•Ph-coated and uncoated steel specimens are:

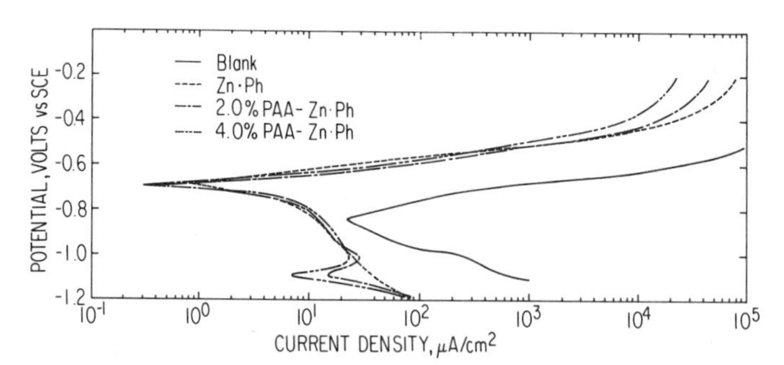

Fig. 12. Polarization curves for untreated Zn•P hand PAA-Zn•Ph-coated steel immersed in a 0.5 M NaCl solution.

(1) in the vicinity of the corrosion potential the relative current density of the blank was considerably reduced (approximately by one order of magnitude) by the overlaid Zn·Ph conversion coating, (2) the short-term steady-state current value for the Zn·Ph specimens was significantly less than that for the blank in the potential region between -1.1 and -1.0 V, thereby suggesting that the evolution of hydrogen is greatly inhibited on Zn·Ph-coated steels, and (3) a large reduction in corrosion potential to less negative potentials was achieved with the Zn·Ph specimens. It is concluded that the rate of corrosion of steel in a deaerated NaCl solution is significantly reduced by the presence of the Zn·Ph.

On the other hand, the differences between the curves (Fig. 12) for the PAA-modified Zn·Ph and the unmodified Zn·Ph specimens are: (1) the cathodic peak at -1.09 V grows with increased PAA concentration, (2) the current density of Zn·Ph at the corrosion potential is reduced by incorporating PAA macromolecules, and (3) the addition of PAA to Zn·Ph greatly reduces the current density in the anodic potential range from -0.3 to -0.2 V. The cathodic peak at -1.09 V (result No. 1, above) may be attributed to the presence of a less stable PAA complex film at the outermost surface sites. This unstable layer may be asociated with the hydrolytic transformation from a PAA-solid to gel state. A possible

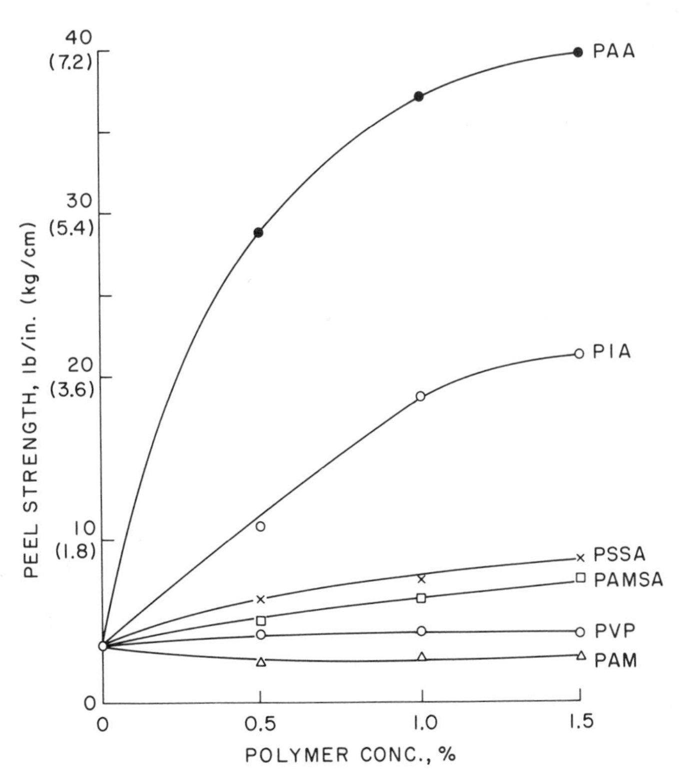

Fig. 13. Changes in the 180°-peel strength at PU-to-modified Zn·Ph joints as a function of polyelectrolyte and non-polyelectrolyte macromolecule concentrations.

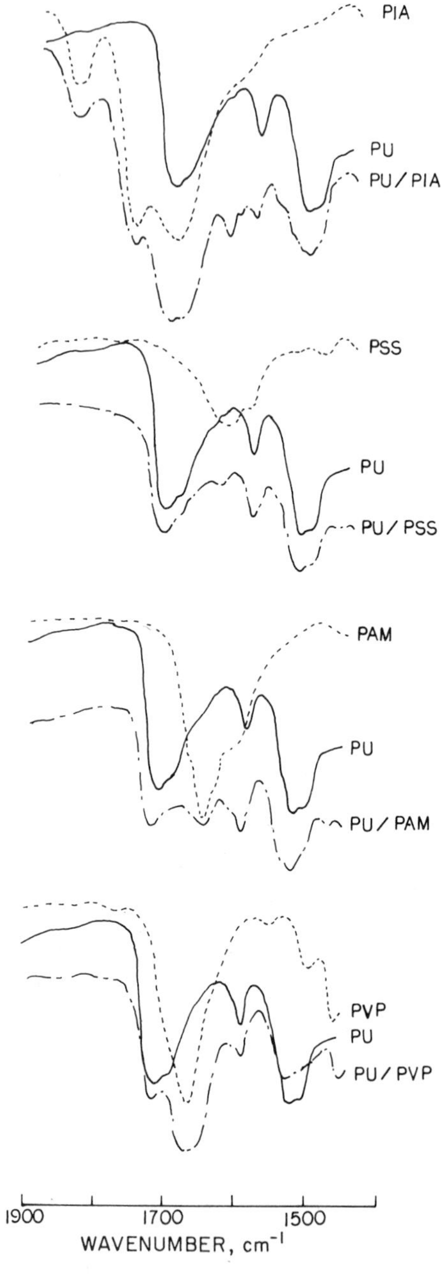

Fig. 14. Specular reflectance IR spectra at the interface of Pu/PIA, PSSA, PAM and PVP joints.

interpretation for the second result is that the presence of PAA macromolecule chemisorbed on the Zn·Ph crystal faces produce a less effective catalyst for the cathodic reaction at a temperature of 25°C. The third result suggests that a more readily passivated surface of Zn·Ph can be formed by PAA. Therefore, the PAA macromolecules appear to be an effective admixture to further enhance of the corrosion resistance of Zn·Ph itself at 25°C.

Adhesion

The presence of the functional organic groups at the outermost surface sites of the conversion coatings serves to promote the adhesive force between the organic polymer topcoat and the crystal precoat. Therefore, the adhesive ability of the polymer-Zn·Ph composite surface was evaluated from the 180°-peel strength of polyurethane (PU) topcoat films overlaid on the composite surfaces. In Fig. 13, test results are given that show the variations in peel strength at PU-to-PAA, PIA, PSSA, PAMSA, PVP, and PAM-modified and unmodified Zn·Ph interfacial joints. The degree of improvement in adhesive bonds at PU-modified Zn·Ph interfaces appears to depend mainly on the species of the functional groups existing at the surface sites of the Zn·Ph. The PU/PAA and PIA/Zn·Ph joint systems exhibited a great enhancement of peel strength as the concentration of these polyelectrolytes was increased.

In this series, the highest strength value of 7.1 kg/cm was attained at the adhesive joint between the PU and the 1.5% PAA-Zn·Ph composite. This superior bond force corresponds to an improvement of 11 times over that measured for the PU-to-unmodified Zn·Ph joints. The intrinsic adhesion observed at PU-to-single Zn·Ph joints is attributed to a mechanical interlocking bond that anchors the polymer as a result of the penetration of PU liquid resin into the open spaces and surroundings of the rectangular-like Zn·Ph crystals. The PIA-Zn·Ph composites also displayed a highly reactive surface, although the maximum peel strength was only ~50% of that from the PAA-Zn·Ph. The modification of Zn·Ph by PSSA and PAMSA macromolecules resulted in considerably smaller increases in peel strength. In contrast, the interfacial bond strengths for the surfaces of PVP- and PAM- Zn·Ph composite systems were not enhanced by the addition of these macromolecules.

It is useful to assess the interfacial bond mechanism responsible for the great enhancement in polymer-polymer adhesion. To gain this information, specular reflectance IR analyses were made. The samples were prepared by spin-casting the PU resin onto polyelectrolyte and nonpolyelectrolyte-coated aluminum mirrors at 4000 rpm. The PU film with $\sim 10^4$ nm thickness overlaid on the films. The resultant IR spectra, shown in Fig. 14, were recorded in the frequency range of 1900 to 1400 cm^{-1}. The spectrum from the PU-coated aluminum mirror in the absence of polyelectrolyte exhibited absorption bands at about 1715, 1590 and 1530 cm^{-1}. These indicate the C=O in amide I, the phenyl and the C=O in amide II, respectively. Interestingly spectra obtained from samples having a PIA layer showed the formation of new bands at 1680 and 1630 cm^{-1} which can be assigned to newly formed amide I and II groups. Although not shown, the intensity of this new peak rises conspicuously as the PIA concentration is increased, whereas that of the isocyanate, -N=C=O, at 2260 cm^{-1} tends to decrease. This fact proves that isocyanate groups in PU polymer adjacent to the PIA macromolecules can form chemical bonds with the carboxylic acid of PIA. This interfacial chemical reaction of isocyanate with carboxylic acid induces the formation of the amide I and II.

An IR spectrum quite similar to that for the PU/PIA interface was obtained for the PU/PAA boundary regions. Therefore, it is believed that the chemical intermolecular reaction is a very important mechanism which greatly enhances the adhesive strength between PU and PAA or PIA having a carboxylic acid side group. On the other hand, no obvious signs of chemical intermolecular reactions were identified on the IR spectra at the interfce of PU/PSSA, PAM, and PVP joints. The spectra do not show the growth of any new bands and nor any marked shifting of frequencies in the range of 1900 to 1400 cm^{-1}. Thus, the reason for the slight increase in peel strength of the PSSA-modified system may be the result of interdiffusion of PU molecules in the absence of strong chemical interactions.

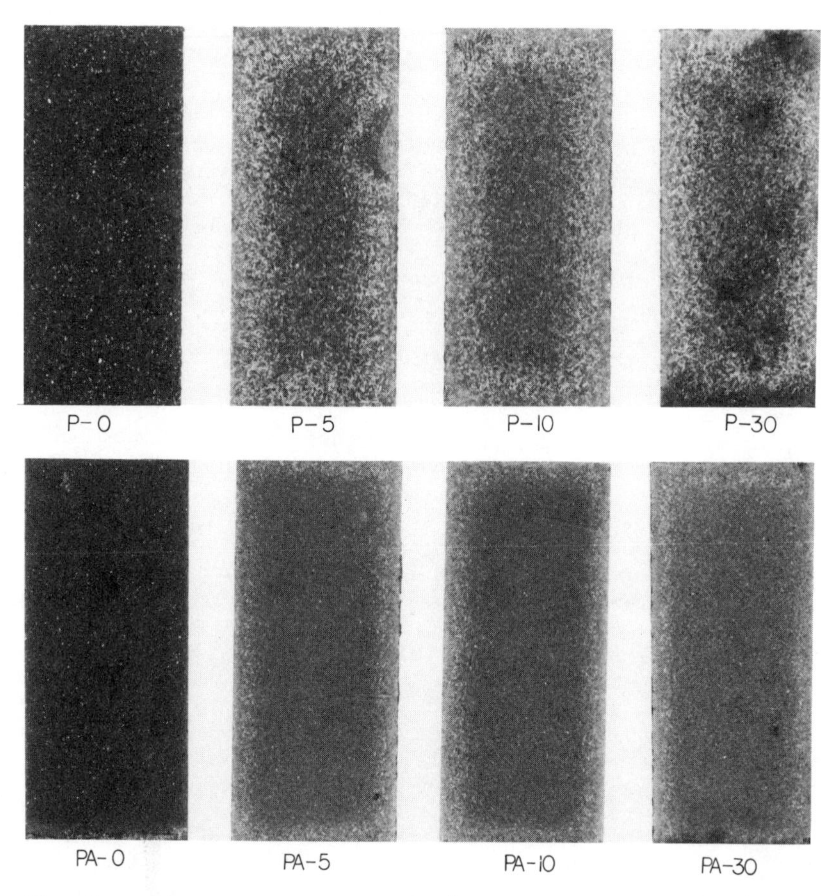

Fig. 15. Comparison of unmodified (P-) and PAA-modified (PA-) precoat surface sides after peeling the topcoat from specimens after exposure to 0.1 M H_2SO_4 at 80°C.

Durability of Adhesive Bond

In order to assess how the surface nature and structure of PAA-chemisorbed Zn•Ph crystal layers affect the durability of adhesive bonds at crystalline precoat-to-polyurethane (PU) topcoat interfacial joints, silica-filled PU polymer was overlaid onto the crystal-precoated steel substrates, and then the PU-coated specimens were subjected to a 0.1 M H_2SO_4 solution at 80°C. The PU films exhibit excellent chemical resistance to the hot acidic solution, but the conventional Zn•Ph layers are rapidly attacked. All of the edges on the PU-coated plate specimens used in the tests to estimate the bond durability of the PU-to-Zn•Ph adhesive joints were unprotected.

Figure 15 shows comparisons between unmodified and PAA-modified Zn•Ph surface sites after PU topcoats were peeled from the interfaces of samples that had been exposed to hot acid solutions for up to 30 days. In Fig. 15, the unmodified and modified Zn•Ph surfaces are designated as samples No. P and PA, respectively, and the exposure periods in days are signified by the numbers which follow the identification letters. For instance, P-30 corresponds to an unmodified specimen after exposure for 30 days. As is evident from the photographs, the extent of the failure of the interfacial adhesion structure can be visually confirmed from the changes in color of the crystals on the edges of the precoat sites. For the unmodified Zn•Ph surfaces, the dark areas that represent adhesively bonded interfaces are reduced markedly by extended times. It appears that the chemical degradation of the Zn•Ph crystal itself, brought about by penetration of the acid solution to the interfacial regions, was progressively promoted as the exposure period increased. On the other hand, the degradation failure for the PAA-chemisorbed Zn•Ph precoat surfaces after peeling the PU films was considerbly less. Although the 30-day exposed specimen exhibits some color change at the edges of the substrate surfaces, the interfacial bonding areas were conspicuously larger than those of the unmodified Zn•Ph surface sites at the same exposure ages. These results apparently demonstrate that complex conversion layers derived from a mix solution of zinc phosphate and polyelectrolyte have a high potential for improving the bond durability of polymer-to-metal adhesive joints.

In the crystalline precoat-to-polymer topcoat adhesion studies, identification of the failure mode can lead to a better understanding of the mechanisms contributing to strong bonds. To obtain information regarding the adhesion failure mode and locus at the topcoat/precoat boundary for the specimens before and after exposure to hot acid solutions, the failure surfaces generated by peeling were explored using scanning electron microscopy (SEM) and energy dispersive X-ray spectrometry (EDX). EDX coupled with SEM has a high potential for the quantitative analysis of any selected elements which exist at solid composite material subsurfaces. This feature can greatly enhance the results, as well as facilitate the interpretation of SEM studies. The adhesive PU polymer surface sites of the peeled specimens were inspected before and after exposure for 30 days to the acid environment.

Figure 16 shows SEM microprobe and EDX elemental analysis of a PU surface that was removed from an unmodified Zn•Ph surface prior to acid exposure. The SEM image is characterized by a concave structure with pits and cavities on the interfacial surfaces. The presence of a large number of rectangular-shaped pits can be explained in terms of the adhesive failure mechanism. The adhesion loss in this bonding system probably occurs near the precoat-to-topcoat interfaces. This failure mode is representative of the weak adhesive bonding at the PU/Zn·Ph interfacial joints. In the EDX spectrum, the abscissa is the X-ray energy characteristic of the elements present, and the intensity of a peak is related to the amount of each element present. EDX data to substantiate the failure locus revealed that the predominant element, which is represented by the highest peak intensity, is silica. Other elements present, but with low peak intensities, are P, Zn and Fe atoms. The Si atom is associated with the silica flour used as a filler in the PU polymer. The notable P, Zn and Fe elements probably come from the zinc- and iron-based phosphate crystal compounds. The presence of these elements indicates a cohesive failure occurring in the precoat subsurface. An appreciable amount of the cohesively failed precoat is, therefore, present on the peeled PU film surfaces.

The magnitude of cohesive failure in the precoat layers can be estimated by a quantitative analysis of the selected elements. The quantitative data involve the intensity (counts/50 sec) of P and Zn atoms and the element ratio of P-to-Zn peak counts. As seen in Fig. 16, the resultant intensity counts for P and Zn atoms were 218 and 243, respectively, and the intensity ratio of P/Zn was 0.90. These values were found to be much lower than those for the adherend surface sites of the precoat. From the above results, the failure occurring at PU/Zn·Ph joints is probably through a mixed-mode of cohesive (in the precoat) and adhesive failures. In contrast, at the PU-to-PAA-complexed

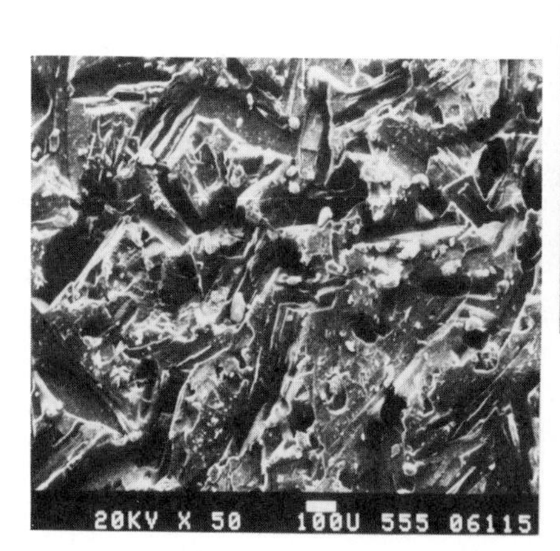

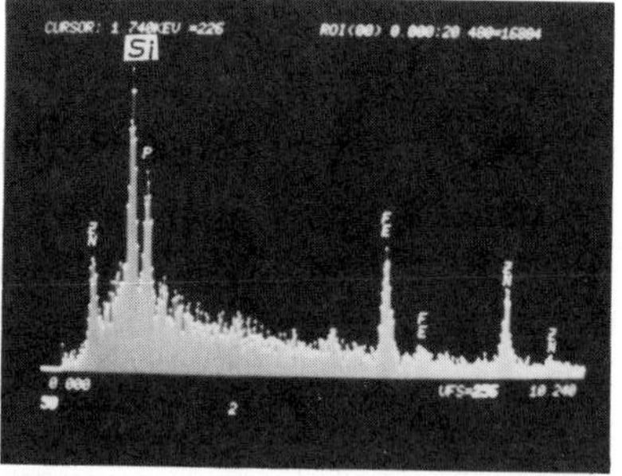

Elements	Intensity counts/50 sec.	Intensity ratio/Zn
P	218	0.90
Zn	243	1.00

Fig. 16. SEM micrograph and EDX of surface side of PU topcoat bonded to Zn·Ph surface before exposure to hot acidic solution.

Zn·Ph joints, the topographical feature of the interfacial PU adhesive sites for the peeled unexposed specimens, was microscopically observed to present a quite different image, as shown in Fig. 17. A large size pit with a rectangular-like shape, which was detected on the PU surface bonded to the Zn·Ph, could not be confirmed on the topcoat sites removed from the PAA-modified Zn·Ph precoat surfaces. EDX peaks indicate that the major chemical constituents on the polymer surface are more likely to be associated with the presence of P, Fe and Zn, rather than that of Si. In support of these facts, the intensity counts for both P and Zn atoms and the P/Zn intensity ratio were found to be considerably higher than those obtained from the interfacial topcoat in the PU-Zn·Ph joint systems. It is of particular interest that the ratio value of 2.16 was very close to that computed for the complex precoat surface after peeling the polymer topcoat. These experimental data, in conjunction with the SEM image of the rough surface texture for the peeled precoat interfaces, indicate that the locus of the adhesion loss is clearly cohesive in the conversion coat layer as a result of a considerable overlayer of the crystal precoat remaining on the PU adhesive surfaces. Hence, the presence of the functional organic species, which occupy the most active sites on the precoat, appears to play a major role in promoting a good chemical bond with the polymer topcoat systems. The observed defective mode, which breaks down the crystal precoat system, suggests tht the intrinsic strength of the crystal itself is less than that of the adhesion force at the topcoat/precoat joints. The formation of stronger conversion crystals, therefore, is requisite for producing further improvements in the interfacial adhesion bonds.

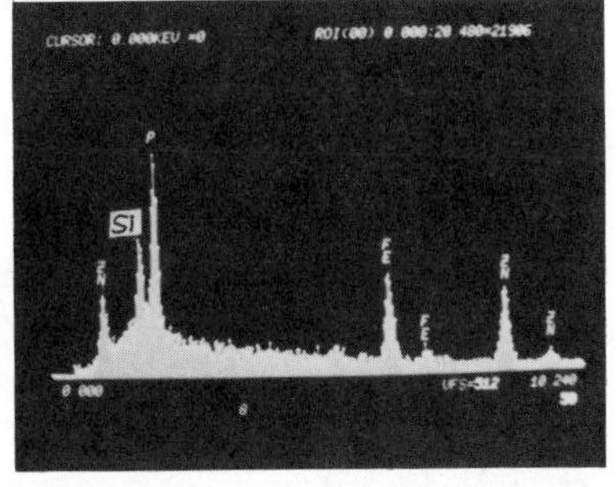

Elements	Intensity counts/50 sec.	Intensity ratio/Zn
P	803	2.16
Zn	371	1.00

Fig. 17. Peculiarity of adhesive side of unexposed PU film removed from the PAA-completed Zn·Ph surface.

For the PU/Zn·Ph joint specimens exposure for 30 days, SEM micrographs and EDX analyses at the edge and center portions of the adhesive PU interfacial surface are given in Fig. 18. As shown in the figure, the microstructure feature disclosed on the edge area of the PU surface sites (Fig. 18-A) is appreciably different from that at the center area (Fig. 18-B). The principal difference is that the number of rectangular-like shaped pits at the edge is less than those in the center view. The micromorphological image from the center area resembles the surface microstructure of the adhesive sites for the unexposed specimens discussed previously. Although not shown in the figure, the EDX spectrum for the edge indicated that the dominant element present is P rather than Si.

A

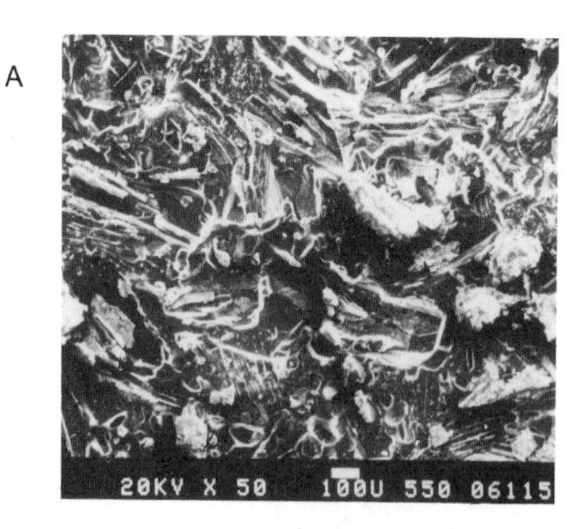

Elements	Intensity counts/50 sec.	Intensity ratio/Zn
P	549	2.09
Zn	263	1.00

B

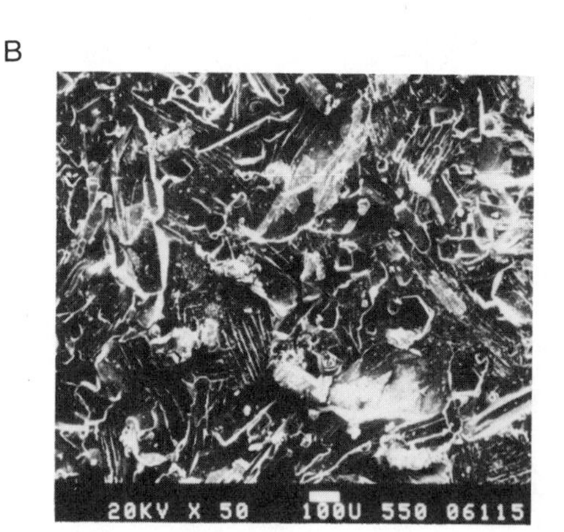

Elements	Intensity counts/50 sec.	Intensity ratio/Zn
P	394	1.71
Zn	230	1.00

Fig. 18. Edge (A) and center (B) portions of PU film site removed from Zn·Ph precoat surface after immersion for 30 days in acidic solution at 80°C.

It is inferred from these data that some crystal precoat remains on the PU interface after peeling. The presence of the joined pieces of the crystal was verified by comparing the elemental quantitative values obtained at different locations of the interfacial polymer surface sites. The resultant P and Zn intensity counts for the edge samples were found to be significantly higher than those for the center portions. This means that the quantity of Zn•Ph crystal, which is transferred to the polymer adhesive sites from the edge region of the precoat interface, is relatively larger than that transferred from the center areas. The reason for the large crystal pieces being left on the polymer interfaces may be as follows: when the PU-to-Zn•Ph adhesion joints come in contact with the hot acid solutions, the corrosive acid fluid readily penetrates through the weak bonding regions at the interfaces. The penetrated fluid reacts particularly with the Zn•Ph layers, whereas the PU topcoat at the interface is durable to acidic chemicals. The significant progression of the reaction leads to the chemical dissolution of the Zn•Ph and the locus of the mechanical disbondment occurs within the highly dissolved precoat layers of Zn•Ph adjacent to the topcoat. This results in the presence of the residual crystal on the interfacial PU film surfaces. Since the concept of dissolution failure within the precoating is related directly to the magnitude of susceptibility of the crystal formation to the chemically

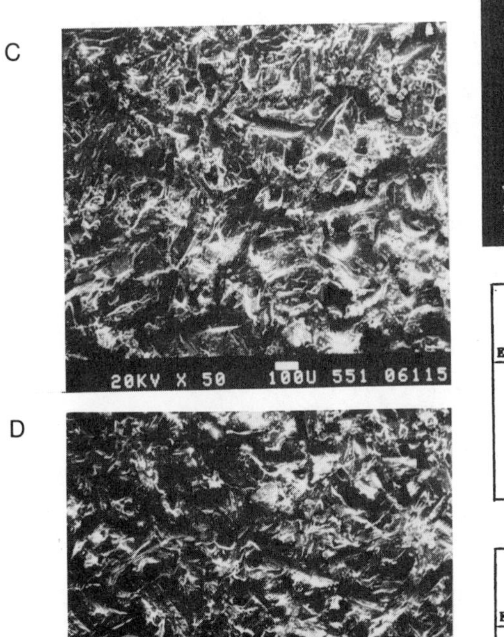

Elements	Intensity counts/50 sec.	Intensity ratio/Zn
P	644	2.09
Zn	308	1.00

Elements	Intensity counts/50 sec.	Intensity ratio/Zn
P	745	2.38
Zn	313	1.00

Fig. 19. Edge (C) and center (D) parts of PU surface site bonded to PAA-Zn•Ph complexes after 30-day exposure.

aggressive fluid, the unmodified Zn·Ph appears to be soluble in the hot acidic solution. The dissolution-induced precoat failure beneath the topcoat seems to be the major factor in the chemical delamination process.

Figure 19 illustrates SEM images and EDX analyses taken on the edge and the center part of the PU interfaces peeled from PU-to-PAA-complexed Zn·Ph joints that were exposed to the hot acid solution for 30 days. The topographical features of the surfaces at either the edge (Fig. 19-C) or the center (Fig. 19-D) had a strong resemblance to that of the PU removed from the unexposed specimens. However, the intensity counts of both P and Zn at the edges were somewhat lower than those at the center. These counts at the center portions are almost equivalent to those of the peeled PU interface for the unexposed specimens (Fig. 17). Hence, it appears that even though a high degree of interfacial bond was developed at the PU/PAA-Zn·Ph interfaces, the precoat layers near the edges might be slightly affected by the acid solutions, thereby inducing an appreciable adhesion loss at the joints. From the above results, the adhesion failure mode and locus at the interfacial boundary for the peeled specimens before and after exposure, were consequently proposed in Fig. 20. As seen in the figure, the failure at the PU/Zn·Ph joint before exposure was through a mixed-mode of adhesive and (in the precoat) failures. In contrast, the locus of adhesive loss of PU/PAA-Zn·Ph joint systems was identified to be clearly cohesive in the precoat, thereby increasing the interfacial bond strength. For the specimens after 30-day exposure, the adhesive loss for the PU/Zn·Ph systems was due mainly to the dissolution of the Zn·Ph layer adjacent to the interface by the penetration of acid solutions. However, the failure mode for the exposed PU/PAA-Zn·Ph was confirmed to be cohesive, similar to that for the unexposed specimens. This suggests that the hydrophobic structure formed by the interfacial chemical reaction between the PU and the functional organic species plays an important role in achieving a long-term bond durability at metal adhesive joints that will be subjected to chemically corrosive environments.

CONCLUSIONS

In deposition processes for polyelectrolyte-Zn·Ph complex conversion precoats on high-strength cold-rolled steel surfaces, the positive surface sites of multiple phosphate crystal embryos at the beginning of the precipitation were strongly chemisorbed by the anionically charged segments of polyelectrolytes having functional carboxylic acid and sulfonic acid pendant groups. The electrostatically segmental chemisorption of polyanions either on newly precipitated nuclei or on growth sites during the primary crystallization processes acts to suppress and delay the crystal growth. In contrast, the magnitude of susceptibility of pendant amide groups to chemisorption on crystal growth sites was very small.

The major conversion product in the well-formed crystals was identified to be the zinc phosphate dihydrate responsible for the corrosion-inhibiting ability of the precipitated layers. Comparisons between electrochemical polarization curves for untreated steel and Zn·Ph-deposited steel indicate that the corrosion rate in a deaerated NaCl solution was reduced considerably by the presence of dense Zn·Ph crystal arrays derived from a zinc orthophosphate dihydrate-based phosphate solution. The presence of polyelectrolyte macromolecules in the precipitated crystal layers act to further enhance the corrosion resistance at 25°C of Zn·Ph itself.

BEFORE EXPOSURE TO HOT ACID SOLUTION

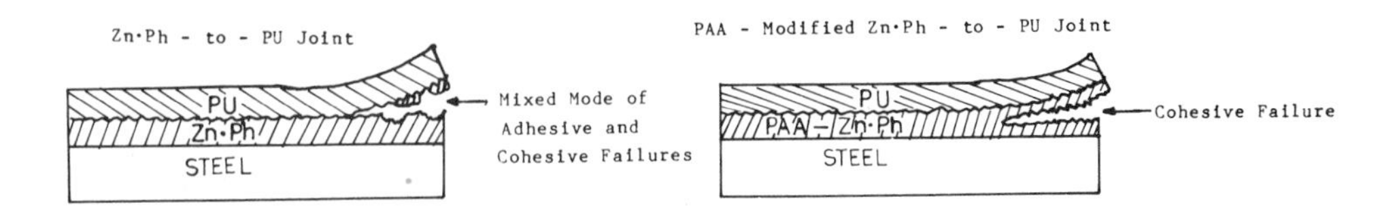

AFTER EXPOSURE TO HOT ACID SOLUTION

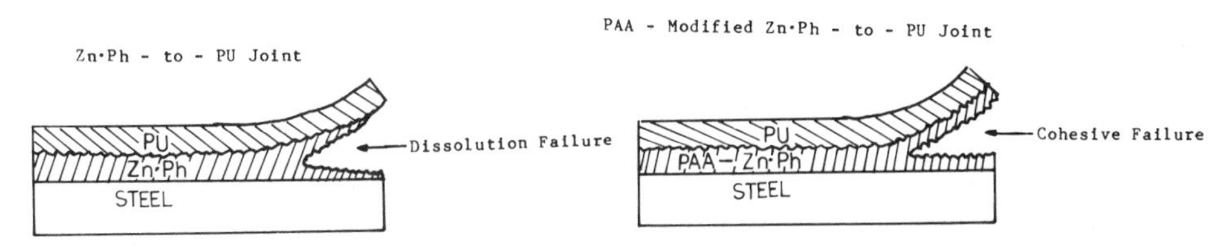

Fig. 20. Proposed failure modes for PU/PAA-Zn·Ph and PU/Zn·Ph joint systems before and after exposure to 0.1 M H_2SO_4 at 80°C for 30 days.

The chemical state at the outermost surface sites of the PAA-chemisorbed Zn·Ph precoat layers was composed of the zinc phosphate hydrates and functional organic groups. The presence of functional organic species such as ionic carboxylate and carboxylic acid groups contributed to the dramatic improvement in the adhesive force at the crystalline precoat-to-polyurethane (PU) topcoat interfacial joints. The major role in promoting a good interfacial bond performance was due to the interfacial chemical reactions between the isocyanate groups in PU and carboxylic acid groups existing at the outermost surface sites of PAA and PIA-modified Zn·Ph in the Zn·Ph-to-PU adhesive joint systems.

The inherent stability of the chemical bonding at the polymer-to-complex Zn·Ph interfaces was found to play the key role in enhancing the bond durability of metal adhesive joints that will be subjected to chemically corrosive environments. The failure mode for the PU/PAA-Zn·Ph exposed to a hot H_2SO_4 solution at 80°C for 30 days was confirmed to be cohesive, (in the Zn·Ph layer) similar to that for the unexposed specimens. This seems to suggest that the hydrophilic COOH groups of PAA at the interfaces were converted into hydrophobic by chemical reaction with the isocyanate groups in the PU topcoat.

REFERENCES

1. T. Sugama, L.E. Kukacka, N. Carciello, N., and J.B. Warren, Polyacrylic acid macromolecule-complexed zinc phosphate crystal conversion coatings, J. Appl. Polym. Sci., 30, 4357 (1985).
2. E.L. Ghali and R.J.A. Potvin, The mechanism of phosphating of steel, Corrosion Sci., 12, 583 (1972).

3. J.B. Lakeman, D.R. Gabe, and M.O.W. Richardson, The physical properties of phosphate coatings and their correlation with potential/time measurements, Trans. Inst. Metal Finishing, 55, 47 (1977).
4. T. Sugama, L.E. Kukacka, N. Carciello and J.B. Warren, Factors affecting improvement in the flexural modulus of polyacrylic acid-modified crystalline films, J. Appl. Polym. Sci., 32, 3469 (1986).
5. R.A. Iezzi and H. Leidheiser, Jr., Surface characteristics of cold-rolled steel as they affect paint performance, Corrosion, 37, 28 (1981).
6. R.P. Wenz, Steel surface cleanliness: Preparation, characterization, and paint performance, in: Organic Coatings Science and Technology, Vol. 6, G.D. Parfitt and A.V. Patsis, eds., Marcel Dekker, Inc., New York (1984).
7. D.A. Jones and N.R. Nair, Electrochemical corrosion studies on zinc-coated steel, Corrosion, 41, 357 (1985).
8. Joint Committee on Powder Diffraction Standards, Card 30-1491 (1984).
9. M.L.Miller, The structure of polymers, Reinhold Publishing Corporation, New York (1966).
10. J.A. Kargol, D.L. Jordan and A.R. Palermo, The influence of high strength cold-rolled steel and zinc coated steel surface characteristics on phosphate pretreatment, Corrosion, 39, 213 (1983).

EFFECTS OF MECHANICAL DEFORMATION ON THE PHOTO-DEGRADATION OF ACRYLIC-MELAMINE COATINGS

Truc-Lam H. Nguyen and Charles E. Rogers

Center for Adhesives, Sealants and Coatings
Case Western Reserve University
Cleveland, Ohio 44106

ABSTRACT

The effects of bending deformations on the photodegradation of cured films of the systems: n-butyl acrylate, styrene or methyl methacrylate, hydroxyethyl acrylate, and melamine coated on aluminum mirror substrate have been studied using several experimental methods. The key characterization method was reflectance absorption Fourier transform infrared spectroscopy to measure the various chemical changes in the polymer caused by UV and moisture exposure. This method was used both for flat (undeformed coating) substrate and for substrates with cured coatings subsequently bent to different radii of curvature to impose different levels of stress on the coatings during exposure. The results showed a pronounced enhancement of degradation with increasing surface stress. The actual imposed stress is limited by the degradation-induced formation of microcracks and crazes. The results are interpreted in terms of dilational voiding under the imposed stress which increases oxygen accessibility, plus the effects of stress, per se, on the oxidation mechanism and its micro-reversibility.

INTRODUCTION

Polymeric coatings on metallic substrates are often subjected to mechanical stress/strain under application conditions during exposure to ultraviolet radiation. It has been shown[1,2] that such static or cyclic deformations affect the magnitude and, possibly, the mechanism of photodegradation processes. Of particular interest are deformation-induced changes in oxygen accessibility and effects of imposed stress, per se, which may affect the micro-reversibility of degradation reactions such as chain scission.

In this study we have used a nondestructive technique to follow the photodegradation and hydrolysis reactions of melamine-acrylic coatings on an aluminum mirror substrate. Transmission infrared spectroscopy has been used[3-8] to follow the photodegradation and hydrolysis of free-acrylic-melamine coatings. Fourier transform infrared reflection absorption spectroscopy (FTIR-RA) is a valuable tool for such analysis since it allows a quantitative study to be made of the coating on the metal substrate, the condition of actual applications. We have extended the FTIR-RA method to allow studies to be made using curved metal substrates.

EXPERIMENTAL

The clear coating formulation was obtained from the Glidden Company. The melamine was partially alkylated. The acrylic copolymers were butyl acrylate/styrene/hydroxyethyl acrylate (BA/S/HEA) and butyl acrylate/methyl methacrylate/hydroxyethyl acrylate (BA/MMA/HEA). Aluminum mirrors (2.5 x 5.0 x 0.2 cm) were mechanically polished with No. 5 chromeoxide, ultrasonically washed with acetone, then dried in air. The as-received coating formulation was dissolved in methyl ethyl ketone. Samples were solution cast onto the aluminum substrates and air dried. The coated substrate was cured at 175°C for 30 minutes in air. Film thickness was calculated based on the concentration of the solution, density of the coating and the surface area of the aluminum substrate.

The cured coatings on the metal substrate were deformed by bending the coated substrate over support hemispheres with different radii. The surface stress, S, applied to the thin coatings was calculated using Euler's approximation: $S = \pi^2 EHY/2L^2$, where H is thickness, L is length, E is Young's modulus and Y is the deflection at the center of the specimen.

The coated substrate was photodegraded at 40°C ± 2°C in air using the QUV weatherometer with ambient humidity. In this study, film thicknesses from 0.5 to 2.0 micrometer were used. The reflection-absorption attachment (Harrick Scientific), along with a gold wire grid polarizer, were used in a Digilab FTS-20 Fourier Transform Infrared spectrometer. The angle of incidence was set at 75° with respect to normal (near grazing angle). The spectrometer was equipped with a narrow band pass, liquid nitrogen-cooled mercury cadmium telluride (MCT) detector and was constantly purged with nitrogen gas to remove moisture. Spectra were obtained at room temperature with 4 cm^{-1} resolution, 400 scans for sample and 200 scans for reference.

RESULTS AND DISCUSSION

FTIR-transmission and FTIR-RA spectra of cured melamine-BA/S/HEA coatings show that there is no band shape distortion when comparing a RA to a transmission spectra. However, there is band shifting from 2 to 5 wavenumbers for the strongly absorbing bands of the benzene ring from 700 cm^{-1} to 702 cm^{-1} melamine-acrylic crosslink or ether linkage from 1096 cm^{-1} to 1101 cm^{-1} and carbonyl to 1735 cm^{-1}. In order to use FTIR-RAS as a quantitative method, calibration curves must be established. A linear relationship is observed between the integrated intensity and the thickness of the coating from 0.5 to 2.0 micrometer. To use FTIR-RAS as a quantitative method for the melamine-acrylic coatings, the film thickness should be below 2 micrometer, as was done for this study.

In the earlier part of this study[9] it was established that in the unstressed (nonbent substrate) condition, photodegradation resulted in a decrease in intensity of the melamine triazine (816 cm^{-1}), benzene, melamine-acrylic crosslink or ether linkage and also that of the -CH stretching (2900 cm^{-1}). In addition with UV exposure, there were formations of new carbonyls (1780 cm^{-1} and 1680 cm^{-1}) as well as amines

(3350 cm^{-1}) and hydroxyls (3250 cm^{-1}). The loss of melamine triazine ring indicates a loss of material due to the photo-oxidation process. There was no indication of formation of the melamine-melamine crosslink (1350 cm^{-1}).

Figure 1 compares the FTIR-RA spectra of melamine-BA/S/HEA coating of a stressed and degraded sample (spectrum AA) to that of an unstressed and degraded sample (spectrum BB). Both samples were photodegraded for 70 hours under UV light. The difference spectrum between the two is illustrated in spectrum CC. The difference spectra across the wavelength region from 3800 to 600 cm^{-1} (not all shown in Figure 1) show an increase in amine and carbonyl compounds and a decrease in -CH stretching, phenyl ring and melamine triazine ring. Figure 2 shows the fraction of melamine triazine remaining as a function of imposed surface stress for melamine-BA/S/HEA samples degraded for 70 hours and for melamine-BA/MMA/HEA samples degraded for 180 hours. The loss of melamine triazine is accelerated by the imposed stress. Likewise, the acceleration in the rate of formation of amines and hydroxyls with increasing stress is shown in Figure 3.

The leveling-off with increasing nominal stress of the effect of imposed stress on the acceleration of photodegradation is attributed, at least in part, to the observed acceleration of microcracking, gross mechanical embrittlement and failure, with increasing imposed surface stress. The formation of cracks (and probable precursor crazes, voids, etc., which were not determined, per se, in this study) serve to act as stress relievers for the coating such that the actual stress experienced is less than the nominal calculated stress from the Euler equation. The plateau stress would tend to that characterizing the threshold cohesive crazing/crack stress of the coating.

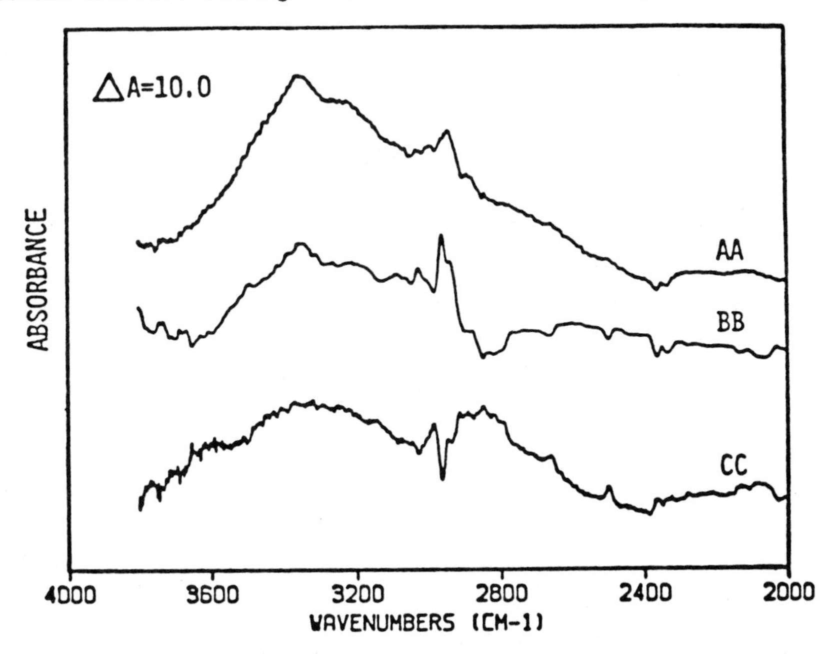

Fig. 1. FTIR-RA spectra of a melamine-BA/S/HEA coating, which had been photodegraded for 70 hours. The samples were stretched at 1.5-inch curvature (spectrum AA) and unstretched (spectrum BB). The difference spectrum is represented by CC.

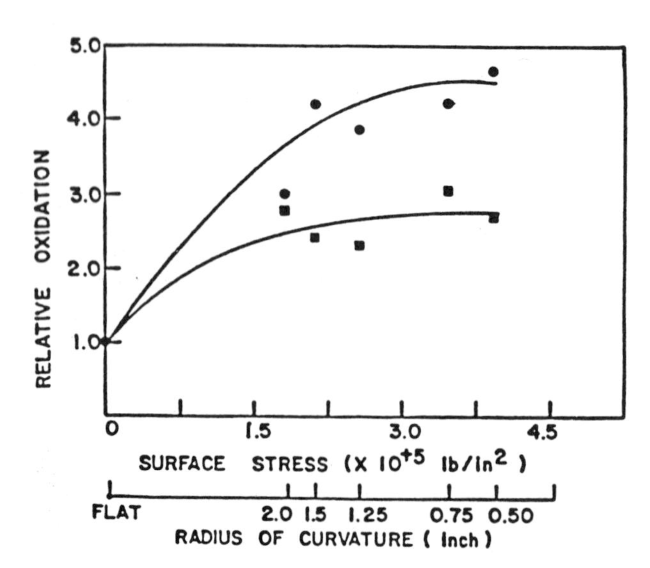

Fig. 2. Fraction of melamine triazine remaining on the aluminum surface for melamine-BA/S/HEA, photodegraded for 70 hours (●) and for melamine-BA/MMA/HEA, photodegraded for 180 hours (▲) as a function of stress.

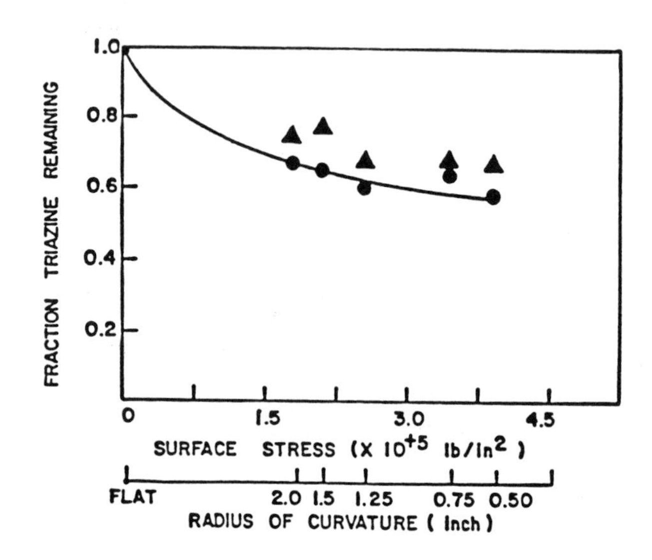

Fig. 3. Relative oxidation (i.e., rate of formation) of amines (■) and hydroxyls (●) for melamine-BA/S/HEA coating photodegraded for 70 hours as a function of stress.

The acceleration of degradation due to imposed mechanical stress can be attributed in part to dilation, including void, craze, and crack formation, which would increase oxygen accessibility to the coating material. A more basic change in the reactions may be due to the imposed stress, per se, as it affects the micro-reversibility of the elementary reactions. As an example, a chain scission in an unstressed state has a finite probability of micro-reversal to reform the original chain bond. Under stress, the equilibrium of the reaction may be shifted to favor the scission product. The results of this present study can neither confirm nor deny this possible mechanistic aspect.

CONCLUSIONS

The imposition of mechanical stress on melamine-acrylic coatings on aluminum substrate caused a pronounced acceleration of photodegradation. The rate of loss of melamine triazine ring increased with increasing stress as did the rate of formation of amine and hydroxyl groups. The leveling-off of the rate of acceleration with increasing stress can be attributed to cohesive failure (cracks, crazes, voids, etc.) in the coating which served to limit the actual stress experienced by the coating to the threshold cohesive failure stress. The acceleration of photodegradation is considered to be due to enhanced oxygen accessibility and, possibly, the more direct effects of stress on the reaction mechanism and its microreversibility.

ACKNOWLEDGEMENTS

The authors appreciate the financial support of the Center for Adhesives, Sealants and Coatings (CASC) at Case Western Reserve University. We also thank F. Louis Floyd of the Glidden Company for kindly providing the coating formulations.

REFERENCES

1. D. Benachour and C.E. Rogers, ACS Symposium Series, 151, 263 (1981); 220, 307 (1983).
2. T.-L.H. Nguyen and C.E. Rogers, Polym. Mater. Sci. Eng. Preprints, 53, 292 (1985).
3. D.R. Bauer and L.M. Briggs, Characterization of Highly Crosslinked Polymers, S.S. Labana and R.A. Dickie, Eds., ACS Symposium Series, 243, 185th ACS Meeting, Washington, D.C., 271 (1984).
4. M.G. Lazarra, J. Coatings Technol., 56 (710), 19 (1984).
5. A.D. English, D.B. Chase and H.J. Spinelli, Macromolecules, 16, 1422 (1983).
6. W.J. Blank, J. Coatings Technol., 51 (656), 61 (1979).
7. D.R. Bauer, J. Appl. Polym. Sci., 27, 3651 (1982).
8. a. D.R. Bauer and R.A. Dickie, J. Polym. Sci. Polym. Phys., 18, 1997 (1980).
   b. ibid., p.2015 (1980).
9. T.-L.H. Nguyen and C.E. Rogers, ACS Polymer Preprints, 27, No. 2, 228 (1986).

Discussion

On the Paper by C.L. Bauer, M. Vratsanos and R.J. Farris

Jwohuei Jou (IBM-Almaden Research Center, San Jose, CA): How do you characterize the interfacial shear stress at the edge?

R.J. Farris (University of Massachusetts): The most practical method is via engineering stress analyses. This requires a knowledge of the stress-strain constitutive relationship for the coating and substrate as well as the associated geometry. The shear stresses at the edges are very sensitive to how the coating is terminated, i.e., how it tapers in thickness as it approaches the edge.

J. Jou: Would the shear stress change with time?

R.J. Farris: For stresses induced by solvent removal or slow thermal excursions, experimental data suggests little time dependence.

Y. Ersun-Hallsby (Dow Chemical Co., Midland, MI): What are the effects of mechanical interlockings on the stresses developed between the plane of the film and substrate?

R.J. Farris: We are looking at flat substrates. Surface asperities would cause multiaxial stresses and stresses through the bonded interface. No such stresses exist away from the edges of flat films.

Y. Ersun-Hallsby: What are the effects of fillers? Could the results presented on films be extrapolated to systems with fillers?

R.J. Farris: Fillers represent an interesting twist. Since they increase the modulus more than they decrease expansion or shrinkage effects, they should increase the stresses during solidification.

Tom Dickinson (Washington State University, Pullman, WA): What is the mechanism for the mass uptake due to stress on Kevlar and other polymers?

R.J. Farris: We have no idea. It is clear that swelling and other related phenomena are stress dependent, e.g., constrained swelling of elastomers. Kevlar and related materials represent an extreme in that they are capable of supporting tensile stresses in excess of their shear moduli.

John Chen (Mameco Internional, RPPM, Cleveland, OH):

(a) E&E obtained by the author comparable to mechanical spectrometer's data.

(b) Stress increases due to cyclic heating-cooling, would this happen for amorphous polymer?

R.J. Farris:

(a) Accuracy 1-3%.

(b) Polycarbonate is amorphous, thus the behavior will be the same for heating-cooling cycle.

On the Paper by F.C. Jain, J.J. Rowato, K.S. Kalonia and V.S. Agarwala

L.H. Lee (Xerox Corp.): How can you translate your concept into practical protection between the polymer-metal interface?

F. Jain (University of Connecticut): Active electronic barriers in metal-semiconductor and metal-SiO_2 (thin insulator)-semiconductor configurations can be realized using appropriately doped semiconducting polymers in a manner similar to Al-ITO and Al-SiO_2-ITO systems. Particularly, extensive literature is available on doped polyacetylene, phthalocyanine and chlorophyll. Several investigators have used Schottky barriers (i.e., metal-polymer interfaces) for solar cell applications.

Unlike the solar cells, the metal-semiconductor interfaces used in corrosion protection are not sensitive to doping variations and, therefore, present fewer problems in their adaptation for practical corrosion applications. The recent usage of plasma deposition of semiconducting polymers (e.g., polymerized polyacrylonitrile presented at this Conference session #6) further ensures the increased application of polymers in active barrier format for corrosion protection.

On the Paper by T.H. Nguyen and C.E. Rogers

C.S. Chen (General Dynamics, Fort Worth, TX): How does the formation of voids affect the observed apparent increase in Poisson's ratio at higher deformations?

C.E. Rogers (Case Western Reserve University): An apparent value of Poisson's ratio greater than one-half indicates a decrease in overall volume which can be attributed to orientation of the polymer chains into structures akin to crystalline regions. This is not to say that voids are not formed between the oriented domains. The decrease in volume due to orientation must be greater than any increase due to void formation to give a net decrease in overall volume. However, as I mentioned relating to transport in shear-banded regions, which are highly oriented, the voids may serve as high diffusion-rate paths through the overall structure. The highly oriented domains in the region are themselves essentially impermeable, but the penetrant simply passes around these domains through the porous void regions. Peterlin has shown essentially this for diffusion in fibers.

C.S. Chen: Can that same reason be used to explain observed values of Poisson's ratio of nearly unity for four-bar linkage materials, such as fabrics measured along the 45° direction?

C.E. Rogers: Yes, I suppose so in a qualitative manner. As you indicate, it's like taking hold of the opposite corners of a square, pulling it into a diamond-shape, and then further to form two nearly parallel lines joined at each end. The decrease in area bounded by the lines corresponds in three dimensions to a decrease in volume. However, it is pushing the concept of Poisson's ratio to apply it to such systems. The representation of Poisson's ratio is really applicable only when the strains are infinitesimal, according to elasticity theory.

C.S. Chen: At larger deformations, will voids become smaller or larger?

C.E. Rogers: Good question. It depends on the polymer and the experimental conditions, rate-of-strain, etc. It probably comes down to a matter of definition of what constitutes a void. We must remember that we deform a bulk material. Local fluctuations or defects within the material serve to nucleate voids upon dilative deformation. These voids may coalesce to form craze-like structures, with fibrils drawn across what otherwise would be called a crack. The question then is do the fibrils rupture upon further deformation, or do they continue to elongate, or is the craze growth somehow terminated thereby stabilizing the void structure within the craze? If they rupture, a crack is formed which will grow. Is that a void? If they continue to elongate, which is very unlikely for any individual fibril, then eventually the voids between the fibrils would decrease in volume as the bulk material forms one large fiber. If the individual crazes are terminated, then new voids and crazes would need to be nucleated. This is what apparently happens during moderate deformations of high impact polymers such as rubber-reinforced glassy polymers. A combination of all of the above is probable depending upon the extent of deformation and other factors.

PART SIX:

COATINGS FOR ELECTRONIC AND OPTICAL ENVIRONMENTS

Introductory Remarks

Lieng-Huang Lee

Webster Research Center
Xerox Corporation
Webster, New York 14580

Corrosion problems are not limited to metals. Modern electronic and optical devices also face severe corrosion problems. During this Session, we shall explore some of those problems.

Dr. W.E. Dennis will give us an overview about optical fiber coatings. The applications and capabilities of various types of fibers and their specific coating requirements for harsh environments will be described. The parameters about "ideal coating" will be presented.

There are other types of optical coatings that could be damaged by radiation. Professor J.T. Dickinson will discuss radiation-induced damage of optical coatings. In an excimer laser, optical components are exposed simultaneously to reactive chemical species as well as bombardment by particles and energetic photons. The film materials will include SiO_2 and ThF_4 sputter-deposited on metal substrates. Measurements of chemisorption (of F_2 and XeF_2), sorption, and mass removal will be presented.

For solar reflectors, new silver/polymer films are being developed. Dr. H.H. Neidlinger will cover the subject on stabilized acrylic glazings for solar reflectors. He will discuss UV light to be the major barrier for this type of development. For ^xample, polymethyl methacrylate is a stable polymer in a terrestrial environment, but the polymer does not provide adequate protection for the silver reflector. Polymeric stabilizers will be presented to improve the stability.

For electroconductive coatings, a new development in the PRC will be reported by Professor Lu of Guangzhou Institute of Chemistry, Academia Sinica. The polyconjugated systems are based on the addition products of 2-butynediol-1,4 with ethylene oxide, propylene oxide, and epichlorohydrin. The polymers possess semiconducting properties having a resistance of 10^7-10^{11} ohm-cm at room temperature. The oligomers could react with di-isocyanates to form polyurethane coatings with good mechanical and chemical stability.

Conductive adhesives have found many applications for electronic industries. Mr. Xia Wen-gan from the PRC will discuss durability of the conductive adhesive joints. The application of a coupling agent tends to increase the bond durability; however, the resistivity of the adhesive joint will be shown to increase, too.

Editor's note: Two new contributions submitted after the Symposium are included in this Part. A paper by Dr. T.S. Wei on polymer materials for optical fiber coatings and one by M.G. Allen and S.D. Senturia on microfabricated structures for the measurement of adhesion and mechanical properties are important additions to the subject matter.

OPTICAL FIBER COATINGS

William E. Dennis

Dow Corning Corporation
Midland, MI 46840

ABSTRACT

Optical fibers require protective coatings to prevent chemical attack and mechanical damage in the natural environment. Glass clad silica fibers, the most common type of commercial optical fibers, lose their strength when exposed to moisture and are coated in line as the fiber is drawn. This paper covers the various types of optical fibers, their dimensions, methods of manufacture and the types of coatings used to protect them. The applications and capabilities of the various types of fibers and their specific coating requirements for harsh environments are described.

The characteristics and limitations of current coatings for optical fibers and the parameters which must be met to provide an "ideal coating" are proposed.

INTRODUCTION

Optical fibers are small diameter filaments which guide light down their length by using a gradient of refractive index from the center to the outer edge of the fiber. When light passes from a region of high refractive index to a region of low refractive index some of it is reflected back into the region of high refractive index. Above some critical angle total reflection occurs. The guiding of light waves in optical fibers depends on this property.

The cone of light in a circular cross section defined by this angle is the cone of acceptance of light and is called the numerical aperture (NA) of the fiber. This cone of acceptance depends on the difference in refractive index between the core, the light carrying medium and the cladding, the light guiding interface. The cone half angle is related to the refractive indices by:

$$\textit{sine of the half angle} = NA = \sqrt{n_1^2 - n_2^2},$$

where n_1 is the refractive index of the core and n_2 is that of the cladding. The greater the difference between the two refractive indices the greater the light gathering capability of the fiber. However, the greater the difference in refractive indices the greater the dispersion and consequently the narrower the bandwidths of information capacity. Therefore, a compromise is usually made at a numerical aperture of 0.2.[1]

Standard telecommunication fibers are all glass and have an outside diameter of 125 µm. These fibers are called glass clad silica fibers and have light carrying cores ranging in diameter from 62.5 µm for multimode fibers to 3-8 µm for single-mode fibers.

Profiles of step-index optical fibers and graded refractive index optical fibers are shown in Fig. 1. Large core sizes with a step-index profile are only suitable for short communication links and/or slow data rates because they allow the signal to disperse rapidly. This dispersion is due to intermodal differences in propagation speeds.

The graded refractive index design greatly decreases this dispersion by keeping the speeds of the different modes more uniform. The higher modes travel farther but through a lower refractive index medium and therefore faster than the lower modes which travel down the center of the core. The graded index core is made by doping the center of the core with a higher concentration of germania which gives a higher index of refraction and gradually decreasing the concentration toward the edge of the cladding which can be pure silica.

The single-mode fibers have cores which are so small only one mode of light can propagate in the core at a wavelength greater than the cutoff wavelength. Therefore, intermodal dispersion does not occur and these optical fibers have a step-index profile. These glass fibers are coated with polymeric coatings for environmental and mechanical protection. The outside diameters of the coated fibers range from slightly over 200 to 900 µm.

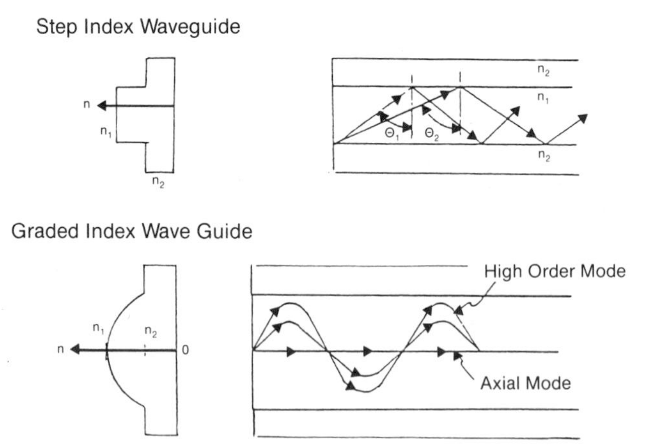

Fig. 1. Step index waveguide and graded index waveguide.

The thin glass fibers are made by drawing them from large diameter preforms. The preforms are usually 1.5" to 4" in diameter and 2 to 4 feet in length. The preforms are presently made by vapor deposition techniques to control the refractive index.

There are three principal methods of using vapor deposition to prepare preforms for drawing optical fibers. They are: outside vapor deposition (OVD), deposition on the inside walls of a quartz tube called modified chemical vapor deposition (MCVD), and axial vapor deposition at the end of a rotating rod.[2]

All of these techniques have unique advantages and disadvantages. All of them utilize the oxidation of silicon tetrachloride and germanium tetrachloride to form the "silica soot" which is later fused to make the glass perform. (See Figures 2-4)

The fibers are formed by melting the end of the preform and pulling the molten glass strand at a rate which gives the desired outside diameter. This process produces a fiber which has the same cross-sectional refractive index profile as the original preform.

The fiber is coated in-line as it is drawn to prevent damage to its surface which lowers its tensile strength. The diameter of the fiber is monitored by a laser, and both the speed of the capstan and the alignment of the coating die are controlled by a feedback mechanism. Drawing and coating speeds in excess of 10 meters/second have been reported.[3]

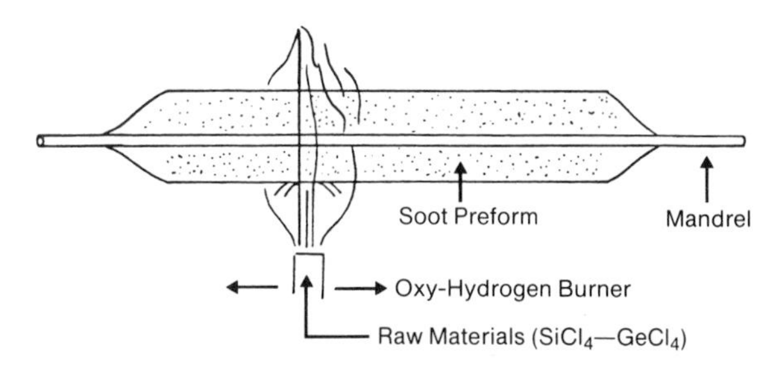

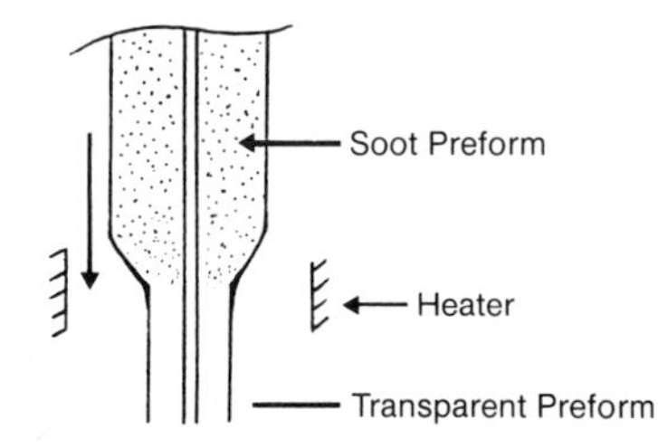

Fig. 2. Outside vapor deposition process.

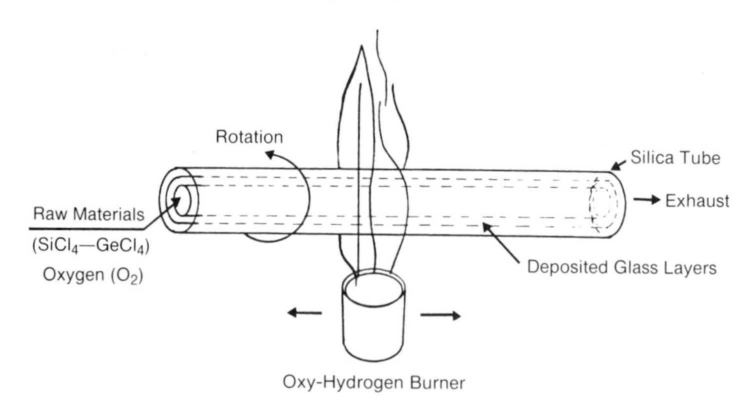

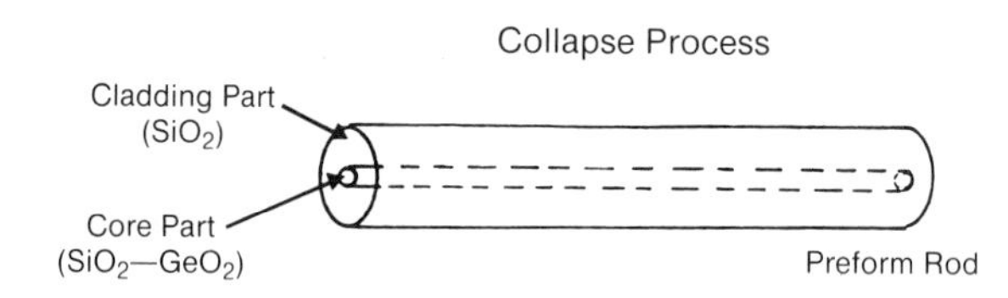

Fig. 3. Modified chemical vapor deposition process.

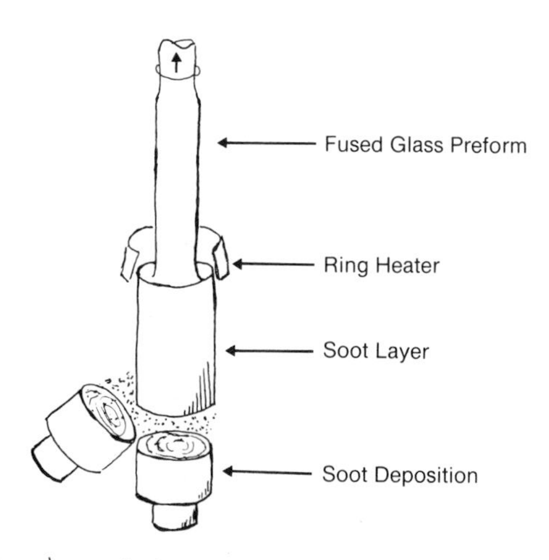

Fig. 4. Axial vapor deposition process.

The rate limiting step in the drawing of optical fibers is the speed with which the protective coatings can be applied and cured. One of the major factors which limit the coating rate is the temperature of the glass. The glass has to cool from the molten state to a temperature which does not cause vaporization or degradation of the coating when it is applied. Consequently, most commercial drawing towers are tall, usually 30 feet or more in height. (See Figure 5)

Glass-clad silica fibers are used in telecommunications because they have greater optical clarity than plastic-clad silica or plastic-clad plastic optical fibers. The light wave interacts with the cladding at the core-cladding interface and therefore the cladding as well as the core must not have absorbing or dispersive impurities.

Transmission of information over 100 kilometers at high bit rates without repeaters has been demonstrated several times and commercial installations routinely have repeater spacings of 20 kilometers or more.

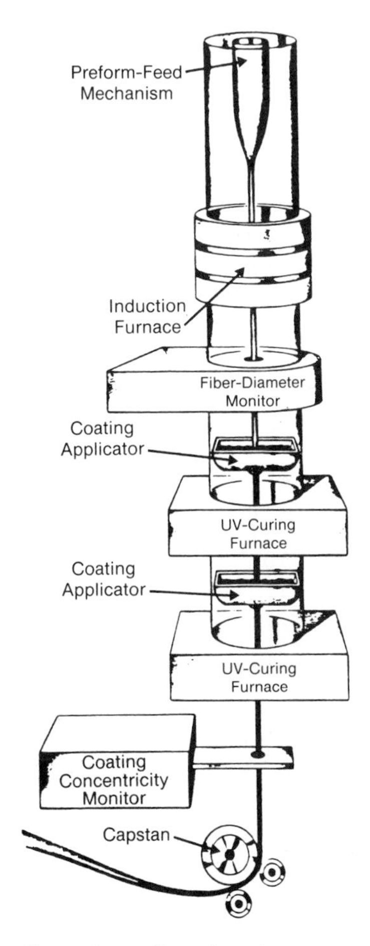

Fig. 5. Drawing towers.

The niche for plastic-clad silica and all plastic fibers are in short data links where high losses can be tolerated. These two fiber types have the advantage of a large core diameter and consequently good light gathering ability and larger outside dimensions making connectorization and handling easier. The core diameters of PCS and plastic fibers are often 200 to 600 µm and the diameter of the cladded fiber can be as high as 1 µm.

The losses of fibers are measured in decibels (dB) per unit length usually kilometers. A decibel is:

$$dB = 10\, log_{10}\left(\frac{light\, received}{light\, sent} \right).$$

The losses in telecommunication fibers are usually 0.2 to 0.5 dB/km. The losses for plastic-clad silica fibers are several dB/km. The losses for plastic-clad plastic fiber are several hundred dB/km.

COATING REQUIREMENTS

The three primary requirements for an optical fiber coating are:[4,5]

1) Providing protection from corrosive environments, specifically water which can attack surface defects and lower the strength of the glass fiber.
2) Providing protection from mechanical forces particularly small radius or point loading forces which bend the fiber sharply causing the signal to escape from the core-cladding interface. These losses are called microbending losses, and
3) Having enough physical and mechanical strength to be handled, spooled, cabled, etc., without losing its physical integrity.

The attributes of a coating which provides protection from chemical degradation particularly moisture are:

1) Good adhesion between the coating and glass fiber,
2) Low water absorption and permeability,
3) Minimal change in physical properties after exposure to moisture.

The attributes of a coating which provides good protection from mechanical stress are:

1) Low modulus,
2) Minimal change in modulus over the temperature range of use.

The properties which enable the coating to be handled and used are:

1) High modulus and tensile strength,
2) Low coefficient of friction,
3) Resistance to moisture and/or fluids used as water blocking agents.

No single coating has all of the characteristics cited above for protection and handleability; therefore, most optical fiber manufacturers apply two coatings.

The first coating is soft and stress-relieving and is called the primary coating and the second coating is hard and tough providing good handling characteristics. This is the secondary or outer coating.

There are two design approaches to cabling and protecting optical fibers.

One of these uses a minimal amount of coating on the fiber and relies on the cable construction and design to prevent stress from being applied to the fiber. The optical fibers in this design are placed in a tube, slot or channel with an excess length of fiber to allow the cable to be bent, twisted, etc., without imparting a stress on the fiber. In most, if not all of these designs, the tube, slot, or chamber containing the fiber is filled with a water blocking compound. This compound fulfills two functions: it prevents the ingress of water and lubricates the fiber so that there is little friction between the fiber and other fibers and/or the walls of the cavity in which it resides. This type of design is called "loose tube", "open channel", or "loose buffer" design.

A second design for protecting the fiber is called tight buffer or tight tube. This design uses a substantially thicker coating on the fiber and the outside diameter of the coated fiber is often 900 µm or greater. The primary coating or coating closest to the fiber is soft and stress-relieving and can be as thick as 400 µm. This soft inner coating is covered with a jacketing which is usually extruded over the coated fiber in a separate step. The outer jacketing is a high modulus, tough coating which provides significant mechanical protection without relying on the cable design to provide all of this mechanical protection.[6]

COATING MATERIALS

Ultraviolet-curing urethane acrylates are commonly used as the soft stress-relieving primary coating for the thinly clad optical fibers. This coating is usually overcoated with a higher modulus urethane acrylate or epoxy acrylate for handleability. The crosslink density of the primary coating is controlled by using chain extending difunctional monomers or oligomers which can be reacted into the coating. This approach is preferable to using non-functional plasticizers because they can migrate out of the primary coating in time and cause the modulus of the primary coating to increase and that of the secondary coating to decrease.

The speed of cure achievable through ultraviolet radiation makes this technology suitable for the coating of optical fibers. This is particularly true for coating formulations which are not also thermally initiated. The conditions for coating the drawn fiber in line must be controlled tightly to maintain a uniform and concentric coating. One of the techniques for controlling the viscosity of the coatings is through temperature. The viscosities of most coatings are highly dependent upon their temperature and this provides a convenient means of regulating them.

One aspect of using UV cure is the sensitivity of some UV curing materials to oxygen inhibition. Since the coatings are quite thin in this application and the surface characteristics critical to the handling and performance of the fibers, any change in the degree of cure caused by oxygen trapping the free radicals and regulating the chain lengths is harmful. An inert atmosphere is usually maintained in the drawing chamber to prevent this undesirable reaction. The draw chamber is often purged with a stream of gas to remove the heat generated by the UV lamps and therefore using an inert gas such as nitrogen for this purpose introduces little expense or complexity.

The relatively thick coatings used on fibers which are cabled in the tight buffer design have traditionally been thermally cured silicone elastomers. A thin layer of high refractive index silicone is used next to the surface of the glass cladding to disperse errant light signals away from the core. The remainder of the buffering coating is polydimethyl siloxane. The jacket which is extruded over this stress relieving combination can be nylon, polyester or any other strong thermoplastic material.

The drawbacks to many UV curing acrylate coating materials are:

1) Their relatively high water-absorption characteristics--3-5% are common.
2) This high water-absorption causes a significant difference in physical properties between wet and dry coatings.
3) They can cause skin sensitization and/or are toxic.

The drawbacks to the thermal curing silicone elastomers are:

1) They cure more slowly than UV cure systems.
2) Their cure mechanism involves silicon hydrides which can generate hydrogen during use. (Hydrogen can cause attenuation of the signal in some optical fiber designs).
3) Silicone elastomers with a high refractive index require special intermediates which are expensive.

UV curing silicone elastomers provide an answer to the drawbacks relating to the cure of thermal curing silicone coatings and for some applications such as plastic-clad silica or plastic fibers their low refractive index is an advantage.

The properties of the cured coatings are critical to the performance and life of the optical fiber. Several characteristics of the cured coating are measured and monitored by the fiber drawers or requested from the suppliers of the coatings.

One of these characteristics is the change in physical properties of the coating as a function of temperature and exposure to moisture. Most of the coating formulations are designed to have a gradual and indistinct glass transition point so that there are no abrupt changes in modulus and coefficient of thermal expansion in the temperature range of use. This is usually achieved by adding modifiers and/or using monomers and oligomers with a variety of glass temperatures. Consequently, a curve from a differential scanning calorimeter of these coatings usually has a broad region of transitions with no sharp discernable glass transition.

Another property of interest to drawers of optical fibers is the change in physical properties of the coating due to exposure to moisture. The percent of water absorbed and the resulting change in physical properties is important to the strength and consequently the life of the fiber if it is used in applications where moisture is present. Fiber drawers often measure the strength and microbending performance of a coated fiber before and after exposure to moisture.

The protection which a coating provides against microbending losses is often determined by placing the coated fiber between rough surfaces at different loading forces. The rough surfaces can be coarse sandpaper, for example, and the loading forces can be several pounds per square inch. The attenuation vs. loading at various wavelengths is graphed to give a picture of the resistance to microbending afforded by various coatings or coating designs.

ATTRIBUTES OF AN IDEAL COATING

An ideal coating for glass-clad silica fibers is one which can be applied in one step and have a low modulus adjacent to the surface of the fiber and increasing modulus with increasing distance from the fiber. The coating would ultimately reach a high modulus and have a hard slick surface.

The coating would have to have good adhesion to the fiber to prevent damage by water but be easily removable for splicing and connecting. The coating should have a refractive index above that of the cladding. It must be free from particles which can abrade and damage the surface of the fiber.

It must have the proper rheology to be applied rapidly and be capable of very fast cure.

All of these requirements have not been met satisfactorily to date and, therefore, most fiber drawers use two coatings or more.

HERMETIC COATINGS

Organic coatings cannot provide true hermeticity because they absorb water and this water can and does diffuse to the coating fiber interface. There are some applications and fibers which require hermetic coatings. Heavy metal fluoride and chalcogenide glasses which are very transparent at longer wavelengths, 2 microns and greater are very susceptible to damage by water. These fibers are potentially capable of transmitting information for hundreds of kilometers without a repeater.

There have been several hermetic coatings proposed and investigated.[7,8] Most of these require high temperature for application and are metal or ceramic. They are usually applied by chemical vapor deposition (CVD) or melting techniques. These coating techniques are very slow compared to coating fibers with organic coatings. The hermetic coatings are very high modulus and do not provide microbending protection, so soft stress relieving silicone and organic coatings are still used with hermetically coated fibers.

REFERENCES

1. Joseph Palais, "Fiber Optic Systems," Information Gatekeepers, Brookline, MA, 7-2 (1982).
2. Tingye Li, ed., Optical Fiber Communications. Volume 1. Fiber Fabrication, Academic Press, NY (1985).
3. M. Ogai, F. Takahashi, N. Sato and M. Nishmura, Technical Digest, Conference on Optical Fiber Communication, Atlanta, GA, 114 (1986).
4. L.L. Blyler, Jr. and F.V. DiMarcello, Proc. IEEE 68, 1194-1198 (1980).

5. L.L. Blyler, Jr. and C.J. Aloisio, "Polymer Coatings for Optical Fibers," American Chemical Society Symposium Series No. 285, Appl. Polym. Sci., p.907 (1986).
6. T. Naruse, Y. Sugawara and K. Masuno, Electron Lett. 13, 153 (1977).
7. R. Chaudhuri and P.C. Schultz, "Hermetic Coating on Optical Fibers," SPIE 717, 27 (1986).
8. M. Sato, O. Fukuda and K. Inada, "Digest of IOOC'81," Paper MG4, San Francisco (1981).

POLYMER MATERIALS FOR OPTICAL FIBER COATINGS

Ta-sheng Wei

GTE Laboratories Incorporated
40 Sylvan Road
Waltham, MA 02254

ABSTRACT

Successful development of polymer materials has met challenges posed by the coating process and the requirements of high mechanical reliability and optical performance for optical fibers. In this overview, an introduction to optical fiber and coating will be given. Diverse physical requirements for fiber coatings in different applications will then be discussed in detail. The new challenge to polymer researchers is to develop polymer coatings that can offer more reliability in a harsh or hostile environment or enhanced performance in special applications.

INTRODUCTION

Silica-based optical fiber was first suggested by Kao and Hockham in 1966 as a potential low-loss transmission medium for optical communications.[1] Since then, vigorous research and development efforts have led to the large-scale, low-cost production of optical fiber with low attenuation, high bandwidth and excellent mechanical strength. It also made the widespread application of optical fiber communication a reality. Today, laboratory systems have demonstrated multigigabit-per-second transmission in a single mode fiber that is well over 100 km long. Such progress is the result of the interdisciplinary exploration of the new optical fiber technology involving physicists, chemists, ceramists, materials scientists and engineers.[2] Among the many major accomplishments in optical fiber technology, the rapid development of polymer materials that meet very diverse and demanding requirements for optical fiber coatings has contributed significantly to the practicality and reliability of optical fiber.[3-4]

Coating is an integral part of optical fiber. In practical use, the polymer coating should, throughout the service life of 25 years, preserve the mechanical strength of the fiber and prevent the deterioration of its optical performance. A single coating usually compromises either the optical or the mechanical performance of optical fiber, because different requirements of coating properties are imposed on fiber coating. On the other hand, dual coatings, which are preferred by most fiber manufacturers and consist of a low-modulus inner coating surrounded by a high-modulus outer coating, can be optimized to satisfy these requirements. In addition, since polymer coating should be compatible with high drawing speed to meet productivity and economic requirements, UV-cured polymers are now being used extensively. In particular, UV-cured urethane acrylates have become the dominant coating materials for optical fiber, due to their design flexibility, diverse properties and advanced development.[5]

For most non-telecommunication applications, very different requirements on fiber coatings are imposed.[6] This is especially true for a large class of optical fiber sensors, where each application may require some unique coating properties to enhance the sensitivity of the sensor.

In this overview a brief introduction to optical fiber and coating will be given. Then various physical requirements for fiber coatings in different applications and the current status of commercial and developmental polymer coating materials will be discussed.

OPTICAL FIBER

Optical fiber, as shown schematically in Figure 1, is a waveguide in which light signals are transmitted. It has a central core surrounding by a cladding. The standard fiber diameter for communication applications is 125 µm. A polymer coating, typically 150-900 µm in diameter, is usually applied over the cladding to protect the fiber. Dual coating designs consisting of primary and secondary coatings (shown in Figure 1) are more popular than a single coating.

The refractive index of the fiber core should be higher than that of the cladding, so the light launched into the core will propagate within it by total internal reflection. The fiber core material must have low optical loss to propagate light efficiently. The cladding material does not need to be as transparent.

Many glassy materials have been used in optical fiber fabrication.[7] Among them, high-silica glass meets the stringent requirements of high transparency and ease of fabrication, because silica has very low intrinsic absorption and scattering losses and can be easily fabricated

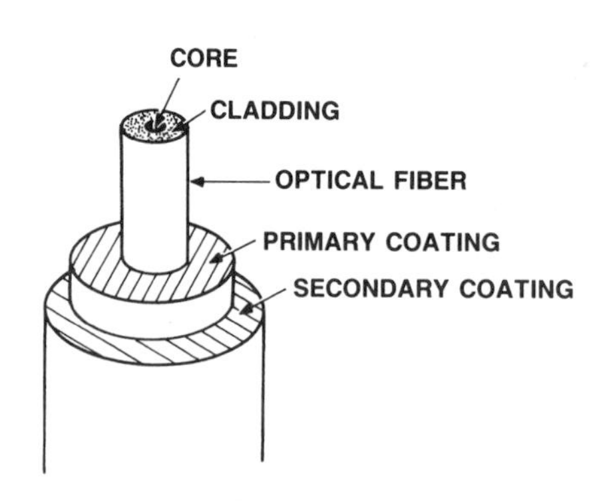

Fig. 1. Schematic of an optical fiber and its coatings.

in long lengths. A minimum loss of about 0.15 dB/km at 1.6 μm has been projected in silica fibers.[8] (Optical fiber loss is normally measured in dB/km and it can be converted into absorbance using 1 $cm^{-1}=10^6$ dB/km). Optical loss as low as 0.154 dB/km near 1.55 μm, which is very close to the intrinsic limit, has been demonstrated in silica fibers of multikilometer lengths.[9]

A large number of organic polymers with good transparency in the visible region are being considered as materials for all-plastic fibers.[10] The best result achieved for all-plastic optical fiber is 20 dB/km at 0.68 μm for a 100 m length, twice the theoretical limit.[11]

Among many ultra-low loss materials, heavy metal fluoride glasses based on ZrF_4 have the most potential. The minimum loss is estimated to be less than 0.01 dB/km at 2.5 μm.[8] The best ZrF_4-based fluoride fiber so far has a loss of 0.7 dB/km at 2.63 μm for a 30 m length.[12]

Depending on the combinations of core and cladding materials, three different types of fibers are commonly available. They are, in the order of increasing losses, all-silica, plastic-clad silica and all-plastic fibers. All-silica fiber typically has either a pure silica core with a fluorine-doped silica cladding or a GeO_2-doped silica core with a silica cladding. These and other dopants, such as P_2O_5 and B_2O_3, have been used to modify the refractive index of silica and facilitate the fabrication.[2] In telecommunication applications, silica based fibers have been used exclusively, because of the superior optical and mechanical properties of silica glasses. On the other hand, plastic-clad silica and all-plastic fibers are being used in short data links and light pipe display applications. Such fibers have either a silica or plastic core with a plastic cladding, which can also be part of the fiber coating. Several publications have discussed optical fiber materials and fabrication techniques extensively.[2,13]

FIBER COATING

Silica fiber is most commonly drawn from a glass preform rod and coated in-line with a polymer coating. The drawing and coating process can impact the optical, dimensional and mechanical properties of optical fiber.[14] A schematic of the apparatus is illustrated in Figure 2. During fiber drawing, the preform is heated in a high-temperature furnace to near 2100 C and drawn into fiber using a capstan to control the drawing speed and fiber diameter. The fiber diameter is normally monitored below the furnace using various high-speed noncontacting diameter measurement techniques. A constant diameter is then maintained by taking a signal from the diameter monitor for feedback control that adjusts the draw speed to compensate for fiber diameter fluctuations.

Although the theoretical strength of fused silica is on the order of 14 GPa ($2x10^6$ psi), the presence of defects or surface flaws can seriously degrade fiber strength due to stress concentration at these defect sites. For example, a flaw size of 2.3 μm decreases fiber strength to 345 MPa (50 kpsi), the standard prooftest level for optical fiber. Because the glass surface is extremely susceptible to abrasion damage, it is necessary to coat the fiber immediately after drawing to avoid contact with any other solid.

In-line primary coating is normally applied using liquid resins, which are cured or solidified rapidly before the fiber reaches the capstan. The concentricity of the coating is monitored continuously and the position of the applicator is adjusted to achieve centered coating. The secondary coating is applied using either UV cured acrylates or extrudable thermoplastics. Other coating techniques, including vapor phase deposition of ceramic materials and deposition of molten metals, provide an impervious hermetic coating. A detailed discussion of this topic has been given in a recent paper on fiber optic applications of hermetic coatings.[15]

COATING REQUIREMENTS

The most prevailing coating materials are liquid prepolymers and thermoplastics, because they meet processing requirements and have the ability to protect and preserve fiber performance. For primary coatings, liquid prepolymers cured by thermal or UV activation or hot-melt materials solidified by cooling are preferred.[3] Due to slow solvent evaporation, solvent-bearing coatings are generally excluded, except in special cases where unique properties of these coatings are required. For secondary coatings, UV cured acrylates and extrudable thermoplastics, fluoropolymers and liquid crystal polyesters are commonly used.[3,16-18]

Table 1 lists polymer materials that are used as optical fiber coatings. Some of the desired coating properties will be discussed according to processing and functional requirements. Table 2 summarizes these requirements and related polymer properties. Detailed discussions on coating criteria are given by considering the coating process, mechanical reliability, optical performance and sensor applications.

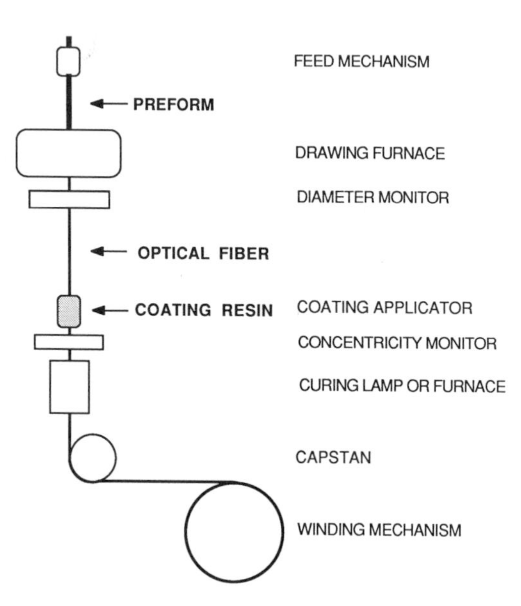

Fig. 2. Schematic of an optical fiber drawing appartus.

Table 1. Polymer Coatings for Optical Fibers

Application/Polymer	Function	Vendor/Reference
UV Curing		
urethane acrylate	primary, secondary	[a],[b]
epoxy acrylate	primary, secondary	(19)
butadiene acrylate	primary	(20-21)
silicone	primary, cladding	[c]
polyene thiol	primary	[d]
acrylate	primary cladding	[e]
Thermal Curing		
silicone	primary,cladding	[c],[f],[g],[h]
Solution		
fluorocarbon	primary, cladding	(10, 22)
organosilicon	primary, cladding	(23)
Hot Melt		
ethylene vinyl acetate	primary	[i]
thermoplastic rubber	primary	(24)
Extrusion		
polyamide	secondary	(3)
polyester	secondary	(16)
fluoropolymer	secondary, cladding	(16)
liquid crystal polyester	secondary	(18)

[a] DeSoto, Inc., Des Plaines, IL
[b] Borden Chemical, Cincinnati, OH
[c] Dow Corning Corp., Midland, MI
[d] W.R. Grace & Co., Columbia, MD
[e] Ensign-Bickford Optics Co., Avon, CT
[f] GE Co., Pittsfield, MA
[g] Petrarch Systems Inc., Bristol, PA
[h] Shin-Etsu Chemical Co., Japan
[i] National Starch Co., Manville, NJ

Table 2. Summary of Polymer Coating Requirements

Property	Functions
Primary Coatings	
elastic moduli	microbending loss, stability, sensor sensitivity
T_g	microbending loss and temperature dependence
particle/gel content	processing, coating concentricity, strength
shelf life	processing, stability
cure/solidification speed	processing, economics
viscosity	processing, draw speed, strength
surface tension	processing
hydrogen evolution	hydrogen-induced loss
UV absorption	UV-induced loss, curing speed
chemical resistance	aging
adhesion-to-glass	strength, fatigue, aging, strippability
water absorption	strength, fatigue
water permeation	strength, fatigue
refractive index	mode stripping, concentricity monitoring
piezoelectricity	sensor sensitivity
Secondary Coatings	
elastic moduli	microbending loss, abrasion resistance
T_g	microbending loss, temperature stability
cure/solidification speed	processing, economics
coefficient of friction	ease of cabling, microbending loss
thermal expansion	microbending loss and temperature dependence
gas evolution	extrudability, hydrogen-induced loss

COATING PROCESS

A primary coating should be applied smoothly and concentrically without damaging the pristine fiber surface. Moreover, the coating resin must solidify prior to contacting the capstan or other solid objects. These requirements imply that the coating resin should have moderately low viscosity (typically 500-6000 cps at the application temperature) and high curing or solidification rate (compatible with a draw speed of 0.5-10 m/s).

Viscosity of the coating liquid is an important property that influence coating concentricity and uniformity. Adjusting the coating viscosity during application is usually done by controlling the temperature of the coating reservoir. The viscosity should be low enough to allow easy filtration prior to application, thereby removing particles which may potentially abrade the fiber surface. For thermoplatic rubber compounds, filtration of particles in its molten form is difficult; this has prevented them from achieving a wider acceptance. On the other hand, higher viscosities may be required if beading of the coating is a problem. Understanding and controlling the temperature dependence of the viscosity help in the proper application of a coating resin.

For large-scale production, economy is the driving force behind the development of coatings that can be applied at a draw speed greater than 2-10 m/s. Using solvent-bearing coatings and thermally-cured silicones usually limits the draw speed to below 1-2 m/s. On the other hand, higher cure speed makes UV cured materials more attractive for primary coatings.

MECHANICAL RELIABILITY

Protecting the optical fiber form abrasion and corrosion in order to preserve its intrinsic mechanical strength is the primary function of a fiber coating. The mechanical failure of silica-based optical fiber is generally attributed to two mechanisms: instantaneous fracture controlled by defects or surface flaws; and delayed failure caused by fatigue, i.e., stress-corrosion-induced flaw propagation. The fracture strength is governed by the statistical distribution of surface flaws or defects in fiber. Fatigue occurs when a stress-dependent chemical reaction between water and the glass network takes place at the defect site. Both failure modes limit fiber strength to values well below the theoretical strength of silica glass ($\sim$14 GPa or $2x10^6$ psi).

Recent progress in improving preform quality and drawing environment has resulted in the production of long-length, high strength optical fibers. High strength fibers, with strength on the order of 4.5-6.2 GPa (650-900 kpsi), are normally characterized by a very narrow, unimodal strength distribution. It has been suggested that high-strength silica fibers are flaw free. Assuming a fiber surface is nearly perfect, even these high measured strengths are less than half the ultimate value. This difference is attributed to fatigue, i.e., flaw initiation and propagation occurring during tensile strength testing.

Polymer coatings should provide adequate protection against mechanical abrasion. For primary coatings, the liquid resin should be free of particle contamination that may damage the fiber surface during drawing and handling. The elastic moduli of secondary coatings should be sufficiently high to provide adequate abrasion resistance. Surface tack and friction of secondary coatings should be low for ease of cabling.

Even though properly applied polymer coatings can provide good protection for fiber strength, polymer coatings cannot provide absolute protection against fatigue because most polymers are highly permeable to water. Although the mechanisms for flaw initiation and propagation are not entirely clear, the influence of a polymer coating on the strength and fatigue behavior of optical fiber has been reported extensively.[25-28] It has been demonstrated that adhesion and water permeation properties of a polymer coating affect fiber strength and fatigue. In particular, the selection of polymer coatings with high adhesion to glass and low water absorption values seems to benefit the mechanical reliability of optical fiber.[27] Nevertheless, many primary coatings of low adhesion values are being used, because splicing is normally done on stripped sections of fibers and mechanical stripping of coating is usually easier and more economical than chemical stripping.

OPTICAL PERFORMANCE

Perhaps the single most important characteristic of any low-loss optical fiber is its attenuation, which is limited by material absorption and scattering. Intrinsic absorption in the host material is caused by electronic transitions in the UV and multiphonon absorption in the IR regions. Intrinsic scattering losses are associated with Rayleigh scattering due to compositional and density fluctuations of the glass structure and waveguide effects near cutoff.

Extrinsic absorption comes mostly from impurities such as transition metal ions and hydroxyl ions. Extrinsic losses are also caused by scattering from bubbles, crystallites and inclusions. In addition, some processing related defects in optical fiber introduce scattering-like losses.

Since most power carried by the guided modes is contained inside the fiber core and the adjacent cladding region, very stringent requirements are placed on OH^{-1} content and impurity levels within the deposited core and cladding materials to eliminate all extrinsic losses. Dramatic improvement in fiber loss over the past twenty years has been accomplished through purification of precursor chemicals, dehydration of deposited glasses, and precise control of preform fabrication and fiber drawing processes. The recently reported loss of 0.154 dB/km at 1.55 μm in silica fibers indicate that the fundamental material limits have been reached and virtually all extrinisic loss mechanisms have been eliminated.[9]

Although extrinsic losses in optical fiber can be controlled during preform fabrication, there is at least one additional loss mechanism, commonly called microbending loss, that is significantly influenced by coating materials. It is caused by scattering resulting from small-scale, random deviations of the fiber axis when a fiber is wound on a spool under tension or placed in a cable. Eccentricity and defects in a primary coating, such as lumps, voids, and beading, may induce microbending losses by producing randomly distributed stresses on the fiber. Such an effect is even more severe at low temperatures, e.g., at -40 C, when the coating moduli increase. In extreme cases microbending can result in very high losses of up to tens of dB/km.

Microbending losses may be minimized through the proper selection of polymer coatings and geometry.[3,29] Soft primary coatings that cushion or buffer the fiber from stresses are usually very effective in reducing the microbending sensitivity. But they cannot provide protection against mechanical abrasion during handling and cabling. Therefore, such coatings are best used in combination with a tough polymer outer coating, thereby forming a dual coating structure. The secondary coating provides abrasion resistance and stress isolation from external forces.

For a dual-coated fiber, elastic properties and geometry of both coatings can be selected to optimize the mechanical and optical performance. In general, polymer materials with low glass transition temperatures (T_g) and low moduli (0.7-6 MPa or 100-900 psi at room temperature) are preferred as primary coatings. Some of the candidate polymers are silicones, hot melt elastomers, butadiene acrylates, urethane acrylates, and epoxy acrylates. However, each has its limitations: low curing speed for some silicones; poor thermal stability for butadiene acrylates and hot melts; high modulus and high T_g for urethane and epoxy acrylates. Most recent developmental work focusses on UV cured primary coating materials with suitable properties.[30-32] The newly developed UV cured silicones seem to offer some interesting features as primary coatings. For secondary coatings, high modulus and high T_g materials including extrudable polyamides, polyesters, fluoropolymers and UV cured epoxy and urethane acrylates have been used. An interesting example is the use of a liquid crystal polyester with an extremely low thermal expansion coefficient to eliminate thermal stresses and microbending losses during temperature cycling.[18]

Besides microbending, there are at least three more important considerations in optical performance for coating materials: hydrogen evolution, UV induced loss, and cladding-mode stripping. The optical reliability of optical fiber is impaired if hydrogen diffuses into the fiber, since considerable absorption is introduced by diffused hydrogen and products of its subsequent reactions with the silica glass. Such a phenomenon has been observed in fibers coated with thermally cured silicones, in which hydrogen is generated from residual silane groups after the crosslinking reaction.[33] The second consideration is related to UV cured coatings, because UV radiation used for curing prepolymer coatings can induce losses in Ge and P doped fibers. Since most damage is caused by short UV wavelengths, coatings with the proper photoinitiator or UV absorber that absorbs strongly below 360 nm can reduce the damage considerably.[34] However, this must be done without affecting the curing rate. The third concern is a practical one, which is related to the propagation of cladding modes if the primary coating has a lower refractive index than the fiber cladding. Cladding modes can cause measurement difficulties at the source and detector ends and in short sections of fibers. The solution is to use a primary coating whose refractive index is higher than that of the cladding (usually 1.46 for silica), so that total reflection can not occur at the interface of fiber and coating. Beside cladding-mode stripping, higher index coatings are also useful in monitoring the coating concentricity.

SENSOR APPLICATIONS

Among a wide variety of optical fiber sensors, interferometric sensors have demonstrated the highest performance.[35] They sense the optical phase change in the fiber that is caused by the physical perturbation to be measured. Such sensors have benefitted most from the growth of fiber coating technology. Fiber coatings often play a critical role in the operation and sensitivity of many interferometric sensors. By changing the fiber coating, the sensing element can be changed from acoustic to magnetic or electric field sensors.

The pressure of acoustic sensitivity of optical fibers is strongly influenced by the elastic coefficients of fiber coatings, since it is related to the acoustically induced phase modulation in optical fibers.[16,36] The soft primary coating has little effect on the acoustic sensitivity. Accordingly, secondary coatings and geometries must be chosen to optimize the fiber response. For a typical coating thickness, high sensitivity requires an outer coating with a high Young's modulus and a low bulk modulus.

Optical fibers coated with piezoelectric polymers have been shown to be sensitive electric-field transducers.[22,37] Normaly, polymers containing polyvinylidene fluoride (PVDF) are used as coatings for such fibers. The strain induced by the electric-field in the coating is transmitted to the fiber to produce an optical phase shift, which can then be detected using interferometric techniques. The electric-field sensitivity depends on the piezoelectric constants and orientation of PVDF molecules.

In sensor development, polymer coatings offer a unique opportunity to sensitize or desensitize sensor fibers to a particular stimulus according to the applications for which they are intended. In most cases, the requirements on optical performance and coating processability can be relaxed, since short lengths can be used. This allows more flexibility in selecting polymer coatings that may not be suited for telecommunication applications.

SUMMARY

The rapid advance in high performance optical communication systems has benefitted from understanding the fundamentals of fiber fabrication processes and the relationships between fiber performance and coating materials. Development of polymer materials has successfully met challenges posed by the coating process and the requirements of high mechanical reliability and optical performance for fibers. The new challenge ahead is to develop new polymer coatings that can offer more reliability in a harsh or hostile environment or more sensitivity in sensor applications.

REFERENCES

1. K.C. Kao and G.A. Hockham, "Dielectric-fibre Surface Waveguides for Optical Frequencies," IEE Proc. 113, 1151-1158 (1966).
2. T. Li, ed., Optical Fiber Communications, Vol. 1, Fiber Fabrication, Academic Press, Orlando, FL (1985).
3. L.L. Blyler, Jr. and C.J. Aloisio, "Polymer Coatings for Optical Fibers," Am. Chem. Soc. Symp. Series, No. 285, 907-930 (1985).
4. G. Kar, "Coatings for Optical Fibers," SPIE 584, 40-43 (1985).
5. K. Lawson and O.R. Cutler, Jr., "UV Cured Coatings for Optical Fibers," J. Rad. Curing, 9 (2), 4-10 (1982).
6. J.M. Martin, "Contributions of Dual Coat UV Cured Coatings to Optical Fiber Strength and Durability," in RADCURE Europe '87 Munich, West Germany, (1987).
7. A.M. Glass, "Optical Materials," Science, 235, 1003-1009 (1987).
8. P.W. France, S.F. Carter, M.W. Moore and C.R. Day, "Progress in Fluoride Fibres for Optical Communications," Br. Telecom. Technol. J. 5 (2), 28-44 (1987).

9. H. Yokota, H. Kanamori, Y. Ishiguro, G. Tanaka, S. Tanaka, H. Takada, M. Watanabe, S. Suzuki, K. Yano, M. Hoshikawa and H. Shimpa, "Ultra-low-loss Pure-silica-core Single-mode Fiber and Transmission Experiment," Tech. Dig., 1986 Opt. Fiber Commun. Conf. Atlanta, GA, paper PD3 (1986).
10. L.L. Blyler, Jr., K.A. Cogan and J.A. Ferrara, "Plastics and Polymers for Optical Transmission," Mater. Res. Soc. Symp. Proc. 88, 3-10 (1987).
11. T. Kaino, K. Jinguji and S. Nara, "Low Loss Poly(methyl methacrylate-d_8) Core Optical Fibers," Appl. Phys. Lett. 42, 567-569 (1983).
12. S. Takahashi, "Low-loss Fluoride Fiber for Mid-infrared Optical Communication," Tech. Dig., 1987 Opt. Commun. Conf. (Reno, NV, p.162 (1987).
13. S.E. Miller and A.C. Chynoweth, eds., Optical Fibert Telecommunications, Academic Press, New York (1979).
14. F.V. DiMarcello, C.R. Kurkjian and J.C. Williams, "Fiber Drawing and Strength Properties," in Optical Fiber Communications (T. Li, ed.), Vol. 1, pp.179-248, Academic Press, Orlando, FL (1985).
15. R. Chaudhuri and P.C. Schultz, "Hermetic Coating on Optical Fibers," SPIE 717, 27-32 (1986).
16. R.N. Capps, "Optical Fiber Coatings. An Overview With Regard to Opto-acoustic Underwater Detection Systems," Ind. Eng. Chem. Prod. Res. Dev. 20, 599-608 (1981).
17. D. Chung and H.F. Finelli, "Fluoropolymers Boost Communication Line Quality," Res. & Dev. 28 (3), 79-82 (1986).
18. S. Yamakawa, Y. Shuto and F. Yamamoto, "Transmission Loss of Thermotropic Liquid-crystal Polyester-jacketed Optical Fibre at Low Temperature," Elect. Lett. 20, 199-201 (1984).
19. H. Schonhorn, S. Torza, R.V. Albarino and H.N. Vazirani, "Organic Polymeric Coatings for Silica Fibers. I. UV-curing Epoxy Acrylate (VIF)," J. Appl. Polym. Sci. 23, 75-84 (1979).
20. T. Kimura and S. Yamakawa, "New UV-curable Primary Coating Material for Optical Fibre," Elect. Lett. 20, 201-202 (1984).
21. T. Kimura and S. Yamakawa, "Effects of Chemical Structure on Low-temperature Modulus for UV-curable Polybutadiene Acrylates as an Optical Fiber Coating Material," J. Appl. Polym. Sci. 24, 1161-1171 (1986).
22. L. J. Donalds, W. G. French, W. C. Mitchell, R. M. Swinehart and T. Wei, "Electric Field Sensitive Optical Fibre Piezoelectric Polymer Coating," Elect. Lett. 18, 327-328 (1982).
23. B.G. Bagley, C.R. Kurkjian and W.E. Quinn, "The Use of an Organosilsesquioxane for the Coating/Cladding of Silica Fibers," Mater. Res. Soc. Symp. Proc. 88, 35-39 (1987).
24. L.L. Blyler, Jr., A.C. Hart, Jr., A.C. Levy, M.R. Santana and L.L. Swift, "A New Dual-coating System for Optical Fibres," Proc., 8th Europ. Conf. Opt Commun. Cannes, France, pp.245-249 (1982).
25. J.E. Ritter, K. Jakus and D.S. Cooke, "Predicting Failure of Optical Glass Fibers," Proc., 2nd Int. Conf. Environ. Degrad. Eng. Mater. Aggreg. Environ., pp.565-575 (1981).
26. J.T. Krause, C.R. Kurkjian and M.J. Matthewson, "Effect of Environment on the Strength of Fibers," Ext. Abst., 87th Ann. Meet. Am. Ceram. Soc . Cincinnati, OH , p.145 (1985).
27. T.S. Wei, "Effect of Polymer Coatings on Strength and Fatigue Properties of Fused Silica Optical Fibers," Adv. Ceram. Mater. 1, 237-241 (1986).
28. B.J. Skutnik, B.D. Munsey and C.T. Brucker, "Coating Adhesion Effects on Fiber Strength and Fatigue Properties," Mater. Res. Soc. Symp. Proc. 88, 27-34 (1987).

29. D. Gloge, "Optical-fiber Packaging and Its Influence on Fiber Straightness and Loss," Bell Syst. Tech. J. 54, 245-262 (1975).
30. W.E. Dennis and D.W. Burke, "Elastomer Protects Communications Fibers," Res. & Dev. 28 (1), 70-71 (1986).
31. H.A. Aulich, N. Douklias and W. Rogler, "Modified UV-curable Epoxy Silicones and Urethane Acrylates as Coating Materials for Optical Fibers," Proc., 9th Europ. Conf. Opt. Commun. Geneva, Switzerland, Oct. pp.377-380 (1983).
32. D.J. Broer and G.N. Mol, "Fast Curing Primary Buffer Coatings for High Strength Optical Fibers," J. Lightwave Tech. LT-4, 938-941 (1986).
33. T. Kimura and S. Sakaguchi, "Transmission Loss of UV-curable Silicone-coated Optical Fibre," Elect. Lett. 20, 315-317 (1984).
34. L.L. Blyler, Jr., F.V. DiMarcello, J.R. Simpson, E.A. Sigety, A.C. Hart, Jr. and V.A. Foertmeyer, "UV-radiation Induced Losses in Optical Fibers and Their Control," J. Non-Cryst. Solids, 38,39, 165-170 (1980).
35. T.G. Giallorenzi, J.A. Bucaro, A. Dandridge, G.H. Sigel, Jr., J.H. Cole, S.C. Rashleigh and R.G. Priest, "Optical Fiber Sensor Technology," IEEE J. Quant. Elect. QE-18, 626-665 (1982).
36. N. Lagakos, E.U. Schnaus, J.H. Cole, J. Jarzynski and J.A. Bucaro, "Optimizing Fiber Coatings for Interferometric Acoustic Sensors," IEEE J. Quant. Elect. QE-18, 683-689 (1982).
37. J. Jarzynski, "Frequency Response of a Single-mode Optical Fiber Phase Modulator Utilizing a Piezoelectric Plastic Jacket," J. Appl. Phys. 55, 3243-3250 (1984).

CONSEQUENCES OF EXPOSURE OF OPTICAL COATINGS TO REACTIVE GASES AND ENERGETIC PARTICLES

J.T. Dickinson[a], M.A. Loudiana[b] and A. Schmid[c]

[a]Department of Physics
Washington State University
Pullman, WA 99164-2814

[b]The Boeing Aerospace Co.
Seattle, Washington

[c]University of Rochester
Laboratory for Laser Energetics
Rochester, NY 14627

ABSTRACT

In an excimer laser, optical components are simultaneously exposed to reactive chemical species as well as bombardment by particles and energetic photons. The effects of exposure to this hostile environment can degrade components through etching reactions, sorption of gases, and particularly, in optical coatings, the formation of color centers. In this paper, studies of simultaneous exposure of dielectric thin films to fluorine containing gases and a flux of energetic electrons or ions are presented. The film materials include SiO_2 and ThF_4 sputter deposited on metal substrates. The reacting gases include F_2 and XeF_2. Techniques to study the subsequent damage include quartz crystal microbalance methods, mass spectroscopy, Auger electron spectroscopy and scanning electron microscopy. Measurements of mass gain and removal as well as chemical species concentration on the surface and in the gas phase are presented. Mechanisms for the observed damage will be discussed.

INTRODUCTION

Radiation induced and enhanced chemistry at solid surfaces is an area of considerable current interest. In an excimer laser, optical components may be simultaneously exposed to reactive species, low-energy ions, energetic electrons, as well as a large flux of UV photons. Thus, the lifetime of materials such as optical coatings and windows depends strongly on radiation/reactive gas phenomena at surfaces related to a number of damage mechanisms. Furthermore, in thin film fabrication technology, there is potential for the use of such etching in highly controlled (nonisotropic) and selective removal of metal, semiconductor and insulating material from a surface.

In this paper, we present results on the effects of electron bombardment on thin films of ThF_4 and SiO_2 as well as exposure of these materials to a source of fluorine (XeF_2 or F_2) under various sequences of electron bombardment. Other experiments involving simultaneous gas and radiation exposure are briefly mentioned.

EXPERIMENTAL

The experiments were done in a stainless steel UHV chamber equipped with an ion pump and cryopump, the latter being used to pump XeF_2 and F_2. XeF_2 was obtained from PCR Research Chemicals and vacuum distilled at -64°C before use. The gas was introduced to the chamber through a narrow stainless steel tube directed at the sample. Following determination of the conductance of the tube for the gases, measurement of the pressure behind the tube using a capacitance manometer allowed the flux to be calculated.

The chamber was equipped with a Varian Auger electron spectrometer with a cylindrical mirror analyzer (CMA) and a UTI 100C quadrupole mass spectrometer. Auger spectra were obtained with a 2-keV, 0.5-μA primary electron beam. The axial electron gun in the CMA was also used as an electron source during irradiation experiments. Surface concentrations were computed from published sensitivities.

Mass changes in the thin films were measured with a sensitive, thermally stabilized, 5 MHz quartz crystal microbalance. For this particular microbalance, an increase in the frequency of 1 Hz corresponds to a mass decrease of 1.26×10^{-8} g.[1,2] 1000Å ThF_4 and SiO_2 films were evaporated onto gold electrodes of microblance crystals at the Naval Weapons Center and shipped under dry nitrogen to Washington State University. The thin films were transferred to the UHV system and used without cleaning.

RESULTS AND DISCUSSION

Effect of Electron Bombardment on ThF_4. ThF_4 rapidly loses mass during electron bombardment, as indicated by microbalance measurements. The microbalance measures the total mass loss which includes the total mass of ions and neutral particles. Simultaneous with this mass loss, the quadrupole mass spectrometer registered a substantial increase in the 38 AMU (F_2) peak. Since the quadrupole is not in line of sight with the ThF_4 surface, it is unlikely that it would detect emitted atomic fluorine due to wall collisions. Although no increase in the 19 AMU (F) peak could be measured, it cannot be ruled out that at least part, if not all, of the desorbing fluorine is atomic.

The surface composition was monitored with AES during an extensive exposure to 1-keV electrons. Surface concentrations are plotted in Fig. 1 as a function of electron fluence. The fluorine concentration decreased from 75 to 50% during the first 0.01 C (3×10^{17} electrons/cm^2) of bombardment, with little change in fluorine concentration thereafter. The thorium concentration initially increased due to the loss of fluorine and then decreased as the film surface slowly adsorbed carbon and oxygen containing gases from the background. The sticking probability of the residual gases on the ThF_4 surface was found to increase considerably with electron bombardment.

The damage caused by electron bombardment is not limited to the near surface layers. Microbalance measurements indicate that during extensive bombardment with a 1-keV, 50-μA electron beam, the mass loss was equivalent to the mass of all the fluorine contained in 27 monolayers of ThF_4. The presence of fluorine on the surface detected by AES indicates that the damage actually extended much deeper than 27 monolayers. This bulk damage was accompanied by a visible change in the ThF_4 film. The film developed a dark blue color where the electron beam was incident upon it. This blue color is due to bulk defects, or color centers, which presumably are most likely associated with fluorine vacancies, although no information on ThF_4 color centers could be found in the literature. This blue color is visibly diminished after the film re-adsorbs a portion of the mass lost during fluorine exposure and heals fluorine vacancy defects.

For the ThF_4 film to lose many monolayers of fluorine very quickly under electron bombardment in vacuum, electron bombardment must be stimulating one or more of the following processes: ThF_4 dissociation, fluorine bulk diffusion in the near surface layers, bulk to surface diffusion (surface segregation), and fluorine desorption. As we shall show, electron bombardment in the presence of XeF_2, under the appropriate conditions, can lead to the uptake of fluorine in ThF_4.[3]

<u>Exposure of Electron Damaged ThF_4 to XeF_2</u>. When "virgin" ThF_4 is exposed to XeF_2, a fraction of a monolayer of fluorine (typically 0.7 monolayer) is adsorbed on the film surface. The sticking probability is small and can vary considerably from one sample to the next due to surface contamination. Following exposure, Xe could not be detected with AES. The surface fluorine concentration, as monitored by AES, was seen to rise by only about 5%, indicating that some of the adsorbed fluorine had diffused into the bulk.

We have found that fluorine readily adsorbs onto a fluorine-depleted ThF_4 film produced by extended electron bombardment.[4] When such a surface is exposed to XeF_2, microbalance data show that the surface rapidly gains a monolayer of fluorine with unity sticking probability. After one monolayer is adsorbed, the fluorine adsorption continues at a very slow but constant rate with less than an additional monolayer adsorbed after several hours. The points plotted on the right-hand-side of Fig. 1 show the increase in surface sample was exposed to XeF_2.

<u>Simultaneous Exposure of ThF_4 to Electrons and XeF_2</u>. When a ThF_4 surface is simultaneously exposed to XeF_2 and high energy electrons (50μA), the film loses less total fluorine than it would if the XeF_2 were not present. Under the identical 1-keV electron beam used before, the maximum amount of fluorine lost by a ThF_4 film subjected to a flux of 1×10^{16} XeF_2 molecules/s was 13 equivalent monolayers or less than half the mass loss when the XeF_2 was not present. Figure 2 shows the concentration of surface species as determined by AES during simultaneous exposure to XeF_2 and a high energy (1 keV) electron beam. Note the rapid initial loss of carbon and accompanying increase in the thorium concentration. The fluorine concentration on the surface increases to 80% at a total electron dose of 0.008 C. Further exposure led to a slow decrease in fluorine surface concentration.

Based on the relative fluxes of electrons and XeF_2, electron stimulated loss of fluorine from ThF_4 must still be occurring under these circumstances. However, an empty surface site is quickly filled from the gas phase. After some time, electron stimulated diffusion of fluorine depletes the bulk fluorine until equilibrium is established where fluorine is desorbed from the surface at the same rate as it is adsorbed from the gas phase. At <u>low</u> incident electron energies (<150 eV), the rate at which fluorine is absorbed into the ThF_4 greatly exceeds the rate at which fluorine is desorbed. During simultaneous bombardment with low-energy electrons and XeF_2, the ThF_4 film gains

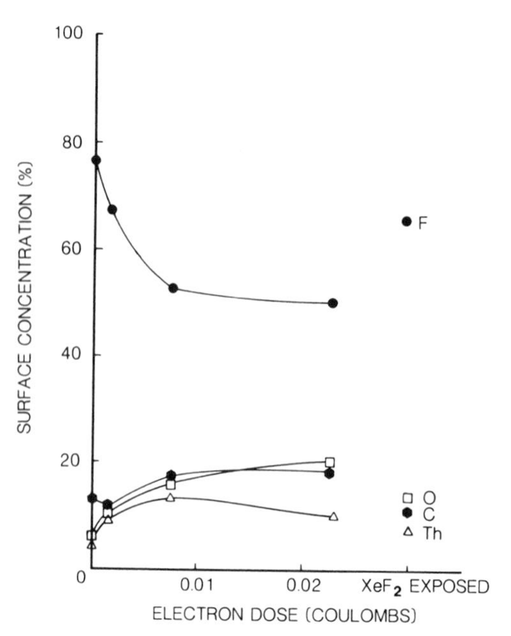

Fig. 1. Concentration of fluorine, oxygen, carbon and thorium on the surface as determined by AES during electron stimulated desorption of ThF_4 with 1-keV electrons. Effect of XeF_2 exposure shown on right.

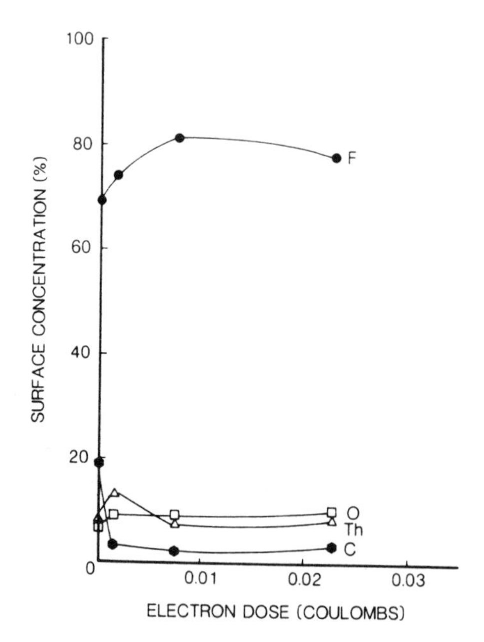

Fig. 2. Surface concentration of fluorine, oxygen, carbon and thorium as determined by AES during simultaneous exposure of ThF_4 to 1-keV electrons and 3.5×10^{15} XeF_2 molecules/cm^2/s.

mass at a steady rate depending on electron energy, XeF_2 flux, and electron flux.[3] After 2 hours of bombardment with a 7-μA, 100-eV electron beam and 1×10^{16} XeF_2 molecules/s, the mass gain rate showed no change.

From our microbalance measurements we can calculate a "sorption yield" which is equal to the rate of mass gain/electron. Figure 3 shows the dependence of the sorption yield on incident electron energy measured by the microbalance during bombardment with 1-μA electron beams. The mass gain rate appears to peak at a relatively low energy (50 eV), suggesting that this is the energy at which the difference between the rate of fluorine diffusion from the surface into the bulk and desorption of fluorine into the gas phase is maximized. At this energy, the rate of fluorine segregation into the bulk is a maximum, and the fluorine flow into the bulk is regulated at the surface by the incoming gas phase fluorine (XeF_2) and the electron current and not by the bulk concentration gradient. Field stimulated effects due to charging of the insulator surface may also play a role since the secondary electron yield decreases at these low energies resulting in increased surface charging.

It should be noted that the results obtained with F_2 as a gas phase fluorine source were almost identical to those obtained with XeF_2 when care was taken to match the fluxes for the two gases.[3]

One particularly striking observation involves the uptake of fluorine by ThF_4 thin films on top of Ag substrates. Microbalance measurements indicate that large quantities of fluorine (100's of monolayers in 1000Å films) can be absorbed by such a system during low-energy electron bombardment. The process can be explained by an electron bombardment driven sorption of fluorine which results in reactions between the fluorine and the silver substrate. At the ThF_4-Ag interface there can also occur substantial segregation of fluorine which can physically damage the thin film. Scanning electron micrographs, such as Fig. 4, indicate the presence of gas underneath such fluorine-rich films. Figure 4 is a photograph made near the edge and outside of the electron

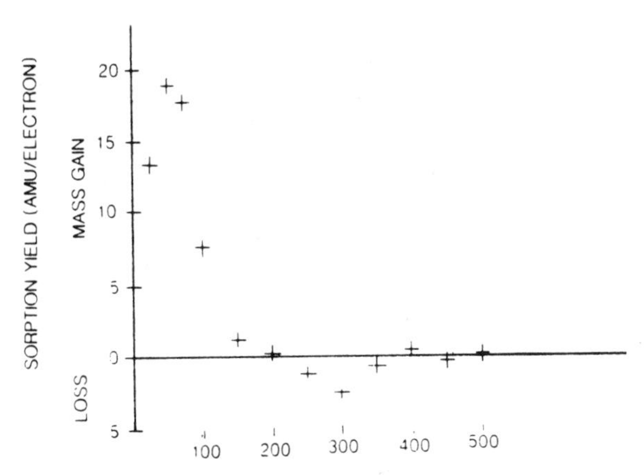

Fig. 3. Yield (mass change/electron current) dependence on electron energy during simultaneous exposure of ThF_4 to XeF_2 and electrons.

bombarded region. It is apparent that gaseous F_2 can form under such a film and produce bubbles such as those in Fig. 4. It is interesting to note that no bubbles were seen inside the electron illuminated area of the film. This suggests a diffusion mechanism that may be driven by electron-stimulated or field-stimulated processes. The low-energy of the incident electrons suggests the latter.

Effect of Electron Bombardment on SiO_2. Microbalance measurements indicate that SiO_2 loses mass slowly during electron bombardment. A loss of oxygen during electron bombardment is indicated by AES. This is in agreement with the results of Carriere and coworkers.[4] Our results indicate that the mass equivalent of one monolayer of oxygen is desorbed by a 1-keV, 20-μA electron beam over a period of 36 minutes. An increase in the partial pressure of oxygen in the chamber could not be measured which is not surprising considering the low electron stimulated desorption rate of oxygen from SiO_2. We saw no evidence of multilayer loss of oxygen from the SiO_2 films.

Exposure of Electron Damaged SiO_2 to XeF_2. Fluorine is adsorbed on fresh SiO_2 during exposure to XeF_2 or F_2 as indicated by AES. Microbalance measurements indicate that one monolayer of fluorine is adsorbed rapidly from XeF_2 or F_2 by SiO_2.[2] SiO_2 stops adsorbing fluorine after one monolayer as determined by both AES and crystal microbalance measurements. We observed no spontaneous etching reaction between XeF_2 or F_2. Microbalance measurements indicate that an oxygen depleted SiO_2 surface (due to electron bombardment) increases mass when exposed to either XeF_2 or F_2 at the same rate as a "virgin" SiO_2 surface.

Simultaneous Exposure of SiO_2 to Electrons and XeF_2. As shown by Coburn and Winters,[5] an electron beam will etch SiO_2 in the presence of XeF_2. The microbalance results shown in Fig. 5 demonstrate the loss of

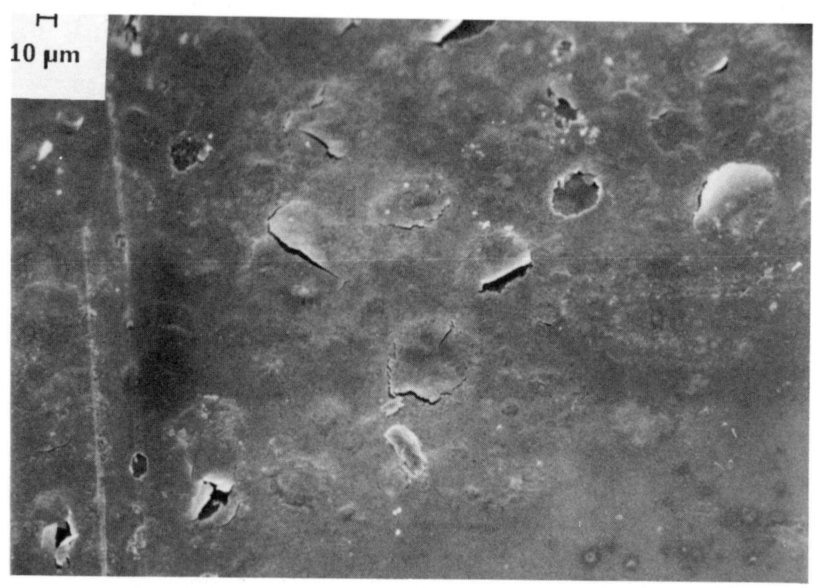

Fig. 4. Scanning electron micrograph near the edge but outside of the electron bombarded region of a ThF_4 film exposed to both electrons and XeF_2

mass (increase in QCM resonant frequency) from simultaneous bombardment of SiO_2 with 1-keV, 20-μA electron beam and $1x10^{16}$ XeF_2 molecules/s. A large increase in QCM frequency occurs when the electron gun is allowed to hit the sample and a large decrease occurs (not shown in figure) when it is turned off due to thermal effects in the quartz crystal microbalance. The dashed line in Fig. 5 indicates the frequency change after subtracting out this thermal effect. Using the equilibrium before and after values of the frequency, the rate of mass loss under these conditions was found to be $2.7x10^{-9}$ g/s. Our preliminary measurements indicate that this etch rate is linear in XeF_2 flux and electron current over a small range of the values given above.

The neutral gaseous species during exposure of SiO_2 were monitored with a quadrupole mass spectrometer. A few of the masses monitored are shown in Fig. 6. Before exposure to the XeF_2 (Fig. 6a), we see the release of neutral atomic oxygen due to electron-stimulated desorption (ESD). Upon simultaneous exposure to XeF_2 (Fig. 6b) we observe a two order of magnitude increase in mass 16 signal with electron bombardment as well as substantial mass 32 (Fig. 6c). Our preliminary cracking fraction analysis indicates that both O atoms and O_2 molecules are being released from the surface during the combined XeF_2/e-beam exposure. In addition, both the mass 16 and mass 32 signals (Figs. 6e and 6f) show a decay under electron bombardment after the XeF_2 beam is turned off, showing that the electrons remove from the surface the necessary surface intermediates. We also show (Fig. 6d) the SiF_4 signal during the combined XeF_2/e-beam exposure, which Winters and co-workers[6] have shown to be the major silicon containing product.

In comparison to this behavior, SiO_2 did not etch rapidly during simultaneous exposure to $1x10^{16}$ F_2 molecules/s and a 1-keV, 20-μA electron beam. There was only a slight mass loss observed in this environment which was approximately the same as that due to the ESD of oxygen from the surface alone.

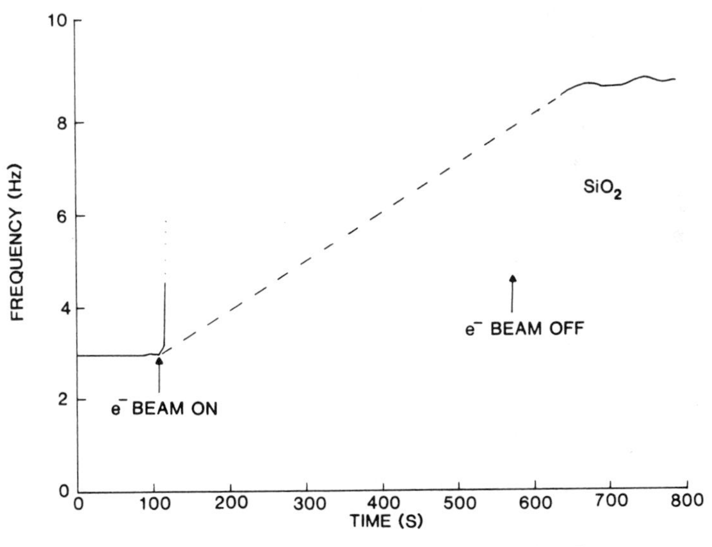

Fig. 5. QCM Response XeF_2 + e^-

This suggests that a more reactive surface intermediate is formed, perhaps atomic fluorine, from XeF_2, as opposed to a molecular fluorine species from F_2. We might expect the surface intermediate from F_2 to be converted to a more reactive species by the incident electrons. We were not able to see sufficient etching with F_2 to verify such a conversion. Another possible difference may involve a more rapid removal of oxygen from the SiO_2 surface during electron bombardment in the presence of the XeF_2 derived surface intermediate in comparison to a much slower rate with the F_2 intermediate.

In addition to these studies, we have also investigated the consequences of simultaneous exposure of dielectric thin films to energetic ions and fluorine containing gases as well as the interaction of silver surfaces with and without particle bombardment. The latter system exhibits enhanced AgF_2 formation in the presence of electron bombardment then can lead to rapid deterioration of such a film.

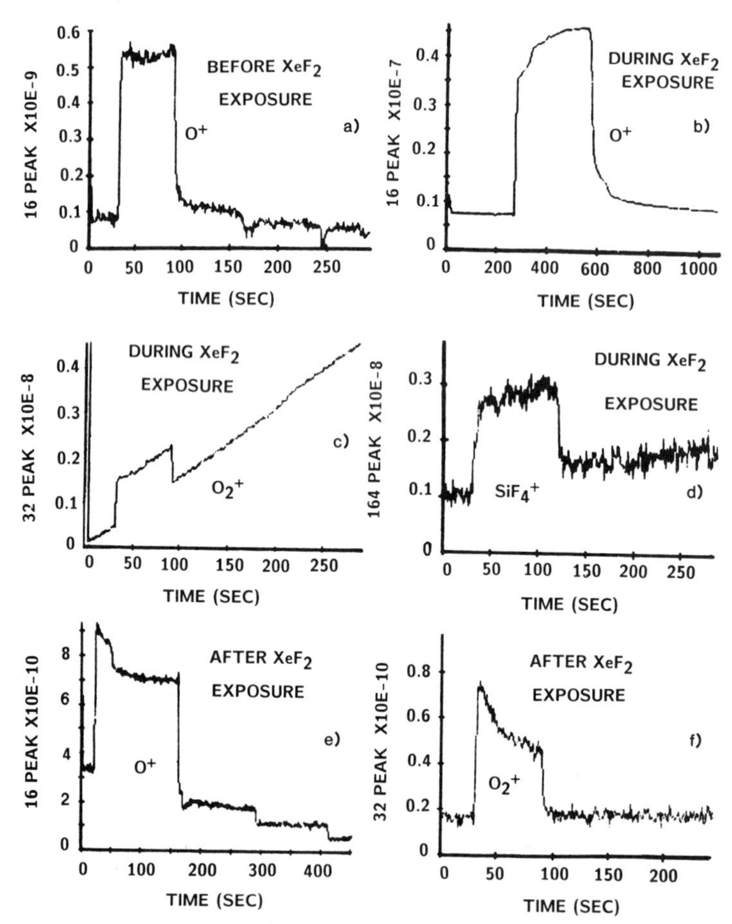

Fig. 6. Quadrupole mass spectrometer outputs corresponding to various neutral gaseous products created during the electron/XeF_2 induced etching of SiO_2 thin films.

ACKNOWLEDGEMENTS

The authors wish to thank Ed Ashley and Jim Stanford of the Naval Weapons Center for providing the ThF_4 and SiO_2 thin films. This work was supported by the Defense Advanced Research project Agency, ARPA order No. 3343 under the Office of Naval Research, Contract No. N00014-80-C-0798, and the National Science Foundation (Ceramics and Electronic Materials Division) DMR 8210406.

REFERENCES

1. A.D. Warner and C.D. Stockbridge, J. Appl. Phys., 34, 437 (1963).
2. M.A. Loudiana, A. Schmid, J.T. Dickinson and E.J. Ashley, Surf. Sci., 141, 409 (1984).
3. M.A. Loudiana, J.T. Dickinson and E.J. Ashley, J. Vac. Sci. Technol., A 3, 647 (1985).
4. B. Carriere and B. Lang, Surf. Sci., 64, 209 (1977).
5. J.W. Coburn and H.F. Winters, J. Appl. Phys., 50, 3189 (1979).
6. H.F. Winters, J.W. Coburn and T.J. Chuang, J. Vac. Sci. Technol., 1, 1157 (1983).

STABILIZED ACRYLIC GLAZINGS FOR SOLAR REFLECTORS

H.H. Neidlinger and P. Schissel

Solar Energy Research Institute
Golden, Colorado 80401

ABSTRACT

Polymer-protected silver reflectors offer one means to decrease the cost of solar concentrators by providing a flexible, light-weight mirror. However, degradation of the optical performance by ultraviolet (UV) light is currently one of the principal barriers in the development of silver/polymer films for solar thermal applications. Empirical evidence has demonstrated that poly(methyl methacrylate) can be a stable polymer in a terrestrial environment, but the polymer does not provide adequate protection for the silver reflector. We have demonstrated that low molecular weight UV stabilizers added to the polymer improve the weatherability of mirrors. We have found that the permanence of the stabilizers may limit mirror durability and that the stabilizer performance slowly diminishes because of photolysis and/or leaching and removal from the host polymer. Polymeric stabilizers are of interest on the basis that they will not leach from the host. Various polymeric UV-absorbing stabilizers have been synthesized from derivatives of 2-hydroxybenzophenone and 2-hydroxyphenylbenzotriazole. The relative effectiveness of the stabilizers in acrylic glazings will be discussed in terms of the weathering modes and retention of optical properties.

INTRODUCTION

In current central receiver systems that utilize concentrated solar energy, first and second generation heliostats have used silver/glass mirrors to reflect solar radiation onto a target receiver that may be several hundred meters away.[1] The central receiver systems require mirrors with high specularity (> 90% within an acceptance angle of 4 mrad) and with a durability of performance of up to 20 years. The first condition is met by silver/glass mirrors, since silver reflects about 97% of the radiation in the 0.3-2.6 µm wavelength range. However, the long-term durability of silver/glass mirrors currently available remains questionable because of fundamental mechanisms of degradation, a problem which has been discussed in depth elsewhere.[2,3] Furthermore, glass, because of its brittle nature and its weight, requires a stiff supporting structure. To reduce weight and cost of structures and to allow design flexibility, the use of silvered polymer materials as the mirror and a thin polymer film to protect the silver is envisioned as a possibility.

A new generation of heliostats based on silver/polymer mirrors must have not only a lower cost, but also adequate durability while maintaining high optical performance. Although some polymer-protected silver mirrors have excellent initial optical properties, further developments in protective glazings are necessary to meet the needs of solar thermal systems for increased durability and maintenance of specularity. Among the desired properties of the silver/polymer films, as identified by system studies,[4] are a solar-weighted hemispherical reflectance of at least 90%, a specularity resulting in 90% of the solar flux being contained within a solid cone with apex angle of a few milliradians, a useful life of at least 5 years, and resistance to UV, pollutants, and environmental degradation.

In this paper, we report our initial investigations on the degradation effects of weathering on silver/polymer mirrors. Furthermore, the paper indicates effective means for inhibiting weathering effects through modification of the polymer and the preparation of durable polymer glazings. In all cases in this paper, the acrylic host polymer is poly(methyl methacrylate) (PMMA). Empirical evidence has demonstrated that PMMA can be a stable polymer in a terrestrial environment, and that it can protect aluminum in that environment. It is also relatively inexpensive and has excellent optical characteristics. We have chosen to emphasize PMMA for silver mirrors, recognizing that unaltered PMMA probably will not protect silver for a sufficiently long time.[5]

EXPERIMENTAL PROCEDURES

An initial set of stabilizers (Table 1) has been identified and prepared. Details of the synthetic chemistry of stabilizers (IV) and (V) are published elsewhere.[6] Stabilizer/PMMA formulations were solution-cast onto silvered glass or glass substrates.[7] While solution casting may not be used in large-scale production, we have observed some factors, such as variables in silver metallization and casting procedures, that influence optical performance.[8] These factors are all relevant for our tests, and they may be relevant where extruded films are used.

Table 1. Representative UV-Stabilizers Studied in Reflector Glazings

Product	Class	Produced By
(I) Tinuvin P	Benzotriazole	Ciba-Geigy
(II) Univul 408	Benzophenone	BASF-Wyandotte
(III) National Starch 78-6121	Copolymeric Benzophenone	National Starch
(IV) PG-1-27-3	Copolymeric Benzotriazole	SERI[6]
(V) MS-1-13-2	Copolymeric Benzophenone	SERI[6]

Optical characterization and testing consisted primarily of measuring hemispherical and specular reflectance, as well as specular transmittance. Details have been previously described.[7]

After initial reflectance measurements, mirrors were exposed to environmental degradation using three techniques. Weather-Ometers (WOM)[9] and QUV[10] accelerated weathering devices provide data for comparison with those from real-time, outdoor exposure (racks facing south with 45° tilt). The spectral distribution for the irradiation in the three weathering modes is given in Fig. 1. Transparent films are exposed only to the two accelerated techniques (WOM and QUV).

The Weather-Ometer specimens were subjected to UV, 60°C, and air at 80% relative humidity. A xenon arc lamp with filter cutoff to match the terrestrial solar spectrum supplied the UV light. The QUV test cyclically used 4 h of UV exposure (from fluorescent lamps) at 60°C and 4 h of condensed water exposure at 40°C (ASTM G53-77). The accelerated weathering devices were used to obtain comparative information on the relative durability of the specimens.

RESULTS AND DISCUSSION

It has been known for many years that light has a deleterious effect upon the properties of polymers. "Weathering", the technological term used to describe the deterioration of polymer properties in the outdoor environment, involves a complex interaction of physical and chemical processes, and photo-oxidation is recognized as the most important of those.[11]

The characteristics of photodegradation permit application of stabilization methods to prevent such degradation by involving photon absorption, transfer of energy, and antioxidant mechanisms.[12] For the protection of solar reflectors, however, the stabilization of the polymer glazing is just one facet of the total problem, which also

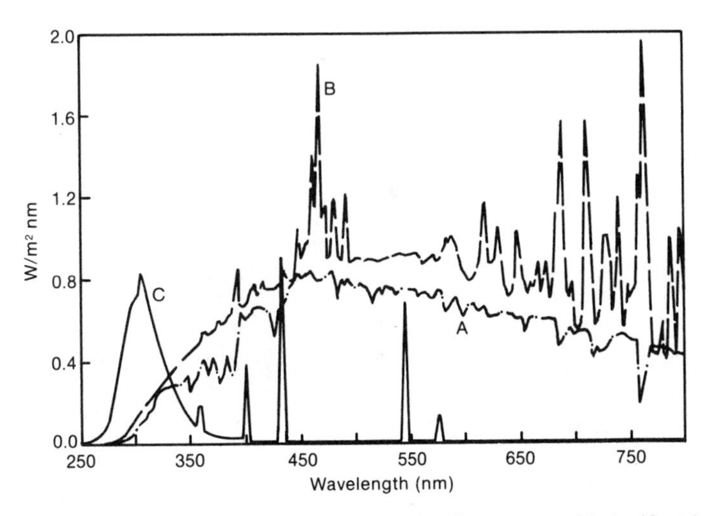

Fig. 1. A comparison of relative spectral energy distribution of Sunlight (A), WOM (B) and QUV (C) artificial light sources.

includes the stabilization of the polymer/silver interface, as well as the stabilization of a potential polymeric backing if no precautions are taken to eliminate the UV transparency of the silver layer. Since UV absorbers decrease the amount of solar light reflected from a mirror, the necessity of the presence of such additives to protect and extend the lifetime of the reflector is of high practical interest.

The phenomena affecting the durability of silvered polymer mirrors have been addressed in prior publications.[8,13] We identified and/or prepared a series of representative examples within the different groups of UV stabilizers (Table 1) and incorporated them into PMMA to enhance its protective effectiveness. The initial emphasis in our studies was on using UV absorbers in solution-cast thin glazings (5-20 μm thickness). Tests performed with these thin glazings were meant to guide later formulations for extruded, thicker films (ca 90 μm thickness).

Solution-cast Stabilized PMMA/Silver/Glass Mirrors

Mirrors made at SERI to test advanced stabilizer systems used glass as a substrate onto which a thin silver film was deposited, and the modified polymeric films were then cast from solution onto the silver. The glass substrate was used simply to simulate a collector substrate of smooth surface quality. In our initial studies, we found that such modified thin glazings prevented or diminished the optical degradation of silver mirrors.[7,13] However, mirror performance was found to depend on the weathering mode (i.e., WOM, QUV, outdoors). None of the mirror samples tested degraded significantly during short-term exposures outdoors (64 weeks) and in the accelerated QUV tests (52 weeks) as determined by comparing hemispherical reflectances. Accelerated Weather-Ometer tests, on the other hand, demonstrated an enhanced optical degradation of unstabilized PMMA/silver mirrors (Fig. 2), while stabilizers added to PMMA impeded the degradation. However, we noticed that the observed optical degradation is not due to increased absorption within the polymer glazing, but to a photo-darkening of the silver near the silver/polymer interface. Moreover, physical aspects of the silver surface (bulk plasmon and surface plasmon absorption), in addition to silver surface degradation phenomena, interfere with the direct evaluation of thin polymer glazings on silver mirrors.[7] Therefore, to evaluate changes in the polymer, we are emphasizing, presently, studies with results obtained from transparent, non-metallized PMMA films.

Solution-cast Stabilized Polymer Films

Based on the above considerations, we studied the effects of polymer additives on weathering of non-metallized PMMA films. We used PMMA in most cases without further purification, realizing that impurities in the polymer may contribute to photodegradation in the polymer films. However, such impurities are expected to be normally present in the commercial materials used for film production, and it is of interest to us to evaluate the effectiveness of stabilizer systems in such technical grade materials. Elimination of impurities in PMMA may lead, of course, to a more photostable product.

A. Optical Properties

Figure 3 shows some representative changes in specular transmittance (acceptance angle 4°) which were observed during accelerated weathering for six weeks in the QUV and WOM for unstabilized PMMA and PMMA stabilized with (I) or (IV). The irradiation of all samples caused some optical degradation during QUV exposure in comparison to the WOM-mode. The increased absorbance changes observed during QUV weathering may be due to the severity of the QUV testing mode, where the UV lamps used emit a significant amount of UV in wavelengths shorter than the solar cut-off of 295 nm (Fig. 1).

Optical losses and mechanical film failure in stabilized films are, however, considerably reduced in comparison to unstabilized PMMA, the lifetime of which did not exceed 6 weeks QUV exposure. Figure 4 summarizes representative results by displaying the change in light absorbance of stabilized PMMA films at 400 nm vs. exposure time in a QUV. (Optical changes of these films during WOM exposure are negligible for the time tested.) All stabilizers tested performed reasonably well in preventing optical degradation of the films except for stabilizer (V), which could not prevent a substantial early surface crazing of this particular film during QUV testing. This phenomenon may be due to a considerably accelerated deactivation of stabilizer (V) (see also Section C). The slight increase in absorbance for most of the other stabilized films indicates that primary photoproducts are still formed in the films,

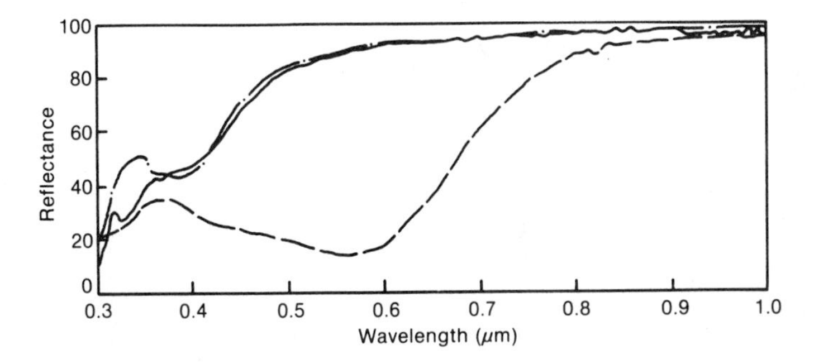

Fig. 2. Hemispherical reflectance of a PMMA glazed mirror as a function of wavelength after 16 weeks of Weathering in a Weather-Ometer (---) and QUV (-.-).

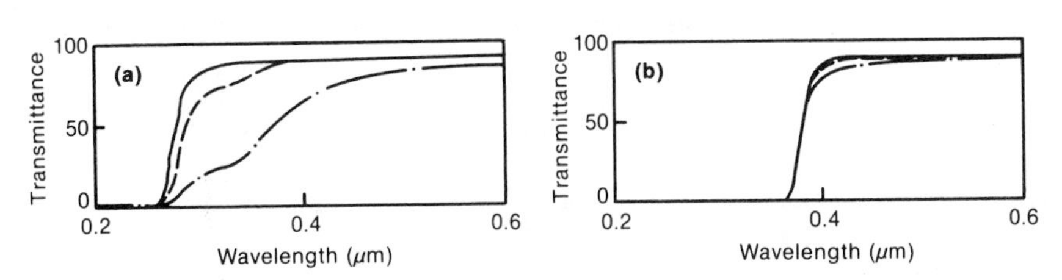

Fig. 3. Changes in the transmittance spectrum between 0.2 and 0.6 μm of (a) unstabilized PMMA, and (b) PMMA stabilized with (I) or (IV) (initial ---) after 6 weeks Weathering in a Weather-Ometer (---) and QUV (-.-).

however, at a much reduced rate. Since the increase in light absorption due to the formation of photoproducts is limited mainly to wavelengths <400 nm, optical performance of these stabilized films is rather unaffected by it. However, main chain scission of PMMA (see Section B), especially in the surface layer, causes a rapid decrease in molecular weight in unprotected or incompletely protected PMMA films, which subsequently is responsible for the formation of surface crazes that can cause considerable light scattering.[6] These results also point out the need for arranging the UV stabilizer as close to the film/air surface as possible.

Furthermore, we have observed in our extraction studies (Section C) that it is not practical to get the solution-cast films solvent free in a reasonable amount of time. Therefore, it is possible that solvent residues (in most cases toluene) in the films initiate or participate in the photo-oxidative degradation of the polymer. However, the presence of stabilizers considerably retards any such participation of the solvent in the photodegradation process.

B. Molecular Weight Changes

PMMA, if not unique, is unusual among polyalkylacrylates in that crosslinking does not occur during photolysis of films in air.[14] PMMA is unusual also in that the number of chain scissions per molecule tends to be linear with irradiation time. The effect of stabilizers on chain scission was studied by gel permeation chromatography (GPC) after irradiation. Values of the number average molecular weight, $\bar{M}_n$, show a rapid decrease in the case of PMMA without stabilizer, more so during QUV exposure than during WOM exposure (Table 2). The rate of decrease is, however, considerably slower in the case of stabilized PMMA. Further analysis of the GPC-eluate at $\lambda > 0.3$ µm indicates that the chain scission of PMMA is preceded by the formation of chromophores absorbing in the range up to around 0.4 µm.[6] Chain scission is observed to be stronger in exposed unprotected or incompletely protected film/air surfaces as compared to the more shielded film interiors. This can result in considerable surface crazing that contributes to a decrease in optical performance.

Table 2. Representative Number-Average Molecular Weight and Chain Scission Data for Unstabilized and Stabilized PMMA After 1000 Hrs. Exposure

Sample	Weathering Mode	$\bar{M}_n \times 10^{-3}$	Number of Scissions $S=(\bar{M}_{n,o}/\bar{M}_n)-1$
PMMA	Initial	174.0	0
	QUV	7.5	22.2
	WOM	79.0	1.2
PMMA+1.5 w%(I)	QUV	125.0	0.39
	WOM	129.0	0.35

C. <u>Permanence of Polymer Additives</u>

The permanence of the stabilizer in the PMMA film is another important factor for preventing long-term degradation of PMMA/silver mirrors. Despite the high inherent UV stability of the selected stabilizers, their concentrations do fall steadily during UV irradiation. The change in the absorption maximum of UV-screening additives was chosen to monitor this factor. The consumption or deactivation of stabilizers is generally higher during QUV exposure and lower during WOM exposure. The additives tested are normally consumed during irradiation according to linear kinetics (Fig. 5). According to the data (Figs. 4 and 5), there is clearly an inverse correlation between the efficiency and deactivation rate of the stabilizers. Polymeric stabilizers decompose in a fashion similar to low molecular weight species, which is evident through losses in their characteristic absorption bands. However, no noticeable molecular weight changes can be observed during the stabilizer deactivation. This may indicate that only cleavage and/or modifications of the pendant stabilizer groups are involved in the deactivation process. A detailed investigation of the reactions leading to stabilizer loss is required to indicate which reactions are an essential part of the protective process, and which are side reactions.

Although molecular weights of the stabilizers do not affect their screening properties in the systems studied, their different solubilities and distributions throughout the polymer film may be a factor of concern. Since photodegradation proceeds from the surface of a film into the bulk, it is desirable to locate UV-absorbing species as close to the surface as possible. Polymeric stabilizers are of particular interest on the basis that they will not leach from the host matrix.

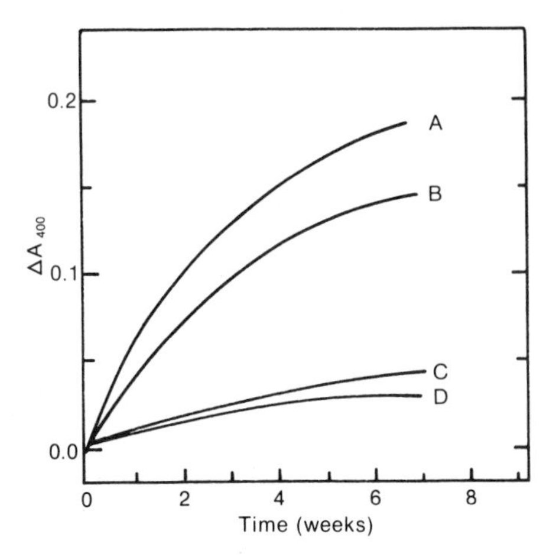

Fig. 4. Changes in absorbance of stabilized PMMA films at 400 nm versus exposure time in the QUV mode: PMMA+V (A), unstabilized PMMA (B), PMMA+I (C), PMMA+IV (C), PMMA+II (D) and PMMA+III (D).

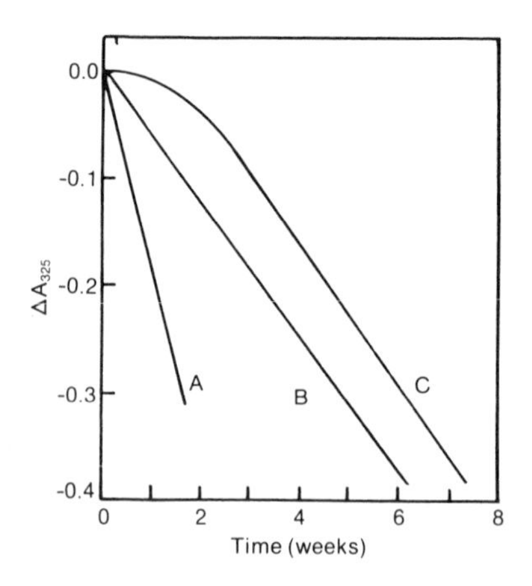

Fig. 5. Changes in absorbance of stabilized PMMA films at 325 nm versus exposure time in the QUV mode: PMMA+V (A), PMMA+IV (B), PMMA+III (C) and PMMA+II (C).

In accelerated extraction experiments,[6] we found no noticeable loss of our polymeric stabilizers (IV, V) from PMMA. This is in contrast to an almost complete extractability of the low molecular weight stabilizers (I) and (II), and a considerable leaching of a commercially available high molecular weight UV absorbing agent (III) from PMMA.

GENERAL CONCLUSIONS

It is apparent from this study that UV light plays a crucial role in the degradation of silver/polymer mirrors, A number of UV-absorbing stabilizers were examined for effectiveness in acrylic glazings and were found to impede the degradation of polymer glazings and of silver/polymer mirrors. Mirror and glazing performance as well as deactivation of UV stabilizers were found to depend on the weathering mode, probably due to variations in the UV-wavelength distribution of the different irradiation sources. Both optical performance degradation and deactivation of UV stabilizers were also markedly affected by solvent residues remaining in the glazings. Changes in optical characteristics were noticeable either through absorbance changes limited to wavelengths below 0.4 µm and/or through increased light scattering over the total solar wavelength range caused by surface crazing of unprotected or insufficiently protected surface layers. These results point out the need for arranging the UV stabilizers as close to the glazing surface as possible. Polymeric stabilizers are of interest here because they are not expected to leach from the glazing. Accelerated extraction experiments demonstrated the physical permanence of polymeric stabilizers in contrast to the extractability of their low molecular weight analogues.

ACKNOWLEDGEMENTS

The authors thank R. Goggin, P. Gomez and M. Steffeck for fabricating the samples, for synthesizing the polymeric stabilizers, and for their careful laboratory measurements.

REFERENCES

1. Solar Thermal Energy Systems, Annual Technical Progress Report FY 1980, DOE/CS.4040-2 (July 1981) p.35 (available from NTIS).
2. Nine papers, Solar Energy Mater. 3 117 (1980).
3. T.M. Thomas, J.R. Pitts and A.W. Czanderna, Appl. Surf. Sci., 15, 75 (1983).
4. L.M. Murphy, J.V. Anderson, W.D. Short and T.J. Wendelin, SERI/TR-253-2694 (December, 1985), Solar Energy Research Institute, Golden, CO (available from NTIS).
5. S.K. Brauman, D.B. MacBlane and F.R. May, Reactivity of Polymers with Mirror Materials, Subcontract No. XP-9-8127-1, Golden, CO: Solar Energy Research Institute (28 September 1982), SRI International, Menlo Park, CA.
6. P. Schissel and H. Neidlinger, SERI/PR-255-3057, (1987), Solar Energy Research Intitute, Golden, CO (available from NTIS).
7. H.H. Neidlinger and P. Schissel, Solar Energ. Mater., 14 327 (1986).
8. P. Schissel, H.H. Neidlinger and A.W. Czanderna, SERI/PR-255-2493 (June 1985), Solar Energy Research Institute, Golden, CO (available from NTIS).
9. Atlas Electric Devices, Chicago, IL.
10. The Q-Panel Co., Cleveland, OH.
11. G. Scott, Atmospheric Oxidation and Antioxidants, Vol. 92, Elsevier, NY (1965).
12. B. Ranby and J.F. Rabek, Photodegradation, Photooxidation, and Photostabilization of Polymers, Wiley NY, p.362 (1975).
13. H.H. Neidlinger and P. Schissel, SERI/PR-255-2590 (July 1985), Solar Energy Research Institute, Golden, CO (available from NTIS).
14. K. Morimoto et al., J. Appl. Polym. Sci., 16 294 (1972).

ELECTROCONDUCTIVE COATING WITH POLYCONJUGATED SYSTEMS

Lu Zhonghe, Chen Jinghong, Li Yaohuan and Yang Yingsong

Guangzhou Institute of Chemistry
Academia Sinica, Guangzhou
People's Republic of China

ABSTRACT

Sublimed zinc chloride and triphenyl phosphine dichloropalladium were found to be effective catalysts for the polymerization and copolymerization of propargyl alcohol and butynediol-1,4 or their derivatives. The polymers obtained in this way can be applied as electroconductive coatings.

INTRODUCTION

The subject of electroconductive polymers has been extensively studied[1,2] and reviewed.[3] It is a field in which the complexities of solid state physics and synthetic chemistry intermingle. Many polymers have been synthesized to produce good conductors or semiconductors which retain the desirable polymeric attributes of moldability, flexibility and toughness.

In the manufacture of many articles with nonconductive substrates as components, it is often desirable to reduce the surface electrical resistivity of the substrate to a value below 10^{14} ohms at 10% relative humidity. For example, paper with electroconductivity may be used to distribute electrical stresses in insulating products. Also, the support normally used in electrographic, electrophotographic, electrostatic and other nonimpact printing processes is conductive, or is coated with a conductive layer, thus it can play a role in the formation of the image or record.[4]

We found that propargyl alcohol and 2-butynediol-1,4 and their derivatives could be catalytically polymerized or copolymerized to form a material with polyconjugated systems in the presence of zinc chloride or bis-triphenylphosphine dichloro-palladium. A nonconductive substrate, such as paper, could be coated with these polymers and copolymers to provide an article with decreased surface electrical resistivity and an increased solvent resistance. These coatings are potentially useful in the manufacture of electrographic printing papers.

It should be pointed out that the polymers obtained in our laboratory fall in the range of semiconductor (Table 1). The comparisons in conductivity are made in Table 1 in terms of σ defined at a reference temperature (frequently near ambient).

Synthesis of monomers The synthesis of a part of the monomers (i.e., the propargyl alcohol(I) derivatives) is discussed in this section. The catalyst for these reactions was basic ion-exchange resin. The reactions were conducted in an autoclave, under a pressure of about 4 kg/cm^2 (nitrogen) for 4-6 hours.

A series of adducts were obtained from these reactions, with the following structure:

$$CH{\equiv}C\text{-}CH_2\text{-}O\text{-}CH_2CH_2OH \qquad \text{(II)}$$

$$CH{\equiv}C\text{-}CH_2\text{-}O\text{-}CH_2\underset{\displaystyle CH_3}{\underset{|}{C}}HOH \qquad \text{(III)}$$

$$CH{\equiv}C\text{-}CH_2\text{-}O\text{-}CH_2\underset{\displaystyle CH_2Cl}{\underset{|}{C}}HOH \qquad \text{(IV)}$$

Table 1. Electrical Conductivities of Materials

Substance		σ, $(ohm.cm)^{-1}$	T(K°)[a]
Class	Material		
Insulator	Quartz Sulphur Polystyrene Nylon-6,6	$10^{-17}\sim10^{-20}$ $\sim10^{-16}$ $3x10^{-14}$ $\sim10^{-14}$	(a) (a) (a) (a)
Semi-conductor	[b]Poly-(Pro)[c] [b]Poly-(Bu)[d] Polyacetylene Selenium Silicon	$\sim10^{-9}$ $\sim10^{-10}$ $\sim10^{-10}$ $(2\sim4)x10^{-6}$ 12~50	298 298 298 273 273
Conductor	Graphite Copper	$10^3\sim10^4$ $5.9x10^5$	273 273

a) Insensitive to small temperature changes, ambient quoted.
b) Prepared by us.
c) Poly-(Pro)=polypropargyl alcohol based polymer.
d) poly-(Bu)=poly(2-butynediol)-1,4 based polymer.

The synthesis for a part of monomers follows the following step:

$$CH\equiv C\text{-}CH_2OH + \begin{cases} CH_2\text{-}CH_2 \text{ (epoxide, -O-)} & \text{(II)} \\ CH_2\text{-}CH\text{-}CH_3 \text{ (epoxide, -O-)} & \text{(III)} \\ CH_2\text{-}CH\text{-}CH_2Cl \text{ (epoxide, -O-)} & \text{(IV)} \end{cases} \xrightarrow[\text{pressure } (N_2)]{\text{basic ion-exchange resin}}$$

Table 2 shows mass-to-charge ratio (MS), elemental analysis and infrared absorption (IR) bands for monomers II, III and IV.

III. POLYMERIZATION

The polymerization procedures are shown in the following scheme: wherein the monomer, $CH\equiv C$ with R attached ($CH\equiv C\text{-}R$), is propargyl alcohol or its derivatives.

R represents the above monomers II, III and IV. When using the sublimed zinc chloride as a catalyst for these monomers, bulk polymerization was preferred. Solution polymerization was chosen in

Table 2. Elemental Analysis, IR and MS Data for Monomers II, III & IV.

Pro-derivatives	MS (m/e)	Elemental analysis						IR(cm^{-1})
		Observed (%)			Calcd (%)			
		C	H	O	C	H	O	
II	101 (M+1)	60.15	8.03	31.72	60.00	8.00	32.00	1030, 1060, 1100, 1130 ($-CH_2-O-CH_2-$) 2100 ($C\equiv C-$) 3290 ($C\equiv CH$) 3400-3450 (-OH)
III	115 (M+1)	63.38	8.93	27.79	63.15	8.77	28.07	1020, 1060, 1100, 1140 ($-CH_2-O-CH_2$) 2100 ($C\equiv C$) 3290 ($C\equiv CH$) 3400-3450 (-OH) 2980 ($-CH_3$)
IV	149 (M+1)	48.35	6.06	21.83	48.64	6.08	21.62	1030, 1070, 1100, 1160 ($-CH_2-O-CH_2-$) 2100 ($C\equiv C$) 3290 ($C\equiv CH$) 3400-3450 (-OH) 750 (C-Cl)

**Elemental analysis: found: Cl, 23.76%, calcd: Cl, 23.64%.

the palladium complex-pyridine catalyst system. However, zinc chloride and triphenyl phosphine dichloro-palladium-pyridine were effective catalysts for the polymerization or copolymerization of propargyl alcohol and its derivatives, and also for butynediol-1,4.

Polymerization Procedure:

1. Polymerization

$$\text{(a)}\quad \underset{\displaystyle R}{\overset{}{CH{\equiv}C}} \xrightarrow[(Ph_3P)_2PdCl_2\text{-pyridine}]{ZnCl_2 \text{ or }} \left(\underset{\displaystyle R}{CH{=}C} \right)_n$$

$$\text{(b)}\quad \underset{R\;\;\;R}{C{\equiv}C} \xrightarrow[(Ph_3P)_2PdCl_2\text{-pyridine}]{ZnCl_2 \text{ or }} \left(\underset{R\;\;R}{C{=}C} \right)_n$$

2. Copolymerization

$$\text{(c)}\quad \underset{CH_2OH\;\;CH_2OH}{C \equiv C} + \underset{CH_2OH}{CH{\equiv}C} \xrightarrow[\text{pyridine}]{(Ph_3P)_2PdCl_2} \left(\underset{CH_2OH\;\;CH_2OH}{C = C} - \underset{CH_2OH}{CH = C} \right)_n$$

where R represents $-CH_2OH$ (I); $-CH_2-O-CH_2CH_2OH$ (II);

$-CH_2-O-CH_2CH(CH_3)-OH$ (III); or $-CH_2-O-CH_2CH(CH_2Cl)-OH$ (IV).

Structure and Properties

Table 3 shows the IR absorption bands for polymers and copolymers obtained by the above methods. We see in Table 3 that the absorption (2210 cm^{-1}) of acetylenic bond of monomers disappeared. IR spectra indicated that the polymers possess a conjugated system, which was characterized by an absorption band at about 1605 cm^{-1} in the IR spectra. It is well established that a conjugated double bond in polymer chains causes an increase in conductivity.

Table 4 gives the percent polymerization, thermal characteristics, electroconductivity and solubility in acetone of the polymers. Under our conditions, the conversions for the series of monomers were very good. From thermogravimetric analysis (TGA) results, we see that an onset temperatures T_i, of weight loss for polymers were higher than 250°C, up to about 300°C, and polybutynediol-1,4 (poly-Bu) had the highest T_i, because the active α-hydrogen in the monomer was substituted by $-CH_2OH$. All polymers had a conductivity of 10^{-10} (ohms. cm)$^{-1}$, while the copolymer of butynediol and propargyl alcohol derivative (IV) had an even higher conductivity to 10^{-8} (ohms. cm)$^{-1}$. The surface electrical resistance of each test strip was measured according to ASTM-D-257-78. In addition, all polymers were soluble in acetone or DMF from 30wt% to 100%. The solutions were very useful for electroconductive coatings.

Table 3. IR Absorption bands for Polymers and Copolymers (cm^{-1}).

Polymer	–OH	$-CH_2$	CH = C or C = C	C–O–C
Poly-(Pro)	3380-3420	2918, 2862	1600	----
Poly-(Bu)	3340-3420	2918, 2850	1605	----
Copoly-(Bu/II)	3340-3420	2918, 2850	1605	1040, 1060 1100, 1130
Copoly-(Bu/III)	3350-3418	2918, 2860	1605	1040, 1060 1100, 1130
Copoly-(Bu/IV)	3360-3420	2915, 2865	1605	1040, 1060 1100, 1130

Table 4. Properties of Newly Synthesized Polymers.

Polymer	Conversion %	T_i (°C)	$\sigma \times 10^{10}$ $(ohm.cm)^{-1}$	Solubility in acetone (%)
Poly-(Pro)[a]	83	251	2.7	~ 30
-(II)[a]	70	260	3.0	~ 45
-(III)[a]	53	253	1.9	100
-(IV)[a]	77	258	4.6	100
-(Bu)[b]	80	313	3.2	~ 80
Copoly-(Bu/IV)[b]	48	301	344	~100

a) $ZnCl_2$, 0.5% moles, 180°C for 22 h.
b) $(Ph_3P)_2PdCl_2$, 0.5% moles, 150°C for 10 h.

Figure 1 shows the current-voltage characteristics of polymers. All the polymers have the characteristics of an organic semiconductor, and usually the copolymers have higher conductivities than the corresponding homopolymers.

Figure 2 shows the effect of water on the resistance of polypropargyl alcohol. When the polymer was heated, the water on the surface of the detected plate caused a decrease in resistance. As heating continued, the water absorbed in the polymers and on the plates evaporated, thus increasing the resistance. At about 250-300°C, the polymer began to decompose, and the resistance also changed. As a result, an "S"-type curve was obtained.

Electroconductive Coatings Since these polymers and copolymers contain reactive hydroxy groups, low molecular weight polymers and copolymers can be crosslinked with both aliphatic and aromatic diisocyanates, such as tolylenediisocyanate (TDI), 4,4-diphenylmethane diisocyanates (MDI), or 1,6-hexamethylene diisocyanate (HDI) to form a film that has high hardness, resistance to marring, gloss, solvent resistance and electroconductivity.

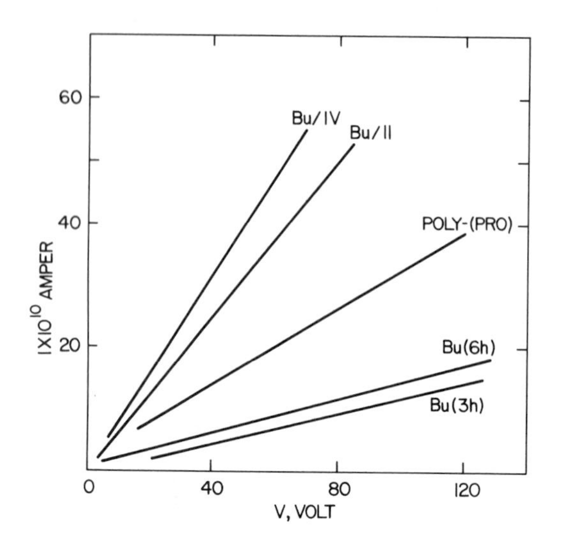

Fig. 1. Volt-ampers characteristics of polymers.

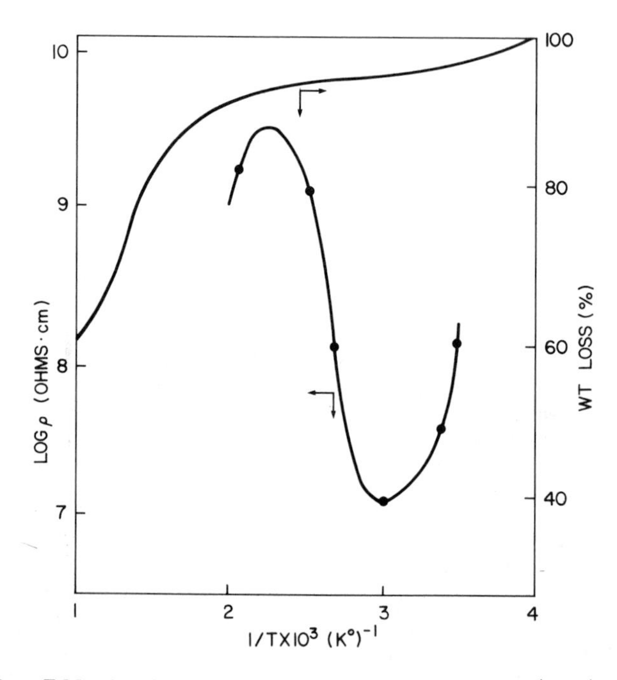

Fig. 2. Effect of water on resistance of poly-(pro) polymer.

For soluble, high molecular weight polymers or copolymers, solvent-based coating systems were prepared. They were dissolved in suitable organic solvents, such as acetone or DMF. The films of such surface coatings were formed by both solvent evaporation and curing reaction between-NCO and -OH in the presence of an organo-tin compound at 100°C for 8-16 h. The coating properties depended on the molecular weight of the polymers and the density of crosslinks of the cured films.

The scheme of formation of polymeric coating is

1. For low MW (liquid) polymers:

$$\left(-CH=C-\right)_n \text{ with side chain } CH_2-O\left(-CH_2-CH_2-O-\right)_n H \quad + \quad O=C=N-C_6H_3(CH_3)-N=C=O$$

100°C for 8~16h. — organo-tin compounds

$$\left(-CH=C)-\right)_n \text{ with side chain } CH_2-O\left(-CH_2-CH_2-O-\right)_m-\overset{O}{\overset{\|}{C}}-NH-C_6H_3(CH_3)-NH-C-(-O-CH_2-CH_2-)-O-CH_2 \text{ attached to } \left(-CH=CH_2-\right)_n$$

2. For soluble, high MW Polymers:

$$\text{Polymer + Solvent} \xrightarrow[\text{evaporated solvent}]{\text{TDI}} \text{Films}$$

CONCLUSIONS

Sublimed zinc chloride and triphenyl phosphine dichloro-palladium were found to be effective catalysts for the polymerization and copolymerization of acetylenic alcohols, such as propargyl alcohol and 2-butynediol-1,4 or their derivatives.

The polymers obtained by these methods contained a polyconjugated structure in their main chain; they were electroconductive, having a resistivity of 10^9-10^{10} ohm. The polymers, whether low or high molecular weight, were soluble in organic solvents; therefore they can be applied as electroconductive coatings for insulators, for example, papers.

REFERENCES

1. G. Manecke, Pure Appl. Chem., 1, 155 (1974).
2. Z.H. Lu, J.H. Chen, Y.H. Li & Y.S. Yang, The 2nd SPSJ International Polymer Conference, Tokyo, Japan, August 18-21, p. 120 (1986).
3. E.P. Goodings, Chem. Soc. Rev., 5, 95 (1976).
4. M. Kryszewski, J. Polym. Sci., Polym. Symp. 50, 359 (1975).

DURABILITY OF THE CONDUCTIVE ADHESIVE JOINTS

Xia Wen-gan and Wang Lu

Central Laboratory
The Huang-He Machinery Factory, Xi'an
The People's Republic of China

ABSTRACT

The bonded joints of aluminum, copper and test sheets coated with silver and gold respectively were tested, using seven kinds of the conductive adhesives with different compounds under five different environmental conditions. The bond strength remained unchanged after high temperature and UV test, but the joints deteriorated faster in high humidity and especially under the water-immersion test. The bonded joints of test pieces coated with silver and gold respectively were not good at high temperature and high humidity, and consequently their resistivities somewhat increased, in the case of aluminum joint the resistivity increased more. When a coupling agent was properly added, the bond strength increased.

The measurement data and theoretical analysis are given in this paper.

INTRODUCTION

With the rapid progress in adhesive technology, the increase in its varieties and the wide application are being found. It is more important today to study durability of adhesive joints than ever. For example, some papers described durability of aluminum joints;[1-2] an advanced study of durability of adhesive joints with metals of different kinds was conducted,[3] and other papers also described the durability of conductive adhesive joints.[4-5] We consider that a thorough study of conductive adhesive joints has to be conducted, because problems can occur in them in the course of its application; for example, the adhesive strength for the conductive adhesive for bonding aluminum alloy decreased significantly under higher temperature and high humidity environment for one year, and some corrosive mass appeared at the adhesive interface; also leakage occurred in the sealing parts, using the epoxy conductive adhesive, made from copper with gold coating under hot and dry outdoor environment for one year. The above mentioned phenomenon contradicts those stated by E. Andrews and others. They hold that the eopxy group is able to set up a reliable covalent bond on the metal surface, it may possibly exist on the gold surface.[6] We consider that the covalent bond is hardly destroyed. For this reason, this paper briefly describes the variation of strength and resistivity for 7 varieties of conductive adhesives for bonding aluminum, copper, gold coating and silver coating parts respectively under different environments.

MAIN COMPONENTS TO FORM SEVEN DIFFERENT CONDUCTIVE ADHESIVES

Table 1 presents the main components for seven different conductive adhesives with the silver powder filler. The silver powder content is 65-75%. These conductive adhesives have been widely applied to following fields:

1. bonding the aluminum wave guide-to-its flange,
2. bonding the gold coating bolt used for adjustment for microwave components,
3. bonding the flat-slot antenna made from aluminum alloy,
4. bonding dynamo brushes,
5. bonding the cavity resonator, and
6. bonding other electronic components.

These adhesives are being developed and produced in China.

E-51, as shown in Table 1, stands for the average epoxy value of bisphenol-A diglycidyl ether. W-95 is 300-400 epoxy; its average epoxy value is 0.95. AG-80 is diamino-diphenylmethantetraglycidyl, and its average epoxy value is 0.80.

RESULTS AND DISCUSSION

Under the action of heat The copper alloy adhesive joints using adhesive No. 1 had been put under hot air at 60°C for one year. As a result its bond strength remained unchanged. For aluminum alloy joints using adhesive No. 2 after 200 hours at 90°C, its degree of conservation of the bond strength was 100%. From the above test result, we can conclude that such conductive adhesives have good thermal stability.

Under the action of ultraviolet light The aluminum alloy joints using adhesive No. 5 were under the action of the light from the long arc xenon lamp at 45-50°C for 3249 hours. As a result, their bond strengths increased by 27%. For copper alloy joints using adhesive No. 3 under showering test at the duration of 15 min. an hour, under coal-arc UV, at 50°C for 500 hours, at last their bond strengths decreased

Table 1. Compositions of Conductive Adhesives

Adhesives	No. 1	No. 2	No. 3	No. 4	No. 5	No. 6	No. 7
Main resin	E-51 Epoxy	E-51 Epoxy	E-51 Epoxy	AG-80 Epoxy	E-51 Epoxy	E-51 Epoxy, W-95 Epoxy	E-51 Epoxy, Polyure-thane
Curing Agent	TEOA	HDA, EMA	MPDA	E.M.I. -24	E.M.I. -24	MOCA	DDM, MOCA

by 45.3%. We can conclude that ultraviolet light has no affect on property of those adhesives.

Atmospheric exposure test Both aluminum alloy and copper alloy joints using adhesive No. 1 respectively were exposed to the atmosphere in Xi'an and Bao-Ji (local yearly average temperature and relative humidity are 17-20°C and 70% respectively) for 4 years. As a result, the bond strength of these two joints decreased by 11.6% and 21.6% respectively. As the above mentioned aluminum alloy joints with adhesive No. 2 was exposed for 3 years, the bond strength decreased by 20.21%. In a few tested joints, there were corrosions found at the boundary. For copper alloy joints with adhesive No. 3, after 3 years the bond strength decreased by 22.4%. For aluminum alloy joints with adhesive No. 5, after 6 years the bond strength decreased by 34.4%

From the above tests, we can see the properties of the adhesive boundary layer for the copper alloy joints gradually degraded. And some corrosion at the aluminum boundary was found.

Under the action of heat and humidity The aluminum alloy joints using adhesive No. 1 were tested under 95% relative humidity at 55°C for 180 days, as a result the bond strength decreased by 34.9%; for 270 days it decreased by 44.4%. The copper alloy joints with adhesive No. 1 were tested under the same condition as mentioned above for 60 days; the bond strength decreased by 41.1%. The test result indicated that adhesive No. 1 used for aluminum joints was more resistant to heat and humidity than that for copper joints. The test results for the joints using adhesives No. 1, No. 4 and No. 6 respectively are similar to one another. When using MPDA as a curing agent for copper alloy joints (such as for adhesive No. 3) after 72 day's heat and humidity test its bond strength decreased only by 14%.

As stated above, MPDA was more resistant to heat and humidity than the fatty amine. Comparing the test results of heat and humidity to that of atmospheric exposure, we can conclude that the adhesive joints which were poorly resistant to heat and humidity environment were also not stable against the atmospheric exposure.

Water immersion test Table 2 shows the test results for the adhesive joints which were immersed in water for 200 days at 50°C. It can be seen that the silver, gold and acid-washed copper alloy joints had poor resistance to water after sand blast, and the conductive adhesive joints after water immersion test had poorer resistance to heat, humidity and atmospheric exposure. The joints using adhesives No. 2 and No. 7 respectively gave similar test results to those as above mentioned after the immersion.

If the coupling agent was properly added to the conductive adhesive, or the surface to be bonded was treated with the solution, then the adhesive property improved. For example, adding KH-560 (Union Carbide A-187) to adhesive No. 1 with which to bond the copper alloy joints after 6 months water immersion the degree of bond strength retention increased by 85% in comparison with those without adding KH-560 to adhesive No.1.

The effect of the coupling agent, in terms of the adhesive bond strength, depends upon the types of adhesive, bonded metals, the method of surface treatment and the type of the coupling agent.

Table 2. Water Immersion Test of Adhesive Joints

Adhesives	No. 1	No. 3		No. 3		No. 3	No. 3
Bonded Material	Copper alloy	Aluminum alloy	Aluminum alloy	Copper alloy	Copper alloy	silver	gold
Surface treatment	Sand blasting	Sand blasting	Base-washing	Sand blasting	Acid washing	Solvent washing	Solvent washing
Strength decreased by (%)	44.4	58.0	10.0 (120 days)	35.5	100 (90 days	100 (60 days)	100 (120 days)

Table 3 presents the degree of the bond strength retention for two kinds of adhesive after 200 days water immersion at 50°C.

<u>Resistivity variation</u> The resistivity of test joints increased by 11.9% after 240 hours of water immersion at 50°C and increased by 12.5% after 750 hours of the same test as mentioned above. To measure the resistivity of different kinds of the conductive adhesives by the method as described in the paper,[7] the resistance of the conductive adhesive using aluminum leads for measuring was at least ten times as strong as that using copper leads for measuring. For example, the resistance of the phenolic aldehyde conductive adhesive using the aluminum leads was 2.51-3.68Ω, and it was 0.15-0.22Ω using the copper leads.

This result indicates that some kinds of conductive adhesives are liable to corrode aluminum alloy especially under the circumstance of heat and humidity. In the boundary area of adhesive No. 2 for the aluminum alloy joints much of the corrosion in white colour appeared after one year in the mountainous area of Guizhou province in Southern China. The resistivity for adhesive No. 1 increased only by less than one order of magnitude after 3 years atmospheric exposure in Bao-ji of Shannxi province in Northwestern China.

As stated above, heat and humidity and especially water affected much of the properties of adhesives.

The reason for the decrease in bond strength and the increase in resistivity is the penetration of water into the adhesive gap of the joints. In this case, the relationship among J, t, A and Q is expressed as follows:

$$J = Q/A\ t, \qquad (1)$$

were J = rate of diffusion,
t = time,
A = adhesive gap area,
Q = quantity of diffusion of water to the interface.

Figure 1 illustrates the diffusion of water at the interface, so:

$$J = -Kdc/dx, \qquad (2)$$

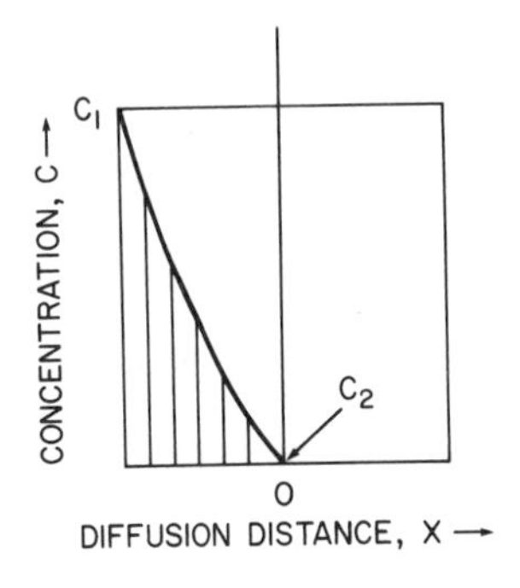

Fig. 1. Diffusion Distance, X—.

where c is water molecular concentration, and x is the distance of diffusion. After integration, the following equation is obtained:

$$Q = K\ A\ t(c_1 - c_2)/L, \tag{3}$$

were K - the penetrating coefficient of water into the adhesive,

c_1, c_2 - water molecular concentration at some points in the adhesive outside and inside parts respectively,

L - depth of the adhesive gap,

In order to increase durability of the bonded joints we must decrease the quantity of the penetrated water Q. This indicates that using an adhesive with a lower penetrating coefficient, decreasing the gap's area in order to prevent the water molecules from condensation on the outside of the adhesive's gap, decreasing the time of the contact with water and at the same time increasing the depth of the adhesive gap, are all beneficial.

Table 3. Bond Strength Retention After 200 Days Immersion in H_2O

Bonded material Surface treatment		Aluminum alloy Sand blasting			Aluminum alloy Base washing			Copper alloy Sand blasting			Copper alloy Acid washing		
Coupling agent		0	KH-550	ND-42	0	KH-550	NE-42	0	KH-550	ND-42	0	KH-550	ND-42
No.2	(%)	42.0	48.0	58.5	58.0	52.0	68.0	38.0	39.5	51.5	10.1	9.4	35.5
No.3 (Ad.)	(%)	42.0	1.5.0	111.0	90.0	115.0	117.0	65.0	84.0	68.5	0 90 days	0 90 days	0 90 days

CONCLUSIONS

From the above results we can conclude that:

1. Water acting at the conductive adhesive joints was the main factor which caused the decrease in bond strength and the increase in resistivity.
2. The durability for the conductive adhesive joints of gold, silver and acid-washed copper was poor by the action of humidity, and the same result was obtained for the aluminum alloy joints with some sorts of the conductive adhesives. Besides, corrosion appeared at the boundary area of the aluminum alloy joints. Consequently their electrical resistance increased, among them the resistance of aluminum alloy joints increased more than others.
3. The durability varied with the composition of the conductive adhesive and surface treatment of the bonded materials.
4. After a coupling agent was added to the conductive adhesive, the durability increased.

REFERENCES

1. N.J. Delollis, Adhes. Age, 20, No. 9 (1977).
2. P.M. Stifel, Durability Testing of Adhesive Bonded Joints, 19th National SAMPE Symposium, Vol. 19.
3. Xia Wen-Gan and Wang Lu, "Preliminary Study of the Durability of Metal Adhesive Joints." Electronic Technology (in Chinese) 8, (1981).
4. Cai Wu-Feng "Environmental-Resistance of the Conductive Adhesive." Electronic Technology (in Chinese) 8, (1982).
5. Zhao Guei-Fang and Zeng Ling-Kuang, "A Study Report on Some Factors Affecting the Electrical and Mechanical Property of the Conductive Adhesive With the Addition Agent." Presented to the First Annual Meeting of China's Adhesive Association 6, (1982).
6. D.T. Clark and W.J. Feast, Editors, Polymer Surface, John Wiley & Sons, Chichaster, New York, Brisbane, Toronto (1978).
7. Bei You-Wei, Synthetic Adhesive and its Property Test, (in Chinese) pp. 236, pp. 241. Fuel and Chemical Industrial Publishing House (1974).

MICROFABRICATED STRUCTURES FOR THE MEASUREMENT OF ADHESION AND MECHANICAL PROPERTIES OF POLYMER FILMS

Mark G. Allen and Stephen D. Senturia

Microsystems Technology Laboratories
Masschusetts Institute of Technology
Cambridge, MA 02139

ABSTRACT

Two microfabricated structures for the measurement of adhesion and mechanical properties of polymer films, suspended membranes and 'island' structures, are described. Measurement of the load-deflection characteristic of the suspended membrane allows determination of the film's Young's modulus and residual stress. Increasing the load on the suspended membrane until the film debonds allows a blister test measurement of the adhesion of the film. The island structure is a modification of the blister test which allows measurement of film adhesion at lower loads, thus overcoming the tensile strength limitation of conventional blister or peel tests.

INTRODUCTION

Determination of the mechanical properties and adhesive strength of thin films in microelectronic devices is important both during fabrication and in evaluation of long-term device reliability. Many tests of these properties are available,[1-5] but few combine the advantages of an _in-situ_ measurement technique and compatibility with standard integrated circuit processes. Several types of microfabricated structures for the measurement of the mechanical properties of polymer films have been fabricated in our laboratory;[6,7] the structures discussed in this work are suspended (free-standing) square membranes of a polymer film supported on an oxidized silicon wafer. In this paper, we report the use of these structures for the _in-situ_ measurement of adhesion of polymer films.

SAMPLE FABRICATION

Suspended membranes are made by first fabricating a square diaphragm 5 microns in thickness in an oxidized silicon wafer using photolithography and anisotropic etching techniques.[6] The wafer is then spin-coated with the polymer of interest and the diaphragm is removed with a backside plasma etch to create a free-standing polymer membrane. Square membranes of a BTDA-ODA/MPDA polyimide (cast without adhesion promoter) from 2 to 10 millimeters (mm) on a side and ranging from 6 to 10 microns in thickness have been fabricated using this technique.

Alternate structures which are based on the suspended membranes are 'island' structures. To fabricate this structure, the square silicon diaphragm is defined so as to leave a small island of thick silicon at the center. The polymer is then spin-cast and cured as for the suspended membranes. Upon removal of the silicon diaphragm, the suspended membrane is left with a small silicon island adhered to its center. This structure is then used as the basis for further adhesion tests.

MECHANICAL PROPERTY MEASUREMENTS

The residual stress and Young's modulus of the film can be determined by measurement of the load-deflection behavior of the membrane (see Fig. 1).[3,7,8] The wafer is secured to a substrate using a commercial epoxy (Tra-Con F-156), which seals the cavity under the membrane. The wafer and substrate are then placed in a chuck which permits the application of differential pressure by use of either a pressure source or a microliter syringe. The differential pressure is measured using a Kulite silicon pressure transducer mounted in the chuck. The entire assembly is placed on a microscope stage with a calibrated z-axis and the deflection d of the film at the center of the membrane is measured.

A theoretical analysis of the load-deflection behavior has been performed using membrane mechanics (the energy minimization approach of Timoshenko),[9] modified to account for the presence of residual stress.[10] This leads to the following relation:

$$\left(\frac{Et}{a^4}\right)d^3 + \left(\frac{1.66t\sigma_o}{a^2}\right)d = 0.547p, \qquad (1)$$

where p is the applied pressure, E is Young's modulus, σ_o is the residual stress in the film, 2a is the site size, t is the film thickness and d is the deflection at the center of the membrane. Using this approach, we have previously found values of E=3 GPa and σ_o=30 MPa for this polyimide.[7] Knowledge of both the above equation and the mechanical property data are necessary for the study of polymer adhesion using these structures.

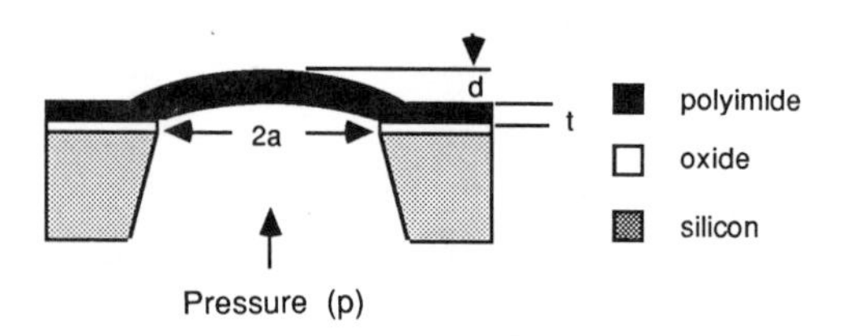

Fig. 1. Membrane parameters.

ADHESION MEASUREMENTS - SUSPENDED MEMBRANES

The suspended membranes can be used for a measurement of the work of adhesion γ_a of the polymer film to the silicon dioxide substrate. By increasing the differential pressure on the test site, the film will peel off the substrate, forming a blister. Computer simulations of the blister volume or radius as a function of critical pressure at constant γ_a indicate that the blister pressure-volume characteristic is unstable; once peel has been initiated at a fixed pressure, the blister will grow without bound. This phenomenon has been observed experimentally in previous applications of the blister test.[11] Our experiment uses a controlled-volume loading to initiate and limit peel. This is accomplished injecting pressurizing fluid (air) into the space under the blister using a calibrated microliter syringe. The PV work necessary to peel (and stretch) the blister from its initial to final radius is measured. The injected fluid is then withdrawn, and the portion of the PV work that went into stretching the blister is measured. This procedure is illustrated graphically in Fig. 2; the shaded area is the average γ_a times the total area peeled.

We have also carried out a theoretical derivation of the critical pressure necessary to initiate peeling of thin films of various geometries under lateral loading and residual tensile stress.[10] This will be useful as a counterpart to the PV analysis described above, and is used to calculate a lower bound for γ_a should it be impossible to nucleate blisters without film failure. The approach utilizes a fracture energy balance and requires knowledge of the pre-peel load-deflection behavior of the suspended membrane. From an energy minimization analogous to the derivation of Eq. (1), the load-deflection behavior of films of several geometries with residual tensile stress can all be described by the equation:

$$p = k_1 d^3 + (k_2 + k_3) d, \tag{2}$$

where k_1, k_2 and k_3 are functions of the geometry of the test site and the type of film (plate or membrane). Assuming a thin film on an infinitely rigid substrate, an energy balance applied to a virtual increment of blister size yields the relationship between the blister deflection and the work of adhesion of the film γ_a:

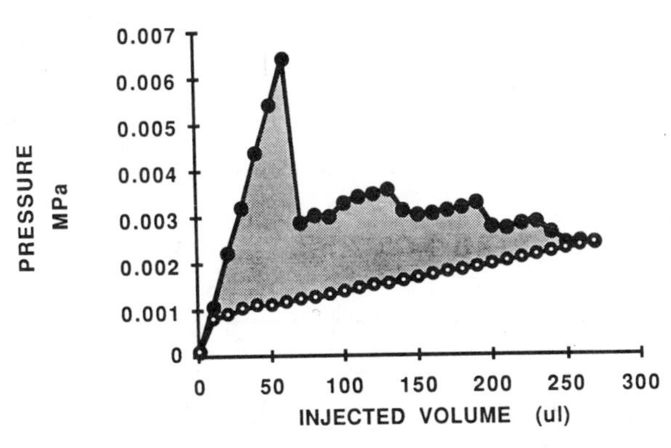

Fig. 2. Adhesion PV data. Solid circles--during peel; open circles--after peel.

$$\gamma_a = \left[\frac{5}{2} \frac{\Delta^4 c_1}{a^{11}} + \frac{3\Delta^2 c_2}{a^7} + \frac{2\Delta^2 c_3}{a^5} \right] \left[da/dA \right], \tag{3}$$

where Δ is the generalized load-point displacement of the blister (in this case, blister volume), da/dA is the incremental dependence of blister size on blister area, and c_1, c_2 and c_3 are mechanical-property-dependent constants. Table 1 gives the values of the various parameters in Eqs. (2) and (3) for three blister geometries: the clamped circular plate, the circular membrane and the square membrane. By substitution of the appropriate constants into Eq. (3), and simultaneous solution of (3) with the corresponding load-deflection relation (2), a value for γ_a can be determined as a function of the critical debond pressure p_c. Thus, experimental measurement of the critical debond pressure can be related to γ_a once the mechanical properties of the film have been accurately determined.

Under certain conditions, Eq. (3) reduces to special cases which have been previously reported in the literature.[11,12] For example, in the case of a clamped circular plate with zero residual stress undergoing small deflections, (c_1=c_3=0) the peel criterion from Eq. (3) is:

$$\gamma_a = 0.5\, p_c d_c, \tag{4}$$

which has been obtained by Williams.[11] Alternatively, for the case of a circular membrane undergoing large deflections with zero residual stress, (c_2=c_3=0) the peel criterion from Eq. (3) is:

$$\gamma_a = 0.625\, p_c d_c. \tag{5}$$

Table 1. Geometric constants for adhesion model

	Square Membrane	Clamped Circular Plate	Circular Membrane
k_1	$1.83\ E\ t/a^4$	$2.77\ E\ t/a^4$	$3.56\ E\ t\ /a^4$
k_2	0	$5.68\ E\ t^3/a^4$	0
k_3	$3.04\ t\ \sigma_o/a^2$	$4\ t\ \sigma_o/a^2$	$4\ t\sigma_o/a^2$
c_1	$0.429\ E\ t$	$2.42\ E\ t$	$0.917\ E\ t$
c_2	0	$5.44\ E\ t^3$	0
c_3	$1.88\ t\ \sigma_o$	$3.82\ t\ \sigma_o$	$2.55\ t\ \sigma_o$
Δ	$16a^2d/\pi^2$	$a^2d\pi/3$	$a^2d\pi/2$
da/dA	$1/2a$	$1/2\pi a$	$1/2\pi a$

Gent[12] has also analyzed this case assuming a slightly different load-deflection profile and has obtained a value of 0.65 for the premultiplying factor in Eq. (5).

For square test sites under residual stress such as the suspended membranes, the relation between γ_a and the critical center deflection d_c at which debond initiates is given by:

$$\gamma_a = 3.70Et(d_c/a)^4 + 4.94\sigma_o t(d_c/a)^2. \qquad (6)$$

The relation between γ_a and p_c can be obtained by simultaneous solution of Eqs. (1) and (6), allowing determination of γ_a from a measurement of either p_c or d_c during peel.

The upper limit of γ_a which can be measured using this technique is limited by the tensile strength of the film (this effect is common in many standard adhesion tests; for example, the 90° peel test is also tensile-strength limited). It was determined that membranes fabricated by the standard process and cure schedule could not be peeled; films always ruptured before blisters were formed. For these samples a lower bound for γ_a was calculated by using the pressure at which the film ruptured as the p_c value for the square membrane.

In order to observe blister nucleation, it was necessary to degrade the adhesion of the PI/SiO_2 interface. This was done by immersing the test sites in 90°C H_2O for varying lengths of time. Table 2 gives a summary of the adhesion data obtained from the suspended membranes.

Table 2. Adhesion data

#	Sample	Environmental Conditions	Analysis method	γ_a (J/m2)
1	6x6 blister site	none	Eq. (6)	>360
2	3x3 blister site	18 hr. H_2O 90°C	PV	1.5×10^{-4}
		18 hr. H_2O 90°C +1 hr. dry 60°C	Eq. (6)	4.4×10^{-3}
			PV	3.5×10^{-3}
3	6x6 blister site	20 hr. H_2O 90°C 6hr. dry 25°C	PV	2.0
			Eq. (6)	3.1
4	10x10 blister site	8 hr. H_2O 90°C	PV	1200 (*)
			Eq. (6)	320
5	10x10 blister site	14.5 hr. H_2O 90°C	Eq. (6)	64
6	10x10 blister site	14.5 hr. H_2O 90°C	Eq. (6)	43

(*) Extensive plastic deformation observed.

Sample 1 was not subjected to any degradative processing and burst before blister nucleation; a lower bound for γ_a was calculated from Eq. (6) to be 360 J/m^2. As expected, adhesive strength generally decreased with increased immersion time (samples 2-6) although this effect was not investigated quantitatively. Upon drying, an increase in adhesive strength was observed (samples 2,3). Agreement between PV method and equation (6) was observed to be within 50% except for sample 4, which showed significant plastic deformation, invalidating the PV analysis. Equation(6) uses only one data point to calculate γ_a, while the PV method is averaged over the entire wafer. This may account for differences in the two approaches. Furthermore, Eq. (6) implicitly assumes an incrementally symmetric peel, which is not strictly correct for the square membrane. However, as peel of the square membrane continues, a circular blister is formed, allowing application of the appropriate form of Eq. (3). These details are presently under study.

ADHESION MEASUREMENTS - ISLAND STRUCTURES

The suspended membrane blister test, like other peel tests, is limited by the tensile strength of the film. One way to overcome this problem is to use thicker films.[13] In the blister test, we have additional flexibility. Different geometries for the microfabricated site are possible which can facilitate peel of thinner films even in systems with very good adhesion.

Equation (3) suggests that if a geometry can be found in which da/dA can be increased, larger values of γ_a may be measured at the same load. For simple blisters, this derivative is inversely proportional to the membrane size (Table 1). Decreasing the membrane size fails since the deflection Δ will also decrease. This problem is overcome in the island structure shown in Fig. 3, where the polymer film will be peeled only off the center island. The deflection Δ is a function of the difference a_2-a_1, where $2a_2$ is the characteristic size (edge length or diameter) of the entire suspended membrane, while $2a_1$ is the characteristic size of the island. However, the derivative da/dA is inversely proportional only to a_1. Thus, a large geometric advantage can be obtained by decreasing a_1 while keeping a_2-a_1 large.

Although the critical pressure analysis for the island structures is considerably more complicated than for the simple blisters, an approximate relation between p_c and γ_a based on a circular geometry can be developed. For a membrane whose load-deflection behavior is dominated by residual tensile stress and which is suspended over a circular annulus of inner radius a_1 and outer radius a_2, γ_a is related to p_c by:

$$\gamma_a = \frac{p_c^2 a_1^2}{32\sigma_o t}\left[\frac{\beta^2-1}{ln\beta} - 2\right]^2, \quad (7)$$

where β is defined as the annular ratio a_2/a_1. Although approximate, it is instructive to examine the limiting behavior of Eq. (4). As β approaches unity, γ_a approaches zero (since no film is exposed, no adhesion can be measured even at infinite pressure), while as β approaches infinity, γ_a gets large for any pressure p_c. Thus, it is theoretically possible to measure large γ_a values at pressures less than the ultimate tensile stress of the film by making the center island sufficiently small.

Concentric square island structures have been fabricated with an outer size ($2a_2$) of 10 mm and inner size ($2a_1$) of 1 and 2 mm. Smaller a_1 values can be obtained by underetching the film on a 1 mm island until only 0.25 mm or even 0.125 mm sections of the film remain adhered to the center island (Fig. 3b). Peel has been achieved using these underetched structures. Although finite-element analysis of the square island structure will be required to generate accurate values of γ_a from observed debond pressures, order of magnitude values of γ_a can be obtained from Eq. (7). Preliminary experiments indicate that such values are in the range of 200-800 J/m^2, in fair agreement with values obtained from application of the peel test to thicker films.[13]

CONCLUSIONS

Microfabricated test structures for the in-situ measurement of adhesion of thin films have been described. Young's modulus and residual tensile stress were determined from pre-peel measurement of the load-deflection behavior of suspended membranes. On systems of weak adhesion, a blister test using suspended membranes has been carried out, while for systems of good adhesion, an island test structure has been developed allowing even thin films to be tested. Mechanical models for all three structures were described.

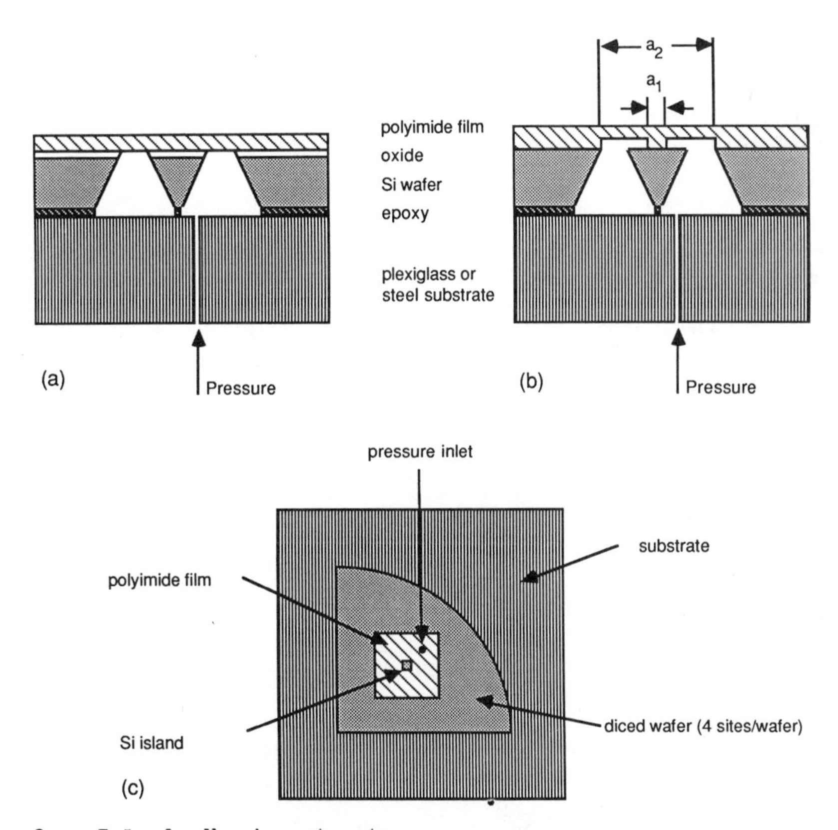

Fig. 3. Island adhesion structures.
(a) side view; (b) side view--undretched film; (c) top view.

ACKNOWLEDGEMENTS

This work was supported in part by E.I. DuPont de Nemours & Co. and the Office of Naval Research. Microfabrication was carried out in the Micro-systems Technology Laboratories, and in the Microelectronics Laboratory of the MIT Center for Materials Science and Engineering, which is supported in part by the National Science Foundation under Contract DMR-84-18718. The development of the fabrication process for suspended membranes was carried out by Mehran Mehregany. We also wish to acknowledge Professor David Parks for discussions on fracture energetics. Professor Alan Gent for providing a preprint of his work and Dr. E. Yuan of Dupont for providing samples of polyimides.

NOMENCLATURE

a blister radius, half length of square suspended membrane
a_1 island inner radius
a_2 island outer radius
A crack area
β a_2/a_1
c constants in load-deflection model
d deflection of membrane or blister at its center
d_c critical deflection of blister at its center
Δ generalized load-point displacement (blister volume)
E Young's modulus
γ_a work of adhesion
k constants in adhesion model
p presssure
p_c blister critical pressure
σ_o residual tensile stress
t film thickness

REFERENCES

1. R.W. Hoffman, in Physics of Nonmetallic Thin Films, ed. Dupey and Cachard, NATO Advanced Study Institutes Series B, Vol. 14 Plenum Press, New York, 1976.
2. L.B. Rothman, J. Electrochem. Soc., 127, 2216 (1981).
3. J.W. Beams, in Structure and Properties of Thin Films, eds. C.A. Neugebauer, J.B. Newkirk, and D.A. Vermilyea, Wiley, 1959.
4. R.J. Jensen, J.P. Cummings and H. Vora, IEEE Transactions on Components, Hybrids and Manufacturing Technology, 7, 384-393 (1984).
5. P. Geldermans, C. Goldsmith and F. Bendetti, in Polyimides, Synthesis, Characterization and Applications, ed. K.L. Mittal, Plenum Press, Vol. 2, 695-711 (1984).
6. M. Mehregany, S.M. Thesis, Department of Electrical Engineering and Computer Science, Massachusetts Institute of Technology, May 1986.
7. M.G. Allen, M. Mehregany, R.T. Howe and S.D. Senturia, Appl. Phys. Lett., in press.
8. M. Mehregany, M.G. Allen and S.D. Senturia, Proceedings of the 1986 Solid-State Sensors Workshop, Hilton Head, S.C., June 1986.
9. S. Timoshenko, Theory of Plates and Shells, McGraw-Hill, 1940, Chapter 9.
10. M.G. Allen, S.M. Thesis, Department of Chemical Engineering, Massachusetts Institute of Technology, May, 1986.
11. M.L. Williams, J. Appl. Polym. Sci., 13, 29 (1969).
12. A.N. Gent and L. Lewandowski, J. Appl. Polym. Sci., in press.
13. D. Suryanarayana and K.L. Mittal, J. Appl. Polym. Sci., 29, 2039 (1984).

Discussion

About the Entire Symposium

Alfred W. Blum Rexnord Chemical Products Inc., Mattawan, Mich.: Pursuant our conversation, I would like to make the following comments regarding the April Denver International Symposium on Adhesives, Sealants & Coatings for Space & Harsh Environments:

The topics for all six sessions were well balanced in terms of past, present and future interests, covering a broad area of interdisciplinary activities. Due to this fact, the attendees gained valuable insights in either direct or indirect related work areas.

As a major achievement, this Symposium brought together representatives from academia and industry, thus providing an excellent link between theoretical aspects and practical implications.

Although very knowledgeable in their fields, some speakers were better prepared than others in clearly presenting their findings within the given time frame. This was a limiting factor which shortened the time available for the question and answer period. Nevertheless, all sessions were well attended and the flow of presentations smooth. The banquet was a memorable experience, highlighted by an excellent speaker, Dr. J. Scott Thornton, President of the Texas Research Institute.

Below are some suggestions for future similar symposia:

- Mid-morning coffee breaks to allow for more intimate contacts.
- An afternoon panel discussion on all topics presented during sessions held in different rooms.
- Better screening of papers as to novelty and/or prior presentations.
- Broader international representation including possibly a speaker from ISO (Int. Standard Org.)

As discussed, I would be glad to assist you in reviewing some pertinent manuscripts prior to publication.

CONTRIBUTORS

Lieng-Huang Lee is a Senior Scientist at the Webster Research Center of Xerox Corporation. He received his B.S. in chemistry from Xiamen (Amoy) University, China and his M.S. and Ph.D. degrees in chemistry from Case Institute of Technology, Cleveland, Ohio. He is the editor of nine books on adhesion, friction and wear of polymers. Currently he is editing a new book on "Fundamentals of Adhesion" to be published by Plenum. In addition, he has published over sixty technical papers and holds twenty-six U.S. patents. He is a member of the Board of Editors for the Journal of Adhesion. In 1983, he was chosen as one of the distinguished scholars by the National Academy of Sciences and the Chinese Academy of Sciences. In 1986, he was awarded a visiting professorship in Chemistry by Xiamen University. Dr. Lee is listed in World's Who's Who in Science.

Charles E. Rogers received the Ph.D. in Physical Chemistry in 1957 from Syracuse University and the State University of New York College of Forestry. In 1965 he was appointed to the faculty of Case Institute of Technology. He is now Professor of Macromolecular Science in the Department of Macromolecular Science, Case Western Reserve University. He also serves as Associate Director of the CWRU Center for Adhesives, Sealants, and Coatings (CASC).

S.J. Shaw obtained his MSc in polymer technology from the Institute of Polymer Technology, Loughborough University (1975), and his Ph.D. in 1985. Currently he is a Senior Scientific Officer at the Royal Armament Research and Development Establishment, engaged on research into various aspects of structural adhesives.

Xia Wen-gan is the Head of the Central Test Department of the Huang-He (Yellow River) Machinery Factory, Xi'an. He has been active in the Xi'an Section of the Chinese Adhesion Society.

Xie Ju-niang studied at Wushang University in 1955. Since 1959, she joined Guangzhou Institute of Chemistry, Academia Sinica. Research interests have been grouting materials, adhesive for wet concrete floor, etc.

Ruan Chuan-liang graduated from the Chemical Engineering Department of Beijing Industrial Institute in 1957. At present, he is Vice-chairman of the Adhesion Society of China. He is also the Board Chairman of the Adhesion Society of Xi'an, Vice-Chairman of the editorial board of "Technology on Adhesion and Sealing" (in Chinese) and a member of the editorial board of "Chemistry and Adhesion" (in Chinese). His major interest is medical adhesives.

Tom Dickinson received his Ph.D. in Physics from the University of Michigan in Molecular Physics. He has been at Washington State University since 1968 where he is a Professor of Physics and Chemical Physics. His current research is in three areas of Materials Physics and Chemistry:

a) The emission of particles accompanying the deformation and fracture of solids (fracto-emission) as a probe of failure mechanisms;

b) The consequences of simultaneous application of mechanical stress and radiation to materials; and

c) Radiation enhanced and induced chemistry at solid surfaces.

Mohammad Parvin received his Ph.D. in Mechanical Engineering from the Imperial College of Science and Technology, London. He was an associate Professor in the Mechanical Engineering Department, Tehran University of Technology. Presently, he is a research fellow in the Graduate Aeronautical Laboratories, California Institute of Technology.

Chi-Tsieh Liu, Ph.D. is a Senior Material Research Engineer and a task manager in the Solid Rocket Division of Air Force Astronautics Laboratory. He has 22 years experience in the study of structural behavior and materials response for metallic, composite materials, and solid propellants. Much of his work has concentrated on fatigue and fracture mechanics. Dr. Liu received his Ph.D. in Engineering Mechanics from Virginia Polytechnic Institute and State University.

R.A. Pike received his BS in chemistry (1949) and an MS degree (1950) from the University of New Hampshire. He was awarded a Ph.D. in organic chemistry from the Massachusetts Institute of Technology in 1953. He has been with the United Technologies Research Center as a Senior Research Scientist from 1967 and Senior Materials Scientist from 1975 to 1984. Dr. Pike is presently Manager, Polymer Science Group of the Materials Laboratory at UTRC. He recently received the Delmonte Award for Excellence presented by the Society for the Advancement of Material and Process Engineering (SAMPE) for work in the area of materials and processes.

Charles Q. Yang obtained a B.S. degree in Chemistry in 1969 from Peking University, Beijing, P.R. of China. He received a M.S. degree in polymer chemistry in 1981 from Nanjing University, Nanjing. He came to the United States in 1983 and obtained his Ph.D. degree under Professor William G. Fateley at Kansas State University in 1987. He joined the Chemistry faculty as an assistant professor at Marshall University, West Virginia in the Fall of 1987.

Randall G. Schmidt obtained his Ph.D. degree in 1987 at the University of Connecticut. He achieved a Bachelor of Science degree in Chemical Engineering from the University of Wisconsin in 1983. Currently he is working at Dow Corning Corporation, Midland, Michigan.

James P. Bell is Professor of Institute of Materials Science and Chemical Engineering Department. He holds a Sc.D. degree from MIT. He is the author or co-author of over 100 journal articles, book chapters, etc. His specific research interests include the relations between various properties and physical structures of polymers, particularly epoxy resins, composite materials and biomedical cement.

Willard Bascom worked in the Naval Research Laboratory for many years. Dr. Bascom joined Hercules Aerospace in 1980. Since 1986, he became a member of the Department of Materials Science and Engineering of University of Utah.

Patrick Cassidy obtained his Ph.D. degree from Iowa State University and was a post-doctorate under Professor C.S. Marvel at the University of Arizona. Currently, he is a professor at Southwest Texas State University, and the Chief Scientist and Vice President of Texas Research Institute, Austin, Texas.

Y. Okamoto is a Scientist in the New Technology Group of Loctite Corp., and his main interest is in polymer synthesis, characterization and mechanism. He received a Ph.D. under Prof. Bailey at the University of Maryland and his post-doc with Prof. H.K. Hall, Jr. at the University of Arizona. Prior to his current position, he was employed at B.F. Goodrich R&D Center.

J.W. Holubka received his B.S. degree in chemistry from the University of Detroit in 1972 and his Ph.D. in Physical Organic Chemistry from Wayne State University in 1977. He has been employed at Ford Motor Company since 1977 as a Staff Scientist, working in the Polymer Science Department. Dr. Holubka's research interests include polymer synthesis and design, surface chemistry, adhesion chemistry of adhesives and coatings, and effects of environment on durability of polymeric systems. He holds 32 patents and published 30 papers.

Chang Chih Ching received his B.S. degree in chemistry in 1953 from Peking University, Beijing, P. R. of China. He is currently a professor of polymer chemistry at Beijing Institute of Aeronautics and Astronautics (BIAA). His research has been the syntheses and applications of high temperature or noninflammable adhesives. He received the China National Award on the subject of strain gauge adhesives in 1982. He spent two years, 1980-1982, at Swarthmore College and Polytechnic Institute of New York as a visiting professor.

Charles Carraher, Jr. is currently Dean and Head of the College of Science, Florida Atlantic University. Dr. Carraher is author or editor of about 30 books in general and polymer chemistry and author of about 400 articles. His research has led to the synthesis of over 60 new families of polymers. These products have wide uses such as flame retardants, antifungal and antibacterial agents, anticancer and antidiabetes agents, UV stabilizers, nonlinear optical materials, conductors, superconductors, semiconductors, adhesives, high thermal applications and insecticides. He is also active in the general field of science education.

Iskender Yilgor received all his degrees from the Middle East Technical University, Ankara, Turkey, where he also served as an Assistant Professor for two years. In 1980 he joined Virginia Tech Polymer Group (Blacksburg, VA) as a Research Associate and then became the Director of Instructional Polymer Laboratories. In 1985, he joined Mercor Incorporated (Berkeley, CA), a specialty polymer company, as the Director of Research. Presently he is the Vice President of Research and Development of Mercor, Inc. His research interests include synthesis and characterization of reactive (siloxane) oligomers, block and segmented copolymers, polymer blends and surface modification of polymeric systems through the use of siloxane containing copolymers.

D. Brenton Paul obtained his Ph.D. degree in 1967 from the University of Adelaide. He joined Materials Research Laboratories in the same year and worked for some time in the area of mechanistic organic chemistry, with particular emphasis on the chemiluminescence of peroxyoxalates. Subsequently he changed his area of interest to polymer chemistry and established a section to investigate advanced polysulfide sealants. He has also been involved in the evaluation of foam materials for use in naval vessels. Currently he is a Principal Scientist directing a research group with primary interests in elastomers.

Mike Owen is a native of Wales. He obtained his B.Sc. and Ph.D. degrees from the University of Bristol, England, He has been with Dow Corning Corporation for 22 years, the last 12 at the Corporate Center in Midland, Michigan. He has had a variety of R&D positions and presently is a Scientist leading the Adhesion Research Group. Most of his activities have focused on the surface chemistry of silicones. He has published over 30 papers and holds patents on this subject.

Waldemar Mazurek completed his MSc degree (1972) and Ph.D. degree (1983) in inorganic chemistry at LaTrobe University, Melbourne. Since 1982 he has worked with aircraft sealants at the Materials Research Laboratories, Melbourne. During this time, he concentrated on metal-based oxidative curing of polysulfides, modification of the polysulfide prepolymers and the dissolution of sealants in reactive media. He is also involved in the application of NMR and electrochemical methods to the study of the chemistry of sealants.

Sung Gun Chu, Chemist, received a Ph.D. degree in Physical Polymer Chemistry from the University of Texas at Austin in 1978. His main interests are in the property-structure processing relations of polymeric materials, and the rheology of polymers. Since he began work at Hercules in April 1981, he has been involved in the study of adhesives and sealants. He is also involved in the development of new, tough matrix resins for graphite composite applications. He has published more than 20 papers and has several patents on adhesives and composite resins. He is presently a Project Leader in Adhesives at the Hercules Research Center.

Robert M. Evans obtained his BS degree from Antioch College and his Ph.D. degree from Case Institute of Technology. He was chairman of ASTM D1.47 Committee on high performance architectural coatings and of the Cleveland Society for Coatings technology. He won two Roon Awards for work on abrasion resistance. In ASTM C-24 committee on Building Sealants he has been the chairman of numerous subcommittees, and until 1987 was Chairman of the US delegation to ISO SC59/8. He has received ASTM's Award of Merit (ASTM Fellow) and was elected to its Sealant Hall of Fame.

Lu Jiqing is doing research on protection of soil with sealants at the Northwest Institute, China Academy of Railway Sciences, Lanzhou, Gansu, People's Republic of China.

Richard Farris received his Ph.D. from the University of Utah in Civil Engineering. He spent 15 years in industry working on the characterization and applied mechanics of composites. He has been with the University of Massachusetts for 13 years. Menas Vratsanos is a Chemical Engineering graduate of Columbia University and Charles Bauer is a Chemistry graduate from Carnegie Mellon University. Both are near completion of their Doctoral studies in the University of Massachusetts.

V. Rascio is Head of the Research and Development Center for Paint Technology, at La Plata, Argentina. Dr. Rascio is a member of the "Comité International Permanent pour la Recherche sur la Preservation des Matériaux en Milieu Marin," "Society for Underwater Technology," "Argentine Committee of Engineering of Oceanic Resources," etc. Specialist in subjects related with corrosion and corrosion protection by means of paints, and more than 100 contributions were published in different scientific journals in different countries. Horacio Damianovich 1972 Award was granted by the "Asociación Química Argentina" for its contributions for the knowledge of antifouling paints formulation, preparation and behavior.

Faquir Jain received his Ph.D. degree from the University of Connecticut in 1973. Since then he has been on the faculty of the School of Engineering at the University of Connecticut, where he is currently the Acting Chairman and Professor of Electrical and Systems Engineering Department. His current research interests include metal-semiconductor and semiconductor heterointerfaces, microwave and optoelectronic semiconductor devices, and integrated circuits. He has published over 45 papers in his field of interest and holds 3 patents.

T. Sugama is an Associate Chemist in the Department of Applied Sciences at the Brookhaven National Laboratory, where he joined in 1978. He holds a Ph.D. degree and has authored 36 scientific and technical articles, and five patents in the areas of ceramic, polymer, and composite materials. His current interests include the research on the surface and interface characteristics of crystalline and fibrous materials modified by polyelectrolyte macromolecules for improving adhesion and corrosion.

William E. Dennis received his B.S. degree from Alma College in 1962 and a Ph.D. degree in synthetic organic chemistry from Wayne State University in 1966. He has worked for Dow Corning Corporation since 1966 in various research and development laboratories. He is currently a Senior Specialist in the High Technology Commercialization Unit working with silicone materials designed for the optical fiber and cable industry.

Ta-sheng Wei received his B.S. degree in physics from National Tsing-Hua University, Hsinchu, Taiwan, ROC, in 1969 and his Ph.D. degree in physics from the University of Pennsylvania, Philadelphia, PA, in 1976. From 1978 to 1984 he was a research specialist with 3M Company, St. Paul, MN, where he engaged in research on fiber optics and polymer physics. In 1984 he joined GTE Laboratories, Waltham, MA, where he is currently a Principal Member of the Technical Staff. At GTE Laboratories he has been involved in research on fabrication and reliability study of optical fibers.

Hermann H. Neidlinger, Senior Scientist in Materials Research Branch of the Solar Energy Research Institute (SERI), has been engaged in research on polymers for conversion, storage, and control of solar energy since 1982. He received his Ph.D. in polymer science at the University of Mainz (Germany) and did post-doctoral work at Stanford University with the late Paul Flory. From 1976 to 1982 he served as a Professor of Polymer Science at the University of Southern Mississippi.

Lu Zhong-he is the President of Academia Sinica, Guangzhou Branch, People's Republic of China. He is also a professor of Guangzhou Institute of Chemistry, Academia Sinica. Currently, he is the President of Guangdong Academy of Sciences. He did some graduate study at Rensselaer Polytechnic Institute, Troy, New York.

Mark G. Allen is currently a Ph.D. candidate in the Department of Chemical Engineering at MIT, having received a B.A. in Chemistry and a B.S.E. in Chemical Engineering from the University of Pennsylvania in 1984, and an S.M. in Chemical Engineering from MIT in 1986.

Stephen D. Senturia is a Professor of Electrical Engineering at MIT, having received an S.B. in Physics from Harvard in 1961 and a Ph.D. in Physics from MIT in 1966.

AUTHOR INDEX

C

D

L

M

N

O

P

Q

R

S

T

SUBJECT INDEX

A

B

C

F

N

O

P

Q

R

S

T

U

V

W

X

Y

Z